Numerical Methods in Finance and Economics

STATISTICS IN PRACTICE

Advisory Editor

Peter Bloomfield
North Carolina State University, USA

Founding Editor

Vic Barnett
Nottingham Trent University, UK

Statistics in Practice is an important international series of texts which provide detailed coverage of statistical concepts, methods and worked case studies in specific fields of investigation and study.

With sound motivation and many worked practical examples, the books show in down-to-earth terms how to select and use an appropriate range of statistical techniques in a particular practical field within each title's special topic area.

The books provide statistical support for professionals and research workers across a range of employment fields and research environments. Subject areas covered include medicine and pharmaceutics; industry, finance and commerce; public services; the earth and environmental sciences, and so on.

The books also provide support to students studying statistical courses applied to the above areas. The demand for graduates to be equipped for the work environment has led to such courses becoming increasingly prevalent at universities and colleges.

It is our aim to present judiciously chosen and well-written workbooks to meet everyday practical needs. Feedback of views from readers will be most valuable to monitor the success of this aim.

A complete list of titles in this series appears at the end of the volume.

Numerical Methods in Finance and Economics
A MATLAB-Based Introduction

Second Edition

Paolo Brandimarte
Politecnico di Torino
Torino, Italy

A JOHN WILEY & SONS, INC., PUBLICATION

Copyright © 2006 by John Wiley & Sons, Inc. All rights reserved.

Published by John Wiley & Sons, Inc., Hoboken, New Jersey.
Published simultaneously in Canada.

No part of this publication may be reproduced, stored in a retrieval system, or transmitted in any form or by any means, electronic, mechanical, photocopying, recording, scanning, or otherwise, except as permitted under Section 107 or 108 of the 1976 United States Copyright Act, without either the prior written permission of the Publisher, or authorization through payment of the appropriate per-copy fee to the Copyright Clearance Center, Inc., 222 Rosewood Drive, Danvers, MA 01923, (978) 750-8400, fax (978) 750-4470, or on the web at www.copyright.com. Requests to the Publisher for permission should be addressed to the Permissions Department, John Wiley & Sons, Inc., 111 River Street, Hoboken, NJ 07030, (201) 748-6011, fax (201) 748-6008, or online at http://www.wiley.com/go/permission.

Limit of Liability/Disclaimer of Warranty: While the publisher and author have used their best efforts in preparing this book, they make no representations or warranties with respect to the accuracy or completeness of the contents of this book and specifically disclaim any implied warranties of merchantability or fitness for a particular purpose. No warranty may be created or extended by sales representatives or written sales materials. The advice and strategies contained herein may not be suitable for your situation. You should consult with a professional where appropriate. Neither the publisher nor author shall be liable for any loss of profit or any other commercial damages, including but not limited to special, incidental, consequential, or other damages.

For general information on our other products and services or for technical support, please contact our Customer Care Department within the United States at (800) 762-2974, outside the United States at (317) 572-3993 or fax (317) 572-4002.

Wiley also publishes its books in a variety of electronic formats. Some content that appears in print may not be available in electronic format. For information about Wiley products, visit our web site at www.wiley.com.

Library of Congress Cataloging-in-Publication Data:

Brandimarte, Paolo.
  Numerical methods in finance and economics : a MATLAB-based introduction / Paolo Brandimarte.—2nd ed.
    p. cm.
  Rev. ed. of: Numerical methods in finance. 2002.
  Includes bibliographical references and index.
  ISBN-13: 978-0-471-74503-7 (cloth)
  ISBN-10: 0-471-74503-0 (cloth)
   1. Finance—Statistical methods. 2. Economics—Statistical methods. I. Brandimarte, Paolo. Numerical methods in finance. II. Title.
  HG176.5.B73 2006
  332.01'51—dc22   2006045787

Printed in the United States of America.

10 9 8 7 6 5 4

This book is dedicated to Commander Straker, Lieutenant Ellis, and all SHADO operatives. Thirty-five years ago they introduced me to the art of using both computers <u>and</u> gut feelings to make decisions.

Contents

Preface to the Second Edition   xvii

From the Preface to the First Edition   xxiii

Part I Background

1 Motivation   3
   1.1 Need for numerical methods   4
   1.2 Need for numerical computing environments:
        why MATLAB?   9
   1.3 Need for theory   13
        For further reading   20
        References   21

2 Financial Theory   23
   2.1 Modeling uncertainty   25
   2.2 Basic financial assets and related issues   30
        2.2.1 Bonds   30
        2.2.2 Stocks   31

vii

	2.2.3	Derivatives	33
	2.2.4	Asset pricing, portfolio optimization, and risk management	37
2.3	Fixed-income securities: analysis and portfolio immunization	42
	2.3.1	Basic theory of interest rates: compounding and present value	42
	2.3.2	Basic pricing of fixed-income securities	49
	2.3.3	Interest rate sensitivity and bond portfolio immunization	57
	2.3.4	MATLAB functions to deal with fixed-income securities	60
	2.3.5	Critique	64
2.4	Stock portfolio optimization	65
	2.4.1	Utility theory	66
	2.4.2	Mean-variance portfolio optimization	73
	2.4.3	MATLAB functions to deal with mean-variance portfolio optimization	74
	2.4.4	Critical remarks	81
	2.4.5	Alternative risk measures: Value at Risk and quantile-based measures	83
2.5	Modeling the dynamics of asset prices	88
	2.5.1	From discrete to continuous time	88
	2.5.2	Standard Wiener process	91
	2.5.3	Stochastic integrals and stochastic differential equations	93
	2.5.4	Ito's lemma	96
	2.5.5	Generalizations	100
2.6	Derivatives pricing	102
	2.6.1	Simple binomial model for option pricing	105
	2.6.2	Black–Scholes model	108
	2.6.3	Risk-neutral expectation and Feynman–Kač formula	111
	2.6.4	Black–Scholes model in MATLAB	113
	2.6.5	A few remarks on Black–Scholes formula	116
	2.6.6	Pricing American options	117
2.7	Introduction to exotic and path-dependent options	118
	2.7.1	Barrier options	119
	2.7.2	Asian options	123
	2.7.3	Lookback options	123

2.8	An outlook on interest-rate derivatives		124
	2.8.1	Modeling interest-rate dynamics	126
	2.8.2	Incomplete markets and the market price of risk	127
	For further reading		130
	References		131

Part II Numerical Methods

3 **Basics of Numerical Analysis** — 137
 3.1 Nature of numerical computation — 138
  3.1.1 Number representation, rounding, and truncation — 138
  3.1.2 Error propagation, conditioning, and instability — 141
  3.1.3 Order of convergence and computational complexity — 143
 3.2 Solving systems of linear equations — 145
  3.2.1 Vector and matrix norms — 146
  3.2.2 Condition number for a matrix — 149
  3.2.3 Direct methods for solving systems of linear equations — 154
  3.2.4 Tridiagonal matrices — 159
  3.2.5 Iterative methods for solving systems of linear equations — 160
 3.3 Function approximation and interpolation — 173
  3.3.1 Ad hoc approximation — 177
  3.3.2 Elementary polynomial interpolation — 179
  3.3.3 Interpolation by cubic splines — 183
  3.3.4 Theory of function approximation by least squares — 188
 3.4 Solving non-linear equations — 191
  3.4.1 Bisection method — 192
  3.4.2 Newton's method — 195
  3.4.3 Optimization-based solution of non-linear equations — 198
  3.4.4 Putting two things together: solving a functional equation by a collocation method — 204

       3.4.5   Homotopy continuation methods   204
       For further reading   206
       References   207

4 Numerical Integration: Deterministic and Monte Carlo
   Methods   209
   4.1 Deterministic quadrature   211
        4.1.1   Classical interpolatory formulas   212
        4.1.2   Gaussian quadrature   214
        4.1.3   Extensions and product rules   219
        4.1.4   Numerical integration in MATLAB   220
   4.2 Monte Carlo integration   221
   4.3 Generating pseudorandom variates   225
        4.3.1   Generating pseudorandom numbers   226
        4.3.2   Inverse transform method   230
        4.3.3   Acceptance–rejection method   233
        4.3.4   Generating normal variates by the polar
                approach   235
   4.4 Setting the number of replications   240
   4.5 Variance reduction techniques   244
        4.5.1   Antithetic sampling   244
        4.5.2   Common random numbers   251
        4.5.3   Control variates   252
        4.5.4   Variance reduction by conditioning   255
        4.5.5   Stratified sampling   260
        4.5.6   Importance sampling   261
   4.6 Quasi-Monte Carlo simulation   267
        4.6.1   Generating Halton low-discrepancy
                sequences   269
        4.6.2   Generating Sobol low-discrepancy
                sequences   281
        For further reading   286
        References   287

5 Finite Difference Methods for Partial Differential
   Equations   289
   5.1 Introduction and classification of PDEs   290
   5.2 Numerical solution by finite difference methods   293
        5.2.1   Bad example of a finite difference scheme   295

	5.2.2	Instability in a finite difference scheme	297
5.3	\multicolumn{2}{l	}{Explicit and implicit methods for the heat equation}	303

	5.2.2	Instability in a finite difference scheme	297
5.3		Explicit and implicit methods for the heat equation	303
	5.3.1	Solving the heat equation by an explicit method	304
	5.3.2	Solving the heat equation by a fully implicit method	309
	5.3.3	Solving the heat equation by the Crank–Nicolson method	313
5.4		Solving the bidimensional heat equation	314
5.5		Convergence, consistency, and stability	320
		For further reading	324
		References	324
6	**Convex Optimization**		**327**
6.1		Classification of optimization problems	328
	6.1.1	Finite- vs. infinite-dimensional problems	328
	6.1.2	Unconstrained vs. constrained problems	333
	6.1.3	Convex vs. non-convex problems	333
	6.1.4	Linear vs. non-linear problems	335
	6.1.5	Continuous vs. discrete problems	337
	6.1.6	Deterministic vs. stochastic problems	337
6.2		Numerical methods for unconstrained optimization	338
	6.2.1	Steepest descent method	339
	6.2.2	The subgradient method	340
	6.2.3	Newton and the trust region methods	341
	6.2.4	No-derivatives algorithms: quasi-Newton method and simplex search	342
	6.2.5	Unconstrained optimization in MATLAB	343
6.3		Methods for constrained optimization	346
	6.3.1	Penalty function approach	346
	6.3.2	Kuhn–Tucker conditions	351
	6.3.3	Duality theory	357
	6.3.4	Kelley's cutting plane algorithm	363
	6.3.5	Active set method	365
6.4		Linear programming	366
	6.4.1	Geometric and algebraic features of linear programming	368
	6.4.2	Simplex method	370

	6.4.3		Duality in linear programming	372
	6.4.4		Interior point methods	375
6.5			Constrained optimization in MATLAB	377
	6.5.1		Linear programming in MATLAB	378
	6.5.2		A trivial LP model for bond portfolio management	380
	6.5.3		Using quadratic programming to trace efficient portfolio frontier	383
	6.5.4		Non-linear programming in MATLAB	385
6.6			Integrating simulation and optimization	387
S6.1			Elements of convex analysis	389
	S6.1.1		Convexity in optimization	389
	S6.1.2		Convex polyhedra and polytopes	393
			For further reading	396
			References	397

Part III Pricing Equity Options

7 Option Pricing by Binomial and Trinomial Lattices 401
 7.1 Pricing by binomial lattices 402
  7.1.1 Calibrating a binomial lattice 403
  7.1.2 Putting two things together: pricing a pay-later option 410
  7.1.3 An improved implementation of binomial lattices 411
 7.2 Pricing American options by binomial lattices 414
 7.3 Pricing bidimensional options by binomial lattices 417
 7.4 Pricing by trinomial lattices 422
 7.5 Summary 425
   For further reading 426
   References 426

8 Option Pricing by Monte Carlo Methods 429
 8.1 Path generation 430
  8.1.1 Simulating geometric Brownian motion 431
  8.1.2 Simulating hedging strategies 435
  8.1.3 Brownian bridge 439
 8.2 Pricing an exchange option 443

8.3	Pricing a down-and-out put option	446
	8.3.1 Crude Monte Carlo	446
	8.3.2 Conditional Monte Carlo	447
	8.3.3 Importance sampling	450
8.4	Pricing an arithmetic average Asian option	454
	8.4.1 Control variates	455
	8.4.2 Using Halton sequences	458
8.5	Estimating Greeks by Monte Carlo sampling	468
	For further reading	472
	References	473
9	**Option Pricing by Finite Difference Methods**	**475**
9.1	Applying finite difference methods to the Black–Scholes equation	475
9.2	Pricing a vanilla European option by an explicit method	478
	9.2.1 Financial interpretation of the instability of the explicit method	481
9.3	Pricing a vanilla European option by a fully implicit method	482
9.4	Pricing a barrier option by the Crank–Nicolson method	485
9.5	Dealing with American options	486
	For further reading	491
	References	491

Part IV Advanced Optimization Models and Methods

10	**Dynamic Programming**	**495**
10.1	The shortest path problem	496
10.2	Sequential decision processes	500
	10.2.1 The optimality principle and solving the functional equation	501
10.3	Solving stochastic decision problems by dynamic programming	504
10.4	American option pricing by Monte Carlo simulation	511
	10.4.1 A MATLAB implementation of the least squares approach	517

	10.4.2 Some remarks and alternative approaches	519
	For further reading	521
	References	522

11 Linear Stochastic Programming Models with Recourse — 525

11.1 Linear stochastic programming models — 526

11.2 Multistage stochastic programming models for portfolio management — 530

 11.2.1 Split-variable model formulation — 532

 11.2.2 Compact model formulation — 540

 11.2.3 Asset and liability management with transaction costs — 544

11.3 Scenario generation for multistage stochastic programming — 546

 11.3.1 Sampling for scenario tree generation — 547

 11.3.2 Arbitrage free scenario generation — 550

11.4 L-shaped method for two-stage linear stochastic programming — 555

11.5 A comparison with dynamic programming — 558

For further reading — 559

References — 560

12 Non-Convex Optimization — 563

12.1 Mixed-integer programming models — 564

 12.1.1 Modeling with logical variables — 565

 12.1.2 Mixed-integer portfolio optimization models — 571

12.2 Fixed-mix model based on global optimization — 576

12.3 Branch and bound methods for non-convex optimization — 578

 12.3.1 LP-based branch and bound for MILP models — 584

12.4 Heuristic methods for non-convex optimization — 591

For further reading — 597

References — 598

Part V Appendices

Appendix A Introduction to MATLAB Programming 603
   A.1 MATLAB environment 603
   A.2 MATLAB graphics 614
   A.3 MATLAB programming 616

Appendix B Refresher on Probability Theory and Statistics 623
   B.1 Sample space, events, and probability 623
   B.2 Random variables, expectation, and variance 625
       B.2.1 Common continuous random variables 628
   B.3 Jointly distributed random variables 632
   B.4 Independence, covariance, and conditional expectation 633
   B.5 Parameter estimation 637
   B.6 Linear regression 642
       For further reading 645
       References 645

Appendix C Introduction to AMPL 647
   C.1 Running optimization models in AMPL 648
   C.2 Mean variance efficient portfolios in AMPL 649
   C.3 The knapsack model in AMPL 652
   C.4 Cash flow matching 655
       For further reading 655
       References 656

Index 657

Preface to the Second Edition

After the publication of the first edition of the book, about five years ago, I have received a fair number of messages from readers, both students and practitioners, around the world. The recurring keyword, and the most important thing to me, was *useful*. The book had, and has, no ambition of being a very advanced research book. The basic motivation behind this second edition is the same behind the first one: providing the newcomer with an easy, but solid, entry point to computational finance, without too much sophisticated mathematics and avoiding the burden of difficult C++ code, also covering relatively non-standard optimization topics such as stochastic and integer programming. See also the excerpt from the preface to the first edition. However, there are a few new things here:

- a slightly revised title;
- completely revised organization of chapters;
- significantly increased number of pages.

The title mentions both Finance *and* Economics, rather than just Finance. To avoid any misunderstanding, it should be made quite clear that this is essentially a book for students and practitioners working in Finance. Nevertheless, it *can* be useful to Ph.D. students in Economics as well, as a complement to more specific and advanced textbooks. In the last four years, I have been giving a course on numerical methods within a Ph.D. program in Economics, and I typically use other available excellent textbooks covering advanced algorithms[1] or offering well-thought MATLAB toolboxes[2] which can be used to solve a wide array of problems in Economics. From the point of view of my students in such a course, the present book has many deficiencies: For instance, it does not cover ordinary differential equations and it does not deal with computing equilibria or rational expectations models; furthermore, practically all of the examples deal with option pricing or portfolio management. Nevertheless, given my experience, I believe that they can benefit from a more detailed and elementary treatment of the basics, supported by simple examples. Moreover, I believe that students in Economics should also get

---

[1] K.L. Judd, *Numerical Methods in Economics*, MIT Press, 1998.
[2] M.J. Miranda and P.L. Fackler, *Applied Computational Economics and Finance*, MIT Press, 2002.

at least acquainted with topics from Operations Research, such as stochastic programming and integer programming. Hence, the *"and Economics"* part of the title suggests potential use of the book as a complement, and by no means as a substitute.

The book has been reorganized in order to ease its use within standard courses on numerical methods for financial engineering. In the first edition, optimization applications were dealt with extensively, in chapters preceding those related to option pricing. This was a result of my personal background, which is mainly Computer Science and Operations Research, but it did not fit very well with the common use of a book on computational finance. In the present edition, advanced optimization applications are left to the last chapters, so they do not get into the way of most financial engineering students. The book consists of twelve chapters and three appendices.

- Chapter 1 provides the reader with motivations for the use of numerical methods, and for the use of MATLAB as well.

- Chapter 2 is an overview of financial theory. It is aimed at students in Engineering, Mathematics, or Operations Research, who may be interested in the book, but have little or no financial background.

- Chapter 3 is devoted to the basics of classical numerical methods. In some sense, this is complementary to chapter 2 and it is aimed at people with a background in Economics, who typically are not exposed to numerical analysis. To keep the book to a reasonable size, a few classical topics were omitted because of their limited role in the following chapters. In particular, I do not cover computation of eigenvalues and eigenvectors and ordinary differential equations.

- Chapter 4 is devoted to numerical integration, both by quadrature formulas and Monte Carlo methods. In the first edition, quadrature formulas were dealt with in the chapter on numerical analysis, and Monte Carlo was the subject of a separate chapter. I preferred giving a unified treatment of these two approaches, as this helps understanding their respective strengths and weaknesses, both for option pricing and scenario generation in stochastic optimization. Regarding Monte Carlo as a tool for integration rather than simulation is also helpful to properly frame the application of low-discrepancy sequences (which is also known under the more appealing name of quasi-Monte Carlo simulation). There is some new material on Gaussian quadrature, an extensive treatment of variance reduction methods, and some application to vanilla options to illustrate simple but concrete applications immediately, leaving more complex cases to chapter 8.

- Chapter 5 deals with basic finite difference schemes for partial differential equations. The main theme is solving the heat equation, which is the prototype example of the class of parabolic equations, to which Black–Scholes equation belongs. In this simplified framework we may understand the difference between explicit and implicit methods, as well as the issues related to convergence and numerical stability. With respect to the first edition, I have added an outline of the Alternating Direction Implicit method to solve the two-dimensional heat equation, which is useful background for pricing multidimensional options.

- Chapter 6 deals with finite-dimensional (static) optimization. This chapter can be safely skipped by students interested in the option pricing applications described in chapters 7, 8, and 9. However, it may be useful to students in Economics. It is also necessary background for the relatively advanced optimization models and methods which are covered in chapters 10, 11, and 12.

- Chapter 7 is a new chapter which is devoted to binomial and trinomial lattices, which were not treated extensively in the first edition. The main issues here are proper implementation and memory management.

- Chapter 8 is naturally linked to chapter 4 and deals with more advanced applications of Monte Carlo and low-discrepancy sequences to exotic options, such as barrier and Asian options. We also deal briefly with the estimation of option sensitivities (the Greeks) by Monte Carlo methods. Emphasis is on European-style options; pricing American options by Monte Carlo methods is a more advanced topic which must be analyzed within an appropriate framework, which is done in chapter 10.

- Chapter 9 applies the background of chapter 5 to option pricing by finite difference methods.

- Chapter 10 deals with numerical dynamic programming. The main reason for including this chapter is pricing American options by Monte Carlo simulation, which was not covered in the first edition but is gaining more and more importance. I have decided to deal with this topic within an appropriate framework, which is dynamic stochastic optimization. In this chapter we just cover the essentials, which means discrete-time and finite-horizon dynamic programs. Nevertheless, we try to offer a reasonably firm understanding of these topics, both because of their importance in Economics and because understanding dynamic programming is helpful in understanding stochastic programming with recourse, which is the subject of the next chapter.

- Chapter 11 deals with linear stochastic programming models with recourse. This is becoming a standard topic for people in Operations Research, whereas people in Economics are much more familiar with dynamic programming. There are good reasons for this state of the matter, but from a methodological point of view I believe that it is very important to compare this approach with dynamic programming; from a practical point of view, stochastic programming has an interesting potential both for dynamic portfolio management and for option hedging in incomplete markets.

- Chapter 12 also deals with the relatively exotic topic of non-convex optimization. The main aim here is introducing mixed-integer programming, which can be used for portfolio management when practically relevant constraints call for the introduction of logical decision variables. We also deal, very shortly, with global optimization, i.e., continuous non-convex optimization, which is important when we leave the comfortable domain of easy optimization problems (i.e., minimizing convex cost functions or maximizing concave utility functions). We also outline heuristic principles such as local search and genetic algorithms. They are useful to integrate simulation and optimization and are often used in computational economics.

- Finally, we offer three appendices on MATLAB, probability and statistics, and AMPL. The appendix on MATLAB should be used by the unfamiliar reader to get herself going, but the best way to learn MATLAB is by trying and using the online help when needed. The appendix on probability and statistics is just a refresher which is offered for the sake of convenience. The third appendix on AMPL is new, and it reflects the increased role of algebraic languages to describe complex optimization models. AMPL is a modeling system offering access to a wide array of optimization solvers. The choice of AMPL is just based on personal taste (and the fact that a demo version is available on the web). In fact, GAMS is probably much more common for economic applications, but the concepts are actually the same. This appendix is only required for chapters 11 and 12.

Finally, there are many more pages in this second edition: more than 600 pages, whereas the first edition had about 400. Actually, I had a choice: either including many more topics, such as interest-rate derivatives, or offering a more extended and improved coverage of what was already included in the first edition. While there is indeed some new material, I preferred the second option. Actually, the original plan of the book included two more chapters on interest-rate derivatives, as many readers complained about this lack in the first edition. While writing this increasingly long second edition, I switched to plan B, and interest-rate derivatives are just outlined in the second chapter to point out their peculiarities with respect to stock options. In fact, when planning this new edition, many reviewers warned that there was little hope to cover interest-rate derivatives thoroughly in a limited amount of pages. They require a deeper understanding of risk-neutral pricing, interest rate modeling, and market practice. I do believe that the many readers interested in this topic can use this book to build a solid basis in numerical methods, which is helpful to tackle the more advanced texts on interest-rate derivatives.

Interest-rate derivatives are not the only significant omission. I could also mention implied lattices and financial econometrics. But since there are excellent books covering those topics and I see this one just as an entry point or a complement, I felt that it was more important to give a concrete understanding of the basics, including some less familiar topics. This is also why I prefer using MATLAB, rather than C++ or Visual Basic. While there is no doubt that C++ has many merits for developing professional code, both in terms of efficiency and object orientation, it is way too complex for newcomers. Furthermore, the heavy burden it places on the reader tends to overshadow the underlying concepts, which are the real subject of the book. Visual Basic would be a very convenient choice: It is widespread, and it does not require yet another license, since it is included in software tools that almost everyone has available. Such a choice would probably increase my royalties as well. Nevertheless, MATLAB code can exploit a wide and reliable library of numerical functions and it is much more compact. To the very least, it can be considered a good language for fast prototyping. These considerations, as well as the introduction of new MATLAB toolboxes aimed at financial applications, are the reasons why I am sticking to my original choice. The increasing number of books using MATLAB seems to confirm that it was a good one.

Acknowledgments. I have received much appreciated feedback and encouragement from readers of the first edition of the book. Some pointed out typos, errors, and inaccuracies. Offering apologies for possible omissions, I would like to thank I-Jung Hsiao, Sandra Hui, Byunggyoo Kim, Scott Lyden, Alexander Reisz, Ayumu Satoh, and Aldo Tagliani.

Supplements. As with the first edition, I plan to keep a web page containing the (hopefully short) list of errata and the (hopefully long) list of supplements, as well as the MATLAB code described in the book. My current URL is:

- http://staff.polito.it/paolo.brandimarte

For comments, suggestions, and criticisms, my e-mail address is

- paolo.brandimarte@polito.it

One of the many corollaries of Murphy's law says that my URL is going to change shortly after publication of the book. An up-to-date link will be maintained both on Wiley Web page:

- http://www.wiley.com/mathematics

and on The MathWorks' web page:

- http://www.mathworks.com/support/books/

PAOLO BRANDIMARTE
Turin, March 2006

From the Preface to the First Edition

Crossroads are hardly, if ever, points of arrival; but neither are they points of departure. In some sense, crossroads may be disappointing, indeed. You are tired of driving, you are not at home yet, and by Murphy's law there is a far-from-negligible probability of taking the wrong turn. In this book, different paths cross, involving finance, numerical analysis, optimization theory, probability theory, Monte Carlo simulation, and partial differential equations. It is not a point of departure, because although the prerequisites are fairly low, some level of mathematical maturity on the part of the reader is assumed. It is not a point of arrival, as many relevant issues have been omitted, such as hedging exotic options and interest-rate derivatives.

The book stems from lectures I give in a Master's course on numerical methods for finance, aimed at graduate students in Economics, and in an optimization course aimed at students in Industrial Engineering. Hence, this is not a research monograph; it is a textbook for students. On the one hand, students in Economics usually have little background in numerical methods and lack the ability to translate algorithmic concepts into a working program; on the other hand, students in Engineering do not see the potential application of quantitative methods to finance clearly.

Although there is an increasing literature on high-level mathematics applied to financial engineering, and a few books illustrating how cookbook recipes may be applied to a wide variety of problems through use of a spreadsheet, I believe there is some need for an intermediate-level book, both interesting to practitioners and suitable for self-study. I believe that students should:

- Acquire *reasonably* strong foundations in order to appreciate the issues behind the application of numerical methods

- Be able to translate and check ideas quickly in a computational environment

- Gain confidence in their ability to apply methods, even by carrying out the apparently pointless task of using relatively sophisticated tools to pricing a vanilla European option

- Be encouraged to pursue further study by tackling more advanced subjects, from both practical and theoretical perspectives

The material covered in the book has been selected with these aims in mind. Of course, personal tastes are admittedly reflected, and this has something to do with my Operations Research background. I am afraid the book will not please statisticians, as no econometric model is developed; however, there is a wide and excellent literature on those topics, and I tried to come up with a complementary textbook.

The text is interspersed with MATLAB snapshots and pieces of code, to make the material as lively as possible and of immediate use. MATLAB is a flexible high-level computing environment which allows us to implement non-trivial algorithms with a few lines of code. It has also been chosen because of its increasing potential for specific financial applications.

It may be argued that the book is more successful at raising questions than at giving answers. This is a necessary evil, given the space available to cover such a wide array of topics. But if, after reading this book, students will want to read others, my job will have been accomplished. This was meant to be a crossroads, after all.

PS1. Despite all of my effort, the book is likely to contain some errors and typos. I will maintain a list of errata, which will be updated, based on reader feedback. Any comment or suggestion on the book will also be appreciated. My e-mail address is: `paolo.brandimarte@polito.it`.

PS2. The list of errata will be posted on a Web page which will also include additional material and MATLAB programs. The current URL is

- `http://staff.polito.it/paolo.brandimarte`

An up-to-date link will be maintained on Wiley Web page:

- `http://www.wiley.com/mathematics`

PS3. And if (what a shame ...) you are wondering who Commander Straker is, take a look at the following Web sites:

- `http://www.ufoseries.com`
- `http://www.isoshado.org`

PAOLO BRANDIMARTE
Turin, June 2001

Part I
Background

1
Motivation

Common wisdom would probably associate the ideas of numerical methods and number crunching to problems in science and engineering, rather than finance. This intuitive view is contradicted by the relatively large number of books and scientific journals devoted to computational finance; even more so, by the fact that these methods are not confined to academia, but are actually used in real life. As a result, there has been a steady increase in the number of academic programs devoted to quantitative finance, both at Master's and Ph.D. level, and they usually include a course on numerical methods. Furthermore, many people with a quantitative or numerical analysis background have started working in finance, including engineers, mathematicians, and physicists.

Indeed, as the term *financial engineering* may suggest, computational finance is a field where different cultures meet. Hence, a wide array of students and practitioners, with diverse background, will hopefully be interested in a book on numerical methods for finance. On the one hand, this is good news for the author. On the other one, the first difficult task is to get everyone on common ground as far as financial theory and the basics of numerical analysis are concerned; if treatment is too brief, there is a significant risk of losing a considerable subset of readers along the way; if it is too detailed, another subset will be considerably bored. The aim of the first three chapters is to "synchronize" readers with a background in Finance and readers with a scientific background, including students in Engineering, Mathematics, and Physics. In chapter 2, we will give the second subset of readers an overview of concepts in finance, with an emphasis on asset pricing and portfolio management. The first subset of readers will find a reasonably self-contained treatment on classical topics of numerical analysis in chapter 3.

In this introductory chapter we want to give a preview of the problems we will deal with, along with some motivation. The reader who is unfamiliar with some topics just outlined here should not be worried, as they are not taken for granted and will be treated thoroughly in the next chapters. We want to make three points:

1. In financial engineering we need numerical methods (section 1.1).

2. We need sophisticated and user-friendly numerical computing environments, such as MATLAB[1] (section 1.2), even if this does not prevent at all the use of (relatively) low-level languages such as Fortran or C++ or spreadsheets such as Microsoft Excel.

3. Whatever software tool we select, we need a reasonably strong theoretical background, as we must often select among competing methods and many things may go wrong with them (section 1.3).

1.1 NEED FOR NUMERICAL METHODS

Probably, the best-known result in financial engineering is the Black–Scholes formula to price options on stocks.[2] Options are a class of *derivatives*, i.e., financial assets whose value depends on another asset, called the underlying. The underlying can also be a non-financial asset, such as a commodity, or an arbitrary quantity representing a risk factor to someone, such as weather, so that setting up a market to transfer risks makes sense. Options are *contracts* with very specific rules for issuing, trading, and accounting. For instance, a European-style call option on a stock gives the holder the right, but not the obligation, of buying a given stock at a given time (maturity, denoted by T), for a prespecified price (the strike price, denoted by K). Similarly, a put option gives the right to sell the underlying asset at a predetermined strike price. In European-style derivatives, the right specified in the contract can only be exercised at maturity T; in American-style derivatives, one can exercise her right at *any* time *before* T, which in this case plays the role of the expiration date of the option.

In the case of a European-style call option, if the asset price at maturity is $S(T)$, then the payoff is $\max\{S(T) - K, 0\}$. The rationale here is that, under idealized assumptions on financial markets, the option holder could purchase

---

[1] MATLAB is a registered trademark of The MathWorks, Inc. For more information, see http://www.mathworks.com.
[2] The formula was published by Fisher Black and Myron Scholes in 1973. A similar research line had been pursued by Robert Merton, and in fact Scholes and Merton were awarded the Nobel prize in Economics in 1997. By that time, unfortunately, Fisher Black was deceased.

the underlying asset at the prevailing price $S(T)$ and immediately sell it at price K. Clearly, the option holder will do so only if this results in a positive profit. Actually, market imperfections, such as transaction costs or bid–ask spreads, prevent such an idealized trade: even if $S(T)$ is the last quoted price, there is no guarantee that the option holder can actually buy the stock at that price. In the book we will neglect such issues, which are related to the *micro-structure* of financial markets.

If we are at a time instant $t < T$, we would like to assign a value, or a fair price, to the option. However, what we know is only the current price $S(t)$ of the underlying asset, whereas its price $S(T)$ at maturity is not known. If we build some mathematical model for the dynamics of the price $S(t)$ as a function of time, we may regard $S(T)$ as a random variable; hence, the payoff is random as well, and there seems to be no trivial way to price this contract. Let $f(S(t), t)$ be the price of the option at time t if the current price of the underlying asset is $S(t)$; to ease the notation burden we will usually write it as $f(S, t)$. It can be shown that, under suitable assumptions, the value of the contract really depends only on t and S, and it satisfies the following partial differential equation (PDE):

$$\frac{\partial f}{\partial t} + \frac{1}{2}\sigma^2 S^2 \frac{\partial^2 f}{\partial S^2} + rS \frac{\partial f}{\partial S} - rf = 0, \tag{1.1}$$

where r is the risk-free interest rate, i.e., the rate of interest one can earn by investing her money in a safe account, and σ is a parameter related to the volatility of the price of the underlying asset, which is a risky asset. Typically, we are interested in the current value $f(S_0, 0)$, where $S_0 = S(0)$. Equation (1.1), with the addition of suitable boundary conditions linked to the type of option, may be solved analytically in some cases. For instance, if we denote the cumulative distribution function[3] for the standard normal distribution by $N(z) = \mathrm{P}\{Z \leq z\}$, where Z is a standard normal variable, the price C_0 for a European call option at time $t = 0$ is

$$C_0 = S_0 N(d_1) - K e^{-rT} N(d_2), \tag{1.2}$$

where

$$d_1 = \frac{\ln(S_0/K) + (r + \sigma^2/2)T}{\sigma\sqrt{T}},$$

$$d_2 = \frac{\ln(S_0/K) + (r - \sigma^2/2)T}{\sigma\sqrt{T}} = d_1 - \sigma\sqrt{T}.$$

This formula is easy to evaluate, but in general we are not so lucky. The complexity of the PDE or of some additional conditions, which we must impose to fully characterize a specific option, may require numerical methods. We will

---

[3] See appendix B for a refresher on Probability and Statistics.

6 MOTIVATION

cover relatively simple numerical methods for solving PDEs, based on finite differences, in chapter 5, and applications to option pricing will be illustrated in chapter 9. Using finite differences, in turn, may call for the repeated solution of systems of linear equations, which is among the topics of chapter 3 on numerical analysis.

Apart from the obvious computational advantage, analytical formulas are of great importance in gaining insights into how different factors affect option prices. They also allow quick calculation of price sensitivities with respect to such factors, which are relevant for risk management. In the book, we will use analytical formulas quite often in order to validate numerical methods, by comparing the numerical result with the theoretically correct one. This is of no practical value by itself, but it is very instructive. Finally, we will also see that when a complex option cannot be priced analytically, knowing an analytical pricing formula for a related simpler option can be of great value. In option pricing by Monte Carlo simulation (see below), analytical pricing formulas may yield control variates useful to reduce variance in the estimate of price.

Nevertheless, we should note that the distinction between numerical and analytical methods is sometimes a bit blurred. It may happen that analytical formulas are quite complicated. As an example, let us consider the following formula, which we give without much explanation[4]:

$$C_J = \sum_{n=0}^{\infty} \frac{e^{-\lambda T}(\lambda T)^n}{n!} \mathrm{E}\left\{C_{\mathrm{BLS}}\left[S_0 X_n e^{-\lambda \chi T}, T, K, \sigma^2, r\right]\right\}.$$

This is a formula for the price of a European-style call option when price jumps are included in the model. The Black–Scholes model assumes continuous paths for prices, and this formula by Robert Merton generalizes to a model in which jumps occur according to a compound Poisson process. Here $C_{\mathrm{BLS}}(S, T, K, \sigma^2, r)$ is the standard Black–Scholes formula with the usual input arguments; λ is related to the rate of jumps, i.e., the expected number of jumps per unit time; X_n is a random variable related to the size of jumps, and expectation in the formula is with respect to this variable; χ is a number which is also related to the probability distribution of jump sizes. Even without fully understanding this formula, which goes beyond the scope of this introductory book, it is clear that evaluating it is not so trivial and calls for some computational approximation. Nevertheless, it gives an explicit representation of the effect of each factor affecting price, whereas in a purely numerical approach this important information is lost.

Even in the simple case of equation (1.2), some numerical method is actually applied, since we have to evaluate the function:

$$N(z) = \frac{1}{\sqrt{2\pi}} \int_{-\infty}^{z} e^{-y^2/2}\, dy,$$

---

[4]See [5, page 320] for details.

where the integral cannot be solved in closed form. Here, we may evaluate the integral by quite efficient ad hoc approximation formulas, rather than by general-purpose methods for numerical integration. Sometimes, however, we have to compute or approximate integrals in multiple dimensions. In fact, thanks to a result known as Feynman–Kač formula, the solution of a PDE such as (1.1) can be expressed as an expected value. This and other pricing arguments imply that option prices may be expressed as expected values, which boil down to an integral. Unfortunately, when expectation is taken with respect to *many* random variables, standard methods to compute integrals in low-dimensional spaces fail.

In other problem settings, we have to approximate a *function* defined by an integral. For instance, consider a function $g(x, y)$ and define a function of x by

$$F(x) = \int_a^b g(x, y) f_Y(y) \, dy.$$

Such a situation occurs often in stochastic optimization, when x is a decision variable influencing the result, which is only partially under our control because of the effect of a random "disturbance" Y, whose density is $f_Y(y)$ over the support $[a, b]$ (possibly $(-\infty, +\infty)$). The function $F(x)$ can be considered as the expected cost or profit resulting from our decisions. We will see concrete examples in chapters 10 and 11.

Since computing integrals is so important, chapter 4 is entirely devoted to this topic. Apart from deterministic integration methods, we will also deal extensively with random sampling methods known as Monte Carlo integration or Monte Carlo simulation. Monte Carlo simulation has a incredibly wide array of applications, including option pricing and risk management. For instance, it can be shown that the price of a European call option at time $t = 0$ is given by the following expected value:

$$C = \mathrm{E}^{\mathbb{Q}} \left[e^{-rT} \max\{S_T - K, 0\} \right],$$

where S_T is the (random) price of the underlying asset at maturity, and the expected value is taken under a suitably chosen probability measure (denoted by $\mathbb{Q}$). In other words, the option value is the expected value of the payoff, discounted back to time $t = 0$, under a certain probability measure. If we are able to generate M independent random samples $S_T^{(j)}$, $j = 1, \ldots, M$, of the asset price, under probability measure $\mathbb{Q}$, then by the law of large numbers we could estimate the expected value by the sample mean

$$\hat{C} = \frac{1}{M} \sum_{j=1}^{M} e^{-rT} \max\{S_T^{(j)} - K, 0\}.$$

This is the essence of Monte Carlo simulation, and a number of tricks of the trade are needed in order to obtain a reliable and computationally efficient estimate.[5] Variance reduction methods and alternative integration approaches based on low-discrepancy sequences will be introduced in chapter 4, and applications to option pricing are illustrated in chapter 8.

Another widely applied approach to option pricing is based on binomial or trinomial lattices. These can be regarded as a sort of clever discretization of the underlying stochastic process. From this point of view, they are a deterministic way to generate sample paths, whereas Monte Carlo is based on random sample path generation. Another point of view is that certain finite difference approaches can be regarded as generalization of trinomial lattices. We will see applications of these methods in chapter 7.

Another major topic of the book is optimization, which is introduced in chapter 6. Optimization models and methods play many different roles in finance. In the option pricing context, optimization is at the core of pricing American-style options. Since American-style options may be exercised at any time before expiration, optimal exercise strategies must be accounted for in pricing. For instance, in an American-style call option, it would be tempting to exercise the option as soon as it gets *in-the-money*, i.e., when $S(t) > K$ for a call option and you could earn an immediate profit. However, one should also wonder if it could be better to wait for a better opportunity. This is not a trivial problem; indeed, it can be shown that it is *never* optimal to exercise an American-style call option on a stock, unless it pays dividends before expiration.

An older type of application of optimization methods is portfolio management. Given a set of assets in which one can invest her wealth, we must decide how much should be allocated to each one of them, given some characterization of the uncertainty in assets return. The best-known portfolio optimization model is based on the idea of minimizing the variance of portfolio return (a measure of risk), while meeting a constraint on its expected value. This leads to mean-variance portfolio theory, a topic pioneered by Harry Markowitz in the 1950s. While somewhat idealized, this model had an enormous practical and theoretical impact, eventually earning Markowitz a Nobel prize in Economics in 1990.[6] Since then, many different approaches to portfolio optimization have been developed, and they will be illustrated in chapters 10, 11, and 12.

---

[5] As we mentioned, option pricing by solving a partial differential equation or by computing an expectation are theoretically equivalent approaches, via Feynman–Kač formula. However, they can be quite different in computational terms. It is interesting to note that, historically, Black–Scholes formula was first obtained by solving the pricing PDE analytically, whereas the recent tendency is to use expectation based approaches because of their generality.

[6] Markowitz shared the prize with Merton Miller and William Sharpe. What is probably less known is that he was among the developers of SimScript, one of the first programming languages for discrete-event simulation. By the way, Robert Merton had a background in engineering. This shows how artificial the barriers between Economics and Engineering may be.

It is also important to note that asset pricing and portfolio optimization are not necessarily disjoint topics. Many Financial Economics theories are based on portfolio optimization models which in turn lead to asset pricing models. We will not cover these topics, however, both because of space limitations and because they are not strictly related to numerical methods.

There are still other kinds of application of optimization methods, which may more instrumental, such as parameter fitting or model calibration. In complex markets, asset prices may depend on a set of unobservable parameters, and one would like to introduce and price a new asset, in a way which is coherent with observed prices for other traded assets. To do so, a typical approach is the following. First we build a theoretical pricing model, depending on such parameters. Then we try to find values for these parameters, which are as coherent as possible with observed prices. Let $\boldsymbol{\alpha}$ be the vector of unknown parameters; according to the asset pricing model, the theoretical price of asset j should be $\hat{P}_j(\boldsymbol{\alpha})$, whereas the observed price is P_j^o. We would like to get a vector of parameters yielding the best fit. A standard way to do so is solving the following optimization model:

$$\min_{\boldsymbol{\alpha}} \sum_j \left(P_j^o - \hat{P}_j(\boldsymbol{\alpha}) \right)^2.$$

Then, given the optimal set of parameters, we may proceed to price new assets using the theoretical model. This type of approach is essential in pricing interest-rate derivatives. Interest-rate derivatives are considerably more difficult to analyze than options on stocks and are outside the scope of this book; we will just outline the related issues in section 2.8.

As expected, some simple optimization models may be solved analytically, yielding quite useful insights. However, as a rule, very sophisticated computational approaches are needed.

1.2 NEED FOR NUMERICAL COMPUTING ENVIRONMENTS: WHY MATLAB?

MATLAB is an interactive computing environment, providing both basic and sophisticated functions. You may use built-in functions to solve possibly complex but standard problems, or you may devise your own programs by writing them as M-files, i.e., as text files including sequences of instructions written in a high-level matrix-oriented language. Moreover, MATLAB has a rich set of graphical capabilities, which we will use in a very limited fashion, including the ability of quickly developing graphical user interfaces. The unfamiliar reader is referred to appendix A for a quick tour of MATLAB programming.

Some classical numerical problems are readily solved by MATLAB functions. They include:

- Solving systems of linear equations

10 MOTIVATION

- Solving non-linear equations in a single unknown variable (including polynomial equations as a special case)

- Finding minima and maxima of functions of a single variable

- Approximating and interpolating functions

- Computing definite integrals (in low-dimensional spaces)

- Solving ordinary differential equations, as well as some simple PDEs

This and much more is included in the basic MATLAB core. More complex versions of these problems may be solved by other MATLAB ready-to-use functions, but you have to get the appropriate toolbox. A toolbox is simply a set of functions written in the MATLAB language, and it is usually provided in source form, so that the user may customize or use the code as a starting point for further work. For instance, the Optimization toolbox is needed to solve complex optimization problems, involving several decision variables and possibly complex constrains, as well as to solve systems of non-linear equations. Another relevant toolbox for finance is the Statistics toolbox, which includes many more functions than we will use. In particular, it offers functions to generate pseudorandom numbers that are needed to carry out Monte Carlo simulations. Based on the Statistics and Optimization toolboxes, a Financial toolbox was first devised a few years ago, which included different groups of functionalities. Some were low-level functions aimed at date and calendar manipulation or finance-oriented charting, which are building blocks for real-life applications; others dealt with simple fixed-income assets, portfolio optimization, and derivatives pricing.

After this first toolbox, others were introduced which are directly related to finance:

- GARCH toolbox

- Financial time series toolbox[7]

- Financial derivatives toolbox

- Fixed-income toolbox

We will not deal with such toolboxes in the book, but information can be obtained by browsing The MathWorks' Web site (http://www.mathworks.com). We should also mention that other toolboxes, which were not specifically developed for financial applications, could be useful, such as the PDEs toolbox

---

[7] At the time of writing, the functionalities of this toolbox have been included in the Financial toolbox.

or the genetic and direct search toolbox.[8] Other more instrumental tools are useful to develop professional applications, such as Excel link, Web server, the compiler, or the Datafeed module enabling web connections to different financial web sites.

Now the question is: Why choose MATLAB for this book? Indeed, there are different competitors, at different levels:

- User-friendly spreadsheets, such as Microsoft Excel. In fact, there are spreadsheet-based books showing how optimization and simulation methods may be applied to financial problems. Spreadsheets are equipped with solvers able to cope with small-scale mathematical programming problems, and extensions are available to run Monte Carlo simulations or optimization by genetic algorithms.

- On the opposite side of the spectrum, one could use low-level languages such as C++ or Fortran. C++ seem a favorite, if you look at the number of books on computational finance based on this language, but there are people maintaining that the recent versions of Fortran do still have some advantages. C++ or Fortran may be used either to implement the algorithms directly or to call available scientific computing libraries.

- There are also specialized libraries or environments, such as statistical or optimization tools.

How does MATLAB compare against such alternatives? The obvious answer is that the choice is largely a matter of taste, and it depends on your aim.

Sure, when you have to carry out simple computations, there's little point in resorting to a full-fledged computing environment, and probably spreadsheets are the best choice. However, the extra effort in learning a programming language pays off when you have to program a complex numerical method which goes beyond what is standard and readily available. Actually, there is no way to really learn numerical methods without some knowledge of a programming language, and in any case, even if you use a spreadsheet as the front end, it is quite likely that you have to write some code in Visual Basic or C++.

Compiled languages such as Fortran and C++ are certainly the most efficient option, in terms of execution speed.[9] If you have to write really lightning-fast code, this is the best choice.

---

[8] Genetic algorithms and direct search methods are optimization methods which do not require computing derivatives of the objective function. This makes them very flexible for some types of optimization models, as we will see in chapters devoted to optimization.

[9] A compiled language is based on the translation of source level code to machine level language. You need a compiler to do that; optimized compilers are able to obtain extremely fast code. An interpreter does not translate to machine level code, but to some internal form which is then executed. Usually an interpreter has some advantage in terms of debugging and flexibility, which is paid in terms of execution speed.

12   MOTIVATION

MATLAB is an interpreted language, and even if it is quite efficient, there is some difference. However, the performance gap is being bridged by increasingly fast MATLAB versions. Furthermore, executable libraries can be generated from MATLAB code by using the MATLAB compiler; these libraries can then be linked within the application just as any C++ code. But the most important advantage of MATLAB is that it is a very simple, yet powerful, programming language. Unlike C++, you may avoid bothering with issues such as memory allocation, variable declaration, etc. MATLAB is an excellent rapid prototyping tool: You may implement a quite complex algorithm with a very limited amount of lines. Simple code means less time to develop and less chances for programming bugs. Then, if it is really needed, you may go on by translating the prototyped code to, e.g., C++. This is obviously important in a practical setting, but it is not really essential in a didactic book like the present one. When learning a numerical method, being distracted by too many programming details is certainly bad.

MATLAB can be thought of as a suitable compromise between conflicting requirements. The increasing number of toolboxes and books using MATLAB is a good proof of that. Needless to say, this does not imply that MATLAB has no definite limitations. When one has to deal with large-scale optimization problems, it is necessary to resort to specialized packages such as CPLEX,[10] against which MATLAB is unlikely to be competitive (it should be noted that the Optimization toolbox is aimed at general non-linear programming, whereas some optimization packages deal only with linear and quadratic programming). Furthermore, mixed-integer programming problems[11] cannot be solved, at present, by MATLAB.[12] Even worse, when you have a large optimization model, loading the data in a form suitable to a numerical library function is a difficult and error-prone task without the support of algebraic modeling languages such as AMPL.[13] This is one of the reasons why, in the chapters on optimization models, we will sometimes solve them using AMPL. This should not place any burden on the reader, since a free demo version can be downloaded from the AMPL web site. See appendix C for a quick tour of AMPL.

By the same token, if one is interested in statistical computing applied to finance, it is quite likely than some of the many econometric packages are

---

[10] CPLEX is a registered trademark of ILOG. See http://www.ilog.com.

[11] Mixed-integer programming models are optimization models in which some decision variables are restricted to integer, rather than real, values. They are dealt with in chapter 12. See also example 1.2 on page 15.

[12] We should mention that the latest release of the Optimization Toolbox does include a solver for certain pure binary (0/1) linear programming. However, this is not suitable to large scale mixed-integer programming.

[13] AMPL (A Mathematical Programming Language) was originally developed at Bell Laboratories. At present it is available in many versions through different sellers, including ILOG, under license from the copyright owner. See http://www.ampl.com.

better suited to the task. The point is that none of these offers the many functionalities of MATLAB within a single integrated environment.

To summarize, we may argue that a product like MATLAB is the best *single* tool to lay down good foundations in numerical methods. Cheap MATLAB student editions are available, and its use in finance is spreading. So we believe that learning MATLAB is definitely an asset for students and practitioners in financial engineering.

A last choice had to be made in writing the book: To which extent should toolboxes be used? On the one hand, using too many toolboxes would place some burden on the reader, who may not have access to all of them. On the other hand, using only the MATLAB core would probably limit what we can do. So, again, a compromise must be reached. Our choice has been to use a very limited subset of functions from the Statistical and Financial toolbox, which can be easily replicated. We will sometimes use functions from the Optimization toolbox, but the same results can be obtained by the free AMPL demo version. We will use neither advanced financial toolboxes nor the Partial Differential Equations Toolbox. This choice is somewhat contradictory: Why use the Optimization toolbox and not the PDEs one? The point is that there is a wide gap between a conceptual statement of optimization methods, and a robust working implementation. It is not the aim of this book to bridge that gap, so we will avoid a detailed treatment of most optimization methods, limiting ourselves to the principles behind them. On the contrary, simple finite difference methods are relatively easy to implement, and can be treated in detail. Finally, we should also note that typical computational finance courses do cover basic finite difference methods for solving PDEs, but not sophisticated optimization methods.

1.3 NEED FOR THEORY

Now that we established that we are going to use MATLAB in the book, another question may arise: Why should we bother learning numerical *methods*, when they are already available in professionally crafted, ready-to-use code? Can we get rid of theory? Although, in most cases, there is no need for a deep knowledge of numerical analysis in order to use MATLAB, there are at least three reasons to gain a basic understanding of the theoretical background of numerical methods.

1. Without a sound background, you cannot go on developing your own solutions when the available methods are not enough.
2. Without a sound background, you cannot choose the most appropriate algorithm when alternatives are given.
3. Without a sound background, you cannot use methods properly and, most important, you cannot understand what is going wrong when results are not reasonable or you get weird error messages.

14   MOTIVATION

In particular, we need some understanding of fundamental issues like "conditioning of a numerical problem" and "stability of an algorithm." These concepts are briefly discussed in chapter 3. Here we give some simple examples of the trouble one can get into without a sound knowledge of the pitfalls of numerical computing.

Example 1.1 Consider the following expression:

$$9 \cdot 8.1 + 8.1$$

Everyone would agree that this is just a complicated way to write $10 \times 8.1 = 81$. Let us try it on a computer, using MATLAB:

```
>> 9 * 8.1 + 8.1
ans =
   81.0000
```

Everything seems right. Now, there is a built-in function in MATLAB, `fix`, which can be used to round a number to the integer nearest to zero.[14] Note that `fix` does *not* round to the nearest integer:

```
>> fix(4.1)
ans =
    4
>> fix(4.9)
ans =
    4
```

Let us try it on the expression above:

```
>> fix(9*8.1 + 8.1)
ans =
   80
```

Now something seems quite wrong. Actually, the point is that the first result is not what it looks like. This may be seen by changing the visualization format of numbers and trying again:

```
>> format long
>> 9 * 8.1 + 8.1
ans =
   80.99999999999999
```

Actually, there was some warning, since MATLAB printed 81.0000 rather than 81, as it happens with

---

[14]The reader is urged to explore the differences between `fix` and the related functions `floor`, `ceil`, and `round`.

```
>> 10 * 8.1
ans =
    81
```

The problem is that an innocent-looking number like 8.1 is not represented exactly on a computer. This is because a computer works with a finite precision arithmetic based on a binary representation, which can represent some numbers only approximately, even if their representation is finite in another system, like the decimal system we are used to. □

In this example we see a large effect of a small error. This happens because of the non-linear character of the `fix` operator. The example may look a bit artificial, and one could be tempted to think that such difficulties do not arise in practice. In the next example we see the relevant effect of similar small errors in a concrete setting.

Example 1.2 Let us consider a trivial model for capital budgeting decisions. We must allocate a given amount W of money to a set of N potential investments. For each investment opportunity, we know

- The initial capital outlay C_i, $i = 1, \ldots, N$

- The revenue R_i that we will get from the investment (which we assume certain)

We would like to select the subset of investments yielding the largest revenue, subject to a budget constraint. This looks like a portfolio optimization model, the key difference being that our decision must be "all-or-nothing." For each investment opportunity we may decide weather we take it or leave it, but we cannot buy a fractional share of it. In typical portfolio optimization models, assets are assumed infinitely divisible, which may often be a reasonable approximation, e.g., for stocks, but not in this case. It may be helpful to think of our investments as projects that can be started or not.

The decision variables must reflect the logical nature of our decision. This is obtained by restricting the decision variables as follows:

$$x_i = \begin{cases} 1 & \text{if we invest in project } i \\ 0 & \text{otherwise.} \end{cases}$$

Now it is easy to build an optimization model:

$$\max \quad \sum_{i=1}^{N} R_i x_i$$

$$\text{s.t.} \quad \sum_{i=1}^{N} C_i x_i \leq W$$

$$x_i \in \{0, 1\}.$$

This model is grossly simplified, but it is a first example of an integer programming model. It is also well known as the *knapsack problem*, as each investment may be interpreted as an object of given value R_i and volume C_i, and we want to determine the maximum value subset of objects that may fit the knapsack capacity W. A model like this looks deceptively simple. However, it cannot be solved by ordinary optimization methods for continuous optimization models. One could think of simply enumerating all of the feasible solutions, which are a finite set, in order to spot the best one. Unfortunately, this is not feasible in general, as the number of feasible solutions may be very large, even though finite. To see this, notice that there are N variables which can take two values; hence, there are 2^N possible variable assignments. Many of them would be ruled out by the budget constraints, but we see that the computational effort of complete enumeration grows exponentially with the size of the problem. A possible solution approach would be ordering the items in decreasing order of their return R_i/C_i and selecting them until the budget allows. This would work with divisible assets, but it does not guarantee the optimal solution in the discrete case. As a counterexample, consider the following problem:

$$\begin{aligned}
\max \quad & 10x_1 + 7x_2 + 25x_3 + 24x_4 \\
\text{s.t.} \quad & 2x_1 + 1x_2 + 6x_3 + 5x_4 \leq 7 \\
& x_i \in \{0, 1\}.
\end{aligned}$$

The returns are, respectively, $5.00, 7.00, 4.17, 4.80$. Hence, according to this logic we would select investment 2 first, then investment 1, and we would stop there, with a revenue 17, because no other investment fits the residual budget. This is a really bad solution, leaving much budget unused. There are two solutions which exploit the whole budget: $[1, 0, 0, 1]$, with total revenue 34, and $[0, 1, 1, 0]$, with total revenue 32. In this trivial case it is easy to see that the first one is optimal. Unfortunately, in general, a problem like this can only be tackled by non-convex optimization methods, such as branch and bound,[15] described in chapter 12; in that chapter we will see that logical decision variables may be useful in capturing various types of constraints in realistic portfolio management models.

The main limitation of the model above is that uncertainty is not considered at all. Another issue is that in general there might be some interaction among different projects. For instance, it could be the case that a given project, say project P_0, may be started only if projects $P_1, P_2, \ldots, P_N$ are started as well. This logical constraint is easily modeled using the binary decision variables we have just introduced. One possibility is to express the constraint in the

---

[15] This problem may be also solved by some form of dynamic programming; see [7, pp. 72–74]. In chapter 10 we only consider dynamic programming for certain stochastic optimization problems, but the principle is much more general and powerful and it can be applied to some combinatorial optimization problems as well.

following form:
$$x_0 \leq \frac{\sum_{i=1}^{N} x_i}{N}.$$

If we start all the N required activities, the right-hand side of this inequality is simply $N/N = 1$, so that we *may* start P_0, since the constraint boils down to the redundant bound $x_0 \leq 1$. If some required project is missing, the constraint amounts to something like $x_0 \leq \alpha < 1$, which, together with the binary requirement $x_0 \in \{0,1\}$, enforces $x_0 = 0$. In principle, the idea is fine, but does it really work on a computer? Well, in many cases it does, but consider what happens with $N = 3$. Project P_0 will never be selected. In fact, in this case, you should read the constraint above as

$$x_0 \leq \frac{1}{3}x_1 + \frac{1}{3}x_2 + \frac{1}{3}x_3,$$

but unfortunately, even if all the x_i variables are set to 1, due to the finite precision of the computer we have something like

$$x_0 \leq 0.3333333 + 0.3333333 + 0.3333333 = 0.9999999 < 1,$$

where the number of decimals depends on the numerical precision of the machine and on the software involved. Actually, sophisticated optimization software for integer programming does not incur this trouble, since some integrality tolerance is introduced, and 0.9999999 is considered just like 1. Similar considerations apply to any high-quality numerical software, such as MATLAB, but the result can be somewhat unpredictable, as the following snapshot shows:

```
>> fix(1/3 + 1/3 + 1/3)
ans =
     1
>> fix(1/7 + 1/7 + 1/7 + 1/7 + 1/7 + 1/7 + 1/7)
ans =
     0
```

Furthermore, if the optimization problem is first written to a text file, which is then loaded by an optimization solver, it may be the case that the number of digits is too small.[16] So it is better to avoid the trouble with division in the first place, by rewriting the constraint as

$$Nx_0 \leq \sum_{i=1}^{N} x_i,$$

---

[16] For instance, if you solve the model within a modeling system like AMPL, calling a solver like CPLEX, there is no trouble. But if you write an MPS file and load the file with CPLEX, the result will not be correct. MPS files are text files representing optimization models according to standard rules; they are read by many optimization software packages.

or, even better, in the disaggregated form

$$x_0 \leq x_i, \quad i = 1, \ldots, N.$$

Why this is the preferred form is counterintuitive: after all, the disaggregated form entails more constraints, and one would think that the less constraints we have, the easier an optimization model is to solve. This need not be true in *computational* optimization, and it also depends on how mixed-integer programming problems are solved by branch and bound methods. More on this in chapter 12. □

Numerical errors may affect the precision in representing numbers, but this issue is not much trouble in itself; after all, a derivative price will not be quoted in millionths of a dollar. But how about the *propagation* of errors within a numerical algorithm? If you have a non-linear operator like fix, a small error gets immediately amplified. The same may happen when you execute a long sequence of operations, such that small errors cumulate, growing without bound. The effect may well be a huge negative price for an option, as we will see in chapter 9. In the next example we consider a well-known example, linked to the solution of a system of linear equations.

Example 1.3 Let us consider a system of linear equations:

$$\mathbf{Hx} = \mathbf{b},$$

for some right-hand side vector $\mathbf{b}$, where $\mathbf{H}$ is a peculiar matrix known as the *Hilbert matrix*:

$$\mathbf{H} = \begin{bmatrix} 1 & \frac{1}{2} & \frac{1}{3} & \cdots & \frac{1}{n} \\ \frac{1}{2} & \frac{1}{3} & \frac{1}{4} & \cdots & \frac{1}{n+1} \\ \frac{1}{3} & \frac{1}{4} & \frac{1}{5} & \cdots & \frac{1}{n+2} \\ \vdots & \vdots & \vdots & \ddots & \vdots \\ \frac{1}{n} & \frac{1}{n+1} & \frac{1}{n+2} & \cdots & \frac{1}{2n-1} \end{bmatrix}.$$

The Hilbert matrix may look a bit artificial, but it may arise in certain function approximation problems (see example 3.20 on page 190).

MATLAB provides us with a function, hilb, to build a Hilbert matrix directly. Now, let us try solving the system for $n = 20$; we cheat a little here, since we assume that the solution is known, and we build the corresponding right-hand side $\mathbf{b}$; then we check if that solution is obtained by solving the system. Let the solution be

$$\mathbf{x} = [1\ 2\ 3\ \cdots\ n]',$$

where we use ' to denote vector (or matrix) transposition. Using MATLAB, we obtain something like[17]

```
>> H = hilb(20);
>> x = (1:20)';
>> b = H*x;
>> H\b
Warning: Matrix is close to singular or badly scaled.
         Results may be inaccurate. RCOND = 1.995254e-019.

ans =

    1.0000
    2.0000
    3.0018
    3.9392
    5.8903
   -1.1035
   41.0915
  -94.0458
  196.5770
 -181.1961
   82.1903
   12.1684
  140.5377
 -265.1117
  309.7053
 -328.9234
  485.5373
 -401.3571
  215.1260
  -17.0274
```

We see that the result doesn't look quite as it should. □

In the last example we see the typical effect of propagation of numerical errors, giving rise to numerical instability. In fact, this is detected by MATLAB, which issues a warning message. However, we need some theoretical background in order to get the meaning of this warning. One could think that similar difficulties arise whenever a matrix is close to singular. Clearly, if you try doing something like $\mathbf{x} = \mathbf{A}^{-1}\mathbf{b}$ in order to solve the system $\mathbf{Ax} = \mathbf{b}$,

---

[17]The actual result may depend on the MATLAB version and the hardware you use. This is not the case for usual problems, but it does happen when numerical instability issues arise.

you are likely to be in trouble if **A** is close to singular. This may be true, but it is somewhat misleading:

1. You may have difficulties even when the matrix is not singular at all (see example 3.8 on page 151). We need to study issues such as problem conditioning in order to understand what really happens.

2. In practice, there is no need to invert a matrix to solve a system of linear equations, as this would be much more work than necessary. Computational mathematics may be quite different from "pencil-and-paper" mathematics.

At this point, the reader will hopefully be convinced that some background in numerical analysis is needed, if we are to solve problems in real life.

For further reading

In the literature

- Another MATLAB-based textbook is [6]. It is more aimed at applications in Economics, but it offers an interesting Computational Economics toolbox which may be downloaded for free.

- Readers interested in details on the development and release of Microsoft Windows components for financial applications may have a look at [4].

- Financial modeling within Microsoft Excel is described, e.g., in [2].

- C++ programmers will find [1] and [3] very useful.

- Many journals devoted to quantitative finance publish papers on computational issues. We should mention at least

    - Journal of Computational Finance
      http://www.thejournalofcomputationalfinance.com
    - Journal of Derivatives
      http://www.iijod.com
    - Quantitative Finance
      http://www.tandf.co.uk

On the Web

- To consult a full and updated listing of MATLAB toolboxes, see http://www.mathworks.com.

- For more information on CPLEX and related software, see

http://www.ilog.com.

- The web page for AMPL is http://www.ampl.com, where you will find a list of vendors and compatible solvers and a free student version for download.

- Two web sites we should also mention are

  http://www.gams.com, where an alternative product to AMPL is described, which has found fairly widespread use among economists,

  and http://www.nag.com where a well-known numerical analysis library is described, for use with programming languages like Fortran and C++.

REFERENCES

1. D.J. Duffy. *Financial Instrument Pricing Using C++*. Wiley, New York, 2004.

2. M. Jackson and M. Staunton. *Advanced Modelling in Finance using Excel and VBA*. Wiley, New York, 2001.

3. M.S. Joshi. *C++ Design Patterns and Derivatives Pricing*. Cambridge University Press, Cambridge, 2004.

4. G. Levy. *Computational Finance. Numerical Methods for Pricing Financial Instruments*. Elsevier Butterworth-Heinemann, Oxford, 2004.

5. R.C. Merton. *Continuous-Time Finance*. Blackwell Publishers, Malden, MA, 1990.

6. M.J. Miranda and P.L. Fackler. *Applied Computational Economics and Finance*. MIT Press, Cambridge, MA, 2002.

7. L.A. Wolsey. *Integer Programming*. Wiley, New York, 1998.

2

Financial Theory

This chapter is a reasonably brief introduction to some basic problems in finance. It is mostly aimed at readers with a scientific or engineering background, but with little previous exposure (if any) to the theory of finance. The complementary set of readers, i.e., those with a background in finance may wish to have a cursory look at the material, or maybe to refer back to this chapter for a quick refresher when needed.

The treatment here is purely instrumental to motivating and stating certain problems to which we may apply numerical methods. So, it is certainly not meant to be a substitute for a good book on finance (see the references at the end of the chapter), and it is not aimed at giving a complete overview of financial theory. Furthermore, many concepts such as bond portfolio immunization, mean-variance efficiency, and Value at Risk have many well-known limitations and have been the subject of quite a bit of controversy. We will point out the limitations of each approach, and we do not suggest that they should be used as they are stated; we use them just to pave the way for further developments.

The main themes in finance are *time* and *uncertainty*. Actually, there is a third one, *information*, which is important in advanced models which are beyond the scope of this book. Time is important since, under normal economic conditions, one dollar now is worth more than one dollar tomorrow. Even if we do not consider inflation, it is reasonable to expect that if we have one dollar now and we do not need it for immediate consumption, we could go to a bank, deposit our dollar, and recover a larger sum later on. If, after one year, we get $1 + r$ dollars, we say that r is the annual interest rate. We may see it the other way around: if we borrow one dollar now, in the future we will have to give back some more. In fact, one function of financial markets is just to shift consumption over time by borrowing or lending money. In practice, the rates for borrowing and lending are not really the same, as there is a bid–ask spread,[1] but for our instrumental purposes we will mostly neglect such issues.

If we are investing money over a relatively short time period, we could assume that we know the interest rate that will be applied for that period. This may not be the case for longer periods, as interest rates are subject to uncertainty. If the interest rate is periodically reset according to prevailing conditions, then the investment is subject to uncertainty which may be considered as a reinvestment risk. Even if a given *nominal* rate is agreed to hold for the entire period, the *real* rate will be subject to inflation. An even larger uncertainty is typically associated to investing in stocks, which are often subject to significant price volatility. Our first task is to introduce different ways to model uncertainty (section 2.1). There is no "best" way to model uncertainty, as this may depend on our aim, but there is no doubt that uncertainty is pervasive in finance.

Uncertainty is strongly linked to risk. Any investor has some implicit risk tolerance. For instance, common wisdom dictates that older investors should invest in relatively safe assets, whereas younger ones may afford the risk of investing in stock. Apart from individual investors, there are institutional investors, such as pension funds, or even non-financial firms which use financial assets to modify their exposure to some risk factors. In fact, another function of financial markets is to transfer risk among market participants, who can be grossly classified as speculators or hedgers. Speculators have some view on how prices will move in the future, and they perceive risk as an opportunity to place bets. Speculation has a somewhat negative connotation, but without speculators, markets would not exist in their present form. The other side of the coin is the set of hedgers, who use certain types of assets as a sort of insurance in order to avoid or reduce uncertainty. In some sense, hedgers sell volatility to speculators.

In modern financial markets, there is huge variety of assets in which we may invest our money. The main assets we will deal with may be classified as bonds, stocks, and derivatives. We will introduce these assets in section 2.2. There, we also introduce the three main problems we are concerned with: asset pricing, portfolio optimization, and risk management. We will also see that these basic problems are strictly related.

After this general introduction, we deal with simple fixed-income instruments (bonds) in section 2.3, where we also consider sensitivity measures related to interest-rate risk, such as duration and convexity. Section 2.4 is

---

[1] The bid price how much a dealer bids (is willing to pay) for an asset; hence, from the point of view of an investor, it is the price at which she may sell. The ask price is the price at which the investor may buy, i.e., the price asked for by a dealer.

dedicated to stock portfolio management. The main concepts we illustrate there are utility theory for decision making under uncertainty, the theory of mean-variance efficient portfolios, and risk measures such as Value at Risk. To deal with derivative pricing, we need first to lay some foundations in modeling by continuous-time stochastic processes: Stochastic integrals and stochastic differential equations are introduced in section 2.5, together with the fundamental Ito's lemma. Then we proceed to illustrate the basics of arbitrage-free pricing in section 2.6, where the celebrated Black–Scholes formula for pricing European-style vanilla options is presented, along with basic issues in pricing American-style options. We expand the treatment of options in section 2.7, where we outline a few types of exotic options which will be used in later chapters to illustrate different numerical methods for pricing. Finally, in section 2.8 we give a very brief introduction to interest rate derivatives and the related problems.

In the course of the exposition we will use short MATLAB snapshots in order to illustrate the material with examples and to make it immediately useful. Sometimes, we will use functions from the Financial toolbox. The reader without access to this toolbox should not worry: these examples are just used for concreteness, but most of the book is just based on the MATLAB core.

A final remark is in order. A large part of modern theory of pricing derivatives is based on the concept of martingale, i.e., a specific type of stochastic process. However, the reader will not find any mention of martingale measures and the like in what follows. Given the increasingly large number of excellent texts covering martingale pricing, we have decided to omit such concepts, which are not strictly necessary to introduce numerical methods. The main consequence of this choice is the lack of coverage of interest-rate derivatives, which cannot be dealt with adequately without solid foundations; but this would require much more space than we can afford.

2.1 MODELING UNCERTAINTY

Before considering "modeling," we must understand what "uncertainty" is. The familiar tools of probability and statistics are what we need to cope with the simplest kind of uncertainty. We assume that a variable, say the price of a stock or a commodity, can be modeled as a random variable, whose probability distribution is known, possibly inferred from available data; the probability distribution encodes the knowledge we have (or think we have) about uncertainty. This may already look complicated, but it is often far worse in practice. To begin with, we will only consider purely exogenous uncertainty. This means that our actions do not influence the distribution of the relevant random variables. This is true if we are small investors or the asset is very liquid and in large supply. In thin markets, however, buying and selling an asset may have a significant impact on its price, and uncertainty

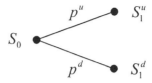

Fig. 2.1 A binomial model for uncertainty.

is partially endogenous. For instance, a trade executed by a large pension fund may have a significant impact on markets; sometimes, to avoid adverse effects, orders are split in different time steps. Another issue is related to "subjective" rather than "objective" uncertainty. We will implicitly assume an objective description of uncertainty, but sometimes an investor has some very specific views, leading to a subjective assessment of uncertainty. The subjective view may be updated whenever we get new information. This is typical of the Bayesian approach to statistics, which has been applied to portfolio management too. Again, given our instrumental point of view, we will avoid such issues. It is important to understand that if we use statistics to identify a probability distribution from past data, and we use that distribution for the future, we are implicitly assuming that, in some sense, history will repeat itself.

To be specific, let us consider possible ways of modeling uncertainty in the price of an asset. The simplest model of uncertainty is the *binomial* model. We know the current price S_0, at time $t = 0$, and we assume that the price S_1, at some future time instant $t = 1$, can take only two values, S_1^u and S_1^d, with probability p^u and p^d, respectively (see figure 2.1). A common choice is to represent uncertainty by a multiplicative shock, i.e., $S_1^u = uS_0$ and $S_1^d = dS_0$, where the letters u and d suggest "up" and "down," respectively (hence, $d < u$). Apparently, this model is very crude, but it is the building block of very useful models.

A more refined model can be built by allowing for more future states. We may consider a sort of tree, like the one depicted in figure 2.2. It is a two-stage tree, in the sense that it represents the world *now* by the single node on the left of the figure, and possible states of the world at one time instant in the future; this structure is sometimes referred to as a *fan*, and it may be used to define a set of discrete scenarios. In this case the random variable S_1 may take values $S_1^{(k)}$, $k = 1, \ldots, m$, with probabilities $p^{(k)}$. An obvious consistency condition is

$$\sum_{k=1}^{m} p^{(k)} = 1; \qquad 0 \leq p^{(k)} \leq 1 \qquad k = 1, \ldots, m.$$

The binomial model or the fan of scenarios are *discrete-state* models, representing uncertainty in a relevant state variable by a discrete probability distribution. The state could be the level of an interest rate, or any underly-

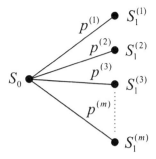

Fig. 2.2 A two-stage tree model for uncertainty.

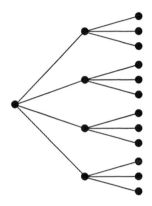

Fig. 2.3 A multistage scenario tree.

ing state variable influencing the price of assets.[2] These models are also the simplest *discrete-time* models, as only two time instants are considered. This may be interesting if we are following a buy and hold strategy, whereby we trade some assets now, and then we just wait for the outcome at some time in the future. If the portfolio will be later rebalanced with some given frequency, we might be interested in a multiperiod model.

A discrete-state, discrete-time, multiperiod model can be depicted as the scenario tree in figure 2.3. This is sometimes called a *bushy* tree. In a bushy tree, the number of nodes following a parent node is called *branching factor*. The larger the branching factor, the more accurate the representation of uncertainty. However, with large branching factors, the number of nodes tends to grow very quickly. Scenario generation is the art of building a suitable tree

---

[2] Strictly speaking, a state variable has the property that knowledge of its value at a time instant is all we need to characterize future evolution. We could have situations in which the whole history of a variable is needed to this purpose. Since this proper use of the term is only relevant for a few topics in the book, we will use the term in the loose sense.

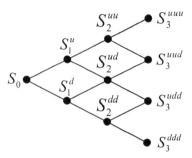

Fig. 2.4 A recombining lattice.

with the minimum number of nodes; note also that there is no need that the branching factor is constant over time, or across nodes. One may use more branches now, and less branches in the future, if it is more important to represent immediate uncertainty. This is important in stochastic programming models (chapter 11). Another point is that the time step involved in a multiperiod model need not be homogeneous. Usually, in a discrete-time model we discretize a time horizon of length T in intervals of length δt, such that $T = M \cdot \delta t$. When we refer to time instant $t = k$, what we really mean is $t = k \cdot \delta t$. However, the time step may change; in such a case, the first time period is short, and time step increases in later periods.

Sometimes, to keep computational effort limited, we prefer using a *recombining lattice*. A recombining binomial lattice is illustrated in figure 2.4. This is obtained if we generalize the binomial model with multiplicative shocks. Since $udS_0 = duS_0$, we see that an up-jump followed by a down-jump is the same as a down-jump followed by an up-jump. In the figure, node S_2^{ud} could also be denoted as S_2^{du}. In the special case $u = 1/d$, we also have, e.g., $S_0 = S_2^{ud}$ and $S_1^u = S_3^{uud}$. The number of nodes grows linearly with the number of periods: We start with one node at $t = 0$, then we have two at time $t = 1$, three at time $t = 2$, and $T + 1$ nodes at time T. In a binary tree we have an exponential growth, as we have 2^T nodes at time T. Note that we are assuming that the multiplicative shocks are always the same, which makes sense if the process is stationary and time step is constant. Lattices may take many different forms, such as trinomial lattices, where each node has three successors. Recombining lattices are very convenient from a computational point of view (see chapter 7). However, they are not always suitable, especially when there are many stochastic factors, calling for larger branching factors and making recombination more difficult to achieve.

Sometimes it is convenient to model uncertainty using a continuous distribution, such as the normal or lognormal distribution. If we think of prices, a continuous distribution is certainly an idealization, since no price is quoted with too many decimal digits. In fact, stock prices are quoted in the USA in fractions of a point, which may be one-eighth or one-sixteenth of a dollar. For

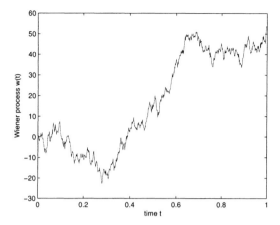

Fig. 2.5 Sample path of a Wiener process.

instance, the price of a stock could be $20\frac{1}{8}$ or $20\frac{1}{4}$, but not $20.19. A similar consideration applies to interest rates. Nevertheless, using a *continuous-state* model may be convenient, if it results in simple modeling of uncertainty and, maybe, in analytical formulas.

By the same token, we may also resort to *continuous-time* models, which may be thought of as the limit of a discrete-time model when the time step tends to zero. In the deterministic case, a standard continuous-time model is a differential equation, like

$$\frac{dB(t)}{dt} = rB(t)$$

with initial condition $B(0) = B_0$. The solution of this equation is $B(t) = B_0 e^{rt}$; in section 2.3.1 we will see that this is the equation of a wealth, initially amounting to B_0, invested at a rate r, with continuous-time compounding of interest. Again, this could be just a convenient approximation. To model uncertainty, differential equations must be extended by introducing a random element, typically represented by some convenient class of stochastic processes. Unlike discrete-time models, we deal in this case with continuous-time stochastic processes (see appendix B). The usual building block is the Wiener process $W(t)$, which is defined later, and is characterized by jagged sample paths like the one depicted in figure 2.5. This process may look fundamentally different from a binomial lattice, but it can be shown that the Wiener process is the continuous-time limit of a certain random walk described by a binomial lattice. By putting Wiener processes and differential equation together in some sensible way, we get stochastic differential equations, which are a rather thorny object to deal with, but are a fundamental tool in financial engineering. We will describe stochastic differential equations in section 2.5.

2.2 BASIC FINANCIAL ASSETS AND RELATED ISSUES

There is a large number of securities in which an investor may be interested. Many of them are standardized, publicly quoted, and traded on exchanges. Some are engineered for a specific need of an investor, or firm, and are traded *over the counter* (OTC); OTC securities are usually less liquid than standardized assets. Despite this virtually infinite variety, we may start by classifying the fundamental securities as

- bonds
- stocks
- derivatives

2.2.1 Bonds

Bonds are one of the instruments that firms and public administrations may use to fund their activities; they are debt instruments which, unlike stocks, do not imply any ownership of a firm on the part of the buyer. Basically, the buyer of a bond lends some money to the issuer, over some time span ending at bond maturity. At maturity the issuer will pay the bond owner an amount of money corresponding to the *face value*, also called the *par value*, of the bond. This could be, e.g., an amount like $100 or $1000. In addition, periodic payments may be made, called *coupons* for historical reasons.[3] In the simplest bonds, coupons are fixed and expressed as a percentage of face value; coupons are usually paid annually or semi-annually. For instance, if the bond has $100 face value, and the coupon rate is 6%, then the bond owner will receive $6 each year, up to and including maturity, when she well receive $106. If coupons are paid semi-annually, the bond owner will receive $3 every six months, up to and including maturity.

There is another class of bonds, which just promise the payment of face value at maturity. They are called *zero-coupon* bonds, and are typically characterized by shorter maturities. We will see that zero-coupon bonds are fundamental in bond pricing. Sometimes, long-term zero-coupon bonds are built by *stripping* coupons from a long-term bond and selling them separately.

The basic type of fixed-coupon bond explains why bonds are usually classified as *fixed-income* securities. Actually, coupons may depend on some underlying variable, but the term "fixed-income" is used for such securities as well. Generally, fixed-income securities are assets whose price depends on the level of interest rates.

It is also important to note that bonds are not necessarily purchased at a price corresponding to face value. This may be the case when bond are first

---

[3] Bonds were physical pieces of paper, and to get the periodic payment the bond owner had to detach a coupon from the document.

issued, and the coupon rate is chosen in order to reflect current interest rates. Since there is a well-developed secondary market for bonds, there is no need to buy a bond right when it is issued, nor to keep it until maturity. If a bond is traded after issue date, we must be able to determine a fair price. This will be the subject of section 2.3.2. Bond prices are quoted as a percentage of the face value, so the actual face value is not so relevant. Assume the face value is 100. If the bond is traded at price larger than 100, we say that it trades *above par*; if the price is smaller, it trades *below par*; otherwise it trades *at par*.

Actually, there are many complicating factors in bond pricing. If the coupon rate is not fixed, but it depends on some random quantity, analyzing a bond may be difficult. Even if the coupon rate is fixed, bond prices may differ depending on the probability of default. Default occurs if the bond issuer is not able to honor his debt and stops paying coupons, or he repays just a fraction of face value. There are different types of default, which represent a risk factor for the investor. This factor is called credit risk. Bonds issued by some governments may be considered risk-free, but corporate bonds cannot; the role of rating agencies is precisely to analyze the financial situation of firms in order to assess how risky their bonds are. Bonds affected by credit risk must sell at lower prices, or promise higher coupon rates. It should also be noted that bonds may be classified in legal terms which are relevant when the firms defaults. We will not consider default issues and credit risk in this book. Furthermore, some bonds have embedded options which complicate the analysis. For instance, a callable bond may be redeemed by the issuer before maturity at a certain price; again, since the issuer may redeem the bond when she finds this advantageous, this must be somehow reflected in the bond price and/or the coupon rate. In this case, the investor is exposed to reinvestment risk, as it is quite likely that she will be forced to reinvest the proceeds from early bond reimbursement in a situation of unfavorable interest rates.

2.2.2 Stocks

Unlike bonds, stocks entitle the owner to a share of the issuing firm. This raises a potentially troublesome legal issue. If you are a stock owner of a firm, and the firm gets involved in a lawsuit, whereby it is liable to pay for some significant damage its products have caused, what is your position? Luckily, stocks are *limited liability* assets; in practice, this means that the worst that may happen is that the stock price goes to zero and you lose all of your investment.

Another difference between stocks and bonds is that the formers do not have a predefined maturity (although the firm can well go out of business). They also entitle the owner to some stream of payments under the form of *dividends*. Unlike fixed bond coupons, dividends are by their very nature stochastic. They depend on how well the firm is faring, and on the dividend policy which is followed by the firm, which may distribute or reinvest its profit. The dividend policy, and the decisions of financing by equity (issuing stocks) or debt (issuing bonds) pertain to a body of knowledge called *corporate finance*.

If you buy a stock share at a price S_0, and then you sell it at a price S_1, you may have a loss or a gain. If you also receive a dividend D, *total return* is

$$\frac{S_1 + D}{S_0},$$

and the *rate of return* is

$$\frac{S_1 + D - S_0}{S_0}.$$

Strictly speaking, we should also consider the timing of dividend payments in order to account for the time value of money, but let us leave this issue aside for now by assuming that dividends are paid exactly when you sell the stock. Since stocks are limited liability assets, the worst-case rate of return is -1. This means that whenever we use a normal distribution to model uncertainty in stock returns, we are committing an error; however, the approximation, *per se*, could be an acceptable one if the probability of an unfeasible return is negligible.[4]

In this book we will not consider pricing issues for stocks. This means that stock prices will be modeled by some stochastic process (see section 2.5) or by some probability distribution, but we will take these as exogenously given. There are "rational" models aimed at suggesting a correct stock value by analyzing the fundamentals of a firm, but they are based on rather uncertain data, and prices may be quite irrational. Nevertheless, such models are useful when trying to assess if some stock is under- or over-priced with respect to other assets, and this is certainly relevant in portfolio management. However, since this is not a matter necessarily dealt with by sophisticated numerical methods, and it calls for integration with qualitative insights, we will leave it aside.

In principle, one would think that an investor buys a stock if she thinks that its price will increase. Actually, with certain limitations, an investor can exploit a strategy called *short-selling* if she thinks the stock price will sink.

Example 2.1 (Short-selling) Suppose a stock is currently selling for $20, and you think that in the near future it will sell for a lower price. In such a case, you may borrow the stock from someone who owns it, and sell it immediately on the market. After a while, you will have to give the stock back to the owner, but if you were right and the price went down to $18, you might buy the stock for this price and close your position. In this case, your return would be $(-18 + 20)/20 = 10\%$. If the stock pays dividend during the

---

[4] Another implicit assumption, when using a normal distribution to model returns, is that these are symmetric, which may not really be the case.

time period over which the stock is lent, dividends must also be paid to the stock lender.

Short-selling is not this easy, as there are several rules constraining it to avoid excessive speculation. Furthermore, it is restricted to certain types of traders; some institutional investors such as pension funds cannot use short-selling because of its speculative nature. Short-selling is very risky: If you are wrong and the price goes up, you may be forced to give the stock back at the worst possible time (this is called *short-squeezing*). □

2.2.3 Derivatives

Derivatives are a broad family of financial contracts, owing their name to the dependency of their payoff on the value of some underlying variable, which may be a stock price, a set of stock prices, an interest rate, an index, or a generic non-financial asset. Suppose that the value of the underlying asset, say a stock which does not pay dividends, is modeled by a stochastic process $S(t)$, depending on time t.

The most common derivatives are forward/future contracts and options. A *forward contract* binds two parties to, respectively, buy and sell a certain asset, in a certain quantity, at a certain date T, and at a fixed forward price F. The party agreeing to buy is said to hold the *long position*, whereas the seller holds the *short position*. By entering a forward contract you basically lock in a fixed price for the underlying asset. You may have two quite different reasons for doing that. You might wish to eliminate, or reduce, risk; in fact, by locking the price for an asset you have to buy or sell, you eliminate the effect of price uncertainty. This does not mean that the final outcome will necessarily be more favorable. If you hold the long position in a forward contract specifying a price F, and the price of the asset when the delivery takes place turns out to be $S(T) < F$, in a sense you have lost an amount $F - S(T)$; if, on the contrary, $S(T) > F$, you have gained a corresponding amount. The point is that if you really need to buy or sell that asset, it may be wise to lock in a certain price rather than taking chances. This type of policy is called *hedging*. Hedging may not be this easy, as you may have difficulties in finding a forward contract for the underlying asset you are interested in, in which case you could settle for a somewhat correlated asset; furthermore, delivery date might differ from the one you would like; finally, one could also decide for a partial hedge, depending on risk attitude. However, you could also be a speculator with a very precise idea of where the price $S(T)$ is going to be, and you may enter a forward contract as a bet. The payoff of a forward contract is depicted in figure 2.6(a) for a long position, in which case it is $S(T) - F$ (it is $F - S(T)$ for the short position). This payoff depends on the random price $S(T)$, and the forward contract is the simplest example of a derivative. Since the payoff is random, we need some way to value a forward contract. We will do this in section 2.6. Here we just note that there is no initial payment with forward contracts; at time $t = 0$ the forward price F is determined in such a way that

34 FINANCIAL THEORY

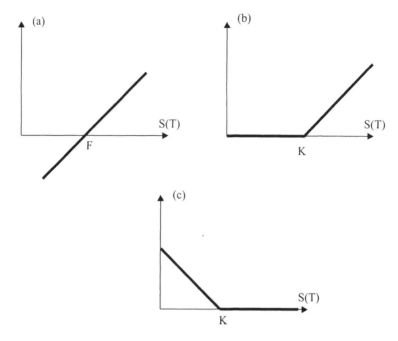

Fig. 2.6 Payoff diagrams for the long position in a forward contract (a), a call option (b), and a put option (c).

the initial value of the contract is zero to both parties. However, at a later time, the value of the contract will not be zero in general.

Derivatives may be private contracts issued by two parties for possibly very peculiar and specific reasons. Alternatively, they may be traded actively on exchanges and quoted on newspapers. In this case, some standardization and regulation is needed to make sure that the derivatives are sufficiently liquid to trade. This is not really the case for forwards, where there is some possibility of default on the part on the part losing money; for this reason future contracts have been devised. A *future contract* is similar to a forward contract; the main difference is that there is an intermediation process such that the detailed working is different. Rather than collecting the payoff at maturity, there is a daily transfer of cash between the two parties, depending on the movement of the underlying asset price. This mechanism is a protection for traders and makes pricing of futures more difficult than forwards, and we refer the reader to references for details on this. It can be shown that prices for futures and forwards are the same if interest rates are deterministic. From a practical point of view, standardized future contracts make trading easier, but hedging more difficult. It may be impossible to find the exact contract you need in terms of time of delivery or underlying asset; in such a case, hedging will eliminate only part of the risk. Nevertheless, futures are a very liquid tool, and it is also interesting to note that, by taking a position with futures, one may also emulate short-selling on assets for which this would be otherwise impossible.

A common feature of forward and future contracts is that the two parties are compelled to buy and sell the asset at delivery (unless you sell the contract to someone else before maturity, as is usually the case with futures). With an *option*, you get the *right*, but not the obligation, to buy or sell a certain asset for a specified price. The two simplest option contracts are the European style call and put options. When you buy a call option, you get the right to buy the underlying asset for a price K, called the *exercise price* (or *strike price*), at a certain date T, called expiration date or maturity. If at maturity the actual price $S(T)$ of the underlying asset is larger than the exercise price K, you would exercise the option and buy the stock, since you may sell the stock immediately and gain $S(T) - K$. If the contrary holds, you would not exercise the option, which expires worthless. Thus, the payoff of this option is

$$\max\{S(T) - K, 0\}$$

and is depicted in figure 2.6(b). If at time t we have $S(t) > K$, we say that the call option is *in-the-money*; this means that we would get an immediate profit by exercising the option. If $S(t) < K$, the call option is said to be *out-of-the-money*. If $S(t) = K$, the option is said to be *at-the-money*.[5] With a put option, you have the right to sell the stock. In this case, you would exercise the option only if the exercise price is larger than the actual price. So the payoff is

$$\max\{K - S(T), 0\}.$$

The payoff diagram for a vanilla European put option is depicted in figure 2.6.(c).

With a European option you may exercise your right only at maturity; an *American option* may be exercised whenever you wish within a prescribed time. European or American call and put options on a single underlying asset are called *vanilla options*, owing their name to their simplicity. A *Bermudan option* is halfway between an American and a European option: It may be exercised at a set of prescribed dates within the horizon. *Asian options* have a payoff depending on the average price of a stock (or some other underlying variable); thus they depend on a set of stock prices. Indeed, quite complex *exotic* options are actually designed and traded; we will describe the simplest exotic options in section 2.7.

Observing the payoff diagrams for the vanilla European call and put, we see that they cannot be negative, unlike a forward contract. Does this imply that you cannot lose money? Well, as you can imagine, the option comes with a price. With a forward contract, you pay nothing when you enter the contract,

---

[5] A simplistic consideration would suggest that an at-the-money option is not worth exercising; however, when considering the transaction costs involved in purchasing a stock, we see that there are circumstances where exercising an at-the-money option may be interesting.

whereas the option has a price depending on several factors including the strike price. Hence, figures 2.6(b) and 2.6(c) are not quite correct, as the payoffs should be shifted down to account for the option price. Indeed, finding this price is the major concern with options, and this is why numerical methods are so important.

Why are options traded? As with futures and forwards, there are two basic reasons. On the one hand, they can be used to control risks. If you hold a stock in your portfolio and you are worried about the possibility of a large drop in its price, you may reduce the risk by buying a protective put. If you hold a portfolio consisting of a stock and a put with strike price K, then the value of the portfolio at option maturity is

$$S(T) + \max\{K - S(T), 0\} = \max\{K, S(T)\}$$

from which we see that the downside risk is limited. This insurance comes with a price, since the option is not free, but in this way you avoid the risk of a large loss. By the same token, you may reduce the interest-rate risk of a fixed-income portfolio by buying interest-rate derivatives. On the other hand, options may also be used for speculation, as shown in the following example.

Example 2.2 Suppose that a stock price is $50, and you believe that it will rise in the near future. You could then buy the stock anticipating a large return. Let's say that you are right and the price rises to $55. Then your rate of return will be

$$\frac{55 - 50}{50} = 10\%.$$

But now imagine that a call option is available with a strike price $50, and that this option costs $5 (this may or may not be a reasonable price, but let us take it as given for the sake of the argument). In this case you will exercise the option, and the rate of return will be much larger:

$$\frac{55 - 50}{5} = 100\%.$$

This effect is called *leverage* or *gearing*. As you may expect, there is another side to the coin. If you are wrong and the stock price drops to $49, then by buying the stock you will lose $1, i.e., 2% of the investment; with the call option you will lose 100%. You are also exposed to other sources of risk if you are interested in selling the option before maturity, as unfavorable movements in the factors determining the option value may have an adverse impact on the value of your portfolio. ▯

Pricing options on stocks is a major topic in the book, and we will see that, depending on the complexity of the model of the underlying asset price dynamics, it may be a rather straightforward task or not. Interest-rate derivatives are definitely more complex, and we will just have an outlook on them

2.2.4 Asset pricing, portfolio optimization, and risk management

We have seen that we need some model to price assets such as bonds and options. In principle, prices are the result of an equilibrium between demand and supply of an asset. *Equilibrium pricing* models are an attempt to capture this equilibrium resulting from the preferences and, possibly, the initial wealth of investors. In the next example we try to illustrate the approach by a very simple example from Microeconomics.

Example 2.3 (Equilibrium pricing in a pure exchange economy) Let us consider a pure exchange economy. In such an economy, we have a set of goods and a set of agents, and production is not considered. Each agent has some endowment of each good, and a preference for consumption of each good. For instance, let us assume that we have two agents, a and b, and two goods. Let the initial endowments for the two agents be, respectively,

$$\mathbf{e}_a = \begin{bmatrix} 1 \\ 0 \end{bmatrix}, \quad \mathbf{e}_b = \begin{bmatrix} 0 \\ 1 \end{bmatrix}.$$

The two agents would probably like to exchange part of the goods they own, at some price which we want to determine. Let p_1 and p_2 be the prices of the two goods. To express the preferences of the two agents, we may introduce a utility function. For instance, let us assume a so-called Cobb–Douglas utility form:

$$u_a(x_{1a}, x_{2a}) = x_{1a}^\alpha x_{2a}^{1-\alpha}, \quad u_b(x_{1b}, x_{2b}) = x_{1b}^\beta x_{2b}^{1-\beta},$$

where x_{ij} is the consumption of good $i = 1, 2$ by agent $j = a, b$, and $\alpha, \beta \in (0, 1)$ are parameters specifying the preferences of the two agents. Nota that this utility function indeed models preference for consumption bundles consisting of *both* goods, thus agents have an incentive to exchange. We have an equilibrium if each agent solves his optimal consumption problem and if markets "clear," i.e., consumption equals availability of each good.

For given prices, agent a will determine optimal consumption by maximizing his utility, subject to a budget constraint. Formally, he should solve the optimization problem:

$$\begin{aligned} \max \quad & x_{1a}^\alpha x_{2a}^{1-\alpha} \\ \text{s.t.} \quad & p_1 x_{1a} + p_2 x_{2a} = W_a, \end{aligned}$$

where $W_a = p_1$ is his initial wealth, i.e., the value of his (unit) endowment of good 1 given price p_1. Strictly speaking, the budget constraint should be written as an inequality, but given the form of utility functions we may assume that non-satiation applies: This means that the two agents are always happier if they can consume some more. By the same token, we should also include non-negativity constraints on consumption ($x_{ij} \geq 0$), but given the form of utility we may assume an interior solution, i.e., a solution in which consumption of each good is strictly positive. The optimal solution[6] is

$$x_{1a}^* = \frac{\alpha p_1}{p_1} = \alpha \qquad x_{2a}^* = \frac{(1-\alpha)p_1}{p_2}.$$

By the same token, agent b solves

$$\max \quad x_{1b}^\beta x_{2b}^{1-\beta}$$
$$\text{s.t.} \quad p_1 x_{1b} + p_2 x_{2b} = W_b$$

where $W_b = p_2$, yielding

$$x_{1b}^* = \frac{\beta p_2}{p_1} \qquad x_{2b}^* = \frac{(1-\beta)p_2}{p_2} = 1 - \beta.$$

However, prices should be compatible with market clearing, i.e., total demand for a good is equal to its total availability. Hence, we must have:

$$x_{1a}^* + x_{1b}^* = \alpha + \frac{\beta p_2}{p_1} = 1 \quad \Rightarrow \quad \frac{p_2}{p_1} = \frac{1-\alpha}{\beta}$$

Requiring market clearing for the second good yields the same condition. This is reasonable as only the ratio of prices matters: a proportional increase in both prices will increase initial wealth without changing the problem. We could normalize prices by setting $p_1 = 1$, i.e., by selecting good 1 as a *numeraire*. □

We see that, in principle, we could find equilibrium prices if we knew the preferences of each agent. Clearly, this does not look very practical. Furthermore, in finance we must also account for time and uncertainty. This means that we should know how investors value immediate consumption relative to future consumption, as well as their attitude towards risk. The task is even more difficult if we take information asymmetries or heterogeneous beliefs into account. Unless very specific hypotheses are made, there is no hope to come up with a feasible pricing approach. However, by making suitable assumptions, interesting equilibrium pricing models have been devised. For stock prices, this leads, e.g., to the Capital Asset Pricing Model (CAPM); equilibrium models have been also proposed for interest-rate dynamics.

---

[6]In this specific case, we could simply get rid of one decision variable by eliminating the equality constraint and enforcing the first-order condition, i.e., by requiring the first-order derivative of the utility function is zero at optimum. We will give a solution by the method of Lagrangian multipliers in chapter 6, page 352.

Nevertheless, in financial engineering a much less ambitious attitude is usually taken. We take the prices of a set of assets as given (and observable in the market), and we try to find the price of other assets in such a way to avoid obvious inconsistencies, like the one illustrated in the following example.

Example 2.4 (Arbitrage in a binomial model) Consider a binomial model of uncertainty, like the one in figure 2.1, and an economy consisting of two assets. The first asset is risk-free, in the sense that its price now is $1, and it will be $1.1 in both future states. We may think of this risk-free asset as a bank account offering a 10% interest rate for the period of time we consider. The second asset is risky: its current price is $1 too, and its future price could be $2 or $3 with equal probability.

It is easy to see that these prices are not consistent. If an investor borrows $1 from the bank in order to buy the risky asset, she will be sure to have a profit: in the worst-case scenario she will gain $(2 - 1.1) = 0.9$, and she will make even more money if the price of the risky asset turns out to be $3. Assuming that unlimited borrowing is allowed, she could make an unbounded amount of money, without incurring any risk. This is an example of an **arbitrage** opportunity. Loosely speaking, an arbitrage opportunity is a money making machine. Such a free lunch is not compatible with economic theory or, for that matter, with common sense.

In general, if we assume a binomial model with multiplicative shocks u and d, and there is a risk-free interest rate denoted by r_f, the following inequalities should apply: $d < 1 + r_f < u$. □

Clearly, the assumptions in the example are not quite reasonable, as unlimited borrowing is not possible and assets are available in limited supply. However, those prices are not reasonable, as they cannot be equilibrium prices, since investors taking advantage of arbitrage opportunity will influence prices. In practice, limited arbitrage opportunities are sometimes available, and there are traders taking advantage of them, but they tend to disappear quickly and are only feasible for very special traders.[7] Hence, typical models for asset pricing are based on the assumption that arbitrage is not possible.

Ruling out arbitrage opportunities leads to arbitrage-free, or *relative*, pricing. We price assets in such a way that their prices are consistent with observed prices for other assets. We will not investigate the relationships between equilibrium and lack of arbitrage, but it is intuitive that arbitrage opportunities are not compatible with equilibrium. The advantage of arbitrage pricing is that it does not rely on too many critical assumptions about the behavior of investors. Their aggregate risk attitude may somehow be taken into account by parameters which are inferred by observing market prices; this model *calibration* concept is fundamental to deal with interest-rate derivatives.

---

[7]Transaction costs may make arbitrage opportunities unprofitable, and so they allow for some slight mispricing; large institutional investors may have to pay very small transaction costs making arbitrage available to them.

40 FINANCIAL THEORY

A large part of the book is devoted to asset pricing under the no-arbitrage hypothesis. The second large body of applications is portfolio optimization. Actually, asset pricing and portfolio optimization, from a theoretical point of view, are not disjoint. After all, allocating wealth to assets in a portfolio generates demand for such assets, and demand contributes to determine asset prices. In financial economics, equilibrium asset pricing models are based on optimization models which are generalizations of the pure exchange economy of example 2.3. However, in everyday portfolio management, it is common to treat uncertainty as purely exogenous. This means that we need first to model uncertainty, and then to select a suitable model for portfolio optimization, together with some computationally feasible way of solving it. Actually, there is much more to that and portfolio optimization is just one part of portfolio management. For instance, risks must be assessed by some sensitivity analysis with respect to the assumed model of uncertainty, which must be somehow stress-tested. Portfolio optimization is only part of a decision process involving different actors with different organizational responsibilities.

In its basic form, portfolio optimization entails some form of stochastic optimization. By selecting a portfolio, we implicitly select a probability distribution for its return or, equivalently, for future wealth. How can we compare probability distributions corresponding to different portfolio choices? One trivial approach would be to maximize the expected value of return. The following examples show that this would result in unreasonable portfolio choices.

Example 2.5 (Putting all of your eggs in one basket) Consider an investor who must allocate her wealth to n assets. The return of each asset, indexed by $i = 1, \ldots, n$, is a random variable R_i with expected value $\mu_i = \mathrm{E}[R_i]$. The asset allocation decision may be modeled by introducing a set of decision variables x_i representing the fraction of wealth invested in asset i. If we rule out short-selling, these decision variables are naturally bounded by $0 \leq x_i \leq 1$. The expected value of return from our portfolio is

$$\mathrm{E}\left[\sum_{i=1}^{n} R_i x_i\right] = \sum_{i=1}^{n} \mathrm{E}[R_i] x_i = \sum_{i=1}^{n} \mu_i x_i.$$

Hence, we should solve the following optimization model:

$$\max \quad \sum_{i=1}^{n} \mu_i x_i$$

$$\text{s.t.} \quad \sum_{i=1}^{n} x_i = 1$$

$$x_i \geq 0,$$

whose solution is quite trivial: we should simply pick up the asset with maximum expected return, $i^* = \arg\max_{i=1,\ldots,n} \mu_i$, and set $x_{i^*} = 1$. It is easy to see that this portfolio is a very dangerous bet; in practice, portfolios are diversified, which means that there must be something else beyond expected values. In practice, one would also have some constraints on portfolio composition, limiting exposure to certain geographical areas or types of industry, and this would make the trivial solution above not feasible. However, if we take only expected return into account, the solution is basically shaped by these constraints. By the way, if short-selling is allowed, the decision variables are unrestricted, and the expected value of future wealth goes to infinity. In fact, one would short-sell assets with low expected return, to make money to be invested in the most promising asset. This is clearly unreasonable. ☐

Example 2.6 (St. Petersburg paradox) Consider the following proposition. You are offered a lottery, whose outcome is determined by flipping a fair and memoryless coin. The coin is flipped until it lands tail. Let k be the number of times the coin lands head; then, the payoff you get is $\$2^k$. Now, how much should you be willing to pay for this lottery? The reader is invited to consider this problem as a pricing problem: the lottery is a sort of derivative with respect to some random outcome. We could consider the expected value of the payoff as the fair price for this rather peculiar asset. The probability of winning $\$2^k$ is the probability of having k consecutive heads followed by one tail, which stops the game, after $k+1$ flips of the coin. Given independence of events, the probability of this sequence is $1/2^{k+1}$, i.e., the product of individual event probabilities. Then, the expected value of the payoff is

$$\sum_{k=0}^{\infty} 2^k \frac{1}{2^{k+1}} = \sum_{k=0}^{\infty} \frac{1}{2} = +\infty.$$

This game looks so beautiful that we should be willing to pay any amount of money to play it! No one would probably do so. Again, we see that expected value does not tell the whole story. ☐

These two examples show that expected values must be complemented by some other information, such as variance or quantiles, in order to take sensible decisions. More generally, we need a way to model decision making under uncertainty, and this calls for a way to model risk aversion. One way to do so is to introduce the concept of expected utility, which is done in section 2.4.1. Expected utility is an interesting concept, with some theoretical and practical pitfalls. In fact, it basically postulates that decision makers are very rational, consistent, and very well informed, all of which is often contradicted. But even if we believe that decision makers are consistently rational, it is difficult to elicit the utility function from any investor. A practical way out is to define suitable risk measures, which can be accounted for in formulating portfolio optimization models. A typical approach is to constrain the expected return of the portfolio, and then to minimize a suitably chosen risk measure. By varying expected return, we can trace a set of reasonable portfolios among which the decision maker may select the best compromise solution, trading off expected return against risk. If we measure risks by the variance of return, we obtain a well-known theory based on mean-variance efficiency (section 2.4.2). Recently, different risk measures have been adopted, such as Value at Risk, which is described in section 2.4.5. This leads to another important body of finance, risk management, which may take advantage from numerical methods as well. We should emphasize again that portfolio optimization models are only a part of the more general portfolio management process, which also includes risk assessment and management.

We have said that asset pricing is somewhat related to portfolio optimization, which in turn is related to risk management. It is also important to understand the link between asset pricing and risk management. On the one hand, we need to understand the sensitivity of asset prices to random fluctuations in underlying factors, so that hypothetical scenarios for the evolution of the underlying factors can be mapped to changes in portfolio value. Furthermore, we would like to devise approaches to design our portfolio in such a way that sensitivity to such changes is minimized. For instance, we may want to understand how interest rates affect bond prices, and to devise portfolios which are at least partially immunized against shocks; this is the subject of the next section.

On the other hand, however, there is a much less obvious link, which will be apparent when we treat option pricing in section 2.6. Consider the point of view of the option writer, i.e., the guy who *sells* an option. Options may be risky for people buying them, but they are even riskier for the party who sells them; in fact the option holder has a *right* to exercise, but the option writer *must* comply with this right. To get the point, consider the extreme case of a call option with strike price $K = 20$ which is exercised when the underlying asset price is $S_T = 80$; this is trouble for the writer if he has to buy the underlying stock at 80 to sell it at 20. Hence, the option writer needs a reliable way to hedge against such risks. We will see that, in an idealized world, the option price is basically the price of a hedging strategy for the option writer.

2.3 FIXED-INCOME SECURITIES: ANALYSIS AND PORTFOLIO IMMUNIZATION

In this section we deal only with "really fixed" income assets, i.e., fixed-coupon and zero-coupon bonds. Even in this simple setting we may introduce several useful concepts.

2.3.1 Basic theory of interest rates: compounding and present value

In order to understand bond pricing, the first concepts we need are related to interest rates and how they are compounded. Assume you have wealth W_0

and you invest it in, say, a bank account for one year. After this period, you will get an amount of money $W_1 > W_0$. Hence, you could measure the rate of return of your investment by

$$r = \frac{W_1 - W_0}{W_0}.$$

In other words, at the end of the investment period you collect an amount of money which is the sum of the *principal*, the original amount you owned, plus *interest*:

$$W_1 = W_0 + rW_0 = (1+r)W_0.$$

The quantity r is referred to as *interest rate* over the time period we are considering. Now assume that you leave your money in the bank account for two years and that the same interest rate r applies for both years. How much money will you get? If the *simple* interest rule applies, you will get twice the interest:

$$W_2 = (1 + 2r)W_0.$$

If the period of your investment is n years, the simple interest rule yields

$$W_n = (1 + nr)W_0.$$

In the general case including fractions of years, one possible rule assumes proportionality:

$$W_t = (1 + tr)W_0,$$

where t is any real number. More often than not, however, you earn interest on interest; after the first year, the interest you earned is added to the original wealth, and the interest rate for the next year will be applied to the new wealth:

$$W_2 = (1+r)W_1 = (1+r)^2 W_0.$$

In this case we speak of *compound* interest, and for n years we have

$$W_n = (1+r)^n W_0.$$

Note that, in the case of compounding, wealth grows more rapidly, according to a geometric progression.

Compounding can occur at any frequency. For instance, let us assume that you get interest every six months. Typically, a nominal interest rate r is quoted yearly, but it is applied dividing it by the number of periods in the year:

$$W_1 = (1 + r/2)^2 W_0.$$

We obtain the *effective* yearly rate by equating wealth at the end of the year:

$$(1 + r/2)^2 W_0 = (1 + r_e)W_0 \quad \Rightarrow \quad r_e = r + r^2/4 > r.$$

44 FINANCIAL THEORY

If interest is compounded m times per year, we have

$$W_1 = (1 + r/m)^m \, W_0.$$

For a given nominal rate, the more frequent the compounding, the faster the growth and the higher the effective yearly rate. What happens if, in the limit, interest is compounded continuously? By taking the limit as m goes to infinity, and using a well-known result from calculus, we get

$$W_1 = \lim_{m \to \infty} (1 + r/m)^m W_0 = e^r W_0.$$

Continuous compounding looks a bit artificial, but in this case many things turn out to be simpler, including the application of an interest rate to an arbitrary period of time t. We may think of dividing the time interval t in small slices of length $1/m$ years, i.e., $t \approx k/m$ for some integer k. Using discrete-time compounding and then taking the limit we get:

$$\left[1 + \frac{r}{m}\right]^k = \left[1 + \frac{r}{m}\right]^{mt} = \left\{\left[1 + \frac{r}{m}\right]^m\right\}^t \to e^{rt}.$$

Again, we may find the effective yearly rate r_e corresponding to the continuously compounded rate r: $r_e = e^r - 1$.

Another fundamental concept in the basic theory of interest rates is the *present value* of a stream of cash flows in time. We will see that absence of arbitrage implies that the price of a bond must be the present value of a cash flow stream. Consider a cash flow stream, i.e., a sequence of periodic payments C_t at discrete-time instants $t = 0, 1, \ldots, n$. Given an interest rate r with discrete compounding, applied over each time period, the present value of the cash flow stream is defined as

$$\text{PV} = \sum_{t=0}^{n} \frac{C_t}{(1+r)^t}.$$

Note that cash flows need not be positive; for instance, in investment analysis we typically have $C_0 < 0$, corresponding to an initial cash outlay. We say that cash flows are discounted, reflecting the fact that the value of \$1 in the future is something less now; the discount factor by which each cash flow is multiplied is smaller for distant periods. When the nominal interest rate is quoted yearly but the payments occur more frequently, the formula may be easily adapted following the previous treatment. If there are m payments per year at regular time intervals, we have

$$\text{PV} = \sum_{k=0}^{n} \frac{C_k}{(1+r/m)^k}, \qquad (2.1)$$

where k indexes the time periods and n is the number of periods, i.e., the number of years times the number of periods within one year.

All of the considerations we have made on compounding apply here. If the interest rate is continuously compounded, present value is

$$PV = \sum_{t=0}^{n} C_t e^{-rt}.$$

Continuous compounding is very convenient when cash flows are not regular in time. Let us denote by t_i, $i = 1, \ldots, n$, the time at which cash flow C_i is received. Then

$$PV = \sum_{i=0}^{n} C_i e^{-rt_i}.$$

In the case of discrete compounding, one possible convention is using fraction of years. For instance, the present value P of cash flow C occurring in nine months could be expressed by

$$P = \frac{C}{(1+r)^{9/12}},$$

if we assume that all months consist of the same number of days.

It is important to note that we have assumed that the same interest rate r, however it is quoted, is applied to any time interval. This need not be the case actually, as we will see later. Furthermore, it is also worth stressing that we have not considered inflation. When inflation is taken into account, we should distinguish between *nominal* and *real* interest rate, but we will always disregard inflation in this book. The calculations above, possibly adjusted to cope with these issues, are very common and have been implemented in a large number of software packages, including MATLAB. Typical functions of this kind have been included in the Financial Toolbox.

Example 2.7 The Financial toolbox includes different functions to analyze cash flow streams, including pvvar, which computes the present value of a stream, given an interest rate. Consider for instance the cash flow stream corresponding to a bond maturing in five years, with face value 100, and a 8% coupon rate. This cash flow can be represented by the following vector:

```
>> cf=[0 8 8 8 8 108]
cf =
     0     8     8     8     8   108
```

The zero in the first position corresponds to an immediate cash flow, which in this case is zero, as the first coupon will be paid in one year (you may think that a coupon have just been paid). What is the present value of this stream if we discount it by an interest rate corresponding to the coupon rate? Not surprisingly, present value is equal to face value:

```
>> pvvar(cf,0.08)
```

```
function pv = mypvvar(cf,r)
% get number of periods
n = length(cf);
% get vector of discount factors
df = 1./(1+r).^(0:n-1);
% compute result
pv = dot(cf,df);
```

Fig. 2.7 Function to compute present value with discrete compounding and regular cash flows.

ans =
    100.0000

If we increase that discount rate, present value is decreased:

```
>> pvvar(cf,0.09)
ans =
    96.1103
```

On the contrary, if the discount rate is decreased, present value is increased:

```
>> pvvar(cf,0.07)
ans =
    104.1002
```

Indeed, we will see that when interest rates rise, bond prices fall, whereas bond prices increase when interest rates drop. A major task in bond portfolio management is to take interest-rate risk into account.

How can we evaluate present value without the Financial Toolbox? Function mypvvar in figure 2.7 is a possible answer. Note that, in computing the vector of discount factors, we must use a vector from 0 to length n minus 1; also note the use of the dot operator both in the division (./) and in the power (.^). The function dot computes the dot product of vectors:

$$\mathbf{x}'\mathbf{y} = \sum_{k=1}^{m} x_i y_i,$$

provided that the vectors have the same number m of elements. The advantage of using dot is that we do not need worrying whether vectors are row or column vectors, as is the case when we use matrix multiplication.

```
>> cf = [0 8 8 8 8 108];
>> mypvvar(cf,0.08)
ans =
```

```
    100.0000
>> mypvvar(cf,0.09)
ans =
    96.1103
>> mypvvar(cf,0.07)
ans =
    104.1002
```

□

Another quite common concept linked to analyzing cash flow streams is the *internal rate of return*. Given a stream of cash flows C_t ($t = 0, 1, 2, \ldots, n$), the internal rate of return is defined as a value ρ such that the present value of the stream is zero. In other words, it is a solution of the non-linear equation

$$\sum_{t=0}^{n} \frac{C_t}{(1+\rho)^t} = 0. \tag{2.2}$$

Clearly, in order to find a solution, we must assume that at least one cash flow is negative. Typically, this is the initial cash flow C_0, which may correspond to an investment or to the price you pay to purchase a bond. MATLAB provides us with useful functions to compute the internal rate of return.

Example 2.8 We will describe methods to solve general non-linear equations in section 3.4. However, the equation defining internal rate of return may be easily transformed to a specific non-linear equation, a *polynomial* equation, which is relatively easy to solve. With the change of variable $h = 1/(1+\rho)$, we may rewrite equation (2.2) as

$$\sum_{t=0}^{n} C_t h^t = 0,$$

which is readily solved by the MATLAB function `roots`. All we have to do is to represent a cash flow stream as a vector, as done in the following MATLAB interaction snapshot.

```
>> cf=[-100 8 8 8 8 108]
cf =
  -100     8     8     8     8   108
>> h=roots(fliplr(cf))
h =
  -0.8090 + 0.5878i
  -0.8090 - 0.5878i
   0.3090 + 0.9511i
   0.3090 - 0.9511i
   0.9259
```

```
>> rho=1./h -1
rho =
   -1.8090 - 0.5878i
   -1.8090 + 0.5878i
   -0.6910 - 0.9511i
   -0.6910 + 0.9511i
    0.0800
```

A few comments are in order. First, we define a variable cf and we associate a cash flow to it. Then, in a single command line, we flip the cash flow from left to right with the function fliplr and we invoke the roots function to assign the roots of the resulting polynomial to the variable h. Flipping the cash flow vector is necessary since roots assumes that a polynomial is represented by a vector in which the first components correspond to the highest power terms in the polynomial, whereas when we represent cash flows we put such terms at the end. After obtaining the solution in terms of h, we go back to the original variable ρ (note that the dot in ./ is necessary since h is a vector of solutions). Since in this example $n = 5$, we have a vector of five roots: four are complex conjugates, and the one we are interested in is the real one, i.e., $\rho = 0.08$. Indeed, it can be shown that for a cash flow stream with $C_0 < 0$ and $C_t \geq 0$ $(t = 1, \ldots, n)$ and $\sum_{t=1}^{n} C_t > 0$, we have a unique real and positive solution of the non-linear equation (see, e.g., [15, chapter 2]).

If we want to devise a function filtering complex roots away, we may use the MATLAB find function, which returns the indexes of the elements in a vector meeting some condition:

```
>> index = find(abs(imag(rho)) < 0.001)
index =
     5
>> rho(index)
ans =
    0.0800
```

What we have done here is finding the indexes of elements in rho such that the absolute value of their imaginary part is less than a specified tolerance; then we get the elements from the vector. It is tempting to think that we should look for elements such that the imaginary part is exactly zero, but this type of "exact thinking" should be avoided when numerical computing is involved. To get the point, consider the trivial equation

$$(x-1)^3 = x^3 - 300x^2 + 30{,}000x - 1{,}000{,}000 = 0$$

and use roots to solve it:

```
>> v = [1 -300 30000 -1000000];
>> h=roots(v)
h =
```

```
  1.0e+002 *
  1.0000 + 0.0000i
  1.0000 - 0.0000i
  1.0000
>> index = find(abs(imag(h)) == 0)
index =
     3
```

The nasty thing occurring here is that multiple real roots may turn out as complex conjugates with a very small imaginary part. This is arguably unlikely to occur when computing internal rates of return of non-pathological cash flow streams, but it is a good example of pitfalls in numerical computing and it points out the care we need to take. All the work above (including filtering complex roots out) is done by the `irr` function available in the Financial toolbox:

```
>> irr(cf)
ans =
    0.0800
```

We urge the reader to try writing a function doing all of this automatically; then, readers having access to the Financial Toolbox may compare their function with `irr`. □

With respect to present value, when computing internal rate of return we are going the other way around, in some sense. Moreover, the present value may be computed using a *set* of discount factors linked to different interest rates applied over time periods differing in length; the internal rate of return is one rate which, applied over *all* of the time periods, would give the same present value.

2.3.2 Basic pricing of fixed-income securities

Pricing a zero-coupon bond Consider a zero-coupon bond, with a face value F, maturing in one year, which is currently sold at price P. If we purchase this security and we keep it until maturity, we will have a total return

$$R = \frac{F}{P}$$

and a rate of return

$$r = R - 1 = \frac{F}{P} - 1.$$

An obvious relationship between r, F, and P is

$$P = \frac{F}{1+r}. \tag{2.3}$$

We may see this relationship the other way around. If we fix F and r, this may be interpreted as a pricing relationship.

What rate r should we use in pricing? If the bond is default-free, as is usually the case with government bonds, this should be the prevailing risk-free interest rate: no more, no less. To see why, we may use a common principle in finance, i.e., the *no-arbitrage* principle. Assume that the bond is underpriced, i.e., it sells for a price P_1 such that

$$P_1 < P = \frac{F}{1+r},$$

and that we may take out a loan at the risk-free interest rate r (we are assuming that borrowing and lending rates are equal). Then we can borrow an amount L and use it to purchase L/P_1 bonds. Note that the immediate net cash flow is zero. Then, at maturity, we must pay $L(1+r)$ to our money lender, and we get an amount FL/P_1 when the face value is redeemed for each bond. But since, by hypothesis,

$$\frac{F}{P_1} > 1 + r,$$

the net cash flow at maturity will be

$$L\frac{F}{P_1} - L(1+r) = L\left(\frac{F}{P_1} - 1 - r\right) > 0.$$

Hence, we pay nothing at the beginning and receive a positive amount in the future; since the bargain is an interesting one, we might well exploit it, in the limit, to ensure an unbounded profit for increasing L. This is a simple example of *arbitrage*. Of course, limitless borrowing is not available; more important, purchasing a huge amount of those bonds would raise their prices, and the arbitrage opportunity would soon disappear. Indeed, a common assumption in many financial problems is that arbitrage opportunities do not exist. Note that this does not imply that they actually do not exist; on the contrary, it is the very fact that many people are out there to exploit those opportunities which tends to eliminate them quickly. The argument may be repeated similarly if the inequality is reversed and the bond is overpriced:

$$P_1 > P = \frac{F}{1+r}.$$

In this case we should borrow the bond itself, rather than the cash needed to buy it. This is accomplished by selling the bond short (see example 2.1 on page 32 for an illustration of short-selling a stock). There are many limitations to short-selling in practice, but for pricing models it is often (not always) reasonable to assume that it is possible. Then we may sell the overpriced bond and invest the proceeds at the risk-free rate; let us assume that we borrow bonds for a total value L, we sell them at price P_1, and we invest the money we obtain. The immediate net cash flow is again zero. At maturity, we get $L(1+r)$ from our investment, and we have to pay the face value F to the owner for each bond that we have borrowed. Hence the net cash flow at maturity is again positive:

$$-L\frac{F}{P_1} + L(1+r) = L\left(-\frac{F}{P_1} + 1 + r\right) > 0.$$

We have also implicitly assumed that transaction costs are negligible and that we may lend or borrow money at the same rate. Again, these assumptions are violated in practice, but they may be close enough to reality, at least for some large investors, to warrant their use. The reader may have the impression that the arbitrage argument is, at least in this case, an unnecessary complication to obtain an almost obvious result: the price is obtained by taking the present value of its future cash flows. However, the no-arbitrage principle is used, with some modification, to price quite complex securities where uncertainty is involved and intuition does not help (as in the case of options; see section 2.6.2).

No-arbitrage and linearity of pricing Before proceeding and considering pricing coupon-bearing bonds, it is useful to point out a couple of important implications of the no-arbitrage principle.

The first implication is the *law of one price*. Different assets cannot sell for different prices, in idealized markets, otherwise an immediate arbitrage opportunity arises. In practice, markets are not perfect, and we all know that the same product may be sold at different prices in different countries. In this case, arbitrage opportunities are eliminated by transportation costs, taxes, etc. Financial markets, also thanks to Internet, are closer to perfect markets, and for modeling purposes we may assume that the law of one price makes sense. We will also see that it makes sense when uncertainty is involved.

Another implication is that pricing is a *linear operator*. To get the point, let us denote by $P(\cdot)$ an abstract pricing operator that maps assets to prices. Linearity means that the price of a portfolio of assets should be the weighted sum of the prices of each single asset. Formally, if we denote an asset by X_i, $i = 1, \ldots, n$, we have

$$P\left(\sum_{i=1}^{n} \alpha_i X_i\right) = \sum_{i=1}^{n} \alpha_i P(X_i),$$

where $P(X_i)$ is the price of asset i and α_i is the number of assets of type i in the portfolio. To see this, let us break the argument in two parts. If we consider one asset, we should have $P(2X) = 2P(X)$. If, for instance, $P(2X) < 2P(X)$, we may make an immediate profit by purchasing two assets and selling them separately. A similar consideration applies if $P(2X) > 2P(X)$. The same reasoning can be applied with an arbitrary number of assets, at least in idealized markets with no friction; in real markets, transaction costs, round lots, etc., make the argument only approximately valid. By the same token, we must have $P(X_1 + X_2) = P(X_1) + P(X_2)$. If, for instance, $P(X_1 + X_2) < P(X_1) + P(X_2)$, we may buy the bundle of two assets and then make an immediate profit by selling them separately. Again, reality is a bit different. Prices may be non-linear when transaction costs are involved or when assets are in limited supply and markets are thin.

Linearity of pricing has an important implication on pricing coupon-bearing bonds; if we regard such a bond as a portfolio of zero-coupon bonds, it is immediate to see that we may price each coupon as a zero-coupon bond and sum the results.

Pricing a coupon-bearing bond Linearity of pricing implies that a bond may be priced by pricing each coupon separately, including payment of face value at maturity. Consider a bond with face value F, paying a coupon C per period. Pricing is very simple, if we assume that the bond is default-free, so that a riskless interest rate may be applied, and that this rate can be applied to any period length (provided that we account for compounding). It is easy to see that the fair bond price may be obtained by computing the present value of its cash flow stream:

$$\mathrm{PV} = \sum_{i=1}^{n} \frac{C}{(1+r)^i} + \frac{F}{(1+r)^n}. \tag{2.4}$$

This is the basic principle, which links present values and prices. As expected, several complications may arise in practice.

- If r is quoted yearly and there is more than one coupon payment per year, the formula could be adjusted in the same vein as equation (2.1). If m coupons are paid in a year:

$$\mathrm{PV} = \sum_{i=1}^{n} \frac{C/m}{(1+r/m)^i} + \frac{F}{(1+r/m)^n}$$

  where n is the number of periods.

- Another fundamental issue is that different interest rates are typically associated to different time horizons. This implies that bond pricing requires knowledge of *several* discount factors. If we denote by r_t the interest rate which applies from now to time t, i.e., the *spot rate*, we should discount each coupon C_t appropriately:

$$\sum_{t=1}^{T} \frac{C_t}{(1+r_t)^t} + \frac{F_T}{(1+r_T)^T}.$$

  The set of rates r_t is related to the *term structure of interest rates*. The idea is depicted in figure 2.8, where we see an upward-sloping structure;

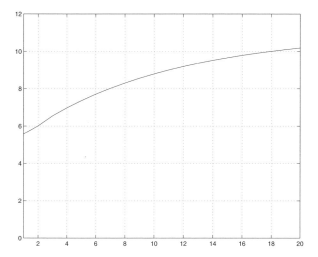

Fig. 2.8 Term structure of the interest rate; years are reported on the horizontal axis, and the corresponding (percentage) spot interest rates are plotted.

this corresponds to the intuitive notion that longer interest rates are usually associated with longer terms. Actually, other shapes are possible in general. A downward sloping curve is usually associated to recession, whereby interest rates are expected to drop in the future. Note that an upward sloping curve does not necessarily imply that interest rates are expected to rise.

- If these simple formulas were generally applicable, any bond with the same coupon rate and maturity date should have the same price, which is actually not the case. A first point is that not all bonds are issued by institutions with the same credit rating. Although a bond issued by some governments may be default-free, a corporate bond may not be of the same quality; hence, all other things being equal, you would require a lower price for it. This difference may be captured by the bond yield, which is introduced in the next section.

Measuring return of a bond: yield to maturity We have seen that the price of a fixed-coupon bond is basically the present value of its cash flow stream, which may depend on a whole set of interest rates. But how can we measure the return of a bond of given price by a *single* number? One possible idea is to compute the internal rate of return of the bond. The internal rate of return of a bond is called the *yield*,[8] and for a bond with price P it is the solution,

---

[8] Actually, there are different concepts of yield (see, e.g., [6] or [7]), but we will stick to this one for the sake of simplicity, even though it may be subject to some criticism.

λ, of the following equation:

$$P = \sum_{i=1}^{n} \frac{C}{(1+\lambda)^i} + \frac{F}{(1+\lambda)^n}.$$

If more than one coupon payment is made during a year, the equation defining yield is immediately adapted:

$$P = \sum_{i=1}^{n} \frac{C/m}{(1+\lambda/m)^i} + \frac{F}{(1+\lambda/m)^n}.$$

From these equations it is easy to see that bond prices will drop if there is an increase in required yield λ, and vice versa. Required yield may increase if bond rating gets worse, which calls for some risk premium, or if the general level of interest rates rises. Analyzing the relationship between price and yield is relatively easy, but it is just an approximation. A full term structure of interest rates should be taken into account, as the curve may not only go up or down, but it may also twist and change its qualitative shape. Nevertheless, an approximate analysis is often valuable, as we will see shortly.

Issues in bond portfolio management: interest-rate risk Intuitively, the higher the required yield, the lower the price, and higher yields must be offered for risky bonds. If the credit rating of the bond issuer changes, the bond price will change accordingly to reflect the new situation. But is credit risk the only source of risk for bonds? Unfortunately, the answer is no. To begin with, coupon rates may depend on some other economic or financial variable, resulting in some uncertainty in the cash flow, so we have a form of financial risk. Another point is that some bonds have embedded options which may be unfavorable for the holder; for instance, the issuer may *call* the bond, that is, redeem it before maturity, which results in reinvestment risk since we would have to reinvest the cash we receive from the bond issuer (bonds with embedded options may be analyzed using techniques we discuss later when we deal with options).

But even if all of these risks are ruled out, there may still be a form of risk, depending on the intended *use* of the security. The point is that any portfolio of bonds has some purpose, and the portfolio risk must be evaluated with reference to this purpose. A common use of a bond portfolio is to enable some institution (e.g., a pension fund) to comply with a stream of future liabilities. To be more concrete, assume that we have to pay a sequence of liabilities over a time horizon which is discretized in T periods and that the liability in period $t = 1, \ldots, T$ is L_t. Now, we could just purchase bonds in such a way as to meet all the liabilities. In fact, this is possible, at least in principle. Consider a set of N bonds, each with a price P_i ($i = 1, \ldots, N$). If the cash flow from a unit of security i at time t is represented by F_{it}, we may consider the following cash flow matching model:

$$\min \quad \sum_{i=1}^{N} P_i x_i$$

$$\text{s.t.} \quad \sum_{i=1}^{N} F_{it} x_i \geq L_t \quad \forall t$$

$$x_i \geq 0.$$

Here the decision variable x_i represents the amount of bond i purchased (rather than the weight in the portfolio). If we neglect the possibility of default and assume that the liabilities are known in advance, the resulting portfolio would certainly meet the obligations; unfortunately, it is likely to be quite expensive. Unless bond maturities are matched to the liabilities, we will have to meet the obligations with coupon payments, requiring a possibly large number of bonds. Note also that liabilities are taken into account by an inequality constraint, which may turn out to be strict, since it is unlikely that a perfect match of cash flows and liabilities may be obtained with a given set of bonds. In the case of a long planning horizon, the lack of suitable long-term bonds may compound these difficulties.

Hence, we must manage our bond portfolio in a more dynamic manner, buying and *selling* bonds along the way. But here comes the trouble. Bond prices are related to interest rates, and these may change in unpredictable ways. For instance, is a five-year zero-coupon bond riskless?

Example 2.9 Consider a five-year zero-coupon bond, with face value 100, sold with required yield $r_1 = 0.08$. Which is the percentage change in its price if the yield is increased immediately after purchase to $r_2 = 0.09$?

```
>> r1=0.08;
>> r2=0.09;
>> P1=100/(1+r1)^5
P1 =
    68.0583
>> P2=100/(1+r2)^5
P2 =
    64.9931
>> (P2-P1)/P1
ans =
    -0.0450
```

We see that we have a 4.5% decrease the value of the bond. Note that this loss occurs only if you have to sell the bond before maturity. No harm is done if you keep the bond to maturity, but this makes sense only if the liability you want to match coincides with maturity. Now what if the maturity is 20 rather than five years?

```
>> P1=100/(1+r1)^20
P1 =
    21.4548
>> P2=100/(1+r2)^20
P2 =
    17.8431
>> (P2-P1)/P1
ans =
    -0.1683
```

We see that the loss is now much larger, almost 17%. Although zero-coupon bonds with long maturities may not be available easily, it is a general rule that the longer the maturity, the more sensitive to yield changes the bond price is. Coupon rates play some role, too. We may compare two bonds with coupon rates of 4% and 8%, respectively.

```
>> cf1=[0 8 8 8 8 8 8 8 8 108];
>> cf2=[0 4 4 4 4 4 4 4 4 104];
>> P1=pvvar(cf1,0.08)
P1 =
    100.0000
>> P2=pvvar(cf1,0.09)
P2 =
    93.5823
>> (P2-P1)/P1
ans =
    -0.0642
>> P1=pvvar(cf2,0.08)
P1 =
    73.1597
>> P2=pvvar(cf2,0.09)
P2 =
    67.9117
>> (P2-P1)/P1
ans =
    -0.0717
```

We see that a lower coupon rate implies a larger sensitivity. ▯

The problem is that the interest rates are not constant over time; they may change, depending, e.g., on inflation or general economic conditions. The changes in interest rates may be complex, as we should take a whole curve of spot rates into account. The curve may shift up or down, but it may also change shape, as it may steepen or flatten. In the example above we have just captured these complex changes with one measure, yield. If rates move up, a higher yield will be required for new bonds of the same characteristics.

For bonds issued in the past and traded on secondary markets, an increase in the yield results in a decrease in the price at which they may be sold. On the contrary, if interest rates drop, we may gain something from the decrease in the required yield, which results in an increase in the price. Depending on the maturity and the coupon rate, we have seen that a bond may be more or less sensitive to yield changes. We need a formal way to measure the interest-rate risk associated with bonds, in order to figure out a way to shape a fixed-income portfolio. A relatively simple answer is represented by the duration and convexity concepts discussed in the next section.

2.3.3 Interest rate sensitivity and bond portfolio immunization

Imagine that you are an investor facing a stream of known liabilities in the future and you want to hold a portfolio of bonds such that you may meet the liabilities. On the one hand, you would like to do it at minimum cost, but you would also like to hold a portfolio that is not likely to get you in trouble in case of changes in the interest rates. As a simple example, imagine that you have one liability L to be paid in five years. If you may find a safe zero-coupon bond maturing in five years, with face value F, you may just buy an amount L/F of these bonds. However, if the bond maturity is less than five years, you will face reinvestment risk; if the bond maturity is more than five years, you will face interest rate risk, as we have seen in example 2.9. Ideally, you would like to find a zero-coupon bond with maturity corresponding exactly to the date of each liability. Unfortunately, it is practically impossible to do so, and we must find another way to protect the bond portfolio against interest rate uncertainty. Immunization is a possible, and simple, solution.

Formally, we have a function $P(\lambda)$ that gives the relationship between the yield and the price of a bond. We may draw this curve (how this may be done in MATLAB is explained in example 2.11), obtaining something like the curve illustrated in figure 2.9. We see that the curve is convex,[9] which is actually the case for usual bonds. Now, consider small movements in the required yield; we would like to find out a way to approximate the change in price with respect to a change in yield. Indeed, there are two concepts, duration and convexity, which can be used to this aim.

Given a stream of cash flows occurring at times $t_0, t_1, \ldots, t_n$, the *duration* of the stream is defined as

$$D = \frac{\text{PV}(t_0)t_0 + \text{PV}(t_1)t_1 + \text{PV}(t_2)t_2 + \cdots + \text{PV}(t_n)t_n}{\text{PV}},$$

where PV is the present value of the whole stream and $\text{PV}(t_i)$ is the present value of cash flow c_i occurring at time t_i, $i = 0, 1, \ldots, n$. In some sense, the

---

[9] Formally, a function f is convex on a set if, for any choice of $\mathbf{x}$ and $\mathbf{y}$ in that set, $f(\lambda \mathbf{x} + (1-\lambda)\mathbf{y}) \leq \lambda f(\mathbf{x}) + (1-\lambda)f(\mathbf{y})$ holds for $0 \leq \lambda \leq 1$; more on this in supplement S6.1.

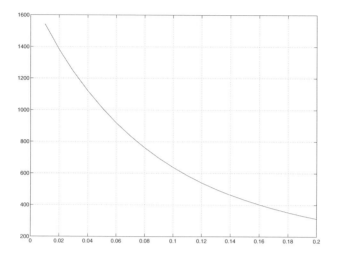

Fig. 2.9 Price–yield curve.

duration looks like a weighted average of cash flow times, where the weights are the present values of the cash flows. Note that for a zero-coupon bond, which has a single cash flow, the duration is simply the time to maturity. When we consider a generic bond and use the yield as the discount rate in computing the present values, we get Macaulay duration:

$$D = \frac{\sum_{k=1}^{n} \frac{k}{m} \frac{c_k}{(1+\lambda/m)^k}}{\sum_{k=1}^{n} \frac{c_k}{(1+\lambda/m)^k}},$$

where it is assumed that there are m coupon payments per year. In order to see why duration is useful, let us compute the derivative of the price with respect to yield:

$$\begin{aligned}\frac{dP}{d\lambda} &= \frac{d}{d\lambda}\left(\sum_{k=1}^{n}\frac{c_k}{(1+\lambda/m)^k}\right) \\ &= \sum_{k=1}^{n} c_k \frac{d}{d\lambda}\left[\frac{1}{(1+\lambda/m)^k}\right] = -\sum_{k=1}^{n}\frac{k}{m}\frac{c_k}{(1+\lambda/m)^{k+1}}. \end{aligned} \quad (2.5)$$

If we define the modified duration $D_M \equiv D/(1+\lambda/m)$, we get

$$\frac{dP}{d\lambda} = -D_M P. \quad (2.6)$$

Thus, we see that the modified duration is related to the slope of the price–yield curve at a given point; technically speaking, it is the price elasticity of the bond with respect to changes in the yield. This suggests the opportunity of using a first-order approximation:

$$\delta P \approx -D_M P \, \delta\lambda.$$

An even better approximation may be obtained by using a second-order approximation. This may be done by defining the *convexity*:

$$C = \frac{1}{P}\frac{d^2 P}{d\lambda^2}.$$

It turns out that, for a bond with m coupons per year,

$$C = \frac{1}{P(1+\lambda/m)^2} \sum_{k=1}^{n} \frac{k(k+1)}{m^2} \frac{c_k}{(1+\lambda/m)^k}.$$

Note that the unit of measure of convexity is time squared. Convexity is actually a desirable property of a bond, since a large convexity implies a slower decrease in value when the required yield increases, and a faster increase in value if the required yield decreases. Using both convexity and duration, we have the second-order approximation

$$\delta P \approx -D_M P \, \delta\lambda + \frac{PC}{2}(\delta\lambda)^2.$$

Example 2.10 We may check the quality of the price change approximation based on duration and convexity with a simple example. Let us consider a stream of four cash flows $(10, 10, 10, 10)$ occurring at times $t = 1, 2, 3, 4$. We may compute the present values of this stream under different yield values using MATLAB function pvvar:

```
>> cf = [10 10 10 10]
cf =
    10    10    10    10
>> p1=pvvar([0, cf], 0.05)
p1 =
   35.4595
>> p2=pvvar([0, cf], 0.055)
p2 =
   35.0515
>> p2-p1
ans =
   -0.4080
```

Note that we have to add a 0 in front of the cash flow vector cf since pvvar assumes that the first cash flow occurs at time 0. We see that increasing the

60 FINANCIAL THEORY

yield by 0.005 results in a price drop of 0.4080. Now we may compute the modified duration and the convexity using the functions `cfdur` and `cfconv`. The function `cfdur` returns both Macauley and modified duration; for our purposes, we must pick up the second output value.

```
>> [d1 dm] = cfdur(cf,0.05)
d1 =
    2.4391
dm =
    2.3229
>> cv = cfconv(cf,0.05)
cv =
    8.7397
>> -dm*p1*0.005
ans =
   -0.4118
>> -dm*p1*0.005+0.5*cv*p1*(0.005)^2
ans =
   -0.4080
```

We see that at least for a small change in the yield, the first-order approximation is satisfactory and the second-order approximation is practically exact.
□

We have defined duration and convexity for a single bond; what about a bond portfolio? If the yield is the same for all the bonds, it can be shown that the duration of the portfolio is simply a weighted average of all the durations (the weight is given by the weight of each bond within the portfolio). This is not exactly true if yields are not the same; however, the weighted average of the durations may be used as an approximation. How can we take advantage of this? In the case of asset liability management, one possible approach is to match the duration (and possibly the convexity) of the portfolio of bonds and the portfolio of liabilities. This process is called *immunization*. To carry out the necessary calculations, we may use the functions available in the Financial toolbox.

2.3.4 MATLAB functions to deal with fixed-income securities

When turning our attention from simple cash flows streams to real-life bonds, various complications arise. The first one is that in order to represent the settlement date and the maturity date of a bond correctly, we must be able to cope with a calendar, taking leap years into account. MATLAB has an internal way of dealing with dates, which is based on converting a date to an integer number. For instance, if we type `today`, MATLAB replies with a number corresponding to the current date; this number may be converted to a more meaningful string by using `datestr`:

```
>> today
ans =
    732681
>> datestr(today)
ans =
04-Jan-2006
```

You may wish to check which date corresponds to day 1. The inverse of datestr is datenum:

```
>> datenum('04-Jan-2006')
ans =
    732681
```

There is a wide variety of string formats that you may use to input a date in MATLAB; the one you see above is only one of them (note that it is necessary to enclose the string between quotes). Dates must be taken into account for different reasons. Consider buying a bond after it is issued; if you buy a bond at a date between two coupon payments, the time elapsed from the last coupon payment date must be taken into account. If not, you would receive a coupon benefit to which the previous owner is partially entitled. Actually, by computing the present value of the cash flow stream you would take it into account; however, the market convention is to quote a bond price without considering this issue. What you read is the *clean price*, to which accrued interest must be added in order to obtain the correct price. Accrued interest may be computed by prorating the coupon payment over the period between two payments. Roughly speaking, if coupons are paid every six months and you buy a bond two months before the next coupon payment, you owe something like two-thirds of the coupon to the previous owner. However, there are different day count conventions to make the necessary calculations. These issues are considered in the **bndprice** function, which is used to price a bond, for a given yield value. To understand the input arguments required, we may use the online help (we have included only the first few lines appearing on the screen):

```
>> help bndprice
 BNDPRICE Price a fixed income security from yield to maturity.
    Given NBONDS with SIA date parameters and semi-annual yields to
    maturity, return the clean prices and the accrued interest due.

    [Price, AccruedInt] = bndprice(Yield, CouponRate, Settle, Maturity)

    [Price, AccruedInt] = bndprice(Yield, CouponRate, Settle, ...
        Maturity, Period, Basis, EndMonthRule, IssueDate, ...
        FirstCouponDate, LastCouponDate, StartDate, Face)
```

We see that, as usual in MATLAB, this function may be called with a minimal set of input arguments, which are required yield, coupon rate, settlement date (i.e., when the bond is purchased), and maturity date. The two output values are the clean price and the accrued interest, which must be summed in order to get the real (dirty) price:

```
>> [clPr accrInt] = bndprice(0.08, 0.1, '10-aug-2007', '31-dec-2020')
clPr =
   116.2366
accrInt =
     1.1141
>> clPr+accrInt
ans =
   117.3507
```

When calling the function this way, all the other arguments take a default value. For instance, the `Period` parameter, which is the number of coupon payments per year, is assumed to be two, and the face value (`Face`) is assumed to be 100. Another possibly important parameter is `Basis`, which controls the day count convention in computing the accrued interest; the default value is 0, which corresponds to the actual/actual convention; if the parameter is set to 1, the convention is 30/360 (i.e., it is assumed that all months consist of 30 days). To appreciate the difference between the day count conventions, we may compute the number of days between two dates by the 30/360 convention and the actual number of days:

```
>> days360('27-Feb-2006', '4-Apr-2006')
ans =
    37
>> daysact('27-Feb-2006', '4-Apr-2006')
ans =
    36
```

Other day count conventions are possible and used for different securities (see, e.g., [7]). The remaining parameters are related to the coupon structure and are described in the Financial toolbox manual.

Example 2.11 To obtain the price–yield curve of figure 2.9, we may use the following code fragment:

```
settle     = '19-Mar-2000';
maturity   = '15-Jun-2015';
face       = 1000;
couponRate = 0.05;
yields = 0.01:0.01:0.20;
[cleanPrices , accrInts] = bndprice(yields, couponRate, settle, ...
    maturity, 2, 0, [] , [] , [] , [], [] , face);
plot(yields, cleanPrices+accrInts);
grid on
```

Note that when we have to provide a function with an optional argument, such as the face value, but we do not want to use optional arguments which should occur before that one, we have to pass empty vectors represented by [] so that the arguments are properly matched. □

For now, we have computed a price given a required yield. We may also go the other way around; we may compute the yield given the price, using another predefined function:

```
>> CleanPrices = [95 100 105];
>> bndyield(CleanPrices, 0.08, datenum('31-Jan-2006'), '31-Dec-2015')
ans =
    0.0876
    0.0800
    0.0728
```

The minimal set of parameters for the bndyield function are: the clean price, with no accrued interest; the coupon rate; the settlement date; and the maturity date. In this case we have used a common feature of MATLAB functions. If a vector is passed as an argument, where a scalar would be used in the simplest case, the output is, typically, the vector of the results obtained by applying the function to each component of the input vector. Here we have used different prices, and we see that a bond selling below par (95) has a yield higher than the coupon rate; yield and coupon rate are equal for a bond selling at par (100); yield is lower for a bond selling above par (105). Optional parameters may be passed to bndyield, which are similar to the parameters of bndprice.

Other useful functions may be used to compute duration and convexity, given the price or the yield of a bond. They are best illustrated by a simple immunization example.

Example 2.12 A common problem in bond portfolio management is to shape a portfolio with a given (modified) duration D and convexity C. Suppose that we have a set of three bonds; we would like to find a set of portfolio weights w_1, w_2, and w_3, one for each bond, such that

$$\sum_{i=1}^{3} D_i w_i = D$$

$$\sum_{i=1}^{3} C_i w_i = C$$

$$\sum_{i=1}^{3} w_i = 1,$$

where C_i and D_i are the bond durations and convexities, respectively ($i = 1, 2, 3$). Note that we have assumed that both the duration and the convexity of the portfolio can be computed as weighted combinations of the bond characteristics; actually, this is not true in general, but for the moment we

64 FINANCIAL THEORY

```
% SET BOND FEATURES (bondimmun.m)
settle     = '28-Aug-2007';
maturities = ['15-Jun-2012' ; '31-Oct-2017' ; '01-Mar-2027'];
couponRates = [0.07 ;  0.06 ; 0.08];
yields = [0.06 ; 0.07 ; 0.075];

% COMPUTE DURATIONS AND CONVEXITIES
durations = bnddury(yields, couponRates, settle, maturities);
convexities = bndconvy(yields, couponRates, settle, maturities);

% COMPUTE PORTFOLIO WEIGHTS
A = [durations'
     convexities'
     1 1 1];
b = [ 10
     160
       1];
weights = A\b
```

Fig. 2.10 Simple code for bond portfolio immunization.

will consider this as a simple approximation. All we have to do is to compute the coefficients C_i and D_i and to solve a system of three equations and three unknowns. This is easily accomplished by the script in figure 2.10. Note that we have assumed a given yield, and that we have used the functions **bnddury** and **bndconvy** to compute durations and convexities. It is possible to carry out a similar computation starting from the clean bond prices; we have just to use functions **bnddurp** and **bndconvp**. By running the script, we obtain the following solution:

```
weights =
    0.1209
   -0.4169
    1.2960
```

Note that we have to sell bond 2 short, which may not be feasible. □

2.3.5 Critique

The naive immunization and cash flow matching models, that we have just discussed, leave room for many criticisms.

To begin with, duration is only an approximate measure of bond price sensitivity. It is a correct measure only if the term structure is flat (i.e., the same rate applies to any period length) or if there is a parallel shift on the term structure. In practice, shape changes are possible, calling for more sophisticated sensitivity measures and immunization approaches.

Another issue is that immunization protects against small changes in the required yield. But after such a change, the duration and convexity are changed and the portfolio is no longer immunized. In fact, we are not paying due attention to the dynamic character of portfolio management. In the limit, consider a portfolio consisting of two bonds, one with a short and the other with a long duration, bracketing the target duration. It may be the case that the first bond has a short maturity; when maturity is reached, we are left with only one bond and a portfolio that is far from immunized. Continuous portfolio rebalancing may lead to nervous trading and high transaction costs. An alternative is to use dynamic optimization models, accounting for uncertainty in the interest rates and for dynamic trading. This leads to stochastic programming models, which are described in chapter 11. With such models, the stochastic nature of liabilities can also be accounted for.

Apart from using more sophisticated models, one can use more sophisticated assets. In fact, the need for interest-rate risk management has produced a vast array of interest-rate derivatives (see section 2.8). Both pricing such derivatives and managing interest-rate risk requires modeling the term structure of interest rates; this is a vast and difficult topic, which is actually beyond the introductory aim of this book.

2.4 STOCK PORTFOLIO OPTIMIZATION

Unlike bonds and derivatives, we do not consider pricing problems for stocks. There are models aimed at finding a "rational" price for a stock share of a firm, but they are beyond the scope of the book. Hence, we will consider stock prices as exogenous and we will only consider stock portfolio management. There is a set of n stocks and we must allocate our wealth among them. For simplicity, we do not consider dividend issues nor consumption, and we tackle a simple single-period problem, leaving multi-period portfolio optimization to later chapters. Our basic assumption is that uncertainty can be modeled by a probability distribution, which we treat as it were objective, and likely built on the basis of historical data. This need not be the case in portfolio management, as one could have some view, or information, which should be reflected in the decision problem. By selecting a portfolio, we select a probability distribution of future wealth, which is a random variable. We have seen in examples 2.5 and 2.6 on page 40 that using plain expected values in decision making under uncertainty may lead to unreasonable results. We must find a sensible way to model preferences under uncertainty, which essentially means that we must express risk aversion. The simplest approach to do so is based on utility theory, which is introduced in section 2.4.1. Since finding the utility function of a decision maker is no trivial task, practical approaches have been proposed based on risk measures. The best-known concept is mean-variance efficiency, which is dealt with in section 2.4.2; in section 2.4.3 we also illustrate a few MATLAB functions to cope with mean-variance portfolio optimization. Alternative risk measures, most notably Value at Risk, are discussed in section 2.4.5.

2.4.1 Utility theory

The idea that most investors are risk averse is intuitively clear, but what does *risk aversion* really mean? A theoretical answer, commonly used in economic theory, can be found by assuming that decision makers order uncertain outcomes by some utility function. To introduce the concept, let us consider simple lotteries, which may be regarded as investments under uncertainty. If a lottery has discrete outcomes, then it corresponds to a random variable X, with possible values x_i and probabilities p_i, and it can be represented by a fan like figure 2.2. The decision maker should select among alternative lotteries or she may also combine them, forming new random variables. For instance, consider an agent who has to choose between the following two lotteries: lottery a_1, which is actually deterministic and ensures a payoff μ, and lottery a_2, which has two equally likely payoffs $\mu + \delta$ and $\mu - \delta$. The two lotteries are clearly equivalent in terms of expected payoff, but a risk-averse agent will arguably select lottery a_1. More generally, if we have a random variable X and we add a *mean-preserving spread*, i.e., a random variable ϵ with $\mathrm{E}[\epsilon] = 0$, this addition is not welcome by a risk-averse decision maker.

Given a set of lotteries, the agent should be able to pick up the preferred one; or, given any pair of lotteries, the agent should be able to tell which one she prefers or to decide that she is indifferent among them. In this case, we would have a preference relationship among lotteries. Since preference relationships are a bit cumbersome and are not easy to deal with, we could map each lottery to a number, measuring the utility of that lottery to the agent, and use the standard ordering of numbers to sort lotteries. For arbitrary preference relationships, a function representing them may not exist, but under a set of more or less reasonable assumptions,[10] such a mapping does exist and it can be represented by a utility function. A particularly simple form of utility function, which looks reasonable but is justified by specific hypotheses on the preference relationship it models, is the Von Neumann–Morgenstern utility:

$$U(a) = \sum_{i=1}^{n} p_i u(x_i)$$

for some function $u(\cdot)$, where a is a lottery with outcomes x_i and probabilities p_i. The function $u(\cdot)$ is the utility of a certain payoff, and $U(\cdot)$ is clearly the expected utility. If $u(x) \equiv x$, then the utility function boils down to the

---

[10] The discussion of these assumptions is best left to books on Microeconomics; we should mention that most of them look rather innocent and reasonable under most circumstances, but they may lead to surprising effects in paradoxical examples.

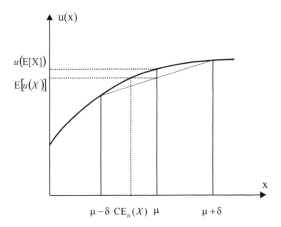

Fig. 2.11 How concave utility functions imply risk aversion; the certainty equivalent is also shown.

expected value of the payoff, but by selecting the utility u we may model different attitudes towards risk. For our problems, it is reasonable to assume that utility $u(\cdot)$ is an increasing function, since we prefer more wealth to less.

In the case of the two lotteries above, preference for a_1 is expressed by

$$U(a_1) = u(\mu) \geq \frac{1}{2}u(\mu - \delta) + \frac{1}{2}u(\mu - \delta) = U(a_2).$$

Since the inequality is not strict, we should say that lottery a_1 is at least as preferred as a_2, as the agent could be indifferent between the two. More generally, if we have two possible outcomes x_1 and x_2, with probabilities $p_1 = p$ and $p_2 = 1 - p$, a risk-averse decision maker would prefer not taking chances:

$$u(\mathrm{E}[X]) = u(px_1 + (1-p)x_2) \geq pu(x_1) + (1-p)u(x_2) = \mathrm{E}[u(X)].$$

This condition basically states that the function $u(\cdot)$ is *concave*. We see that concavity is linked to convexity, as the two concepts are related by a change in the sense of the inequality, and a function $f(\cdot)$ is concave if and only if the function $-f(\cdot)$ is convex (see supplement S6.1). Figure 2.11 illustrates the role of concavity. It can be shown that for a continuous or discrete random variable, the following Jensen's inequality holds for a concave function:

$$u(\mathrm{E}[X]) \geq \mathrm{E}[u(X)]. \tag{2.7}$$

It is fundamental to observe that the exact numerical value of the utility assigned to lotteries is irrelevant; only the relative ordering of alternatives is essential. In fact, we speak of *ordinal* rather than cardinal utility. Given the linearity of expectation, we also see that an affine transformation of utility has no effect, provided it is increasing: if we use $au(x) + b$ instead of $u(x)$, the ordering is preserved, provided that $a > 0$.

How can we say something about the properties of a specific utility function? In particular, we would like to come up with some way to measure risk aversion. We have said that a risk-averse agent would prefer a certain payoff rather than an uncertain one, when the expected values are the same. She would take the gamble only if the expected value of the risky lottery were suitably larger than the certain payoff. In other words, she requires a *risk premium*. The risk premium depends partly on the risk attitude of the agent, partly on the uncertainty of the gamble itself. We will denote the risk premium by $\rho_u(X)$; note that it is a number, which a decision maker with utility $u(\cdot)$ associates to a random variable X. The risk premium is defined by requiring

$$u(\mathrm{E}[X] - \rho_u(X)) = U(X). \tag{2.8}$$

The risk premium implicitly defines a *certainty equivalent*, i.e. a certain payoff such that the agent would be indifferent between the lottery and this payoff:

$$\mathrm{CE}_u(X) = \mathrm{E}[X] - \rho_u(X).$$

Note that the certainty equivalent is smaller than the expected value, and the difference is larger when the risk premium is larger. These concepts may be better grasped by looking again at figure 2.11.

A difficulty with the risk premium concept is that it mixes the intrinsic risk of a lottery with the risk attitude of the agent. We might wish to separate the two sides of the coin. Consider a lottery $X = x + \tilde{\epsilon}$, where x is a given number and $\tilde{\epsilon}$ is a random variable with $\mathrm{E}[\tilde{\epsilon}] = 0$ and $\mathrm{Var}(\tilde{\epsilon}) = \sigma^2$. Assume that the random variable $\tilde{\epsilon}$ is a "small" perturbation, in the sense that each of its realizations ϵ is a relatively small number.[11] Hence, we may approximate both sides of equation (2.8) by Taylor expansions. Consider for instance the expression $u(x + \epsilon)$. Since only numbers are involved here, we may write

$$u(x + \epsilon) \approx u(x) + \epsilon u'(x) + \frac{1}{2}\epsilon^2 u''(x).$$

By writing the same approximation for the random variable $\tilde{\epsilon}$ and taking expected values, we may approximate the right-hand side of (2.8):

$$\begin{aligned}\mathrm{E}[U(X)] &\approx \mathrm{E}\left[u(x) + \tilde{\epsilon}u'(x) + \frac{1}{2}\tilde{\epsilon}^2 u''(x)\right] \\ &= u(x) + \mathrm{E}[\tilde{\epsilon}]u'(x) + \frac{1}{2}\mathrm{E}[\tilde{\epsilon}^2]u''(x)\end{aligned}$$

---

[11] For the sake of convenience, in this section we denote by $\tilde{\epsilon}$ a random variable and by ϵ a realization of that variable. This notation is common in Economics; in Statistics, one typically uses X and x with the corresponding pair of meanings.

$$= u(x) + 0 \cdot u'(x) + \frac{1}{2}\text{Var}(\tilde{\epsilon})u''(x)$$
$$= u(x) + \frac{1}{2}\sigma^2 u''(x).$$

In the second-to-last line we have used $\text{Var}(\tilde{\epsilon}) = \text{E}[\tilde{\epsilon}^2] - \text{E}^2[\tilde{\epsilon}] = \text{E}[\tilde{\epsilon}^2] - 0$. We may also approximate the left-hand side of (2.8), which only involves numbers, by a first-order expansion around $\text{E}[X] = x$:

$$u(\text{E}[X] - \rho_u(X)) \approx u(x) - \rho_u(X)u'(x).$$

Equating both sides and rearranging yields

$$\rho_u(X) = -\frac{1}{2}\frac{u''(x)}{u'(x)}\sigma^2.$$

Since we assume utility is concave and increasing, the right-hand side is positive.[12] We may also see that the risk premium is factored as the product of one term depending on agent's risk aversion and of another one depending on uncertainty. This justifies the definition of the coefficient of absolute risk aversion:

$$R_u^a(x) \equiv -\frac{u''(x)}{u'(x)} \qquad (2.9)$$

We have said that, given the linearity of the expectation operator, an (increasing) affine transformation of a utility function $u(x)$ is inconsequential. The definition of the risk-aversion coefficient is consistent with this observation, as it is easy to see that the coefficients for $u(x)$ and $au(x) + b$ are the same.

Note that $r_u^a(x)$ does not depend on uncertainty, but it does depend on the expected value of the lottery. From an investor's point of view, this implies that risk aversion depends on the current level of wealth. The more concave the utility function, the larger risk aversion.

By the same token, we may define a coefficient of *relative* risk aversion. This is motivated by considering a multiplicative, rather than additive, shock on an expected value x: $X = x(1 + \tilde{\epsilon})$. Using a similar reasoning, we get:

$$\rho_u(X) = -\frac{1}{2}\frac{u''(x)}{u'(x)}x\sigma^2,$$

which motivates the definition

$$R_u^r(x) \equiv -\frac{u''(x)x}{u'(x)}. \qquad (2.10)$$

---

[12] A useful property of differentiable concave function of one variable is $u''(x) \leq 0$; see supplement S6.1.

Example 2.13 (A few standard utility functions) A typical utility function is logarithmic utility[13]:

$$u(x) = \log(x).$$

Clearly this makes sense only for positive values of wealth. It is easy to check that for the logarithmic utility we have

$$R_u^a(x) = \frac{1}{x}, \qquad R_u^r(x) = 1.$$

Hence, logarithmic utility has decreasing absolute risk aversion, but constant relative risk aversion. We say that logarithmic utility belongs to the families of DARA (decreasing absolute risk aversion) and CRRA (constant relative risk aversion) utility functions. We will see that this has important implications in portfolio optimization.

Another common utility function is quadratic utility:

$$u(x) = x - \frac{\lambda}{2}x^2. \tag{2.11}$$

Note that this function is not monotonically increasing and makes only sense for $x \in [0, 1/\lambda]$. Another odd property of quadratic utility is that it is IARA (increasing absolute risk aversion):

$$R_u^a(x) = \frac{\lambda}{1 - \lambda x} \Rightarrow \frac{dR_u^a(x)}{dx} = \frac{\lambda^2}{(1 - \lambda x)^2} > 0.$$

This is usually considered at odds with typical behavior of investors. Nevertheless, we may also see that quadratic utility emphasizes the role of variance, since for this utility

$$U(X) = \mathrm{E}[X - \frac{\lambda}{2}X^2] = \mathrm{E}[X] - \frac{\lambda}{2}\left(\mathrm{Var}(X) + \mathrm{E}^2[X]\right).$$

A decision maker with quadratic utility is basically concerned only with the expected value and the variance of an uncertain outcome. We will see how quadratic utility is linked to mean-variance portfolio optimization. □

Armed with the utility function concept, we may formalize portfolio optimization problems. In a single period portfolio optimization problem, we have an investor with given initial wealth W_0, which must be allocated to different assets, in such a way to maximize expected utility. Let θ_i be the wealth invested in asset $i = 1, \ldots, n$, and let $\tilde{R}_i$ be the random return of the asset. The

---

[13] In the following we will use the notation log, rather than ln, to denote the *natural* logarithm.

simplest formulation of the portfolio optimization problem is:

$$\max \quad \mathrm{E}\left[u\left(\sum_{i=1}^{n} \tilde{R}_i \theta_i\right)\right]$$
$$\text{s.t.} \quad \sum_{i=1}^{n} \theta_i = W_0. \tag{2.12}$$

The formulation is single-period, in the sense that no rebalancing is involved: a buy and hold strategy is assumed over the time period of interest. If short-selling is ruled out, we should also add non-negativity restrictions $\theta_i \geq 0$. It is common to include in the model a risk-free asset, whose return is deterministic, but this does not affect the *form* of the optimization model (it may affect the solution, of course).

In general, we should not take for granted that the above optimization model has a solution. For instance, if the model of uncertainty is not arbitrage free, we may expect an unbounded solution exploiting the arbitrage opportunity. But for non-pathological cases, an optimal portfolio (not necessarily unique) exists. It is important to note that the optimal portfolio may depend on the initial wealth W_0. Quite often, we may see models in which the decision variables are the weights $w_i \equiv \theta_i/W_0$ of each asset in the portfolio, and the budget constraint (2.12) is rewritten as

$$\sum_{i=1}^{n} w_i = 1.$$

The drawback of such a model formulation is that we do not see clearly the effect of initial wealth on the optimal solution. Since risk aversion depends on wealth, the optimal solution does depend on W_0. There are exceptions, however, as shown by the following example.

Example 2.14 Consider the following portfolio optimization problem:

- Uncertainty is modeled by a binomial distribution: There are two possible states of the world in the future, the up and down state, with probabilities p and q, respectively.

- There are two assets: one is risk-free, the other one is risky.

- The risk-free asset has total return R_f in both states (total return is one plus interest rate).

- Current price for the risky asset is S_0 and its total return is u in the up-state and d in the down-state.

- Initial wealth is W_0 and the investor has logarithmic utility.

In this problem, there is actually one decision variable, which we may take as δ, the number of stock shares purchased by the investor. To get rid of the budget constraint, we observe that δS_0 is the wealth invested in the risky asset, and $W_0 - \delta S_0$ is invested in the risk-free asset. Then, future wealth will be, for each of the two possible states:

$$\begin{aligned} W_u &= \delta S_0 u + (W_0 - \delta S_0) R_f = \delta S_0 (u - R_f) + W_0 R_f \\ W_d &= \delta S_0 d + (W_0 - \delta S_0) R_f = \delta S_0 (d - R_f) + W_0 R_f, \end{aligned}$$

and expected utility is $p \log(W_u) + q \log(W_d)$. The problem is then

$$\max_{\delta} \; p \log \{\delta S_0 (u - R_f) + W_0 R_f\} + q \log \{\delta S_0 (d - R_f) + W_0 R_f\}.$$

A necessary condition for optimality is stationarity (the first-order derivative vanishes):

$$p \frac{S_0(u - R_f)}{\delta S_0(u - R_f) + W_0 R_f} + q \frac{S_0(d - R_f)}{\delta S_0(d - R_f) + W_0 R_f} = 0.$$

In order to solve for δ, we may rewrite the equation a bit:

$$\frac{\delta S_0(u - R_f) + W_0 R_f}{p S_0(u - R_f)} = - \frac{\delta S_0(d - R_f) + W_0 R_f}{q S_0(d - R_f)}.$$

Straightforward manipulations yield

$$\frac{\delta}{p} + \frac{W_0 R_f}{p S_0(u - R_f)} = -\frac{\delta}{q} - \frac{W_0 R_f}{q S_0(d - R_f)}$$

and

$$\delta \left[\frac{1}{p} + \frac{1}{q} \right] = - \frac{W_0 R_f [q(d - R_f) + p(u - R_f)]}{p q S_0 (u - R_f)(d - R_f)}$$

and, finally

$$\frac{\delta S_0}{W_0} = \frac{R_f [up + dq - R_f]}{(u - R_f)(R_f - d)}.$$

This relationship implies that the *fraction* of initial wealth invested in the risky asset does not depend on the initial wealth itself. We have derived this property in a simplified setting, but it holds more generally for logarithmic utility, and is essentially due to its CRRA characteristic. □

Specifying a utility function may be a difficult task, since assessing the trade-off between risk and return is far from trivial. This may be no concern in Economics, if the aim is to build a model explaining some observed behavior and qualitative insights are of interest; however, in Financial Engineering and operational decision making, this is a difficulty. A relatively simple approach is based on the idea of restricting the choice to "reasonable" portfolios. If you fix the expected return you want to get from the investment, you would like to find the portfolio achieving that expected return with minimal risk. By the same token, if you fix the level of risk you are willing to take, you would like to select a portfolio maximizing the expected return. This approach leads to mean-variance portfolio theory, which, despite considerable criticism, underlies quite a significant part of financial theory.

2.4.2 Mean-variance portfolio optimization

Let us go back to the asset allocation problem, when only two risky assets are available. Let us denote by $\tilde{r}_i$, $\bar{r}_i$, and σ_i the random rate of return for asset $i = 1, 2$ and its expected value and standard deviation respectively. It is tempting to say that the problem is trivial when $\bar{r}_1 > \bar{r}_2$ and $\sigma_1 < \sigma_2$. In this case, stock 1 has a larger expected return than stock 2, and it is also less risky; hence, a naive argument would lead to the conclusion that asset 2 should not be considered at all. Actually, this may not be the case, since we have neglected the possible correlation between the two assets. The inclusion of asset 2 may, in fact, be beneficial in reducing risk, if its return is negatively correlated with the return of asset 1. So we see that there is some need for formalization in order to solve the problem.

Assume that we are interested in defining the portfolio weights, w_1 and w_2 in our case. A natural constraint is

$$w_1 + w_2 = 1.$$

Note that we are not considering the initial wealth level W_0, since we deal with the allocation of fractions of wealth. If we want to rule out short-selling, we must also require $w_i \geq 0$. Elementary probability theory tells us that the portfolio rate of return will be

$$\tilde{r} = w_1 \tilde{r}_1 + w_2 \tilde{r}_2,$$

and the expected return will be

$$\bar{r} = w_1 \bar{r}_1 + w_2 \bar{r}_2.$$

More generally, when we must devise a portfolio of n risky assets, the expected return is given by

$$\bar{r} = \sum_{i=1}^{n} w_i \bar{r}_i = \mathbf{w}'\bar{\mathbf{r}}.$$

The variance of $\tilde{r}$ is given, for the two-asset case, by

$$\sigma^2 = \text{Var}(w_1 r_1 + w_2 r_2) = w_1^2 \sigma_1^2 + 2 w_1 w_2 \sigma_{12} + w_2^2 \sigma_2^2,$$

where σ_{12} is the covariance between r_1 and r_2. For n assets we have

$$\sigma^2 = \sum_{i,j=1}^{n} w_i w_j \sigma_{ij} = \mathbf{w}'\Sigma\mathbf{w},$$

where all covariances σ_{ij} have been collected in the covariance matrix Σ.

By choosing the weights w_i, we will get different portfolios characterized by the expected value of the return and by its variance or standard deviation, which we may assume as a risk measure. Any investor would like both to maximize the expected return and to minimize variance. Since these two objectives are, in general, conflicting, we must find a trade-off. The exact trade-off will depend on the degree of risk aversion, which is hard to assess, but it is reasonable to assume that for a given target value $\bar{r}_T$ of the expected return, one would like to minimize variance. This is obtained by solving the following optimization problem:

$$\begin{aligned}
\min \quad & \mathbf{w}'\Sigma\mathbf{w} \\
\text{s.t.} \quad & \mathbf{w}'\bar{\mathbf{r}} = \bar{r}_T \\
& \sum_{i=1}^{n} w_i = 1 \\
& w_i \geq 0.
\end{aligned} \quad (2.13)$$

This is a quadratic programming problem, which may be solved by numerical methods described in chapter 6, where we also show how to use MATLAB functions provided by the Optimization Toolbox. The Financial Toolbox also includes functions to solve mean-variance portfolio optimization problems, which are described in the next section.

By changing the target expected return, one may obtain a set of *efficient* portfolios. Roughly speaking, a portfolio is efficient if it is not possible to obtain a higher expected return without increasing risk. There are infinite efficient portfolios in general, and it is reasonable to assume that the preferred portfolio will be one of them.

2.4.3 MATLAB functions to deal with mean-variance portfolio optimization

MATLAB includes a set of functions based on mean-variance portfolio theory. They rely on the Optimization toolbox to solve optimization problem (2.13) for different values of expected return. The first function we consider is frontcon. In the simplest case, frontcon receives three arguments: the vector of expected rates of return, covariance matrix, and the number of efficient portfolios we wish to find. The last argument is actually the number of risk minimization subproblems we wish to solve; this yields a finite subset of the efficient frontier, which may be enough to trace a good plot. The output arguments are: a vector of expected portfolio risks (standard deviation) for each efficient portfolio; expected rates of return; portfolio weights for each asset in each portfolio. It is instructive to go back to the case of two assets. Assume the following data:

$$\bar{r}_1 = 0.2 \qquad \bar{r}_2 = 0.1$$

$$\sigma_1^2 = 0.2 \qquad \sigma_2^2 = 0.4$$
$$\sigma_{12} = -0.1.$$

Note that asset 2 is apparently useless, but it is negatively correlated with asset 1; hence, when asset 1 performs poorly, we may hope that asset 2 will perform well (and vice versa). Hence, including asset 2 may result in some beneficial diversification. Let us find a set of efficient portfolios:

```
>> r = [0.2 0.1];
>> s = [0.2 -0.1; -0.1 0.4];
>> [PRisk, PRoR, PWts] = frontcon(r,s,10);
>> [PWts, PRoR, PRisk]
ans =
    0.6250    0.3750    0.1625    0.2958
    0.6667    0.3333    0.1667    0.2981
    0.7083    0.2917    0.1708    0.3051
    0.7500    0.2500    0.1750    0.3162
    0.7917    0.2083    0.1792    0.3312
    0.8333    0.1667    0.1833    0.3496
    0.8750    0.1250    0.1875    0.3708
    0.9167    0.0833    0.1917    0.3944
    0.9583    0.0417    0.1958    0.4200
    1.0000         0    0.2000    0.4472
```

Here we display a table showing expected rate of return, in the first column, standard deviation, and portfolio weights. Each line correspond to one of the ten portfolios we wanted to find. The last line correspond to the riskiest portfolio, yielding the largest expected return. As we could expect, return is maximized by investing 100% of our wealth in the first asset, with $\bar{r}_1$ and $\sigma_1 = \sqrt{0.2} = 0.4472$ (recall that we are forbidding short sales in this model). It is interesting to note that it is possible to obtain portfolios whose standard deviation of return is lower than the standard deviation of both assets, which is due to negative correlation between returns in this case. The first portfolio displayed in the first line corresponds to the portfolio of minimal risk. We may also plot the efficient frontier by calling frontcon without output arguments:

```
>> frontcon(r,s,10);
```

We get the plot in figure 2.12.

We may repeat the experiment with more complex portfolios:

```
>> ExpRet = [ 0.15 0.2 0.08];
>> CovMat = [ 0.2 0.05 -0.01 ; 0.05 0.3 0.015 ; ...
             -0.01 0.015 0.1];
>> [PRisk, PRoR, PWts] = frontcon(ExpRet, CovMat, 10);
>> [PWts, PRoR, PRisk]
ans =
```

76 FINANCIAL THEORY

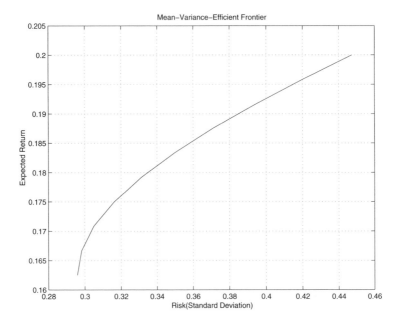

Fig. 2.12 Efficient frontier for a portfolio with two risky assets.

0.2914	0.1155	0.5931	0.1143	0.2411
0.3117	0.1831	0.5052	0.1238	0.2456
0.3320	0.2506	0.4174	0.1333	0.2588
0.3524	0.3181	0.3295	0.1428	0.2794
0.3727	0.3857	0.2417	0.1524	0.3060
0.3930	0.4532	0.1538	0.1619	0.3370
0.4133	0.5207	0.0659	0.1714	0.3714
0.3811	0.6189	0	0.1809	0.4093
0.1905	0.8095	0	0.1905	0.4682
0	1.0000	-0.0000	0.2000	0.5477

By the way, we should not get fooled by the apparent negative weight of an asset in the last portfolio:

```
>> PWts(10,3)
ans =
  -1.4461e-017
```

This is a typical example of small numerical errors that we must expect.

Like any professionally crafted code, `frontcon` is safe in the sense that some consistency checks are carried out on the input arguments. For instance, a covariance matrix must be positive semidefinite. The reader is urged to try `frontcon` with the following covariance matrix:

```
CovMat = [0.2 0.1 -0.1 ; 0.1 0.2 0.15 ; -0.1 0.15 0.2]
```

We have considered trivial portfolio optimization problems with no additional constraints. In real life, it is typical to have some constraints enforcing lower and upper bounds on the allocation to single assets or groups of assets. This may make sense if you want to limit the exposure to certain risky stocks or to market sectors (e.g., telecommunications or energy). The frontcon function is able to cope with such constraints, which may be represented by using additional arguments. However, a richer function, from this point of view, is portopt, which is able to cope with more general constraints.

To illustrate, consider a problem involving five assets. Suppose that you do not want to consider short-selling and that the following upper bounds are given on each asset weight in the portfolio:

$$0.35 \quad 0.3 \quad 0.3 \quad 0.4 \quad 0.5.$$

Furthermore, the assets can be partitioned into two groups, consisting of assets 1 and 2 and of assets 3, 4, and 5, respectively. You might wish to enforce both lower and upper bounds on asset allocation to each group; say the lower bounds are 0.2 and 0.3 and the upper bounds are 0.6 and 0.7. Formally, this would result in a constraint set like the following, which should be added to our quadratic programming problems:

$$0 \le w_1 \le 0.35 \quad 0 \le w_2 \le 0.3 \quad 0 \le w_3 \le 0.3$$
$$0 \le w_4 \le 0.4 \quad 0 \le w_5 \le 0.5$$
$$0.2 \le w_1 + w_2 \le 0.6$$
$$0.3 \le w_3 + w_4 + w_5 \le 0.7.$$

The optimization functions available in MATLAB can easily cope with such constraints, but they must be represented in *matrix form*. In other words, it is customary to specify (linear) constraints as systems of equations $\mathbf{A}_{eq}\mathbf{w} = \mathbf{b}_{eq}$ or inequalities $\mathbf{A}\mathbf{w} \le \mathbf{b}$. Writing constraints in such a form is conceptually simple, but practically difficult. In the past, persons working on numerical optimization had to write matrix generators in order to solve large problems by numerical libraries. Then, to ease a tedious and error-prone task, algebraic languages have been developed, such as AMPL, which is used in chapters 11 and 12 (see also appendix C). Algebraic languages allow us to express an optimization model in a quite natural way. In MATLAB there is no high-level way to express optimization models, but for mean-variance problems there is a sort of specialized matrix generator, called portcons.

For our small example, we would call this function as illustrated in figure 2.13, obtaining the constraint matrix in figure 2.14.[14] Note that we must

---

[14] We should note that frontcon can also be used for such a problem, but we prefer using portcons and portopt to illustrate a more general point related to matrix generators.

```
% CallPortcons.m
NAssets = 5;
AssetMin = NaN;
AssetMax = [0.35 0.3 0.3 0.4 0.5];
Groups  = [1 1 0 0 0 ; 0 0 1 1 1];
GroupMin = [ 0.2 0.3 ];
GroupMax = [ 0.6 0.7 ];

ConstrMatrix = portcons('Default', NAssets, ...
    'AssetLims', AssetMin, AssetMax, NAssets, ...
    'GroupLims', Groups, GroupMin, GroupMax)
```

Fig. 2.13 How to use portcons to build the constraint matrix.

```
ConstrMatrix =
    1.0000    1.0000    1.0000    1.0000    1.0000    1.0000
   -1.0000   -1.0000   -1.0000   -1.0000   -1.0000   -1.0000
   -1.0000         0         0         0         0         0
         0   -1.0000         0         0         0         0
         0         0   -1.0000         0         0         0
         0         0         0   -1.0000         0         0
         0         0         0         0   -1.0000         0
    1.0000         0         0         0         0    0.3500
         0    1.0000         0         0         0    0.3000
         0         0    1.0000         0         0    0.3000
         0         0         0    1.0000         0    0.4000
         0         0         0         0    1.0000    0.5000
   -1.0000   -1.0000         0         0         0   -0.2000
         0         0   -1.0000   -1.0000   -1.0000   -0.3000
    1.0000    1.0000         0         0         0    0.6000
         0         0    1.0000    1.0000    1.0000    0.7000
```

Fig. 2.14 Sample constraint matrix built by portcons.

```
% CallPortopt.m
CallPortcons;
ExpRet = [0.03 0.06 0.13 0.14 0.15];
CovMat = [
    0.01     0       0      0     0
    0        0.04   -0.05   0     0
    0       -0.05    0.30   0     0
    0        0       0      0.40  0.20
    0        0       0      0.20  0.40 ];
[PRisk, PRoR, PWts] = portopt(ExpRet, CovMat, 10, [], ConstrMatrix);
[PRoR, PRisk]
PWts
```

Fig. 2.15 Calling portopt.

include a 'Default' argument in order to specify that the sum of weights does not exceed 1 and short selling is ruled out. This is why we use NaN (not-a-number) as a lower bound on asset allocation AssetMin: otherwise, we would have twice the same constraints $w_i \geq 0$. Also note how the equality constraint $\sum_{i=1}^{5} w_i = 1$ is represented by two inequalities, $\sum_{i=1}^{5} w_i \leq 1$ and $\sum_{i=1}^{5} (-w_i) \leq -1$. This is because portopt assumes inequality constraints only. Then the matrix may be used by calling portopt as illustrated in figure 2.15 (some optional arguments are omitted; see MATLAB online help).

Using that script, we get the following output:

```
ans =
    0.0816    0.1487
    0.0860    0.1620
    0.0904    0.1762
    0.0948    0.1906
    0.0991    0.2054
    0.1035    0.2203
    0.1079    0.2361
    0.1122    0.2526
    0.1166    0.2799
    0.1210    0.3995

PWts =
    0.3000    0.3000    0.2250    0.0875    0.0875
    0.2623    0.3000    0.2309    0.0905    0.1163
    0.2220    0.3000    0.2496    0.0998    0.1286
    0.1816    0.3000    0.2683    0.1091    0.1410
    0.1413    0.3000    0.2870    0.1185    0.1533
```

0.1017	0.3000	0.3000	0.1299	0.1684
0.0639	0.3000	0.3000	0.1463	0.1899
0.0260	0.3000	0.3000	0.1627	0.2113
0.0000	0.3000	0.2650	0.1075	0.3275
0	0.3000	0	0.2000	0.5000

It is useful to check that the maximum return portfolio allocates 50% of wealth to asset 5, which is the maximum return asset; the upper bound $w_5 \leq 0.5$ prevents us from investing all of our wealth in this asset. Then 20% is allocated to asset 4 and nothing to asset 3, because $w_3 + w_4 + w_5 \leq 0.7$. The last 30% is allocated to asset 2.

Another consideration we should point out is that portcons generates a full matrix with many zero entries. Good optimization solvers deal with sparse matrices, which avoid storing zero entries in order to save memory space. Algebraic languages exploit this possibility, which is essential to deal with large-scale problems with special structure.

A last function we describe here may be used to find an optimal portfolio. So far, we have dealt with efficient portfolios, leaving the risk/return trade-off unresolved. We may resolve this trade-off by linking mean-variance portfolio theory to the more general utility theory illustrated in section 2.4.1. Actually, mean-variance theory is not necessarily compatible with an arbitrary utility function: An optimal portfolio for some utility function need not be on the mean-variance efficient frontier. It can be shown that this inconsistency does not arise if the returns are normally distributed or if the utility function is quadratic (see, e.g., [11] or [15]). The last point implies that if may specify a quadratic utility function such as (2.11), the optimal solution will be a mean-variance efficient portfolio. All we have to do is to choose the λ parameter according to our degree of risk aversion. In the Financial toolbox the function portalloc is provided, which yields the optimal portfolio assuming quadratic utility with some risk-aversion parameter; its default value is 3 and suggested alternative values range between 2 and 4. There is still another issue that we have neglected so far. We have considered mean-variance efficient portfolios, assuming that only risky assets were available. However, we may obtain a known return by investing in a bank account with a fixed interest rate or in a safe zero-coupon bond (with maturity equal to our investment horizon, to avoid interest rate risk issues). What is the effect of the inclusion of such a risk-free asset in our portfolio? A detailed analysis of this issue is rich in implications in financial theory, but it would lead us too far. For our purposes it is sufficient to say that the optimal portfolio will be a combination of the risk-free asset and one particular efficient portfolio. The amounts invested in the risk-free asset and in the risky portfolio depend on our risk aversion, but the risky portfolio involved does not. An important implication of this, if we believe in the theory, is that investors could live with just one "mutual" fund, mixing it with the risk-free asset. The portalloc function yields the optimal combination of the risky portfolio and the risk-free asset; it assumes further

```
% CallPortAlloc.m
ExpRet = [ 0.18 0.25 0.2];
CovMat = [ 0.2 0.05 -0.01 ; 0.05 0.3 0.015 ; ...
      -0.01 0.015 0.1];
RisklessRate = 0.05;
BorrowRate = NaN;
RiskAversion = 3;

[PRisk, PRoR, PWts] = frontcon(ExpRet, CovMat, 100);

[RiskyRisk , RiskyReturn, RiskyWts, RiskyFraction, ...
   PortRisk, PortReturn] = portalloc(PRisk, PRoR, PWts, ...
     RisklessRate, BorrowRate, RiskAversion);
AssetAllocation = [1-RiskyFraction, RiskyFraction*RiskyWts]
```

Fig. 2.16 Calling portalloc.

that cash may be borrowed at some rate. Figure 2.16 illustrates a script to call this function.

Some explanation is in order. First, we give the vector of the expected rates of return and the covariance matrix, which are used by frontcon to generate an approximation of the efficient frontier with a given number of points. We also give a riskless rate (for investing) and a risk-aversion coefficient. The borrowing rate is set to NaN since we do not consider the possibility of borrowing. There are several output returned by portalloc: RiskyRisk, RiskyReturn, and RiskyWts are the risk, the expected return, and the composition of the ideal fund. RiskyFraction is the fraction we should invest in the risky portfolio; PortRisk and PortReturn are the risk and return of the portfolio consisting of the risky portfolio and the risk-free asset.

Calling portalloc with these parameters will produce the following output:

```
>> CallPortAlloc
AssetAllocation =
    0.1401    0.2004    0.1640    0.4954
```

One could wonder why we should compute first the efficient frontier. In fact, this is due to the way portalloc is built. We can formulate and solve an optimization problem directly, using the concepts we will illustrate in chapter 6 (see also section C.2).

2.4.4 Critical remarks

Mean-variance portfolio theory leads to relatively simple numerical problems. However, despite its prominent role in financial theory, the approach has been the subject of widespread criticism. We have pointed out that mean-variance portfolio theory is consistent with the utility function framework in the case of normally distributed returns and in the case of a quadratic utility function. Both conditions may be debated.[15]

One important feature of the normal distribution is its symmetry. If the return distribution is symmetric, then using variance or standard deviation as a measure of risk may make sense; in fact, variance takes into account returns that are both higher and lower than the average. The former are actually desirable, but in the case of normal distribution a potential for good performance is exactly counterbalanced by the risk of underperformance. However, if the distribution is not symmetric, we must distinguish the upside potential from the downside risk. While symmetric returns may be assumed for stocks, derivative assets, such as those we shall describe shortly, may lead to more complex distributions. As for the quadratic utility function, we have seen that it implies increasing absolute risk aversion, which is itself a counterintuitive behavior for the usual investor. A solution to both issues would be the use of a carefully chosen utility function, which is hard to come up with, when dealing with real investors. We could also enforce constraints on the probability of large losses; if L is the random variable modeling the portfolio loss, we could require something like

$$P\{L > w\} \leq \alpha,$$

where α is a small probability and w is a threshold parameter; such a probabilistic constraint is known as *chance constraint*. All of these ideas lead to more complex optimization problems, namely stochastic programming problems, which are dealt with in chapter 11.

A further reason for using stochastic programming models is another difficulty in mean-variance theory. The covariance matrix is assumed to be constant over time. Unfortunately, it is likely that correlation may rise when stock market crashes occur, just when diversification should help. So we should use more complex models in describing the uncertainty. Stochastic programming does so by building a set of multiperiod scenarios, like the tree in figure 2.3 on page 27. This also enables us to consider another feature that is disregarded by mean-variance models: the dynamic nature of portfolio management, which is not considered in single-period models. Portfolios are revised in time, and the impact of transaction costs should not be neglected.

Modeling transaction costs exactly may be rather difficult. They depend in a non-trivial way on the amounts traded. For instance, it may be preferable to buy and sell stocks in round lots, since trading in odd lots may increase transaction costs. It might also be advisable to avoid a portfolio with a very

---

[15] See, e.g., [13] for a discussion of alternative utility functions in portfolio optimization. We should also mention that mean-variance theory is justified not only when returns are assumed normally distributed, but in the more general case of elliptic distributions, which include the normal; see [11].

small weight on some assets; the benefit of diversification will probably be lost because of increasing transaction costs. So we could require that if a stock enters the portfolio, it does so with a minimal weight. We may also look for portfolios including no more than a predetermined number of assets. Such constraints require the introduction of integer programming models, which are the subject of chapter 12.

2.4.5 Alternative risk measures: Value at Risk and quantile-based measures

Mean-variance portfolio theory is based on the use of variance or standard deviation as risk measures. We have already pointed out that this may not be always appropriate, but another practical issue is that they may be difficult to interpret by a portfolio manager. This is why alternative risk measures have been proposed and adopted, based on the concept of a portfolio loss. In general, a risk measure is a function mapping a random variable to a number; the larger this number, the riskier the distribution. More specifically, some measures are based on quantiles of the probability distribution of portfolio loss. The most widely known such measure is Value at Risk, or VaR (not to be confused with variance or, for people with a background in Econometrics, with a Vector Auto-Regressive, VAR, model).

The VaR concept was introduced as an easy-to-understand measure of portfolio risk. In fact, measuring, monitoring, and managing risk are fundamental activities for any portfolio manager. Bonds and stocks involve different forms of risk, and derivatives, if used for speculation, may be even riskier. Basically, VaR aims at measuring the maximum portfolio loss one could suffer, over a given time horizon, within a given confidence level. Technically speaking, it is a quantile of the probability distribution of future wealth. Suppose that our initial wealth is W_0 and the future (random) wealth is, at the end of the time horizon,

$$\tilde{W} = W_0(1 + \tilde{r}),$$

where $\tilde{r}$ is the random rate of return. We are interested in characterizing the potential loss, which occurs when the wealth increment

$$\delta W = \tilde{W} - W_0 = W_0 \tilde{r}$$

turns out to be negative. The VaR at confidence level α is implicitly defined by the following condition:

$$P\{\delta W \leq -\text{VaR}\} = 1 - \alpha, \qquad (2.14)$$

which shows that VaR is, disregarding the change in sign to make it positive, a quantile with confidence level α. Typical values for the confidence level could be $\alpha = 0.95$ or $\alpha = 0.99$. To be precise, the definition above holds for a continuous probability distribution, but it can be extended to a discrete probability distribution.

84　FINANCIAL THEORY

Let $f(r)$ be the probability density of the rate of return. Then we should look for a critical rate of return $r_{1-\alpha}$ such that

$$P\{\tilde{r} \leq r_{1-\alpha}\} = \int_{-\infty}^{r^*} f(r)\, dr = 1 - \alpha.$$

The quantile $r_{1-\alpha}$ is obviously linked to a critical wealth $w_{1-\alpha}$, since from equation (2.14) we may deduce

$$w_{1-\alpha} - W_0 = -\text{VaR},$$

which in turn implies

$$\text{VaR} = W_0 - w_{1-\alpha} = -W_0 r_{1-\alpha}.$$

Note that the critical return is usually negative and VaR is positive. Sometimes VaR is defined with respect to the expected future wealth:

$$\text{VaR} = \text{E}[\tilde{W}] - w_{1-\alpha} = -W_0(r_{1-\alpha} - \text{E}[\tilde{R}]).$$

The two definitions may give approximately the same value for a short time horizon, say a few days. In this case volatility dominates drift[16] and $\text{E}[\tilde{W}] \approx W_0$. This assumption is not unreasonable, as regulations suggest using a risk measure in order to set aside enough cash to be able to cover short-term losses.

Computing VaR is easy if one assumes that returns are normally distributed and we are considering short time periods, so that the rate of return over a few successive periods is the sum of returns on each period (i.e., the compounding effect is negligible). For simplicity, assume that we hold N shares of an asset whose current price is S. Let σ be the daily volatility for that asset; hence, for a period of length δt days, volatility is $\sigma\sqrt{\delta t}$, if we assume daily returns are independent on each other. Since, by summing normal random variables, we get another normal variable, the return over the time period δt is normal too, and to get the quantile we need we may standardize as usual. Hence, given a confidence level α, we have to obtain the quantile $z_{1-\alpha}$ of the standard normal distribution by inverting its cumulative distribution function. For instance, if α is 99% and 95%:

```
>> z = norminv([0.01 0.05], 0, 1)
z =
    -2.3263   -1.6449
```

---

[16] The terms "volatility" and "drift" will be clarified in the next sections on stochastic differential equations. Intuitively, drift is related to expected return and volatility is related to standard deviation. On a short time interval of length δt, drift scales linearly with δt, whereas volatility is proportional to $\sqrt{\delta t}$, which means that when the time interval tends to zero, drift goes to zero faster than volatility.

STOCK PORTFOLIO OPTIMIZATION

For the VaR over the time period δt, with a confidence level α, we have

$$\text{VaR} = -z_{1-\alpha}\sigma\sqrt{\delta t}\,NS, \tag{2.15}$$

where the term NS is the current wealth W_0. If the time horizon is longer, we should not neglect the drift due to the expected return. In such a case, we should modify (2.15) as follows:

$$\text{VaR} = NS(\mu\,\delta t - z_{1-\alpha}\sigma\sqrt{\delta t}),$$

where μ is the expected daily return. For a portfolio of assets, computing VaR is again easy if normality is assumed. We have just to evaluate the portfolio risk as in mean-variance theory.

Example 2.15 Suppose that we hold a portfolio of two assets. The portfolio weights are $w_1 = 2/3$ and $w_2 = 1/3$, respectively; the two daily volatilities are $\sigma_1 = 2\%$ and $\sigma_2 = 1\%$, and the correlation is $\rho = 0.7$. Let the time horizon δt be 10 days. To obtain the portfolio risk, we compute the variance:

$$\sigma^2 = [w_1\ w_2] \begin{bmatrix} \sigma_1^2\,\delta t & \rho\sigma_1\sigma_2\,\delta t \\ \rho\sigma_1\sigma_2\,\delta t & \sigma_2^2\,\delta t \end{bmatrix} \begin{bmatrix} w_1 \\ w_2 \end{bmatrix} = 0.0025111;$$

hence $\sigma = 0.05011$. Assuming that the overall portfolio value is \$10 million, and that the confidence level is 99%,

$$\text{Var} = 10^7 \cdot 2.3263 \cdot 0.05011 = \$1{,}165{,}709.$$

The same result can be obtained by using the MATLAB functions `portstats` and `portvrisk`. The first one, given the expected return vector for each asset, the covariance matrix, and the portfolio weights, computes the portfolio risk and the expected return:

[PRisk, PReturn] = portstats (ExpReturn, CovMat, Wts).

The second one computes the VaR, given the expected portfolio return, its risk, the risk threshold $1 - \alpha$, and the portfolio current value:

VaR = portvrisk(PReturn, PRisk, RiskThreshold, PValue)

Using these functions, we get

```
>> format bank
>> s1 = 0.02 * sqrt(10);
>> s2 = 0.01 * sqrt(10);
>> rho = 0.7;
>> CovMat = [ s1^2 rho*s1*s2 ; rho*s1*s2 s2^2];
>> s = PortStats([0 0], CovMat, [2/3 1/3]);
>> var = portvrisk(0,s,0.01,10000000)
```

var =
     1165755.90

Note that the previous result was a bit different because of truncation errors in the pencil-and-paper calculation. □

The general formula for a portfolio of n assets with current price S_i, $i = 1, \ldots, n$, daily volatility σ_i, correlation ρ_{ij} between assets i and j, where we hold a number N_i of shares for each asset is

$$\text{VaR} = -z_{1-\alpha} \sqrt{\delta t \sum_{i=1}^{n} \sum_{j=1}^{n} N_i N_j \rho_{ij} \sigma_i \sigma_j S_i S_j}.$$

Needless to say, this formula holds if normality is assumed. But what if the assumption is not warranted? Indeed, empirical data do not suggest that stock returns are normally distributed. Furthermore, we may have to deal with assets which depend on risk factors, and even if a risk factor is normally distributed, non-linear dependence of the price with respect to the underlying factor will destroy normality. A familiar example is the non-linear dependence of a bond price with respect to required yield. In this case, however, if we recall equation (2.6), we may settle for a duration-based approximation like

$$\delta P \approx D_M P \, \delta \lambda.$$

Hence, if $\delta \lambda$ is normally distributed, δP will be too, and normality holds approximately. Similar considerations apply in the case of derivatives, if we are able to compute suitable sensitivities of the price of the derivative with respect to the price of the underlying asset.

If we look for a better approximation, we must give up normality and deal with the consequences. Indeed, in this case there are many issues. To begin with, we cannot find the quantile of the wealth distribution by looking at the quantile of the standard normal distribution. In this case, a numerical solution can rely on Monte Carlo simulation (see chapter 4). A thornier issue concerns the way we model the dependence among the different risk factors. In fact, correlation tells the whole story when normality is assumed, but not in general. This requires the adoption of more sophisticated statistical models, such as copula theory, which is beyond the scope of this book (see references).

Even if we leave all such modeling and computational issues aside, and we assume that we can compute VaR, there is something wrong with the VaR concept itself. For instance, a quantile cannot distinguish between different distributions. Consider figure 2.17. The plot on the left shows the normal case; if we assume a sort of truncated distribution like the one on the right, VaR will be the same, since the area under the density function to its left is the same. However, the potential loss in the second case is quite different. In particular, it is different the expected value of loss *conditional* on being

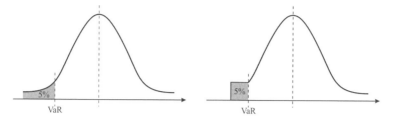

Fig. 2.17 Value at risk can be the same in different cases.

on the left (unlucky) tail of the portfolio value distribution. This has led to the definition of alternative risk measures, such as Conditional Value at Risk (CVaR), which is the expected value of loss, conditional on being to the left of VaR.

Risk measures like VaR or CVaR could be also used in portfolio optimization by solving optimization problems with the same structure as 2.13, with variance replaced by such measures. The resulting problem can be rather complex. In particular it may lack convexity properties that are so important in numerical optimization (see chapter 6). It turns out that minimizing VaR when uncertainty is modeled by a finite set of scenarios (which may be useful to capture complex distributions and dependencies among asset prices) is a nasty non-convex problem, whereas minimizing CVaR is (numerically) easier.

There is one last issue with VaR that deserves mention. Intuitively, risk is reduced by diversification. This should be reflected by any risk measure $\rho(\cdot)$ we consider. A little more formally, we should require a *subadditivity* condition like

$$\rho(A+B) \le \rho(A) + \rho(B),$$

where A and B are two portfolio positions. The following counter-example is often used to show that VaR lack this property.

Example 2.16 Let us consider two corporate bonds, A and B, whose issuers may default with probability 4%. Say that, in the case of default, we lose \$100 (in practice, we might partially recover the face value of the bond). Let us compute the VaR of each bond with confidence level 95%.

Before doing so, we should clarify what VaR is, when uncertainty is modeled by a discrete distribution. Definition (2.14) can be extended by defining VaR as the *smallest* value γ such that

$$P\{\delta W \le -\gamma\} \ge 1 - \alpha.$$

Basically, with a discrete distribution we may not find a value such that equation (2.14) is satisfied and we must resort to an inequality. Since default probability is only 4%, and $1 - 0.04 = 0.96 > 0.95$, we have in our case

$$\text{VaR}(A) = \text{VaR}(B) = \text{VaR}(A) + \text{VaR}(B) = 0.$$

Now what happens if we hold both bonds, and assume independent defaults? We will suffer

- a loss 0, with probability $0.96^2 = 0.9216$;
- a loss 200, with probability $0.04^2 = 0.0016$;
- a loss 100, with probability $2 \times 0.96 \times 0.04 = 0.0768$.

Hence, with that confidence level, $\text{VaR}(A+B) = 100 > \text{VaR}(A) + \text{VaR}(B)$, which means that diversification increases risk, if we measure it by VaR. □

Subadditivity is one of the properties that sensible risk measures should enjoy. The term *coherent risk measure* has been introduced to label a risk measure that meets a set of sensible requirements. VaR is not a coherent risk measure, whereas it can be shown that CVaR is.

2.5 MODELING THE DYNAMICS OF ASSET PRICES

In mean-variance portfolio theory we have considered a buy-and-hold portfolio. Hence, we were not interested in modeling the dynamics of asset prices, but only the distribution of return at the end of a given time interval. For more complex portfolio management models, we do need a dynamic model of asset prices. This is also required to solve option pricing models, as we will see in section 2.6. A model of the dynamics of asset prices must reflect the random nature of price movements, and the asset price $S(t)$ must be described as a stochastic process. This could be a discrete- or a continuous-time process. It turns out that for option pricing purposes, a continuous-time model is most useful, based on random walks. In this section with deal with modeling asset prices as stochastic processes in continuous time, which will lead us to consider stochastic differential equations and stochastic integrals.

2.5.1 From discrete to continuous time

It is a good idea to start with a discrete-time model and then derive a continuous-time model. Consider a time interval $[0, T]$, and imagine that we discretize the interval with a time step δt such that $T = N \cdot \delta t$; we may index the discrete-time instants by $t = 0, 1, 2, \ldots, N$. Let S_t be the stock price at time t. One possible and reasonable model is the multiplicative form:

$$S_{t+1} = u_t S_t, \qquad (2.16)$$

where u_t is a nonnegative random variable and the initial price S_0 is known. If we consider continuous random variables u_t, the model is continuous-state. The random variables u_t are assumed identically distributed and independent.

Independency is an assumption linked to market efficiency. Under this (debatable and debated) assumption, current prices reflect all information available so far.

The multiplicative model is reasonable since it ensures that prices will stay nonnegative, which is an obvious requirement for stock prices. If we used an additive model such as $S_{t+1} = u_t + S_t$, we should admit negative values for the random variables u_t to model price drops, and we would not have the guarantee $S_t \geq 0$. With the multiplicative form, a price drops when $u_t < 1$, but it stays positive. Furthermore, the actual price change depends on the present stock price (a $1 increase is different if the present price is $100 rather than $5), and this is easily accounted for by the multiplicative form.

In order to determine a plausible probability distribution for the random variables u_t, it is helpful to consider the natural logarithm of the stock price:

$$\log S_{t+1} = \log S_t + \log u_t = \log S_t + z_t.$$

The random variable z_t is the increment in the logarithm of price, and a common assumption is that it is normally distributed, which implies that u_t is lognormal.[17] Starting from the initial price S_0 and unfolding (2.16), we get

$$S_t = \prod_{k=0}^{t-1} u_k S_0,$$

which implies that

$$\log S_t = \log S_0 + \sum_{k=0}^{t-1} z_k.$$

Since the sum of normal random variables is still a normal variable (see appendix B), we have that $\log S_t$ is normally distributed, which in turn implies that, according to this model, stock prices are lognormally distributed. Using notation

$$\mathrm{E}[z_t] = \nu, \qquad \mathrm{Var}(z_t) = \sigma^2,$$

we see that

$$\begin{aligned}
\mathrm{E}[\log S_t] &= \mathrm{E}\left[\log S_0 + \sum_{k=0}^{t-1} z_k\right] \\
&= \log S_0 + \sum_{k=0}^{t-1} \mathrm{E}[z_k] = \log S_0 + \nu t \qquad (2.17) \\
\mathrm{Var}(\log S_t) &= \mathrm{Var}\left(\log S_0 + \sum_{k=0}^{t-1} z_k\right) = \sum_{k=0}^{t-1} \mathrm{Var}(z_k) = t\sigma^2, \qquad (2.18)
\end{aligned}$$

---

[17] If X is a normal random variable, then taking the exponential $\exp(X)$ yields a lognormal random variable; see appendix B.

where intertemporal independence of z_t is used in computing variance. The important point to see here is that the expected value and the variance of the increment in the logarithm of the stock price scale linearly with time; this implies that the standard deviation scales with the square root of time.

The next step is to obtain a model in continuous time. In the deterministic case, when you take the limit of a difference equation, you get a differential equation. Informally, in the deterministic case, we may recast what we have seen in discrete time as

$$\delta \log S(t) = \log S(t + \delta t) - \log S(t) = \nu\, \delta t$$

(note that we are basically working with the expected values, since for the moment we do not include randomness). If we take the limit as $\delta t \to 0$, we obtain:

$$d \log S(t) = \nu\, dt.$$

Integrating both differentials over the interval $[0, t]$ yields

$$\int_0^t d \log S(\tau) = \nu \int_0^t d\tau \Rightarrow \log S(t) - \log S(0) = \nu t \Rightarrow S(t) = S(0)e^{\nu t}. \quad (2.19)$$

This is coherent with the discrete time result. Actually, in the deterministic case, it is customary to write the differential equation as

$$\frac{d \log S(t)}{dt} = \nu$$

or, equivalently, as

$$\frac{dS(t)}{dt} = \nu S(t),$$

where we have used calculus to rewrite the differential

$$d \log S(t) = \frac{dS(t)}{S(t)}. \quad (2.20)$$

We see that ν is linked to the continuously compounded return of the asset.

When we include noise, there are a few important changes. The first, is that we should write the equation in the form

$$d \log S(t) = \nu\, dt + \sigma\, dW(t), \quad (2.21)$$

where $dW(t)$ can be considered as the increment of a stochastic process over the interval $[t, t + dt]$. This is a rather tricky object, called a *stochastic differential equation*. It is reasonable to guess that the solution of a stochastic differential equation is a stochastic process, rather than a deterministic function of time. However, this topic is quite difficult to deal with rigorously, as it requires some background in measure theory and stochastic calculus (see the references at the end of the chapter). We will limit ourselves to a reasonably detailed treatment.

The first thing we need is to investigate which type of continuous-time stochastic process $W(t)$ we can use as a building block. In the next section we introduce such a process, called the Wiener process, which plays more or less the same role as process z_t above. It turns out that this process is not differentiable, whatever this may mean for a stochastic process. Hence, we **cannot** write the stochastic differential equation as

$$\frac{d\log S(t)}{dt} = \nu + \sigma \frac{dW(t)}{dt}.$$

Actually, a stochastic differential equation must be interpreted as a shorthand for an *integral* equation much like (2.19), involving increments of a stochastic process. This calls for the definition of a *stochastic integral* and the related stochastic calculus. A consequence of the definition of the stochastic integral is that working with differentials as in equation (2.20) is not possible. We need a way to generalize the chain rule for differentials from the deterministic to the stochastic case. This leads to a fundamental tool of stochastic calculus called *Ito's lemma*.

2.5.2 Standard Wiener process

In the discrete-time model, we have assumed normally distributed increments in logarithmic prices, and we have also seen that the expected value of the increment of the logarithm of price scales linearly with time, whereas standard deviation scales with the square root of time.

In discrete time, we could consider the following process as a building block:

$$W_{t+1} = W_t + \epsilon_t \sqrt{\delta t},$$

where ϵ_t is a sequence of independent standard normal variables. We see that, for $k > j$,

$$W_k - W_j = \sum_{i=j}^{k-1} \epsilon_i \sqrt{\delta t},$$

which implies that

$$\begin{aligned} E[W_k - W_j] &= 0 \\ \text{Var}(W_k - W_j) &= (k-j)\,\delta t. \end{aligned}$$

Passing to continuous time, we may define the standard Wiener process as a continuous-time stochastic process characterized by the following properties.

1. $W(0) = 0$, which is actually a convention.

2. Given any time interval $[s, t]$, the increment $W(t) - W(s)$ is distributed as $\mathcal{N}(0, t-s)$, a normal random variable with zero expected value and

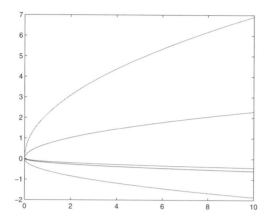

Fig. 2.18 Sample paths of a "degenerate" stochastic process.

standard deviation $\sqrt{t-s}$. Increments are stationary, as they do not depend on where the time interval is, but only on its width.

3. Increments are independent: If we take time instants $t_1 < t_2 \leq t_3 < t_4$, then $W(t_2) - W(t_1)$ and $W(t_4) - W(t_3)$ are independent random variables.

To see the importance of the independent increments assumption, let us compare the sample path of the Wiener process, which was shown in figure 2.5 on page 29, with the sample paths of a process defined as $Q(t) = \epsilon\sqrt{t}$, with $\epsilon \sim N(0,1)$, which are shown in figure 2.18. This is a "degenerate" stochastic process, since knowledge of one point on a sample path implies knowledge of the whole sample path, which makes the process quite predictable. However, if we just look at the marginal distribution of $Q(t)$, it seems just like the Wiener process, since

$$\mathrm{E}[Q(t)] = 0 = \mathrm{E}[W(t)]$$
$$\mathrm{Var}[Q(t)] = t = \mathrm{Var}[W(t)].$$

It is lack of independence that makes the difference. From figure 2.5, we also see that sample paths of the Wiener process look continuous, but not differentiable. This may be stated precisely, but it is not very easy. Introducing continuity and differentiability rigorously calls for specifying some concept of stochastic convergence. In fact, we should say that the Wiener process is nowhere differentiable with probability 1. To get an intuitive feeling for this fact, let us consider the incremental ratio:

$$\frac{\delta W(t)}{\delta t} = \frac{W(t + \delta t) - W(t)}{\delta t}.$$

Given the above properties, it is easy to see that
$$\text{Var}\left(\frac{\delta W(t)}{\delta t}\right) = \frac{\text{Var}\left[W(t+\delta t) - W(t)\right]}{(\delta t)^2} = \frac{1}{\delta t}.$$

If we take the limit for $\delta t \to 0$, this variance goes to infinity. Strictly speaking, this is no proof of non-differentiability of $W(t)$, but it does suggest that there is some trouble in using something like $dW(t)/dt$; indeed, you will never see a notation like this. We only use the *differential* $dW(t)$ of the Wiener process. Informally, we may think of $dW(t)$ as a random variable with distribution $\mathcal{N}(0, dt)$. Actually, we should think of this differential as an increment, which may be integrated as follows:
$$\int_s^t dW(\tau) = W(t) - W(s).$$

This looks reasonable, doesn't it? We may even go further and use $W(t)$ as the building block of stochastic differential equations. For instance, given real numbers a and b, we may imagine a stochastic process $X(t)$ satisfying the equation
$$dX(t) = a\,dt + b\,dW(t).$$

This is a *generalized Wiener process* and straightforward integration yields
$$X(t) = X(0) + at + bW(t).$$

But if we consider something more complicated, like
$$dX(t) = a(t, X(t))\,dt + b(t, X(t))\,dW(t) \tag{2.22}$$

things are not that intuitive. A process satisfying an equation like (2.22) is called an Ito process. We could argue that the solution should be something like
$$X(t) = X(0) + \int_0^t a(s, X(s))\,ds + \int_0^t b(\tau, X(\tau))\,dW(\tau). \tag{2.23}$$

Here the first integral looks like a standard Riemann integral of a function over time, but what about the second one? We need to assign a precise meaning to it, and this leads to the definition of a stochastic integral.

2.5.3 Stochastic integrals and stochastic differential equations

In a stochastic differential equation defining a process $X(t)$, where a Wiener process $W(t)$ is the driving factor, we may assume that the value $X(t)$ depends only on the history of $W(t)$ over the time interval from 0 to t. Technically speaking, we say that process $X(t)$ is adapted to process $W(t)$. Now let us consider a stochastic integral like
$$\int_0^T X(t)\,dW(t).$$

How can we assign a meaning to this expression? To begin with, it is reasonable to guess that a stochastic integral is a random variable. If we integrate a deterministic function of time we get a number; so, it is natural to guess that, by integrating a stochastic process over time, we should get a random variable. Furthermore, the stochastic integral above looks related to the sample paths of process $W(t)$, and an approximation could be obtained by partitioning the integration interval in small subintervals by selecting points $0 = t_0, t_1, t_2, \ldots, t_n = T$ and considering the sum

$$\sum_{k=0}^{n-1} X(t_k) \left[W(t_{k+1}) - W(t_k) \right]. \tag{2.24}$$

It is very important to notice how we select the time instants in the expression above: $X(t_k)$ is a random variable which is independent from the increment $W(t_{k+1}) - W(t_k)$ by which it is multiplied. This is actually *one* possible choice, which may be motivated as follows.

Example 2.17 Consider a set of n assets, whose prices are modeled by stochastic processes $S_i(t)$, $i = 1, \ldots, n$, which are described by stochastic differential equations like (2.22), and assume that we have a portfolio strategy represented by functions $h_i(t)$. These functions represent the number of stock shares we hold in the portfolio. But which functions make sense? An obvious requirement is that functions $h_i(\cdot)$ should not be anticipative: $h_i(t)$ may depend on all the history so far, over the interval $[0, t]$, but clairvoyance should be ruled out. Furthermore, we should think of $h_i(t)$ as the number of shares we hold over a time interval of the form $[t, t + dt)$.

Now, assume that we have some initial wealth that we invest in our portfolio, whose initial value, depending on portfolio strategy h, is

$$V_h(0) = \sum_{i=1}^{n} h_i(0) S_i(0) = \mathbf{h}'(0)\mathbf{S}(0),$$

where we have grouped h_i and S_i in vectors and use notation $\mathbf{h}'\mathbf{S}$ to denote inner vector product. What about the dynamics of the portfolio value? If the portfolio is self-financing, i.e., we can trade assets but we do not invest (nor withdraw) any more cash after $t = 0$, it can be shown that the portfolio value will satisfy the equation

$$dV_h(t) = \sum_{i=1}^{n} h_i(t)\, dS_i(t) = \mathbf{h}'(t)\, d\mathbf{S}(t)$$

This looks fairly intuitive and convincing, but some careful analysis is needed to prove it.[18] In particular, we may guess that the wealth at time $t = T$ will

---

[18] See, e.g., [1, chapter 6].

be:
$$V_h(T) = V_h(0) + \int_0^T \mathbf{h}'(t)\, d\mathbf{S}(t).$$

However, it is fundamental to interpret the stochastic integral as the limit of an approximation like (2.24), i.e.,

$$\int_0^T \mathbf{h}'(t)\, d\mathbf{S}(t) \approx \sum_{k=0}^{n-1} \mathbf{h}'(t_k)\left[\mathbf{S}(t_{k+1}) - \mathbf{S}(t_k)\right].$$

The number of stock shares we hold at time t_k does *not* depend on future prices $\mathbf{S}(t_{k+1})$. First we allocate wealth, and *then* we observe return. This is why Ito stochastic integrals are defined the way they are, and this makes financial sense. □

Now, if we take approximation (2.24) and consider finer and finer partitions of the interval $[0, t]$, letting $n \to \infty$, what do we obtain? The answer is technically involved. We must select some concept of stochastic convergence and check that everything makes sense. Using mean square convergence, it can be shown that the definition makes indeed sense, and we get the so-called stochastic integral *in the sense of Ito*.

The definition of stochastic integral has some important consequences. To begin with, what is the expected value of the integral above? We may get a clue by considering approximation (2.24):

$$\begin{aligned}
\mathrm{E}\left[\int_0^T X(t)\, dW(t)\right] &\approx \mathrm{E}\left\{\sum_{k=0}^{n-1} X(t_k)\left[W(t_{k+1}) - W(t_k)\right]\right\} \\
&= \sum_{k=0}^{n-1} \mathrm{E}\left\{X(t_k)\left[W(t_{k+1}) - W(t_k)\right]\right\} \\
&= \sum_{k=0}^{n-1} \mathrm{E}\left[X(t_k)\right] \cdot \mathrm{E}\left[W(t_{k+1}) - W(t_k)\right] = 0,
\end{aligned}$$

where we have used independence of $X(t_k)$ from the increments of the Wiener process, along with the fact that the expected value of the increments is zero.

The definition of stochastic integral does not yield a precise way to compute it. We may try, however, to consider a specific case to get some intuition. The following example illustrates one nasty consequence of the way we have defined the stochastic integral.

Example 2.18 (The chain rule does not apply to stochastic differentials) Say that we want to "compute" the stochastic integral

$$\int_0^T W(t)\, dW(t).$$

96   FINANCIAL THEORY

Analogy with ordinary calculus would suggest using the chain rule of differentiation to obtain a differential which can be integrated directly. Specifically, we *might guess* that

$$dW^2(t) = 2W(t)\,dW(t).$$

This *would* suggest

$$\int_0^T W(t)\,dW(t) = \frac{1}{2}\int_0^T dW^2(t) = \frac{1}{2}W^2(T).$$

But this cannot be true. We have just seen that the expected value of the integral is zero, but

$$\mathrm{E}\left[\frac{1}{2}W^2(T)\right] = \frac{1}{2}\mathrm{E}[W^2(T)] = \frac{1}{2}\{\mathrm{Var}[W(T)] + \mathrm{E}^2[W(T)]\} = \frac{T}{2} \neq 0.$$

We see that the expected values do not match.   □

The last example shows that the chain differentiation rule does not work in Ito stochastic calculus. To proceed further, we need to find the right rule, and the answer is Ito's lemma which is introduced below.

We close this section by noting that we started from differential equation (2.22) and we ended up studying the equivalent integral form (2.23). Actually, from a mathematical point of view, only the latter makes sense, and we should regard the differential form as a *shorthand* notation for the integral form. An obvious advantage of the differential form is its readability; working on this form helps intuition, which is essential in devising sensible models for asset prices and interest rates.

2.5.4   Ito's lemma

We now give an informal argument (following [10, chapter 10]) to obtain Ito's lemma. Recall that an Ito process $X(t)$ satisfies a stochastic differential equation such as

$$dX = a(X,t)\,dt + b(X,t)\,dW, \qquad (2.25)$$

which is in some sense the continuous limit of

$$\delta X = a(X,t)\delta t + b(X,t)\epsilon(t)\sqrt{\delta t}, \qquad (2.26)$$

where $\epsilon \sim N(0,1)$, i.e., it has a standard normal distribution. Our aim is to derive a stochastic differential equation for a function $F(X,t)$ of $X(t)$. One key ingredient is the formula for the differential of a function $G(x,y)$ of two variables:

$$dG = \frac{\partial G}{\partial x}dx + \frac{\partial G}{\partial y}dy,$$

which may be obtained from Taylor expansion,

$$\delta G = \frac{\partial G}{\partial x}\delta x + \frac{\partial G}{\partial y}\delta y + \frac{1}{2}\frac{\partial^2 G}{\partial x^2}(\delta x)^2 + \frac{1}{2}\frac{\partial^2 G}{\partial y^2}(\delta y)^2 + \frac{\partial^2 G}{\partial x\,\partial y}\delta x\,\delta y + \cdots$$

when $\delta x, \delta y \to 0$. Now we may apply this Taylor expansion to $F(X,t)$, limiting it to the leading terms. In doing so it is important to notice that the term $\sqrt{\delta t}$ in equation (2.26) needs careful treatment when squared. In fact, we have something like

$$(\delta X)^2 = b^2 \epsilon^2 \delta t + \cdots,$$

which implies that the term in $(\delta X)^2$ cannot be neglected in the approximation. Since ϵ is a standard normal variable, we have $\mathrm{E}[\epsilon^2] = 1$ and $\mathrm{E}[\epsilon^2\,\delta t] = \delta t$. A delicate point is the following. It can be shown that, as δt tends to zero, the term $\epsilon^2\,\delta t$ can be treated as non-stochastic, and it is equal to its expected value. A useful way to remember this point is the *formal* rule

$$(dW)^2 = dt. \tag{2.27}$$

Hence, when δt tends to zero, in the Taylor expansion we have

$$(\delta X)^2 \to b^2\,dt.$$

Neglecting higher-order terms and taking the limit as both δX and δt tend to zero, we end up with

$$dF = \frac{\partial F}{\partial X}dX + \frac{\partial F}{\partial t}dt + \frac{1}{2}\frac{\partial^2 F}{\partial X^2}b^2\,dt,$$

which, substituting for dX, becomes the celebrated **Ito's lemma**:

$$dF = \left(a\frac{\partial F}{\partial X} + \frac{\partial F}{\partial t} + \frac{1}{2}b^2\frac{\partial^2 F}{\partial X^2}\right)dt + b\frac{\partial F}{\partial X}dW. \tag{2.28}$$

Although this proof is far from rigorous, we see that all the trouble is due to the term of order $\sqrt{\delta t}$ linked to the Wiener process. Indeed, if we set $b = 0$, i.e., there is no random term due to the Wiener process in the differential equation, Ito's lemma boils down the chain rule for derivatives

$$\frac{dF}{dt} = \frac{\partial F}{\partial x}\frac{dx}{dt} + \frac{\partial F}{\partial t},$$

and thus, given differential equation (2.22) for x,

$$dF = a\frac{\partial F}{\partial x}dt + \frac{\partial F}{\partial t}dt.$$

In Ito's lemma we have an extra term in dW, which is expected given the input stochastic process, and an unexpected term:

$$\frac{1}{2}b^2\frac{\partial^2 F}{\partial x^2}.$$

In the deterministic case, second-order derivatives occur in second-order terms linked to $(\delta t)^2$, which can be neglected; but here we have a term of order $\sqrt{dt}$ which must be taken into account even when it is squared. In order to grasp Ito's lemma, we should try a couple of examples.

Example 2.19 Let us consider again example 2.18. In order to compute the stochastic integral of $W^2(t)$, we may simply apply Ito's lemma to the case $X(t) = W(t)$, by setting $a(X,t) \equiv 0$, $b(X,t) \equiv 1$, and $F(X,t) = X^2(t)$. Hence we have:

$$\frac{\partial F}{\partial t} = 0 \tag{2.29}$$

$$\frac{\partial F}{\partial X} = 2X \tag{2.30}$$

$$\frac{\partial^2 F}{\partial X^2} = 2. \tag{2.31}$$

It is important to point out that in equation (2.29) the partial derivative with respect to time is zero; it is true that $F(X(t), t)$ depends on time through $X(t)$, but here we have no direct dependence on t, thus the *partial* derivative with respect to time vanishes.

Ito's lemma tells us

$$dF = d(W^2) = dt + 2W\,dW.$$

It is instructive to note that dt is the term which we would *not* expect by applying the usual chain rule. But this term allows us to get the correct expected value of $W^2(T)$, since

$$W^2(T) = W^2(0) + \int_0^T dW^2(t) = 0 + \int_0^T dt + \int_0^T W(t)\,dW(t).$$

Taking expected values we get

$$\mathrm{E}[W^2(T)] = T,$$

which is coherent with what we have seen in example 2.18. □

Ito's lemma may be used to find the solution of a stochastic differential equation, at least in relatively simple cases. A most important one is geometric Brownian motion.

Example 2.20 Geometric Brownian motion. Geometric Brownian motion is defined by the stochastic differential equation

$$dS(t) = \mu S(t)\,dt + \sigma S(t)\,dW(t),$$

where μ and σ are constant parameters referred to as drift and volatility, respectively. Intuition would suggest to rewrite the equation as

$$\frac{dS(t)}{S(t)} = \mu\, dt + \sigma\, dW(t),$$

and then to consider the differential of $d \log S$, which would be dS/S in deterministic calculus, to get the integral. However, we know that some extra care is needed. Nevertheless, it is useful to find the stochastic differential equation for $F(S,t) = \log S(t)$. To apply Ito's lemma, we first compute partial derivatives:

$$\frac{\partial F}{\partial t} = 0$$
$$\frac{\partial F}{\partial S} = \frac{1}{S}$$
$$\frac{\partial^2 F}{\partial S^2} = -\frac{1}{S^2},$$

from which we may write

$$\begin{aligned} dY &= \left(\frac{\partial F}{\partial t} + \mu S \frac{\partial F}{\partial S} + \frac{1}{2}\sigma^2 S^2 \frac{\partial^2 F}{\partial S^2}\right) dt + \sigma S \frac{\partial F}{\partial S} dW \\ &= \left(\mu - \frac{\sigma^2}{2}\right) dt + \sigma\, dW. \end{aligned}$$

Now we see that our guess was not that bad, as this equation may be integrated and yields

$$\log S(t) = \log S(0) + \left(\mu - \frac{\sigma^2}{2}\right) t + \sigma W(t).$$

Recalling that $W(t)$ has a normal distribution, as it can be written as $W(t) = \epsilon\sqrt{t}$, where $\epsilon \sim \mathcal{N}(0,1)$, we see that the logarithm of price is normally distributed:

$$\log S(t) \sim \mathcal{N}\left[\log S(0) + \left(\mu - \frac{\sigma^2}{2}\right) t,\ \sigma^2 t\right].$$

We can rewrite the solution in terms of $S(t)$:

$$S(t) = S(0) e^{(\mu - \sigma^2/2)t + \sigma W(t)}.$$

or

$$S(t) = S(0) e^{(\mu - \sigma^2/2)t + \sigma\sqrt{t}\epsilon}.$$

This shows that prices, according to the geometric Brownian motion model, are lognormally distributed. Recalling the relationships between normal and lognormal variables (see appendix B), we may also conclude that

$$E[S(t)] = S(0) e^{\mu t},$$

from which we see that the drift parameter μ is linked to continuously compounded return. The volatility parameter σ is related to standard deviation of the increment of logarithm of price.

The roles of drift and volatility can also be grasped intuitively by considering the following approximation of the equation defining Brownian motion:

$$\frac{\delta S}{S} \approx \mu\, \delta t + \sigma\, \delta W,$$

where $\delta S/S$ is the return of the asset over small time interval δt. According to this approximation, we see that return can be approximated by a normal variable with expected value $\mu\, \delta t$ and standard deviation $\sigma\sqrt{\delta t}$. Actually, this normal distribution is only a local approximation of the "true" (according to the model) lognormal distribution. □

Example 2.21 In the next sections we will apply Ito's lemma to pricing options written on an underlying asset whose price follows geometric Brownian motion. Assuming that the option price at time t is a function of time and price only, i.e., a function $f(S,t)$, let us write a differential equation for the value of an option. Applying again Ito's lemma, with $a = \mu S$ and $b = \sigma S$, yields

$$\begin{aligned} df &= \left(\frac{\partial f}{\partial t} + \mu S \frac{\partial f}{\partial S} + \frac{1}{2}\sigma^2 S^2 \frac{\partial^2 f}{\partial S^2}\right) dt + \sigma S \frac{\partial f}{\partial S}\, dW \\ &= \frac{\partial f}{\partial t}\, dt + \frac{\partial f}{\partial S}\, dS + \frac{1}{2}\sigma^2 S^2 \frac{\partial^2 f}{\partial S^2}\, dt. \end{aligned} \qquad (2.32)$$

This seems an intractable object, since it looks like a partial differential equation involving a stochastic process. Actually, by exploiting the no-arbitrage principle, it can be simplified and transformed to a deterministic partial differential equation, which is amenable to solution by numerical methods. In some cases it may even be solved analytically. □

2.5.5 Generalizations

Geometric Brownian motion is not the only type of stochastic process relevant in finance, and the Wiener process is not the only relevant building block. One of the main features of these processes is the continuity of sample paths. However, discontinuities do occur sometimes, such as jumps in prices. In this case, different building blocks are used, such as the Poisson process, which is used to count events occurring with a certain rate. We should also note that continuous sample paths do not make sense for certain state variables such as credit rating. Another point is that the lognormal distribution, that we get from geometric Brownian motion, is a consequence of the normality associated to the Wiener process. Distributions with fatter tails are typically observed, questioning the validity of the models we have seen so far. However, dealing with sophisticated stochastic processes is beyond the scope of this book.

What we should consider, at least, is generalizing the Wiener process to a multidimensional process; we should also point out different forms of stochastic differential equations, leading to qualitatively different processes, such as mean reverting processes.

Correlated Wiener processes and multidimensional Ito's lemma When an option depends on more than one underlying asset, the simplest model is a generalization of geometric Brownian motion. According to this approach, we assume that the price $S_i(t)$ of asset $i = 1, \ldots, n$ satisfies

$$dS_i(t) = \mu_i S_i(t)\, dt + \sigma_i S_i(t)\, dW_i(t),$$

where the Wiener processes $W_i(t)$ are not necessarily independent. They are characterized by a set of instantaneous correlation coefficients ρ_{ij}, whose meaning can be grasped by an extension of the usual formal rule:

$$dW_i \cdot dW_j = \rho_{ij}\, dt.$$

Another point of view is that when simulating correlated Wiener processes, we must generate standard normal variates ϵ_i which are correlated; how this can be accomplished will be explained in the chapter on Monte Carlo simulation. It is relatively easy to generalize the results of example 2.21 to an option whose price at time t depends on time and a set of asset prices. To generalize Ito's lemma, we write the differential of $f(S_1(t), S_2(t), \ldots, S_n, t)$, using Taylor expansion to get

$$df = \frac{\partial f}{\partial t}\, dt + \sum_{i=1}^{n} \frac{\partial f}{\partial S_i}\, dS_i + \frac{1}{2} \sum_{i,j=1}^{n} \frac{\partial^2 f}{\partial S_i \partial S_j}\, dS_i\, dS_j,$$

where terms have been included or neglected according to the formal multiplication rules:

$$(dt)^2 = 0$$
$$dt \cdot dW_i = 0 \quad \forall i$$
$$dW_i \cdot dW_j = \rho_{ij}\, dt \quad \forall i, j$$

and $\rho_{ii} = 1$.

If we plug the equation of geometric Brownian motion here, we get the multidimensional Ito's lemma:

$$df = \left\{ \frac{\partial f}{\partial t} + \sum_{i=1}^{n} \mu_i \frac{\partial f}{\partial S_i} + \frac{1}{2} \sum_{i,j=1}^{n} \sigma_i \sigma_j \rho_{ij} \frac{\partial^2 f}{\partial S_i \partial S_j} \right\} dt + \sum_{i=1}^{n} \sigma_i \frac{\partial f}{\partial S_i}\, dW_i. \quad (2.33)$$

Mean reverting processes With geometric Brownian motion, the expected value of a price should go to infinity as time goes by, which is not really what happens in practice. In fact, stocks pay dividends, and no-arbitrage arguments show that the stock price should drop when dividends are paid. Other relevant variables, such as interest rates, cannot grow without bound. On the contrary, they tend to swing around long-term values, depending on economic conditions. We say that interest rates are characterized by mean reversion. Modeling interest rates is needed when dealing with interest rate derivatives which are used to control risk in fixed-income portfolios. We will have a brief look at such models in section 2.8. We just note here that we could model interest rates, and any variable showing mean reversion, by a stochastic differential equation like

$$dr = a(\hat{r} - r)\,dt + \sigma\,dW,$$

where $a > 0$. There is much to say about a model like this, since we should investigate consistency with the entire term structure of interest rates and with no-arbitrage properties. Actually, a model like this is only concerned with the short term interest rate. Yet it is easy to see that the process $r(t)$ tends to swing around the value $\hat{r}$. If $r > \hat{r}$, the drift term is negative, and $r(t)$ tends to drop; if $r < \hat{r}$, the drift term is positive and $r(t)$ tends to increase. Variations of such a model may be needed in order to make sure that the output is consistent with observed dynamics and that interest rates stay positive.

Similar considerations hold when modeling a stochastic and time-varying volatility $\sigma(t)$. Indeed, geometric Brownian motion assumes constant volatility, whereas in practice we may observe time periods in which volatility is higher than usual. One possible model for stochastic volatility consists of a pair of stochastic differential equations:

$$dS(t) = \mu S(t)\,dt + \sigma(t)S(t)\,dW_1(t)$$
$$dV(t) = \alpha(\bar{V} - V(t))\,dt + \xi\sqrt{V(t)}\,dW_2(t)$$

where $V(t) = \sigma^2(t)$, $\bar{V}$ is a long-term value, and different assumptions can be made on the correlation of the two driving Wiener processes. According to this model, volatility displays mean reversion, and it can be shown that the square root term prevents negative values of $V(t)$. Complex models may also link volatility to price.

2.6 DERIVATIVES PRICING

There are two basic issues in dealing with derivatives. The first issue is pricing. What is the fair price of a forward or an option contract? The second issue is hedging. Suppose that you are the writer of an option rather than the holder. In some sense the holder is at an advantage, since she is not forced to exercise the option if the circumstances are unfavorable (although example 2.2 on page 36 shows that careless management of an option portfolio may lead to a disaster). If you are the writer of an option and this is exercised, you have to meet your obligation, and in principle there may be no limit to your loss. Thus, you are interested in trading policies to reduce the risk to which you are exposed. We will not pursue real-life hedging in any detail in this book (see, e.g., [26]), but it is worth noting that, at least in theory, hedging is related to pricing.

A key role in pricing is played by the no-arbitrage argument we have already used, in a trivial situation, for bond pricing. This is best illustrated by a couple of examples. In the first one we derive the price of a forward contract. In the second one we derive a fundamental relationship between the price of a call and the price of a put, called *put-call parity*.

Example 2.22 Consider a forward contract for delivery at time T of an asset whose spot price now is $S(0)$. The spot price $S(T)$ at delivery is a random variable; hence, it would seem that randomness is involved in finding the fair forward price F that the holder of the long position of the forward will have to pay to the holder of the short position to purchase the underlying asset. Actually, a simple arbitrage argument shows that this is not the case.

Suppose that we hold the short position in the contract, and consider the following portfolio. We may borrow an amount $S(0)$ at the risk-free interest rate r, assuming continuous compounding, to buy the asset. The net cash flow now is zero. Then, at time T we may deliver the asset at price F, and we must pay back $S(0)e^{rT}$. Despite the randomness in the spot price, the value of our portfolio at T is deterministic and given by $F - S(0)e^{rT}$. But since the portfolio value at time $t = 0$ is zero, the same must hold at time $t = T$. Hence,

$$F = S(0)e^{rT}.$$

Any different forward price would lead to an arbitrage opportunity. If $F > S(0)e^{rT}$, the portfolio above will lead to a safe gain $F - S(0)e^{rT}$, with no initial commitment. If $F < S(0)e^{rT}$, we may reverse the portfolio by short-selling the asset and investing the proceeds. The reasoning assumes that short-selling the asset is possible and that no storage charge is paid for keeping the asset. See [10] for a full account of forward pricing.

It is interesting to note that a simple-minded approach would suggest a guess like $F = \mathrm{E}[S(T)]$, i.e., that the fair forward price is the expected price of the underlying in the future. This could look reasonable, assuming risk neutrality (linear utility function). The trouble with a reasoning like this is that we know most individual decision makers are characterized by some degree of risk aversion, but coming up with the "market" risk aversion, on the basis of individual utility functions, is awkward. Actually, in the idealized case we are considering, risk aversion does not play any role. This does not mean that risk aversion is not important, but that in this case we are using a sort of *relative* pricing, in which the attitude towards risk is irrelevant.

Finally, we should note that we *could* write the forward price as an expected value, if we assume that the underlying asset price $S(t)$ satisfies an equation

like
$$dS(t) = rS(t)\,dt + \sigma S(t)\,dW(t),$$
where the "true" drift has been replaced by the risk free rate. Indeed, in a risk neutral world investors would not care about risk and they would not require a risk premium. Hence, all assets would have the same return r. We begin seeing here a powerful principle: risk-neutral pricing. □

Example 2.23 Consider a call and a put options, both European-style, written on an underlying asset whose current price is $S(0)$, with the same exercise price K and maturity T. For now, we are not able to figure out the fair prices C and P of the two options, but it is easy to see that a precise relationship must hold between them. Consider two portfolios:

1. Portfolio P_1 consists of one European call option and an amount of cash equal to Ke^{-rT}, where r is the risk-free interest rate.

2. Portfolio P_2 consists of one European put option and one share of the underlying stock.

The value of portfolio P_1 at time $t = 0$ is $C + Ke^{-rT}$; the value of portfolio P_2 at time $t = 0$ is $P + S(0)$. At time T, we may have two cases, depending on the price $S(T)$. If $S(T) > K$, the call option will be exercised and the put option will not. Hence, under this hypothesis, portfolio P_1 at time $t = T$ will be worth
$$[S(T) - K] + K = S(T),$$
and portfolio P_2 will be worth
$$0 + S(T) = S(T).$$

If $S(T) < K$, the put option will be exercised and the call option will not. In this case, portfolio P_1 is worth
$$0 + K = K$$
and portfolio P_2
$$[K - S(T)] + S(T) = K.$$

In both cases, the two portfolios have the same value at time T. Hence, their values at time $t = 0$ must be equal; otherwise, there will be an arbitrage opportunity. We have shown that the following put-call parity relationship must hold:
$$C + Ke^{-rT} = P + S(0).$$
This implies that if we are able to find the fair price for one of the two options, the other one is obtained as well. □

We will see that the use of arbitrage arguments leads to pricing equations in the form of partial differential equations. These may sometimes be solved

DERIVATIVES PRICING 105

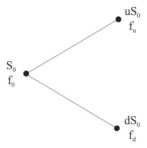

Fig. 2.19 Simple single-period binomial lattice.

analytically to yield a pricing formula in closed form, as in the case of Black and Scholes. In other cases, an analytical approach to option pricing may lead to useful approximate pricing formulas. In general, however, we need to resort to numerical procedures. There are basically three numerical approaches to price a derivative:

- Solving a partial differential equation, e.g., by finite difference approximations

- Monte Carlo simulation

- Binomial or trinomial lattices

All of them will be pursued in later chapters.

The first ingredient of an option pricing model is a model for the dynamics of the underlying asset price. The simplest such model, in continuous time, is geometric Brownian motion, which we have introduced in example 2.20. However, it is best to start with an even simpler representation model of price uncertainty: a one-step binomial model.

2.6.1 Simple binomial model for option pricing

Consider a single time step of length δt. We know the asset price S_0 at the beginning of the time step; the price S_1 at the end of the period is a random variable. The simplest model we may think of specifies only two possible values, accounting, e.g., for the possibility of an increase and a decrease in the stock price. To be specific, let us consider figure 2.19. We start with a price S_0; at the next time instant we assume that the price may take either value $S_0 u$ or $S_0 d$, where $d < u$, with probabilities p_u and p_d, respectively. Note the similarity with the multiplicative model of equation (2.16); this is a discrete-time model as well, but it is also discrete-state. Now, imagine an option whose unknown value now is denoted by f_0. If the option can only be exercised after δt, it is easy to find its values f_u and f_d corresponding to the two outcomes. They are simply the option payoffs, which are determined by the type of contract. How can we find f_0? We may again exploit the no-arbitrage principle. Let us set up a portfolio consisting of two assets: a riskless bond, with initial price $B_0 = 1$ and future price $B_1 = e^{r \cdot \delta t}$, and the underlying asset with initial value S_0. We denote the number of stock shares in the portfolio by Δ and the number of bonds by Ψ. The initial value of this portfolio is

$$\Pi_0 = \Delta S_0 + \Psi,$$

and its future value, depending on the realized state, will be either

$$\Pi_u = \Delta S_0 u + \Psi e^{r \cdot \delta t}, \quad \text{or} \quad \Pi_d = \Delta S_0 d + \Psi e^{r \cdot \delta t}.$$

Now let us try to find a portfolio which will exactly *replicate* the option payoff, i.e.,

$$\Delta S_0 u + \Psi e^{r \cdot \delta t} = f_u$$
$$\Delta S_0 d + \Psi e^{r \cdot \delta t} = f_d.$$

Solving this system of two linear equations in two unknown variables, we get

$$\Delta = \frac{f_u - f_d}{S_0(u - d)}$$
$$\Psi = e^{-r \cdot \delta t} \frac{u f_d - d f_u}{u - d}.$$

But in order to avoid arbitrage, the initial value of this portfolio must be exactly f_0:

$$\begin{aligned} f_0 &= \Delta S_0 + \Psi \\ &= \frac{f_u - f_d}{u - d} + e^{-r \cdot \delta t} \frac{u f_d - d f_u}{u - d} \\ &= e^{-r \cdot \delta t} \left\{ \frac{e^{r \cdot \delta t} - d}{u - d} f_u + \frac{u - e^{r \cdot \delta t}}{u - d} f_d \right\}. \end{aligned} \qquad (2.34)$$

It is important to note that this relationship does *not* depend on the objective probabilities p_u and p_d. In particular, the option price is *not* the, discounted, expected value of the payoff, which could have been a seemingly reasonable guess. If we think again at example 2.22 on forward pricing, we could wonder if we can nevertheless interpret equation (2.34) as an expected value. Indeed, if we set

$$\pi_u = \frac{e^{r \cdot \delta t} - d}{u - d}, \quad \pi_d = \frac{u - e^{r \cdot \delta t}}{u - d},$$

we may notice that

- $\pi_u + \pi_d = 1$
- π_u and π_d are positive if $d < e^{r \cdot \delta t} < u$, which *must* be the case if there is no arbitrage strategy involving the riskless and the risky asset; hence, we may interpret π_u and π_d as probabilities;

- the option price (2.34) can be interpreted as the discounted expected value of payoff under those probabilities:

$$f_0 = e^{-r \cdot \delta t} \hat{E}[f_1] = e^{-r \cdot \delta t}(\pi_u f_u + \pi_d f_d), \tag{2.35}$$

where notation $\hat{E}$ is used to point out that expectation is taken with respect to a different probability measure;

- the expected value of S_1 under probabilities π_u and π_d is

$$\hat{E}[S_1] = \pi_u S_0 u + \pi_d S_0 d = S_0 e^{r \cdot \delta t}.$$

The last observation explains why the "artificial probabilities" π_u and π_d are called *risk-neutral*. What we have found is coherent with pricing of a forward contract and suggests that derivatives can be priced by taking expectations under a risk-neutral measure. The objective probability measure does not play any role here, as the option payoff can be perfectly replicated by the two "primary" assets. When a set of "primary" assets allows us to replicate an arbitrary payoff, we say that the market is *complete*. It can be proved that a risk-neutral measure exists if arbitrage is impossible and it is unique if the market is complete. The risk-neutral valuation principle has far-reaching consequences; we refer the reader to a book like [20] for a deeper, yet readable, analysis.

What we have seen is a typical pricing argument based on *replication*. We may obtain the same result by taking a slightly (but equivalent) view. Assume that we have *written* a call option on a stock. How can we *hedge* against our risk? One possibility would be to purchase one stock share, so that if the holder will exercise the option, our position is covered. However, this strategy may be too conservative and expensive, if the option expires worthless. We could try to find the "right" number of shares to hold. Say that we purchase Δ stock shares to cover the writer's risk for a generic option with payoffs f_u and f_d. If we have written the option, the initial value of our portfolio is

$$\Pi_0 = \Delta S_0 - f_0.$$

Note that the option value, f_0, has a minus sign because we have a short position in the option, whose value in the future is a liability. The possible portfolio values after time period δt are

$$\Pi_u = \Delta u S_0 - f_u$$
$$\Pi_d = \Delta d S_0 - f_d.$$

In the replication argument, we have built a synthetic option using the stock and the riskless asset. Here we may replicate the riskless asset by choosing Δ such that

$$\Pi_u = \Pi_d \Rightarrow \Delta = \frac{f_u - f_d}{S_0(u - d)}$$

must hold. But due to the no-arbitrage principle, if this portfolio is riskless, it must earn the risk-free interest rate r. Assuming continuous compounding, we must have

$$S_0 \Delta - f_0 = (\Delta u S_0 - f_u)e^{-r \cdot \delta t},$$

or

$$f_0 e^{r \cdot \delta t} = \Delta u S_0 - e^{r \cdot \delta t} S_0 \Delta - f_u.$$

Substituting the expression for Δ and rearranging, we obtain equation (2.34) again.

We may interpret Δ as a hedging parameter, in the sense that it is the number of stock shares we should hold in order to hedge risk away. It is also useful to interpret

$$\Delta = \frac{f_u - f_d}{S_0(u - d)} = \frac{f_u - f_d}{S_u - S_d}$$

as a discretized approximation of the derivative of the option value with respect to changes in the underlying price, i.e., $\Delta = \partial f / \partial S$. In the next section we show that, in the continuous-time and continuous-state case, this interpretation is indeed correct.

2.6.2 Black–Scholes model

In the single-step binomial model, we are able to price an option assuming that future prices of the underlying will take one of two values. Hence, using only two assets, we are able to replicate any payoff. But two states make a rather crude model of uncertainty. What if we want to use a better probability distribution? One possibility would be to use more assets for replication, but this may be rather impractical. An alternative is to allow for trading at intermediate times. We should model asset prices not only now and at maturity, but also along the whole way. This can be done by using the binomial scheme recursively and devising a full recombining binomial lattice; this route yields interesting numerical schemes which are treated in chapter 7. Multi-stage binomial lattices are discrete-state and discrete-time models. But what if we want to account for a continuous distribution of future prices, such as the lognormal distribution associated with geometric Brownian motion? The answer is that we should allow for trading at infinite times, which calls for a continuous-time, continuous-state model. Curiously enough, this apparently complex model may yield simple solutions in closed form.

Consider a vanilla option like a European-style call option written on a non-dividend paying stock, whose price $S(t)$ follows a geometric Brownian motion. Since increments in the driving Wiener process are independent, we may say that future history does not depend on the past. And we may also show that the value of the option at a time t before maturity will depend only on time (more precisely, time to maturity) and current price of the underlying. If we denote this value by $f(S(t), t)$, we have seen in example 2.21 that it satisfies the stochastic differential equation:

$$df = \frac{\partial f}{\partial t} dt + \frac{\partial f}{\partial S} dS + \frac{1}{2}\sigma^2 S^2 \frac{\partial^2 f}{\partial S^2} dt. \tag{2.36}$$

What we know is that, at maturity, the option value is just the payoff,

$$F(S(T), T) = \max\{S(T) - K, 0\},$$

and what we would like to know is $f(S(0), 0)$, the fair option price now. Equation (2.36) does not suggest an immediate way to find the option price, but it would look a little bit nicer without the random term dS. Remember that by using no-arbitrage arguments, we have obtained deterministic relationships in examples 2.22 and 2.23, despite the randomness involved. To get rid of randomness, we may try to use options and stock shares to build a portfolio whose value is deterministic, just as we did in the simple binomial setting. Consider a portfolio consisting of a short position in an option and a long position in a certain number, say Δ, of stock shares. The value of this portfolio is

$$\Pi = \Delta \cdot S - f(S, t).$$

Differentiating Π and using equation (2.32), we get

$$d\Pi = \Delta\, dS - df = \left(\Delta - \frac{\partial f}{\partial S}\right) dS - \left(\frac{\partial f}{\partial t} + \frac{1}{2}\sigma^2 S^2 \frac{\partial^2 f}{\partial S^2}\right) dt \tag{2.37}$$

We may eliminate the term in dS by choosing

$$\Delta = \frac{\partial f}{\partial S}.$$

With this choice of Δ, our portfolio is riskless; hence, by no-arbitrage arguments, it must earn the risk-free interest rate r:

$$d\Pi = r\Pi\, dt. \tag{2.38}$$

Eliminating $d\Pi$ between equations (2.37) and (2.38), we obtain

$$\left(\frac{\partial f}{\partial t} + \frac{1}{2}\sigma^2 S^2 \frac{\partial^2 f}{\partial S^2}\right) dt = r\left(f - S\frac{\partial f}{\partial S}\right) dt,$$

and finally

$$\frac{\partial f}{\partial t} + rS\frac{\partial f}{\partial S} + \frac{1}{2}\sigma^2 S^2 \frac{\partial^2 f}{\partial S^2} - rf = 0. \tag{2.39}$$

Now we have a deterministic partial differential equation describing an option value $f(S, t)$. This equation applies to any option whose payoff depends only on the current price of the underlying asset, or its price at maturity. When the payoff depends on the whole history of prices, as in the case of Asian options, we get a slightly more complex equation. Typical partial differential equations need boundary and initial conditions to pin down a specific solution. In our case we have *final* conditions. For a vanilla European call we have a final condition at time T:

$$f(S,T) = \max\{S - K, 0\}.$$

By the same token, the terminal condition for a put is

$$f(S,T) = \max\{S - K, 0\}.$$

A remarkable and counterintuitive feature of equation (2.39) is that the drift μ of the underlying asset does not play any role. Only the risk-free interest rate r is involved. This is not really a surprise, given what we have seen for a forward or for an option under the single-step binomial model, and it is another example of the general and far-reaching principle of risk-neutral pricing.

In general, a partial differential equation is too difficult to use to get a solution in closed form, and it must be solved by numerical approaches; the difficulty stems partly from the equation itself and partly from the boundary conditions. We illustrate rather simple methods in chapter 5, and their application to option pricing is described in chapter 9. However, there are a few cases where equation (2.39) can be solved analytically. The most celebrated case is due to Black and Scholes, who were able to show that the solution for a European call is

$$C = S_0 N(d_1) - K e^{-rT} N(d_2), \qquad (2.40)$$

where

$$d_1 = \frac{\log(S_0/K) + (r + \sigma^2/2)T}{\sigma\sqrt{T}}$$

$$d_2 = \frac{\log(S_0/K) + (r - \sigma^2/2)T}{\sigma\sqrt{T}} = d_1 - \sigma\sqrt{T},$$

and N is the distribution function for the standard normal distribution:

$$N(x) = \frac{1}{\sqrt{2\pi}} \int_{-\infty}^{x} e^{-y^2/2}\, dy.$$

By using put-call parity, it can be shown that the value of a vanilla European put is

$$P = K e^{-rT} N(-d_2) - S_0 N(-d_1). \qquad (2.41)$$

It is also possible to give a value to the number Δ of shares we should sell short to build the riskless portfolio Π:

$$\Delta = \left.\frac{\partial C}{\partial S}\right|_{S=S_0} = N(d_1).$$

For a generic option of value $f(S,t)$,

$$\Delta = \frac{\partial f(S,t)}{\partial S}$$

measures the sensitivity of the option price to small variations in the stock price. Other sensitivities may be obtained, such as

$$\Gamma = \frac{\partial^2 f(S,t)}{\partial S^2}, \qquad \Theta = \frac{\partial f(S,t)}{\partial t}, \qquad \rho = \frac{\partial f(S,t)}{\partial r}, \qquad \mathcal{V} = \frac{\partial f(S,t)}{\partial \sigma}.$$

These sensitivities, collectively nicknamed the *Greeks*, may be used to evaluate the risk involved in holding a portfolio of options. They are known in closed form for some options and must be estimated numerically in general. Δ and Γ play a somewhat similar role to that of duration and convexity in bond portfolios. Θ measures the change in option value as the expiration date is approached, whereas ρ and $\mathcal{V}$ (vega) measure the sensitivity to changes in the riskless rate and in volatility. Δ is particularly significant due to its role in the riskless portfolio we have used to derive the Black–Scholes equation. In fact, the writer of an option might use that portfolio to hedge the option. In principle, this requires a continuous portfolio rebalancing since Δ will change in time; since practical considerations and transaction costs make continuous rebalancing impossible, some hedging error would result. In practice, hedging is not just based on option Δ; furthermore, a whole portfolio of options must be typically hedged.

2.6.3 Risk-neutral expectation and Feynman–Kač formula

In the case of the simple binomial model, we have found that the option value is the discounted expected value of future payoff, under a risk-neutral measure. But in continuous time, so far, we have relied on an apparently different framework, based on partial differential equations. Actually, they are two sides of the same coin, and the gap can be bridged by one version of the Feynman–Kač formula.

THEOREM 2.1 Feynman–Kač representation theorem. *Consider the partial differential equation*

$$\frac{\partial F}{\partial t} + \mu(x,t)\frac{\partial F}{\partial x} + \frac{1}{2}\sigma^2(x,t)\frac{\partial^2 F}{\partial x^2} = rF,$$

and let $F = F(x,t)$ be a solution, with boundary condition

$$F(T,x) = \Phi(x).$$

Then, under technical conditions, $F(x,t)$ can be represented as

$$F(x,t) = E_{x,t}\left[\Phi(X_T)\right],$$

where $X(t)$ is a stochastic process satisfying the differential equation

$$dX_\tau = \mu(X_\tau, \tau)\, d\tau + \sigma(X_\tau, \tau)\, dW_\tau$$

with initial condition $X_t = x$.

The notation $\mathrm{E}_{x,t}$ points out that this is a *conditional* expectation, given that at time t the value of the stochastic process is $X(t) = x$. From a mathematical point of view, the theorem is a consequence of how Ito stochastic integral is defined (see [1] for a clear proof). From a physical point of view, it is a consequence of the connection between Brownian motion (which is a diffusion process) and a certain type of PDEs which can be transformed into the heat equation.[19]

Applying this representation theorem to Black–Scholes equation, for an option with payoff function $\Phi(\cdot)$, immediately yields

$$f(S_0, 0) = e^{-rT} \hat{\mathrm{E}}\left[\Phi(S_T)\right],$$

which is consistent with (2.35). We point out that expectation is taken under a risk-neutral measure, which essentially means that we work as if the stochastic differential equation for $S(t)$ were

$$dS = rS\, dt + \sigma S\, dW.$$

It is interesting to note that changing measure in this case means changing the drift coefficient, whereas volatility is not affected.[20]

We should recall that according to the geometric Brownian motion model, a positive drift means that expected price in the future goes to infinity. This does not happen because dividends are paid, which cause a corresponding decrease in the stock price. It s fairly easy to show by no-arbitrage arguments that the price should fall by an amount corresponding to the paid dividend. Options on stocks paying lump sums at certain time instants can be priced by numerical methods such as binomial lattices. Black–Scholes model is easily extended if we assume that dividends are paid as a continuous stream at a rate q (the rate is applied to the current stock price, just like a continuously compounded interest rate). In this case, the risk neutral dynamics can be described by the equation

$$dS = (r - q)S\, dt + \sigma S\, dW. \tag{2.42}$$

A continuous dividend yield is a useful idealization in many circumstances. We may think of a stock index, which aggregates many stocks: Their discrete dividend cash flows may be aggregated to one dividend yield.

---

[19] We will introduce parabolic PDEs and the heat equation in chapter 5.
[20] Formally, this is a consequence of a theorem due to Girsanov; see [1].

2.6.4 Black–Scholes model in MATLAB

Implementing the Black–Scholes formula in MATLAB is quite easy. We may take advantage of the `normcdf` function provided by the Statistics toolbox to compute the cumulative distribution function for the standard normal distribution. Straightforward translation of equation (2.40) gives

```
d1 = (log(S0/K)+(r+sigma^2/2)*T) / (sigma * sqrt(T));
d2 = d1 - (sigma*sqrt(T));
C = S0 * normcdf(d1) - K * (exp(-r*T)*normcdf(d2));
P = K*exp(-r*T) * normcdf(-d2) - S0 * normcdf(-d1);
```

where the variables `S0, K, R, T, sigma` are self-explanatory. The Financial toolbox function `blsprice` implements these formulas with a couple of extensions. First, it may take vector arguments to compute a set of option prices at once; second, it may take into account a continuous dividend rate q (whose default value is zero). It is easy to adjust the Black–Scholes model and the related pricing formula to cope with a continuous dividend rate (see [28, chapter 5]). The following is an example of calling `blsprice`:

```
>> S0 = 50;
>> K = 52;
>> r = 0.1;
>> T = 5/12;
>> sigma = 0.4;
>> q = 0;
>> [C, P] = blsprice(S0, K, r, T, sigma, q)
C =
    5.1911
P =
    5.0689
```

It is interesting to plot the value of an option, say a vanilla European call, for different values of the current stock price while approaching the maturity. Running the code illustrated in figure 2.20, we get the plot of figure 2.21. We see that as time progresses, the plot approaches the kinky payoff diagram.[21] An important point is that we have to be consistent in specifying the risk-free interest rate, the volatility, and the expiration date. In the snapshot above everything is expressed in a yearly base; hence, the expiration date is in five months. Similar functions are available to compute the Greeks, too; they are best illustrated through a simple example.

Example 2.24 The Greeks may be used to approximate the change in an option value with respect to risk factors, just like duration and convexity for

---

[21] See section A.2 to see how to get a surface, rather a set of plots.

114 FINANCIAL THEORY

```
% PlotBLS.m
S0 = 30:1:70;
K = 50;
r = 0.08;
sigma = 0.4;
for T=2:-0.25:0
   plot(S0,blsprice(S0,K,r,T,sigma));
   hold on;
end
axis([30 70 -5 35]);
grid on
```

Fig. 2.20 Valuing a European call for different current prices of the underlying stock while approaching the expiration date.

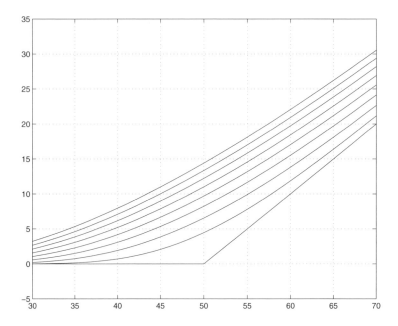

Fig. 2.21 Option value approaching the expiration date.

a bond portfolio, where the main risk factor is interest rate uncertainty. For instance, consider the change in the price of a call option due to an increase in the price of the underlying asset. Using a second-order Taylor expansion, we get the following approximation of this change:

$$C(S_0 + \delta S) \approx C(S_0) + \Delta \cdot \delta S + \frac{1}{2}\Gamma \cdot (\delta S)^2. \tag{2.43}$$

In MATLAB we may use such an approximation by exploiting the functions `blsdelta` and `blsgamma`. It is important to note that, unlike the other two functions, `blsgamma` returns only one argument, as it can be shown that Γ is the same for a call and a put. A simple MATLAB snapshot shows that the approximation is fairly good:

```
>> C0 = blsprice(50, 50, 0.1, 5/12, 0.3)
C0 =
    4.8851
>> dS = 2;
>> C1 = blsprice(50+dS, 50, 0.1, 5/12, 0.3)
C1 =
    6.2057
>> delta = blsdelta(50, 50, 0.1, 5/12, 0.3)
delta =
    0.6225
>> gamma = blsgamma(50, 50, 0.1, 5/12, 0.3)
gamma =
    0.0392
>> C0 + delta*dS + 0.5*gamma*dS^2
ans =
    6.2086
```

☐

Greeks, as we have said, may play a role in hedging, and Δ and Γ play the same role as duration and convexity for bonds. We may come up with strategies to build portfolios of options which are Δ-neutral, which means that the overall value of the portfolio will not change for small changes in the underlying price. Actually, from a practical point of view, *small* changes is not enough, and it is arguably better to have an imperfect hedging for large perturbations than a perfect hedging for infinitesimal perturbations.

Leaving hedging aside, we should note that Greeks also have a role in risk management. Consider estimating Value at Risk for a portfolio of options. Even if we assume that risk factors such as stock price perturbations δS are normally distributed, the pricing formula is non-linear in S_0, and this will destroy normality. However, if we use a Δ-based approximation like $\delta C \approx \Delta \cdot \delta S$ we see that normality is preserved, resulting in easy calculations. Actually, more accurate models and better descriptions of statistical dependence which go beyond correlation require numerical evaluation methods, such as Monte Carlo simulation.

2.6.5 A few remarks on Black–Scholes formula

The Black–Scholes formula has been a remarkable achievement and has played a fundamental role in the development of a huge and increasingly sophisticated market. However, there is a little fly in the ointment. If the Black–Scholes formula were "really correct," there would be *no* market for derivatives. The reason is disarmingly simple: The formula is based on replicating the option with two basic assets, and if this were really that easy, there would be no need for derivatives altogether. A little more formally, in a complete market there is no need for further assets, which would be redundant by definition. But of course, markets are not complete. The replication (or hedging) argument we have used assumes a rather idealized market, whereas, in practice, perfect hedging is made impossible by issues such as transaction costs, stochastic volatility, jumps in asset prices, etc. Geometric Brownian motion does not account for all of these features.

Furthermore, if we assume that perfect replication is feasible, there is no need to consider risk aversion; in fact the machinery we have developed in section 2.4.1 on utility theory does not play any role in simple option pricing models. In fact, several alternative pricing models have been developed, based on more sophisticated models of the dynamics of the underlying asset price. Moreover, while lack of arbitrage implies that a risk-neutral measure exists, market incompleteness implies that it is not unique. Hence, there is a range of prices which are compatible with lack of arbitrage. Which one is the right one? It depends on risk. From a theoretical point of view, we cannot get rid of issues related to decision making under uncertainty.

From a practical point of view, the simplicity and intuitive appeal of the Black–Scholes formula should not be discarded, however. Indeed, rather than resorting to overly complex models, the common practical approach is to use the Black–Scholes framework in a slightly different way, whose aim is to get *relative* prices; in other words, given prices we observe in financial markets, we use the arbitrage-free pricing machinery to price other assets in a way that is consistent with observed prices. Indeed, the Black–Scholes formula is sometimes considered as a sort of "interpolation" formula.

One common way to use the formula is by computing implied volatility. In a naive view, the volatility parameter σ in the formula should be estimated by analyzing the time series of prices of the underlying asset; this is what we mean by *historical* volatility. Implied volatility is computed the other way around: We observe option prices, and compute the volatility that makes the prices from the Black–Scholes formula consistent with the observed prices. This looks a bit like chasing our tail, but it allows to price new instruments in a consistent way. In practice, volatility surfaces are estimated as implied volatility depends on multiple factors, including time to maturity and strike price.

Another way to extend the machinery we have just developed to cope with incomplete market is by calibrating models directly under the risk-neutral measure, which is implicitly chosen by the market. We will motivate the idea in section 2.8, where we see that the Black–Scholes approach can be generalized by introducing a market price of risk. Roughly speaking, for each possible value of the market price of risk there is a risk-neutral measure, and a price under that measure. By observing prices, we may try to recover the market price of risk, or alternatively the risk-neutral measure; then, we may proceed pricing other instruments whose value depends on interest rates. One way to do so is to analyze bond prices to calibrate a model which can be used to price interest-rate derivatives.

2.6.6 Pricing American options

Unlike their European counterparts, American options can be exercised at any date prior to expiration. This seemingly innocent variation makes the analysis of American options much more complex. One easy conclusion is that an American option has a larger value than the corresponding European option, as it gives more opportunity for exercise. From a theoretical point of view, valuing an American option entails the solution of a dynamic stochastic optimization problem. If you hold such an option, you must decide, for each time instant, if it is optimal or not to exercise the option. You should compare the *intrinsic value* of the option, i.e., the immediate payoff you would get from exercising the option early, and the *continuation value*, which is linked to the possibility of waiting for better opportunities.

Formally, the price of an American option can be written as

$$\max_\tau \hat{\mathrm{E}}\left[e^{-r\tau}\Phi(S_\tau)\right], \qquad (2.44)$$

where function Φ is the option payoff, expectation is taken under a risk-neutral measure, and τ is a stopping time. The term "stopping time" has a very precise meaning in the theory of stochastic processes, but here we may simply interpret stopping time as the time at which we exercise the option. The time of early exercise (if this occurs) is a random variable depending only on the history of prices so far.

Clearly, early exercise will not occur if the option is not in-the-money. For a put option, we do not exercise the option at time t if $S(t) > K$. But even if $S(t) < K$, it may be better to keep the option and wait. Early exercise will occur only if the option is "enough" in-the-money; by how much, it will generally depend on time to expiration, and we may expect that when expiration gets closer, we are more willing to exercise early. Qualitatively, for an American put option we would expect an early exercise boundary like the one depicted in figure 2.22. This boundary specifies a stock price $S^*(t)$ such

118 FINANCIAL THEORY

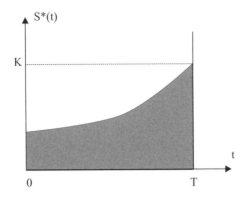

Fig. 2.22 Qualitative sketch of the early exercise boundary for a vanilla American put. The option is exercised within the shaded area.

that if $S(t) < S^*(t)$, i.e., the option is sufficiently deep in-the-money, then we are in the exercise region and it is optimal to exercise the option.[22] If we are above the boundary, we are in the continuation region, and we keep the option.

Finding this boundary is part of the problem and it is what makes it difficult. Unlike European options, we cannot simply compute an expected value, and this makes the use of Monte Carlo methods for pricing American-style options much more difficult. In the past, this was considered impossible, but we will see relatively simple approaches in chapter 10. Within the partial differential equation framework, this reasoning translates to a *free boundary problem*, which is contrasted against typical problems in which boundary conditions are given. However, in the context of finite difference methods of chapter 9, we will see that this essentially boils down to comparing the intrinsic and the continuation value to take a decision.

2.7 INTRODUCTION TO EXOTIC AND PATH-DEPENDENT OPTIONS

The variety of options that have been conceived in the past years seems to have no limit. You have options on stocks, commodities, and even options on options. Interest-rate derivatives play a fundamental role in interest-rate risk management. Some options are rather peculiar and are traded over-the-counter for specific needs.[23]

---

[22] For a detailed treatment of the exercise boundary for American options, see, e.g., [14, chapter 4].
[23] As we mentioned, this means that they are not traded on an organized exchange.

Exotic options on stocks may be designed by introducing a certain degree of path dependency. The idea is that, unlike a vanilla European option, the payoff depends not only on the underlying asset price at expiration, but also on its whole path. In the following we briefly describe barrier, Asian, and lookback options. They are of particular interest in learning and testing numerical methods.

2.7.1 Barrier options

In barrier options, a specific asset price S_b is selected as a barrier value. During the life of the option, this barrier may be crossed or not. In knock-out options, the contract is canceled if the barrier value is crossed at any time during the whole life; on the contrary, knock-in options are activated only if the barrier is crossed. The barrier S_b may be above or below the current asset price S_0: if $S_b > S_0$, we have an up option; if $S_b < S_0$, we have a down option. These features may be combined with the payoffs of call and put options to define an array of barrier options.

For instance, a down-and-out put option is a put option that becomes void if the asset price falls below the barrier S_b; in this case $S_b < S_0$, and $S_b < K$. The rationale behind such an option is that the risk for the option writer is reduced. So, it is reasonable to expect that a down-and-out put option is cheaper than a vanilla one. From the point of view of the option holder, this means that the potential payoff is reduced; however, if you are interested in options to manage risk, and not as a speculator, this also means that you may get cheaper insurance. By the same token, an up-and-out call option may be defined.

Now, consider a down-and-in put option. This option is activated only if the barrier level $S_b < S_0$ is crossed. Holding both a down-and-out and a down-and-in put option is equivalent to holding a vanilla put option. So we have the following parity relationship:

$$P = P_{\text{di}} + P_{\text{do}},$$

where P is the price of the vanilla put, and P_{di} and P_{do} are the prices for the down-and-in and the down-and-out options, respectively. Sometimes a rebate is paid to the option holder if the barrier is crossed and option is canceled; in such a case the parity relationship above is not correct.

In principle, the barrier might be monitored continuously; in practice, periodic monitoring may be applied (e.g., the price could be checked each day at the close of trading). This may affect the price, as a lower monitoring frequency makes crossing the barrier less likely.

Analytical pricing formulas are available for certain barrier options. As an example, consider a down-and-out put with strike price K, expiring in T time units, with a barrier set to S_b. The following formulas are known (see, e.g., [28, pp. 250–251]), where S_0, r, σ have the usual meaning.

$$P = Ke^{-rT}\{N(d_4) - N(d_2) - a[N(d_7) - N(d_5)]\}$$
$$- S_0\{N(d_3) - N(d_1) - b[N(d_8) - N(d_6)]\},$$

where

$$a = \left(\frac{S_b}{S_0}\right)^{-1+2r/\sigma^2}$$

$$b = \left(\frac{S_b}{S_0}\right)^{1+2r/\sigma^2}$$

and

$$d_1 = \frac{\log(S_0/K) + (r + \sigma^2/2)T}{\sigma\sqrt{T}}$$

$$d_2 = \frac{\log(S_0/K) + (r - \sigma^2/2)T}{\sigma\sqrt{T}}$$

$$d_3 = \frac{\log(S_0/S_b) + (r + \sigma^2/2)T}{\sigma\sqrt{T}}$$

$$d_4 = \frac{\log(S_0/S_b) + (r - \sigma^2/2)T}{\sigma\sqrt{T}}$$

$$d_5 = \frac{\log(S_0/S_b) - (r - \sigma^2/2)T}{\sigma\sqrt{T}}$$

$$d_6 = \frac{\log(S_0/S_b) - (r + \sigma^2/2)T}{\sigma\sqrt{T}}$$

$$d_7 = \frac{\log(S_0 K/S_b^2) - (r - \sigma^2/2)T}{\sigma\sqrt{T}}$$

$$d_8 = \frac{\log(S_0 K/S_b^2) - (r + \sigma^2/2)T}{\sigma\sqrt{T}}.$$

A MATLAB code implementing these formulas is given in figure 2.23.

```
>> [Call, Put] = blsprice(50,50,0.1,5/12,0.4);
>> Put
Put =
    4.0760
>> DOPut(50,50,0.1,5/12,0.4,40)
ans =
    0.5424
>> DOPut(50,50,0.1,5/12,0.4,35)
ans =
    1.8481
```

```
% DownOutPut.m
function P = DownOutPut(S0,K,r,T,sigma,Sb)
a = (Sb/S0)^(-1 + (2*r / sigma^2));
b = (Sb/S0)^(1 + (2*r / sigma^2));
d1 = (log(S0/K) + (r+sigma^2 / 2)* T) / (sigma*sqrt(T));
d2 = (log(S0/K) + (r-sigma^2 / 2)* T) / (sigma*sqrt(T));
d3 = (log(S0/Sb) + (r+sigma^2 / 2)* T) / (sigma*sqrt(T));
d4 = (log(S0/Sb) + (r-sigma^2 / 2)* T) / (sigma*sqrt(T));
d5 = (log(S0/Sb) - (r-sigma^2 / 2)* T) / (sigma*sqrt(T));
d6 = (log(S0/Sb) - (r+sigma^2 / 2)* T) / (sigma*sqrt(T));
d7 = (log(S0*K/Sb^2) - (r-sigma^2 / 2)* T) / (sigma*sqrt(T));
d8 = (log(S0*K/Sb^2) - (r+sigma^2 / 2)* T) / (sigma*sqrt(T));
P = K*exp(-r*T)*(normcdf(d4)-normcdf(d2) - ...
    a*(normcdf(d7)-normcdf(d5))) ...
    - S0*(normcdf(d3)-normcdf(d1) - ...
    b*(normcdf(d8)-normcdf(d6)));
```

Fig. 2.23 Implementing the analytical pricing formula for a down-and-out put option.

```
>> DOPut(50,50,0.1,5/12,0.4,30)
ans =
    3.2284
>> DOPut(50,50,0.1,5/12,0.4,1)
ans =
    4.0760
```

We see that the down-and-out put is indeed cheaper than the vanilla put; the price of the barrier option tends to that of the vanilla put as S_b tends to zero. It is also interesting to see what happens with respect to volatility:

```
>> [Call, Put] = blsprice(50,50,0.1,5/12,0.4);
>> Put
Put =
    4.0760
>> [Call, Put] = blsprice(50,50,0.1,5/12,0.3);
>> Put
Put =
    2.8446
>> DOPut(50,50,0.1,5/12,0.4,40)
ans =
    0.5424
>> DOPut(50,50,0.1,5/12,0.3,40)
ans =
```

```
        0.8792
>> DOPut(50,50,0.1,5/12,0.4,30)
ans =
        3.2284
>> DOPut(50,50,0.1,5/12,0.3,30)
ans =
        2.7294
```

For a vanilla put, less volatility implies a lower price, as there is less uncertainty; for the barrier option, less volatility *may* imply a higher price since breaching the barrier may be less likely. We see that the dominating effect depends on the barrier level.

In the formula above, it is assumed that barrier monitoring is continuous. When monitoring discrete, we should expect that the price for a down-and-out option is increased, since breaching the barrier is less likely. An approximate correction has been suggested (see [2] or [14, p. 266]). The idea is using the analytical formula above, correcting the barrier as follows:

$$S_b \Rightarrow S_b e^{\pm 0.5826 \cdot \sigma \sqrt{\delta t}},$$

where the term 0.5826 derives from the Riemann zeta function, δt is time elapsing between two consecutive monitoring time instants, and the sign $\pm$ depends on the option type. For a down-and-out put we should take the minus sign, as the barrier level should be lowered to reflect the reduced likelihood of crossing the barrier. For instance, if we monitor the barrier each day, the prices above change approximately as follows:

```
>> DOPut(50,50,0.1,5/12,0.4,40)
ans =
        0.5424
>> DOPut(50,50,0.1,5/12,0.4,40*exp(-0.5826*0.4*sqrt(1/12/30)))
ans =
        0.6380
>> DOPut(50,50,0.1,5/12,0.4,30)
ans =
        3.2284
>> DOPut(50,50,0.1,5/12,0.4,30*exp(-0.5826*0.4*sqrt(1/12/30)))
ans =
        3.3056
```

We have assumed here that each month consists of 30 days. It should be noted that alternative analytical methods for discrete-time barrier options have been developed, but we will stick to this one because of its conceptual simplicity.

2.7.2 Asian options

Barrier options exhibit a weak degree of path dependency. A stronger degree of path dependency is typical of Asian options, as the payoff depends on the average asset price over the option life.

Different Asian options may be devised, depending on how the average is computed. Sampling may be discrete or (in principle) continuous. Furthermore, the average may be arithmetic or geometric. The discrete arithmetic average is

$$A_{\mathrm{da}} = \frac{1}{n} \sum_{i=1}^{n} S(t_i),$$

where t_i, $i = 1, \ldots, n$, are the discrete sampling times. The geometric average is

$$A_{\mathrm{dg}} = \left[\prod_{i=1}^{n} S(t_i) \right]^{1/n}.$$

If continuous sampling is assumed, we get

$$A_{\mathrm{ca}} = \frac{1}{T} \int_0^T S(t)\, dt$$

$$A_{\mathrm{cg}} = \exp\left[\frac{1}{T} \int_0^T \log S(t)\, dt \right].$$

Given some way to measure the average A, you may use it to define a rate or a strike. An average rate call has a payoff given by

$$\max\{A - K, 0\},$$

whereas for an average strike call we have

$$\max\{S(T) - A, 0\}.$$

By the same token, we may define an average rate put:

$$\max\{K - A, 0\},$$

or an average strike put:

$$\max\{A - S(T), 0\}.$$

Early exercise features may also be defined in the contract.

2.7.3 Lookback options

Lookback options come in many forms, just like Asian options. The basic difference is that a maximum (or a minimum) value is monitored during the option life. Assuming continuous monitoring, we may measure the maximum and the minimum asset price:

$$S_{\max} = \max_{t \in [0,T]} S(t)$$
$$S_{\min} = \min_{t \in [0,T]} S(t).$$

A European style lookback call has a payoff given by

$$S(T) - S_{\min},$$

whereas in the case of a lookback put we have

$$S_{\max} - S(T).$$

Just as in the Asian option case, you may use the maximum and minimum to define rates or strikes, and you may also add early exercise features. Assuming continuous monitoring, some analytical pricing formulas are known for lookback options.

2.8 AN OUTLOOK ON INTEREST-RATE DERIVATIVES

In this book we will only deal with pricing equity options, as this is enough to introduce and motivate the numerical methods we are interested in.[24] However, there is a huge market of interest-rate derivatives, and in this section we would like to point out why they are important and why they are so difficult to deal with. Actually, any bond is an interest-rate derivative, since its value depends on interest rates; if we model interest rates as stochastic processes, we may apply the option pricing machinery to pricing a zero-coupon bond. This may look like "overkill," but it may play a fundamental role in pricing more complex interest-rate derivatives, as we will see.

The following is a non-exhaustive list of the most basic assets that can be classified as interest-rate derivatives.

- **Interest-rate swaps.** A swap is an arrangement between two parties, which agree to exchange cash flows at predetermined dates in the future. In the vanilla swap, one party will pay cash flows given by a fixed interest rate applied to a nominal amount of money (the notional principal). The other party will pay an amount given by a variable interest rate, applied to a given interval of time (the tenure), on the same notional principal. The net cash flow will depend on the level of future interest rates.

---

[24]This section is included for the sake of completeness, but it can be safely skipped by readers just interested in numerical methods.

- **Bond options.** A call option on a bond works more or less like a call option on a stock, with a different underlying asset. In this case we have two maturities: the maturity T of the option, at which the option can be exercised, and the maturity S of the bond. Obviously, we must have $T < S$. The payoff of the option will depend on the bond price at T, which in turn depends on uncertain interest rates. Call options are actually embedded in certain types of bonds. A callable bond can be redeemed before maturity by the issuer, if prevailing interest rates make this choice attractive, i.e., when interest rates drop and the bond issuer may refinance its debt at lower rates. In this case, the investor purchasing the bond implicitly sells a call option to the bond issuer. Hence, the callable bond must cost less than its non-callable counterpart.

- **Interest-rate caps.** A cap offers protection against a rise in interest rates. This may be interesting to someone who wants to borrow money at a variable rate. A cap is a portfolio of caplets, applying to different time intervals in the future. If L is a notional principal and R_K is the cap rate, a caplet applying to a time interval of length δt gives a payoff

$$L \cdot \delta t \cdot \max\{0, R - R_K\},$$

where R is the interest rate prevailing for that interval. Should interest rates rise in the future, the owner of the cap will receive a payoff covering the payment interest above the cap rate. It can be shown that caps are equivalent to portfolios of bond options.

- **Interest-rate floors.** A floor is similar to a cap, but it offers protection against a drop in interest rates. The payoff of a floorlet is

$$L \cdot \delta t \cdot \max\{0, R_K - R\}.$$

The list of available interest-rate derivatives is increasing because of their usefulness as interest-rate risk management tools. They are, at least potentially, more powerful than older-style practices based on immunization.

The elementary interest-rate derivatives we have just described can be priced using fairly simple models, if some assumptions are made. But this does not hold in general, and more sophisticated models are needed, either to account for the complexity in the dynamics of interest rates, or to price complex derivatives. In the following sections we will just offer some intuition about the reasons behind such a complexity. In the Black–Scholes model for stock options, we have assumed constant interest rates and constant volatility for the price of the underlying asset. Of course the first assumption does not make any sense for interest-rate derivatives. But also the second one cannot be reasonable: The bond price, when maturity is approaching, is less and less volatile (the duration gets smaller and smaller).

2.8.1 Modeling interest-rate dynamics

Several models have been proposed over the years to capture the uncertain dynamics of interest rates. They differ in the following basic features:

- The number of stochastic factors. In the simplest models, we describe the dynamics of the *short rate* $r(t)$, which is essentially a rate applying for a very short time span $(t, t + \delta t)$ in the future. However, we know that bond prices depend on a whole term structure of interest rates. If we build a one-factor model, we are essentially assuming that we may capture the dynamics of the whole term structure just by the short rate and its future evolution. Actually, it is difficult to get a realistic model based on one factor only, and more complex models based on a set of factors should be built, with a corresponding increase in difficulty.

- The focus on equilibrium or arbitrage. It is possible to pursue the somewhat ambitious idea of building an economically sound model, which yields interest rates as a consequence of market equilibrium. An alternative idea is trying to build models which match the currently observed term structure. This is less ambitious, but it may better replicate observed prices. In fact, a basic requirement of a credible model is that it replicates the prices of basic assets, which may be observed in the market. In general, arbitrage based approach aim at this idea of *relative* pricing.

As a result, there is a significant variety of models, with advantages and disadvantages, and there is no obvious choice among them. We do not want to venture into this difficult domain but, given our knowledge of Ito processes, we may at least sketch a few models based on stochastic differential equations for the short rate.

The general structure of such models is

$$dr(t) = \mu[t, r(t)]\, dt + \sigma[t, r(t)]\, dW(t), \tag{2.45}$$

where $W(t)$ is a standard Wiener process. Multifactor models use multidimensional Wiener processes. Geometric Brownian motion is a clearly inadequate model, at least in the long term, as interest rates cannot grow without bound. Mean reversion is a common feature of many models, among which we mention:

1. Vasiček:
$$dr = (b - ar)\, dt + \sigma\, dW,$$
where $a > 0$.

2. Cox–Ingersoll–Ross (CIR):
$$dr = a(b - r)\, dt + \sigma\sqrt{r}\, dW.$$

3. Black–Derman–Toy (BDT):

$$dr = \Theta(t) r\, dt + \sigma(t) r\, dW.$$

4. Hull–White (extended CIR):

$$dr = [\Theta(t) - a(t) r]\, dt + \sigma(t) \sqrt{r}\, dW,$$

where $a(t) > 0$.

Vasiček model exhibits mean reversion, but the rate can get negative. Avoiding negative rates is the rationale behind the $\sqrt{r}$ term in the CIR model. The BDT model includes time-varying functions: On the one hand, this makes the model more complicated, but it allows to match the current term structure (which can be done only approximately with simpler models). The Hull–White extension of CIR model, in some sense, puts all of the above ideas together.

In the next chapters, we will see how continuous-time stochastic models may be exploited computationally, either by Monte Carlo simulation or by building discretized approximations such as binomial lattices or trees. The same ideas, with significant complications may be applied to interest rate models. For instance, the MATLAB Financial derivatives toolbox includes functions to build trees for the BDT short rate model and the Heat–Jarrow–Morton (HJM) model, which the best-known multifactor model. Whatever model and computational technique we use, we must calibrate the parameters of the models above. One would think that to accomplish this task, we should gather market data for interest rates and use some numerical procedure to fit model parameters to observed data. The next section shows that this is not really the case.

2.8.2 Incomplete markets and the market price of risk

We have already pointed out that an apparently paradoxical feature of Black–Scholes formula is that it prices an option under the very assumption that options are no use. This is due to the fact the markets are assumed complete, thus options can be replicated using a risk-free asset and the underlying asset. In practice, this is not true for many reasons, including market imperfections (e.g., transaction costs) and stochastic volatility. This does not imply that the theory is useless: On the contrary, it is used to build internally coherent prices by exploiting concepts such as implied volatility and volatility surfaces.

When we consider interest-rate derivatives, however, we are facing an immediate difficulty: The interest rate is not an asset that can be included in a portfolio. Hence, we cannot build a replicating portfolio. A similar difficulty is faced with certain derivatives written on commodities which are not investment goods and which cannot be included in an investment portfolio leading to replication arguments. The fundamental difficulty is that markets are incomplete. Hence, while no-arbitrage conditions imply that a risk-neutral measure exists, market incompleteness implies that it is not unique. All we can do is to build an internally coherent price system, which is consistent with some observed prices and is arbitrage-free. In other words, we need to pin down a risk-neutral measure which is linked to observed prices.

When dealing with interest-rate derivatives, the simplest asset we may work with is a zero-coupon bond. Actually, we need a set of zero-coupon bonds, one for each possible maturity. Let us assume that a market exists for zero-coupon bonds of any maturity. We may work with a short rate model like (2.45) to explore the consequences of no-arbitrage. Let $p(t,T)$ be the price at time t of a zero-coupon bond with maturity T. Given a model for the short rate, it is reasonable to assume that this price is a function of time t and the current short rate $r(t)$:

$$p(t,T) = F(t, r(t); T).$$

As we have seen with pricing stock options, we need some boundary or terminal condition. If we assume that the face value of the bond is \$1, the terminal condition is

$$F(T, r; T) = 1,$$

for any value of r. To ease the notation, we will denote the price of this bond by F^T. Assuming that the short rate is modeled by equation (2.45), application of Ito's lemma yields

$$\begin{aligned} dF^T &= \left(\frac{\partial F^T}{\partial t} + \mu \frac{\partial F^T}{\partial r} + \frac{1}{2} \sigma^2 \frac{\partial^2 F^T}{\partial r^2} \right) dt + \sigma \frac{\partial F^T}{\partial r} dW \\ &= \nu_T F^T dt + \xi_T F^T\, dW, \end{aligned}$$

where, for the sake of convenience, we have introduced

$$\nu^T = \frac{1}{F^T} \left(\frac{\partial F^T}{\partial t} + \mu \frac{\partial F^T}{\partial r} + \frac{1}{2} \sigma^2 \frac{\partial^2 F^T}{\partial r^2} \right)$$

$$\xi_T = \frac{\sigma}{F^T} \frac{\partial F^T}{\partial r}.$$

If we consider another bond, with maturity S, we have

$$dF^S = \nu_S F^S dt + \xi_S F^S dW,$$

where $W(t)$ is the same Wiener process, as both bonds depend on the same underlying factor. Hence, we may eliminate the term dW by forming the following portfolio of bonds:

$$\Pi = (\sigma_S F^S) F^T - (\sigma_T F^T) F^S.$$

It is important to realize that the expressions between parentheses are the amounts of each bond we hold, which do not change over a short time period

δt, whereas the bond value does. Hence, differentiation in the Ito sense yields

$$\begin{aligned}d\Pi &= (\sigma_S F^S)\, dF^T - (\sigma_T F^T)\, dF^S \\ &= \left(\mu_T \sigma_S F^S F^T - \mu_S \sigma_T F^T F^S\right) dt.\end{aligned}$$

But since this is a risk-free portfolio, lack of arbitrage opportunities implies

$$d\Pi = r\Pi\, dt,$$

which in turn gives

$$\mu_T \sigma_S - \mu_S \sigma_T = r\sigma_S - r\sigma_T.$$

This equality must hold for any maturity. This means that if our bond market is arbitrage free, there must exist a process $\lambda(t)$ such that

$$\frac{\mu_T(t) - r(t)}{\sigma_T(t)} = \lambda(t), \tag{2.46}$$

for any maturity T. The process $\lambda(t)$ is called the **market price of risk**. If we write $\mu = r + \lambda\sigma$, we may understand the reason behind this name: the drift μ is the risk free rate plus a compensation depending on volatility and the price of risk. If the price of risk is $\lambda = 0$, as in the usual risk neutral world, we have $\mu = r$, which is exactly the drift we use when pricing options in the Black–Scholes world.

If we substitute μ_T and σ_T in (2.46), we get the following PDE:

$$\frac{\partial F^T}{\partial t} + (\mu - \lambda\sigma)\frac{\partial F^T}{\partial r} + \frac{1}{2}\sigma^2 \frac{\partial^2 F^T}{\partial r^2} - rF^T = 0.$$

This PDE, together with the boundary condition $F^T(T, r) = 1$, is called **term structure equation**. Application of the Feynman–Kač formula to this PDE yields the price of the zero-coupon bond as an expected value:

$$F^T(t, r) = \mathrm{E}^Q_{t,r}\left[e^{\int_t^T r(s)\, ds}\right],$$

where notation $\mathrm{E}^Q_{t,r}$ means that we are taking a conditional expectation given t and $r(t)$, under a risk-neutral measure Q, and the process $r(s)$ satisfies the stochastic differential equation:

$$dr(s) = \{\mu - \lambda\sigma\}\, ds + \sigma\, dW(s),$$

with initial condition $r(t) = r$.

Using a similar procedure we could price other interest-rate derivatives, provided we use the appropriate market price of risk. To spot the right λ, we should calibrate the model, in the sense that we should find the market price of risk that fits the observed prices of zero-coupon bonds. This means that we should find a stochastic differential equation describing the dynamics of the short rate *directly* in the risk-neutral world. Doing so basically requires the solution of an inverse problem: Given bond prices and the term structure equation, we should find the market price of risk. This task may be relatively easy or not, depending on the model we assume for the short rate. Some models result in an analytical solution, some do not. Of course, if a model depends on three numerical parameters (like CIR), we cannot hope to find an exact fit.

In practice, model calibration based on zero-coupon bonds is not that easy because of the lack of enough assets. Actually, what we need is a model enabling coherent pricing with traded assets; hence, any asset related to interest rates is a possible data source for calibration. Recently, many *market models* have been developed which do not claim to be economically motivated models, but aim at making practical pricing easier. In fact, the short rate is a mathematically convenient object, but it is not directly observable. Other rates, such as LIBOR,[25] are more convenient from this point of view.

For further reading

In the literature

- A book dealing with investments in general and their mathematical modeling is [15]. It is comprehensive and quite readable. A higher-level treatment can be found in [11]. Another general reference is [28], which has a sharper focus on derivatives.

- If you are interested in how a stock exchange actually works, see [27].

- More specific references for bond markets and fixed-income-related assets are [6], [7], [8], and [25]. See also [16].

- Portfolio theory is covered in [5]; you might wish to have a look at chapter 10 there to gain a deeper understanding of utility theory.

- Advanced issues in portfolio management are dealt with in [23].

- The classical reference for options and derivatives in general is [10]. For a more formal treatment, see, e.g., [14].

- A good reference on Value at Risk is [12].

- A book dealing extensively with the intricacies of option hedging is [26]; it is not very readable for the uninitiated, but it gives a precise idea of practical option trading.

---

[25] London Inter-Bank Offer Rate

- There is a growing literature on continuous-time stochastic calculus in finance. Many books in this vein are quite hard to read; but if you want to find a good compromise between intuition and mathematical rigor, take a look at [17] or [19]. A more recent text is [24].

- Discrete-time models are dealt with in [20], which is an excellent reference for an understanding of the relationship between risk-neutral probability measures and the no-arbitrage hypothesis.

- Readers interested in a broader view of Financial Economics should consult [4]. Another readable reference is [3].

- Interest-rate derivatives are also covered in books on fixed-income securities such as [16]. A book which is more focused on this class of assets is [21]. Recent market models are described in [22].

- For a mathematically rigorous yet readable treatment of the theoretical background of interest-rate derivatives, see [1].

- Readers interested in the use of derivatives for interest-rate risk management are [9] and [18].

On the Web

- A site where you may find a list many interesting resources for finance is http://fisher.osu.edu/fin/journal/jofsites.htm.

- An academic society that could be of interest to you is IAFE (International Association of Financial Engineers, http://www.iafe.org). Another interesting academic society is the Bachelier Finance Society (http://www.bachelierfinance.com).

REFERENCES

1. T. Björk. *Arbitrage Theory in Continuous Time (2nd ed.)*. Oxford University Press, Oxford, 2004.

2. M. Broadie, P. Glasserman, and S.G. Kou. A Continuity Correction for Discrete Barrier Options. *Mathematical Finance*, 7:325–349, 1997.

3. J. Cvitanić and F. Zapatero. *Introduction to the Economics and Mathematics of Financial Markets*. MIT Press, Cambridge, MA, 2004.

4. J.-P. Danthine and J.B. Donaldson. *Intermediate Financial Theory (2nd ed.)*. Academic Press, San Diego, CA, 2005.

5. E.J. Elton and M.J. Gruber. *Modern Portfolio Theory and Investment Analysis (5th ed.)*. Wiley, New York, 1995.

6. F.J. Fabozzi. *Bond Markets: Analysis and Strategies*. Prentice Hall, Upper Saddle River, NJ, 1996.

7. F.J. Fabozzi. *Fixed Income Mathematics: Analytical and Statistical Techniques (3rd ed.)*. McGraw-Hill, New York, 1997.

8. F.J. Fabozzi and G. Fong. *Advanced Fixed Income Portfolio Management: The State of the Art*. McGraw-Hill, New York, 1994.

9. B.E. Gup and R. Brooks. *Interest Rate Risk Management: The Banker's Guide to Using Futures, Options, Swaps, and Other Derivative Instruments*. Irwin Professional Publishing, New York, 1993.

10. J.C. Hull. *Options, Futures, and Other Derivatives (5th ed.)*. Prentice Hall, Upper Saddle River, NJ, 2003.

11. J.E. Ingersoll, Jr. *Theory of Financial Decision Making*. Rowman & Littlefield, Totowa, NJ, 1987.

12. P. Jorion. *Value at Risk: The New Benchmark for Controlling Derivatives Risk*. McGraw-Hill, New York, 1997.

13. J.G. Kallberg and W.T. Ziemba. Comparison of Alternative Utility Functions in Portfolio Selection Problems. *Management Science*, 29:1257–1276, 1983.

14. Y.K. Kwok. *Mathematical Models of Financial Derivatives*. Springer-Verlag, Berlin, 1998.

15. D.G. Luenberger. *Investment Science*. Oxford University Press, New York, 1998.

16. L. Martellini, P. Priaulet, and S. Priaulet. *Fixed-Income Securities: Valuation, Risk Management, and Portfolio Strategies*. Wiley, Chichester, 2003.

17. T. Mikosch. *Elementary Stochastic Calculus with Finance in View*. World Scientific Publishing, Singapore, 1998.

18. S.K. Nawalkha, G.M. Soto, and N.A. Beliaeva. *Interest Rate Risk Modeling*. Wiley, Hoboken, NJ, 2005.

19. S. Neftci. *An Introduction to the Mathematics of Financial Derivatives (2nd ed.)*. Academic Press, San Diego, CA, 2000.

20. S.R. Pliska. *Introduction to Mathematical Finance: Discrete Time Models*. Blackwell Publishers, Malden, MA, 1997.

21. R. Rebonato. *Interest-Rate Option Models (2nd ed.)*. Wiley, Chichester, 2000.

22. R. Rebonato. *Modern Pricing of Interest-Rate Derivatives: the LIBOR Market Model and Beyond*. Princeton University Press, Princeton, NJ, 2002.

23. B. Scherer and D. Martin. *Introduction to Modern Portfolio Optimization with NuOPT, S-Plus, and S^+ Bayes*. Springer, New York, 2005.

24. S. Shreve. *Stochastic Calculus for Finance (vols. I & II)*. Springer-Verlag, New York, 2003.

25. S.M. Sundaresan. *Fixed Income Markets and Their Derivatives*. South Western College Publishing, Cincinnati, OH, 1997.

26. N. Taleb. *Dynamic Hedging: Managing Vanilla and Exotic Options*. Wiley, New York, 1996.

27. S.R. Veale, editor. *Stocks, Bonds, Options, Futures: Investments and their Markets*. New York Institute of Finance / Prentice Hall, Paramus, NJ, 1987.

28. P. Wilmott. *Quantitative Finance (vols. I and II)*. Wiley, Chichester, West Sussex, England, 2000.

Part II
Numerical Methods

3

Basics of Numerical Analysis

The core of the MATLAB system implements a set of functions to cope with some classical numerical problems. Although there is no need for a really deep knowledge of numerical analysis in order to use MATLAB, a grasp of the basics is useful in order to choose among competing methods and to understand what may go wrong with them. In fact, numerical computation is affected by machine precision and error propagation, in ways that may result in quite unreasonable outcomes. Hence, we begin by considering the effect of finite precision arithmetic and the issues of numerical instability and problem conditioning, which are outlined in section 3.1. This material is essential, among other things, in understanding the pitfalls of pricing derivatives by solving PDEs.

Then we describe methods for solving systems of linear equations in section 3.2; MATLAB provides the user with both direct and iterative methods to this purpose, and it is important to understand the characteristics of the two classes of methods. Section 3.3 introduces the reader to the problems of approximating functions and interpolating data values. Solving non-linear equations is the subject of section 3.4.

Other topics, such as numerical integration and finite difference methods for PDEs are dealt with in specific chapters. With respect to standard textbooks in numerical analysis, a few types of numerical problems have been omitted, most notably the computation of matrix eigenvalues and eigenvectors and the solution of ordinary differential equations. Both problems are solved by methods available in MATLAB, but since they will not be used in the rest of the book, we refer the reader to the references listed at the end of the chapter.

3.1 NATURE OF NUMERICAL COMPUTATION

Real analysis is based on real numbers. Unfortunately, dealing with real numbers on a computer is impossible. Each number is represented by a finite number of bits, taking the values 0 or 1. Hence, we have to settle for binary and finite precision arithmetic. The progress in computing hardware has improved the quality of the representation, since more bits may be used efficiently without resorting to low-level software tricks. Yet some representation error is unavoidable, and its effect may lead to unexpected results. We have seen some examples of what may go wrong in section 1.3. In this section we try to explain why this may happen.

3.1.1 Number representation, rounding, and truncation

The usual way we represent numbers relies on a decimal base. When writing 1492, we actually mean

$$1 \times 10^3 + 4 \times 10^2 + 9 \times 10^1 + 2 \times 10^0.$$

Similarly, when we have to represent the fractional part of a number, we use negative powers of the base 10:

$$0.42 \Rightarrow 4 \times 10^{-1} + 2 \times 10^{-2}.$$

Some numbers, such as $1/3 = 0.\overline{3}$, do not have a finite representation and should be thought as limits of an infinite series. However, on a computer we must use a binary base, since the hardware is based on a binary logic; for instance,

$$(21.5)_{10} \Rightarrow 2^4 + 2^2 + 2^0 + 2^{-1} = (10101.1)_2.$$

How can we convert numbers from a decimal to a binary base? Let us begin with an integer number N. It can be thought of as

$$N = (b_k \cdot 2^k) + (b_{k-1} \cdot 2^{k-1}) + \cdots + (b_1 \cdot 2^1) + (b_0 \cdot 2^0).$$

Dividing both sides by 2, we get

$$\frac{N}{2} = (b_k \cdot 2^{k-1}) + (b_{k-1} \cdot 2^{k-2}) + \cdots + (b_1 \cdot 2^0) + \frac{b_0}{2}.$$

Hence, the rightmost digit in the binary representation, b_0, is simply the remainder of the integer division of N by 2. We may think of N as

$$N = 2 \cdot Q + b_0,$$

where Q is the result of the integer division by 2. Repeating this step, we obtain all the digits of the binary representation. This suggests the algorithm whose MATLAB code is illustrated in figure 3.1. The function DecToBinary

```
function b=DecToBinary(n)
n0 = n;
i=1;
while (n0 > 0)
   n1 = floor(n0/2);
   b(i) = n0 - n1*2;
   n0=n1;
   i = i+1;
end
b=fliplr(b);
```

Fig. 3.1 MATLAB code to obtain the binary representation of an integer number.

takes an integer number n and returns a vector b containing the binary digits[1]:

```
>> DecToBinary(3)
ans =
     1     1
>> DecToBinary(8)
ans =
     1     0     0     0
>> DecToBinary(13)
ans =
     1     1     0     1
```

Similarly, the fractional part of a number is represented in a binary base as

$$R = \sum_{k=1}^{\infty} d_k 2^{-k}.$$

Some numbers, which can be represented finitely in a decimal base, cannot in a binary base; for instance,

$$7/10 = (0.7)_{10} = (0.1\overline{0110})_2.$$

Clearly, in such cases the infinite series is truncated, with a corresponding error. The binary representation of a fractional number R can be obtained by the following algorithm, which is similar to the previous one (int and frac denote the integer and the fractional part of a number, respectively):

1. Set $d_1 = \text{int}(2R)$ and $F_1 = \text{frac}(2R)$.

---

[1] This is not the best implementation, as the output vector b is resized incrementally. We could compute the number of necessary bits and preallocate b.

2. Recursively compute $d_k = \text{int}(2F_{k-1})$ and $F_k = \text{frac}(2F_{k-1})$ for $k = 2, 3, \ldots$.

Knowing how to change the base may seem useless, but we will see an application of these procedures in section 4.6, dealing with quasi-Monte Carlo simulation.

In practice, we have to represent both quite large and quite small numbers. Hence we resort to a floating-point representation like

$$x \approx \pm q \times 2^n,$$

where q is the mantissa and n is the exponent. The exact details of the representation depend on the chosen standard and the underlying hardware. In any case, since only a finite memory space is available to store the mantissa, we will have a *roundoff* error.

Rounding off is not the only source of error in numerical computation. Another one is truncation. This occurs, for instance, when we substitute a finite sum for an infinite sum. As an example, consider the following expression for the exponential function:

$$e^x = \sum_{k=0}^{\infty} \frac{x^k}{k!}.$$

When we truncate a sum like this, a truncation error occurs.

Example 3.1 One typically troublesome situation is when you subtract two nearly equal numbers. To see why, consider the following example[2]:

$$\begin{aligned} x &= 0.3721478693 \\ y &= 0.3720230572 \\ x - y &= 0.0001248121. \end{aligned}$$

If you represent the numbers by five significant digits only (rounding the last one), the actual result will be

$$\hat{x} - \hat{y} = 0.37215 - 0.37202 = 0.00013,$$

with a relative error of about 4% with respect to the correct result. In fact, it is good practice to avoid expressions like

$$\sqrt{x^2 + 1} - 1,$$

which could result in remarkable losses in significance for small values of x. In such cases, it is easy to rewrite the expression above as

$$\sqrt{x^2 + 1} - 1 = \left(\sqrt{x^2 + 1} - 1\right)\left(\frac{\sqrt{x^2+1}+1}{\sqrt{x^2+1}+1}\right) = \frac{x^2}{\sqrt{x^2+1}+1}.$$

---

[2] See [13, pp. 58–59]

Here there is no subtraction involved, but in other cases, there is no easy way to avoid the difficulty. □

3.1.2 Error propagation, conditioning, and instability

Roundoff errors have been mitigated by the increase in the number of bits used to store numbers on modern computers. From a practical perspective, numbers are virtually represented exactly. Nevertheless, such errors may accumulate within the steps of an algorithm, possibly with disruptive effects, as we have seen in example 1.3. Hence, algorithms should be analyzed with respect to their numerical stability properties. We typically have alternative algorithms for the *same* problem, and it may happen that some of them are subject to instability issues and some are not. A typical case we will consider in chapter 5 is the choice between explicit and implicit methods to solve PDEs by finite differences. Sometimes, but not always, there is a trade-off between potential instability and computational efficiency. As an example, an advanced optimization library like ILOG CPLEX offers different interior point solvers to tackle large-scale linear programming problems[3]; in case of numerical difficulties we may switch to more robust but slower options.

We see that stability is a property of a specific *algorithm* to solve a numerical problem. There is still another issue, which is related to the difficulty of solving the problem *per se*, which is called *conditioning*. When we consider the numerical conditioning of a problem, we are not dealing with specific algorithms to compute a solution, but with the intrinsic difficulty of a problem. Hence, it is important to have a conceptually clear view of how stability and conditioning are related.

From an abstract point of view, a numerical problem may be considered as a mapping

$$y = f(x),$$

which transforms the input data x into the output y. An algorithm is a computationally workable approach to computing that function; different algorithms may be used to solve the same numerical problem, possibly with different characteristics with respect to computational effort and stability properties. When using a computer, roundoff errors will be introduced in the representation of the input; we should check the effects on the output of a perturbation δx in the input data. Denoting the actual input by $\bar{x} = x + \delta x$, the output should be $f(\bar{x})$, whereas an algorithm will yield some answer, say y^*. An algorithm is stable if the relative error

$$\frac{\|f(\bar{x}) - y^*\|}{\|f(\bar{x})\|}$$

---

[3]Interior point methods are dealt with in section 6.4.4.

142   BASICS OF NUMERICAL ANALYSIS

is of the same order of magnitude as the machine precision.[4]

By comparing $f(\bar{x})$ with $f(x)$, we analyze a different issue, called the conditioning of the numerical problem. We should compare the error in the output with the error in the input; when the input error is small, the output error should be small, too. Ideally, it would be nice to have a bounding relationship like

$$\frac{\|f(x) - f(\bar{x})\|}{\|f(x)\|} \leq K \frac{\|x - \bar{x}\|}{\|x\|}, \qquad (3.1)$$

where $\|\cdot\|$ is an appropriate norm.[5] The number K is called the *condition number* of the problem. Later, we investigate the condition number for the problem of solving a system of linear equations, but a simple example will illustrate the point.

Example 3.2 Consider the following non-linear equation:

$$\begin{aligned} p(x) &= x^8 - 36x^7 + 546x^6 - 4536x^5 + 22449x^4 - 67284x^3 \\ &\quad + 118124x^2 - 109584x + 40320 = 0. \end{aligned}$$

This is actually a specific type of non-linear equation, as it is a polynomial equation, and it can be solved by special purpose methods, one of which is implemented in the function `roots`[6]:

```
>> p1=[ 1 -36 546 -4536 22449 -67284 118124 -109584 40320];
>> roots(p1)
ans =
    8.0000
    7.0000
    6.0000
    5.0000
    4.0000
    3.0000
    2.0000
    1.0000
```

Note how the polynomial is represented by a vector containing its coefficients. We see a clear pattern in the solution. In particular, we have one root in the interval $[5.5, 6.5]$. Now let us change the second coefficient from -36 to

---

[4] To get an intuitive idea of what machine precision is about, consider the inequalities $1 - \epsilon < 1 < 1 + \epsilon$, which are obviously true for any $\epsilon > 0$. With computer arithmetic, there is a smallest ϵ such that the inequalities hold; below that value, we cannot tell the difference between the two sides of the inequalities.

[5] The reader should be familiar with the norm concept for vectors; anyway, it is recalled in section 3.2.1.

[6] We have already met `roots` when computing the internal rate of return in example 2.8 on page 47.

36.001. This is a small change in the problem data, and one would expect a corresponding slight change in the solution:

```
>> p2=[ 1 -36.001 546 -4536 22449 -67284 118124 -109584 40320];
>> roots(p2)
ans =
   8.2726
   6.4999 + 0.7293i
   6.4999 - 0.7293i
   4.5748
   4.1625
   2.9911
   2.0002
   1.0000
```

Some roots do not move that much, but now there is no root in the interval $[5.5, 6.5]$, and we have a pair of complex conjugate roots, instead. Note again that the conditioning issue is linked to the numerical problem itself, not to the specific algorithm used to solve it: With `roots` we are able to find a very good approximation of the solution, but this is significantly changed by a slight change in the problem data. Indeed finding the roots of a high-degree polynomial is an ill-conditioned problem, and you may imagine the potentially dramatic effects of errors in collecting empirical data to define a numerical problem. □

Putting the two concepts together, we will find a "good" answer to a specific problem when the problem is well-conditioned and the algorithm is stable.

3.1.3 Order of convergence and computational complexity

Sometimes, we are able to find a solution of a numerical problem directly by a relatively straightforward procedure. In other cases, we use iterative algorithms which generate a sequence of approximations. Given an approximate solution $\mathbf{x}^{(k)}$, some transformation is applied to obtain an improved approximation $\mathbf{x}^{(k+1)}$. The minimal requirement of a good algorithm is that the sequence generated converges to the correct solution $\mathbf{x}^*$. Furthermore, one would hope that such convergence is reasonably fast. The speed of convergence may be quantified by a rate. The rate of convergence is at least **linear** if there are a constant $c < 1$ and an integer N such that

$$\| \mathbf{x}_{n+1} - \mathbf{x}^* \| \le c \, \| \mathbf{x}_n - \mathbf{x}^* \|, \qquad n \ge N.$$

The rate of convergence is at least **quadratic** if there are a constant C and an integer N such that

$$\| \mathbf{x}_{n+1} - \mathbf{x}^* \| \le C \, \| \mathbf{x}_n - \mathbf{x}^* \|^2, \qquad n \ge N.$$

In this case we do not require $C < 1$. This can be generalized to an arbitrary order of convergence α:

$$\|\mathbf{x}_{n+1} - \mathbf{x}^*\| \leq C \|\mathbf{x}_n - \mathbf{x}^*\|^\alpha, \qquad n \geq N.$$

The larger the rate q, the better; quadratic convergence $(q = 2)$ is preferred to linear convergence $(q = 1)$. An iterative method need not always converge. Sometimes, convergence depends on the initial estimate $\mathbf{x}^{(0)}$ and its distance from the solution.

When we use an iterative algorithm, we may have no precise idea of the number of iterations we need to get a satisfactory solution. In other cases, some direct method will yield the answer. By *direct method* we mean a procedure which, after a known number of steps, gives the desired solution (if no difficulty due to instability arises). For direct methods, it may be possible to quantify the number of elementary operations (e.g., additions and multiplications) needed to get the answer; this measures the computational complexity of the algorithm. The amount of computation will be a function of the size of the problem. The number of operations may depend on implementation details, and the size of the problem may depend on the type of encoding used to represent the problem. In practice, it is not necessary to be overly precise in this measure as it is usually enough to have an idea of the rate of growth of the computational effort with respect to the increase in problem size. Furthermore, the computational burden of running an algorithm may depend on the specific problem instance at hand, where by *problem instance* we mean a specific problem with specific numerical data. Sometimes, it is possible to analyze the average complexity with respect to the universe of problem instances. Usually, it is easier to quantify the worst-case complexity.

Computational complexity issues are quite important for discrete optimization problems, as they must often solved by potentially time-consuming algorithms.

Example 3.3 Consider again the knapsack model for capital budgeting, which was introduced in example 1.2. Since there is a finite set of possible solutions, in principle one could find the optimal solution by enumerating all of them. However, since each project may or may not be financed, there may be up to 2^N solutions, where N is the number of competing projects and is the essential measure of the problem size. This number is actually only an upper bound on the number of solutions, since many will be infeasible with respect to the budget constraint. Yet we may say that the worst-case complexity of complete enumeration is in the order of 2^N [technically speaking, we say that the complexity is $O(2^N)$].[7] □

Clearly, an exponential growth like this is quite undesirable. Efficient algorithms are usually characterized by a polynomial growth of the computational

---

[7] A function $f(n)$ is $O(g(n))$ if $\lim_{n \to \infty} f(n)/g(n) < \infty$.

effort; their complexity is something like $O(N^p)$ for some constant p. When we find a polynomial algorithm for an optimization problem, we say that the problem has polynomial complexity. However, if we cannot find a polynomial algorithm and only methods with worst-case exponential complexity are available, does this mean that the *problem* has exponential complexity? Actually, this need not be the case: Maybe there is a polynomial algorithm, but we are not smart enough to come up with it. So, while considering the complexity of an algorithm may be relatively easy, doing that for a problem is not trivial in general. We wee here the same problem–algorithm duality that we have seen with stability and conditioning.

3.2 SOLVING SYSTEMS OF LINEAR EQUATIONS

The solution of systems of linear equations is an important problem per se; however, it is also instrumental for a variety of other problems. For instance, Newton's method for solving systems of non-linear equations calls for the repeated solution of linear systems (see section 3.4.2); in chapter 5 we will also see how solving linear systems is needed in certain methods to cope with PDEs.

In pencil-and-paper mathematics, when we have to solve a system of linear equations like $\mathbf{A}\mathbf{x} = \mathbf{b}$, we use matrix inversion to get $\mathbf{x} = \mathbf{A}^{-1}\mathbf{b}$ (provided the matrix is non singular). Although MATLAB offers a function, called inv, to invert a matrix, it may sound surprising to the newcomer that this is *not* used to solve systems of linear equations. More efficient approaches are used.

It is not our aim to dwell too deeply on this subject; we limit ourselves to the basic concepts needed to understand what MATLAB offers to solve linear equations. Methods for solving linear equations can be broadly classified as direct or iterative. Direct methods have a clearly defined computational complexity, as they yield the result directly within a given number of steps; *iterative methods* build a sequence of solutions whose limit is (under some conditions) the desired solution. For iterative methods, the number of steps is not known a priori, as it depends on convergence speed. They are useful for some large systems characterized by sparse matrices (i.e., matrices with a small number of non-zero entries). Both classes are available in MATLAB, and there exist definite situations where application of one class is advantageous over application of the other.

We have seen in example 1.3 that solving linear systems may be a difficult task with certain matrices. One would expect that when a matrix is close to singular, solving the related system may be numerically hard. While this is reasonably true, there are other reasons why numerical difficulties may arise. In order to see why, we need to analyze problem *conditioning*, which in this case amounts to consider the condition number of the matrix. Before doing so, we must introduce preliminary concepts related to the norms of vectors and matrices.

3.2.1 Vector and matrix norms

We are all familiar with the concept of vector length in the Euclidean sense. The norm is a generalization of that idea, which can be extended to matrices and functions, and it is extremely useful in analyzing convergence, stability, and conditioning issues in numerical analysis.

The vector norm is a function mapping vectors $\mathbf{x} \in \mathbb{R}^n$ to real numbers $\|\mathbf{x}\|$ such that:

- $\|\mathbf{x}\| > 0$ for any $\mathbf{x} \neq \mathbf{0}$, and $\|\mathbf{x}\| = 0$ if and only if $\mathbf{x} = \mathbf{0}$;
- $\|c\mathbf{x}\| = |c| \|\mathbf{x}\|$ for any $c \in \mathbb{R}$;
- $\|\mathbf{x} + \mathbf{y}\| \leq \|\mathbf{x}\| + \|\mathbf{y}\|$ for any $\mathbf{x}, \mathbf{y} \in \mathbb{R}^n$.

These properties are the intuitive properties a measure of vector length should satisfy. The most natural way to define a vector length is through the Euclidean norm

$$\|\mathbf{x}\|_2 \equiv \sqrt{\sum_{i=1}^{n} |x_i|^2}.$$

However, there are different notions of vector length, which satisfy the conditions above for a vector norm. The most common ones are:

- $\|\mathbf{x}\|_\infty \equiv \max_{1 \leq i \leq n} |x_i|$, which is known as L_1 norm;
- $\|\mathbf{x}\|_1 \equiv \sum_{i=1}^{n} |x_i|$, which is known as L_∞ norm.

Generally speaking, one may define a vector L_p norm as

$$\|\mathbf{x}\|_p = \left(\sum_i |x_i|^p\right)^{1/p}.$$

Letting p tend to infinity we get L_∞ norm.

Example 3.4 Vector and matrix norms are computed in MATLAB by the norm function.

```
>> v = [2 4 -1 3];
>> [norm(v,1) norm(v,2) norm(v,inf)]
ans =
   10.0000    5.4772    4.0000
```

The function takes two arguments: the vector and an optional parameter specifying the type of norm. The default value for the optional parameter is 2. A call like norm(v,p) corresponds to

sum(abs(v).^p)^(1/p).

The L_∞ norm is computed when the value of the optional parameter is inf.
◻

Example 3.5 Quite often we consider the norm of an "error." In numerical analysis the error can be the distance between the solution of a problem and the current approximation in an iterative algorithm, or an error due to round-off or truncation. Most people in Finance and Economics are familiar with the idea of least squares. In the simplest setting, given a set of experimental data represented by pairs (x_i, y_i), $i = 1, \ldots, n$, we look for a linear law like

$$y = a + bx,$$

which fits the experimental data as best as possible. Since perfect fitting is impossible in practice, one defines an "error" e_i such that for each experimental point $y_i = a + bx_i + e_i$. Typically, the term *residual* is used rather than error, which in any case we would like to keep as low as possible. This can be accomplished by minimizing the norm $\|\mathbf{e}\|$ of the residual by solving

$$\min \quad \sum_{i=1}^{n} e_i^2$$
$$\text{s.t.} \quad y_i = a + bx_i + e_i \quad \forall i.$$

Taking squares makes sense in order to avoid compensation between positive and negative residuals, but we should wonder if there is something wrong in using alternative norms such as L_1 and solving

$$\min \sum_{i=1}^{n} |e_i|$$

or, if we consider the L_∞ norm, solving the min–max problem

$$\min_{a,b} \left\{ \max_{i=1,\ldots,n} |e_i| \right\}.$$

The first case makes perfect sense, as it is related to plain average of residuals in absolute value, whereas using Euclidean norm tends to penalize large errors a bit more. However, given the non-differentiability of absolute value as a function, minimization using the L_1 norm requires numerical solution by linear programming, whereas the least squares problem has a straightforward analytical solution which paves the way to statistical interpretations in the case of linear regression. The L_∞ norm makes sense when we are interested in controlling the worst-case deviation, rather than minimizing a measure related to average residual.
◻

A less familiar concept is the matrix norm, which can be defined by requiring the same properties as above. In the case of square matrices, the norm function maps $\mathbb{R}^{n \times n}$ to $\mathbb{R}$. The required properties are:

148 BASICS OF NUMERICAL ANALYSIS

- $\|\mathbf{A}\| > 0$ for any $\mathbf{A} \neq \mathbf{0}$, and $\|\mathbf{A}\| = 0$ if and only if $\mathbf{A} = \mathbf{0}$.
- $\|c\mathbf{A}\| = |c| \cdot \|\mathbf{A}\|$ for any $c \in \mathbb{R}$.
- $\|\mathbf{A} + \mathbf{B}\| \leq \|\mathbf{A}\| + \|\mathbf{B}\|$ for any $\mathbf{A}, \mathbf{B} \in \mathbb{R}^{n \times n}$.

Sometimes, the following additional condition is required:

$$\|\mathbf{AB}\| \leq \|\mathbf{A}\| \cdot \|\mathbf{B}\|.$$

It may also be important to connect vector and matrix norms. We say that a vector and a matrix norm are **compatible** if the following inequality holds:

$$\|\mathbf{Ax}\| \leq \|\mathbf{A}\| \|\mathbf{x}\|$$

for any matrix $\mathbf{A}$ and vector $\mathbf{b}$ (note that in the left-hand side of the inequality we are using the vector norm).

Typical matrix norms are:

- $\|\mathbf{A}\|_\infty \equiv \max_{1 \leq i \leq n} \sum_{j=1}^n |a_{ij}|$.
- $\|\mathbf{A}\|_1 \equiv \max_{1 \leq j \leq n} \sum_{i=1}^n |a_{ij}|$.
- $\|\mathbf{A}\|_F \equiv \left(\sum_{i=1}^n \sum_{j=1}^n |a_{ij}|^2\right)^{1/2}$, the Frobenius norm.
- $\|\mathbf{A}\|_2 \equiv \sqrt{\rho(\mathbf{A}'\mathbf{A})}$, the *spectral norm*, where $\rho(\cdot)$ is the spectral radius of a matrix, i.e., $\rho(\mathbf{B}) \equiv \max\{|\lambda_k|: \lambda_k \text{ is an eigenvalue of } \mathbf{B}\}$.

The first two norms may look a bit weird, but they are easy to compute. In the first case, for each matrix row we sum absolute values of the elements in each column, and then we take the maximum over the rows. In the second case the two roles are swapped.

Example 3.6 The norm function may be used to compute matrix norms as well. A call like

```
>> A = [ 2 4 -1 ; 3 1 5 ; -2 3 -1];
>> [norm(A,inf) norm(A,1) norm(A,2) norm(A,'fro')]
ans =
    9.0000    8.0000    6.1615    8.3666
```

computes the four matrix norms we have defined, including the spectral and Frobenius norms. For the spectral norm, you may check the result by computing the square root of the eigenvalues of $\mathbf{A}'\mathbf{A}$:

```
>> sqrt(eig(A' * A))
ans =
    2.2117
    5.2100
```

and picking up the largest value. □

The Frobenius norm looks like a straightforward generalization of Euclidean vector norm, but the other three norms look somewhat unnatural. In fact, there is a natural way to introduce a matrix norm, given a vector norm. A square matrix may be considered as an operator transforming vectors: it rotates a vector and it changes its length, making it longer or shorter. We may consider the degree of "amplification" of the vectors as the norm of the matrix. Formally, given a vector norm, we may define its subordinate norm as

$$\|\mathbf{A}\| \equiv \sup_{\mathbf{x} \neq 0} \frac{\|\mathbf{A}\mathbf{x}\|}{\|\mathbf{x}\|} = \max_{\|\mathbf{x}\|=1} \|\mathbf{A}\mathbf{x}\|. \qquad (3.2)$$

In this case we also say that the matrix norm is induced by the vector norm. It is easy to see that in this case the two norms are compatible. Now it can be shown that the vector $\|\cdot\|_\infty$ norm induces the matrix $\|\cdot\|_\infty$ norm and that the same holds for the $\|\cdot\|_1$ norms. A surprising fact is that the Euclidean vector norm does *not* induce the Frobenius norm. In fact it is easy to see that the Frobenius norm is not a subordinate norm, by considering the identity matrix $\mathbf{I}$: From (3.2) we should have $\|\mathbf{I}\| = 1$, but $\|\mathbf{I}\|_F = \sqrt{n}$, for a matrix of order n. The matrix norm induced by the Euclidean vector norm is the spectral norm, and this explains why it is denoted by $\|\cdot\|_2$ (see, e.g., [13]).

A fundamental property of compatible matrix norms is the following.

THEOREM 3.1 *For any matrix norm that is compatible with a vector norm, we have*

$$\rho(\mathbf{A}) \leq \|\mathbf{A}\|.$$

The proof is straightforward. Given a pair of compatible vector and matrix norms, consider any eigenvalue λ of $\mathbf{A}$ and let $\mathbf{v}$ be a related eigenvector of unit length, $\|\mathbf{v}\| = 1$. Then we have

$$|\lambda| = \|\lambda\mathbf{v}\| = \|\mathbf{A}\mathbf{v}\| \leq \|\mathbf{A}\|\|\mathbf{v}\| = \|\mathbf{A}\|.$$

Since this holds for any eigenvalue of the matrix, the theorem follows.

3.2.2 Condition number for a matrix

Now we are ready to start analyzing the effect of numerical errors on the solution of a linear system. Consider the system

$$\mathbf{A}\mathbf{x} = \mathbf{b}$$

and suppose that we perturb $\mathbf{b}$ by adding a term $\delta\mathbf{b}$; such a perturbation may indeed occur due to rounding off. Then the solution will somehow be perturbed, too. We will have

$$A(x + \delta x) = b + \delta b,$$

which implies that

$$A \cdot \delta x = \delta b \quad \Rightarrow \quad \delta x = A^{-1} \delta b.$$

We would like to assess the error in the solution, δx, as a function of the input error δb. If we adopt compatible matrix and vector norms, we may write

$$\|\delta x\| = \|A^{-1} \delta b\| \leq \|A^{-1}\| \cdot \|\delta b\|$$
$$\|b\| = \|Ax\| \leq \|A\| \cdot \|x\|.$$

Dividing term by term these two inequalities yields

$$\frac{\|\delta x\|}{\|A\|\|x\|} \leq \|A^{-1}\| \cdot \frac{\|\delta b\|}{\|b\|} \quad \Rightarrow \quad \frac{\|\delta x\|}{\|x\|} \leq \|A\| \cdot \|A^{-1}\| \cdot \frac{\|\delta b\|}{\|b\|},$$

which is analogous to (3.1). The condition number $K(A) \equiv \|A\| \cdot \|A^{-1}\|$ gives an upper bound on the ratio of the relative error in the solution to the relative perturbation. Generally speaking, the higher the condition number, the more difficult it is to solve a linear system.

Example 3.7 The cond function computes the condition number. An optional parameter may be provided to select a norm; the default value corresponds to the spectral norm.

```
>> cond(hilb(3))
ans =
   524.0568
>> cond(hilb(7))
ans =
   4.7537e+008
>> cond(hilb(10))
ans =
   1.6025e+013
```

Checking these numbers it is easy to see why solving a linear system involving the Hilbert matrix is a difficult task.

Intuitively, we expect that a matrix which is close to singular will be difficult to deal with. The following theorem, due to Gastinel, somewhat supports this view.

THEOREM 3.2 Let A a non-singular matrix of order n. Then for any subordinate matrix norm we have

$$\frac{1}{\text{cond}(A)} = \min\left\{ \frac{\|A - B\|}{\|A\|} \mid B \text{ is a singular matrix} \right\}.$$

The theorem basically states that when condition number is large, the matrix can be well approximated by a singular matrix, which may mean trouble when we deal with that matrix numerically. However, ill-conditioning is *not* necessarily related to singularity, as the following example clearly shows.

Example 3.8 Consider the system[8]

$$x_1 - x_2 - x_3 - \ldots - x_n = -1$$
$$x_2 - x_3 - \ldots - x_n = -1$$
$$x_3 - \ldots - x_n = -1$$
$$\vdots$$
$$x_{n-1} - x_n = -1$$
$$x_n = 1.$$

Note that the matrix

$$\mathbf{A} = \begin{bmatrix} 1 & -1 & -1 & \ldots & -1 & -1 \\ 0 & 1 & -1 & \ldots & -1 & -1 \\ 0 & 0 & 1 & \ldots & -1 & -1 \\ \vdots & \vdots & \vdots & \ddots & \vdots & \vdots \\ 0 & 0 & 0 & \ldots & 1 & -1 \\ 0 & 0 & 0 & \ldots & 0 & 1 \end{bmatrix}$$

is *not* singular, as $\det(\mathbf{A}) = 1$. We have

$$\mathbf{b} = [-1, -1, -1, \ldots, -1, 1]^T,$$

and the solution is easy to find by a process called "backsubstitution." We see $x_n = 1$. Then we may find $x_{n-1} = x_n - 1 = 0$. Knowing x_{n-1}, we find x_{n-2}, and so on. Using this strategy systematically, we get

$$\mathbf{x} = [0, 0, 0, \ldots, 0, 1]^T.$$

We may also "verify" this using MATLAB:

```
>> N=20;
>> A = eye(N);
>> for i=1:N, for j=i+1:N, A(i,j) = -1;, end, end
>> b=-ones(N,1);
>> b(N,1) = 1;
>> A\b
ans =
        0
```

---

[8]See chapter 3 of E.A. Volkov, *Numerical Methods*, MIR Publishers, 1986.

$$\begin{matrix} 0 \\ 0 \\ 0 \\ 0 \\ 0 \\ 0 \\ 0 \\ 0 \\ 0 \\ 0 \\ 0 \\ 0 \\ 0 \\ 0 \\ 0 \\ 0 \\ 0 \\ 0 \\ 0 \\ 1 \end{matrix}$$

Now, assume that we apply a small perturbation to the right-hand side vector b, adding ϵ to the last component. Then we should find a different solution. The first step of backsubstitution shows a small effect of this small perturbation:

$$\widetilde{x}_n = x_n + \delta x_n = 1 + \epsilon.$$

However, if we go on finding the remaining unknown variables, we see that the perturbation gets amplified:

```
>> b(N,1) = 1.00001;
>> A\b
ans =
    2.6214
    1.3107
    0.6554
    0.3277
    0.1638
    0.0819
    0.0410
    0.0205
    0.0102
    0.0051
    0.0026
    0.0013
    0.0006
    0.0003
    0.0002
    0.0001
    0.0000
```

```
      0.0000
      0.0000
      1.0000
```

Thus, a negligible error in the input may result in a large error in the output. Please note that this is due to the structure of the matrix itself, even though it is not singular. We are facing a difficulty with the conditioning of the problem itself, *not* with stability. Indeed, we can try to figure out what's happening analytically. The error vector $\delta \mathbf{x}$ satisfies the system of equations:

$$\begin{aligned}
\delta x_1 - \delta x_2 - \delta x_3 - \ldots - \delta x_n &= 0 \\
\delta x_2 - \delta x_3 - \ldots - \delta x_n &= 0 \\
\delta x_3 - \ldots - \delta x_n &= 0 \\
\vdots \quad \vdots \quad \vdots & \\
\delta x_{n-1} - \delta x_n &= 0 \\
\delta x_n &= \epsilon.
\end{aligned}$$

By backsubstitution we see

$$\begin{aligned}
\delta x_n &= \epsilon \\
\delta x_{n-1} &= \delta x_n = \epsilon \\
\delta x_{n-2} &= \delta x_n + \delta x_{n-1} = \epsilon + \epsilon = 2\epsilon \\
\delta x_{n-3} &= \delta x_n + \delta x_{n-1} + \delta x_{n-2} = \epsilon + \epsilon + 2\epsilon = 2^2\epsilon \\
&\vdots \\
\delta x_{n-k} &= \delta x_n + \delta x_{n-1} + \cdots + \delta x_{n-(k-1)} = 2^{k-1}\epsilon \\
&\vdots \\
\delta x_1 &= \delta x_{n-(n-1)} = 2^{(n-1)-1}\epsilon = 2^{n-2}\epsilon.
\end{aligned}$$

In our case

$$\|\delta \mathbf{x}\|_\infty = 2^{n-2}|\epsilon|, \quad \|\mathbf{x}\|_\infty = 1, \quad \|\delta \mathbf{b}\|_\infty = |\epsilon|, \quad \|\mathbf{b}\|_\infty = 1,$$

and

$$K_\infty(\mathbf{A}) = \|\mathbf{A}\|_\infty \cdot \|\mathbf{A}^{-1}\|_\infty \geq \frac{\|\delta \mathbf{x}\|_\infty / \|\mathbf{x}\|_\infty}{\|\delta \mathbf{b}\|_\infty / \|\mathbf{b}\|_\infty} = 2^{n-2}.$$

In fact,

```
>> cond(A,inf)
ans =
    10485760
>> 2^18
ans =
```

```
        262144
>> 0.00001 * 2^18
ans =
        2.6214
```

3.2.3 Direct methods for solving systems of linear equations

Direct methods for solving linear equations are based on the idea of transforming the matrix into a suitable form. Example 3.8, among other things, shows that if the matrix is in upper triangular form, we may immediately find the last unknown x_n and then the other ones by backsubstitution. Let us make the approach explicit for a system

$$\mathbf{Ax} = \mathbf{b}$$

where $\mathbf{A}$ is an upper triangular matrix:

$$
\begin{aligned}
a_{11}x_1 + a_{12}x_2 + \cdots + a_{1n}x_n &= b_1 \\
a_{22}x_2 + \cdots + a_{2n}x_n &= b_2 \\
&\vdots \\
a_{nn}x_n &= b_n.
\end{aligned}
$$

Backsubstitution starts from the last variable x_n and proceeds backwards as follows:

$$x_n = \frac{b_n}{a_{nn}}$$

$$x_k = \left(b_k - \sum_{j=k+1}^{n} a_{kj} x_j \right) \bigg/ a_{kk}, \qquad k = n-1, n-2, \ldots, 1.$$

Now we should come up with a systematic method to transform a linear system of equations into an equivalent triangular form. Gaussian elimination is such a procedure. In principle, the idea is rather simple; we must form linear combinations of equations in order to eliminate some coefficients from some equations. Since combining equations linearly does not change the solution, the resulting system is equivalent to the original one. Starting from the system in the form

$$
\begin{aligned}
(E_1) \quad & a_{11}x_1 + a_{12}x_2 + \cdots + a_{1n}x_n = b_1 \\
(E_2) \quad & a_{21}x_1 + a_{22}x_2 + \cdots + a_{2n}x_n = b_2 \\
& \qquad \qquad \qquad \vdots \\
(E_n) \quad & a_{n1}x_1 + a_{n2}x_2 + \cdots + a_{nn}x_n = b_n,
\end{aligned}
$$

SOLVING SYSTEMS OF LINEAR EQUATIONS   155

we may try to obtain a column of zeros under the coefficient a_{11}. This is the first step in getting an equivalent triangular system. For each equation (E_k) $(k = 2, \ldots, n)$, we must apply the transformation

$$(E_k) \leftarrow (E_k) - \frac{a_{k1}}{a_{11}}(E_1),$$

which leads to the equivalent system:

$$\begin{aligned}
a_{11}x_1 + a_{12}x_2 + \cdots + a_{1n}x_n &= b_1 \\
a_{22}^{(1)}x_2 + \cdots + a_{2n}^{(1)}x_n &= b_2^{(1)} \\
\vdots &= \vdots \\
a_{n2}^{(1)}x_2 + \cdots + a_{nn}^{(1)}x_n &= b_n^{(1)}.
\end{aligned}$$

Now we may repeat the procedure to obtain a column of zeros under the coefficient $a_{22}^{(1)}$, and so on, until the desired form is obtained, allowing for backsubstitution.

Example 3.9 Consider the following system:

$$\begin{bmatrix} 1 & 2 & 1 \\ 2 & 2 & 3 \\ -1 & -3 & 0 \end{bmatrix} \begin{bmatrix} x_1 \\ x_2 \\ x_3 \end{bmatrix} = \begin{bmatrix} 0 \\ 3 \\ 2 \end{bmatrix}.$$

It is convenient to represent the operations of Gaussian elimination on an augmented matrix:

$$\left[\begin{array}{ccc|c} 1 & 2 & 1 & 0 \\ 2 & 2 & 3 & 3 \\ -1 & -3 & 0 & 2 \end{array}\right] \Rightarrow \left[\begin{array}{ccc|c} 1 & 2 & 1 & 0 \\ 0 & -2 & 1 & 3 \\ 0 & -1 & 1 & 2 \end{array}\right] \Rightarrow \left[\begin{array}{ccc|c} 1 & 2 & 1 & 0 \\ 0 & -2 & 1 & 3 \\ 0 & 0 & \frac{1}{2} & \frac{1}{2} \end{array}\right].$$

From this it is easy to get $x_3 = 1$, $x_2 = -1$, and $x_1 = 1$.   ☐

We will not quantify exactly the number of operations needed for the overall procedure, but it is evident that the algorithm has a quantifiable computational complexity, which is of order $O(n^3)$ for a system of order n.

Actually, what we have explained is only the starting point of Gaussian elimination, as many things may go wrong with this naive procedure. A first point is that we must have $a_{11} \neq 0$ to carry out the first step of the Gaussian elimination; by the same token, we must have $a_{22}^{(1)} \neq 0$, and so on. Fortunately, if the original system is non-singular, this may be accomplished by a suitable permutation of variables (columns) or equations (row).

Example 3.10 Consider the matrix

$$\mathbf{A} = \begin{bmatrix} 5 & 1 & 4 \\ 0 & 0 & 3 \\ 0 & 1 & 2 \end{bmatrix}.$$

156   BASICS OF NUMERICAL ANALYSIS

If we try Gaussian elimination to get rid of element $a_{32} = 1$, we are in trouble since $a_{22} = 0$. However, to avoid the difficulty, we may simply swap the second and the third equation. Formally, permutations may be represented by suitable matrices, called permutation matrices, characterized by the following properties:

- All elements are either 0 or 1.
- For each row, one element is equal to 1.
- For each column, one element is equal to 1.

As an example, consider

$$\mathbf{P} = \begin{bmatrix} 1 & 0 & 0 \\ 0 & 0 & 1 \\ 0 & 1 & 0 \end{bmatrix}.$$

We may check the effect on matrix $\mathbf{A}$:

$$\mathbf{PA} = \begin{bmatrix} 1 & 0 & 0 \\ 0 & 0 & 1 \\ 0 & 1 & 0 \end{bmatrix} \begin{bmatrix} 5 & 1 & 4 \\ 0 & 0 & 3 \\ 0 & 1 & 2 \end{bmatrix} = \begin{bmatrix} 5 & 1 & 4 \\ 0 & 1 & 2 \\ 0 & 0 & 3 \end{bmatrix}.$$

◻

There is another reason why Gaussian elimination should include the possibility of swapping rows or columns: Some care is needed to minimize the effects of finite precision arithmetic. We have seen in example 3.1 that subtraction is a potentially dangerous operation, because of the potential loss of significance. Suitable row and column permutations may help in keeping the trouble to a minimum; such operations are called *pivoting*. Scaling the size of the coefficients may be used, too. These points are well treated in any numerical analysis book, and the details are beyond the scope of this one.

There are alternative ways to see Gaussian elimination. A compact representation is obtained if we see Gaussian elimination as a way of factoring the matrix $\mathbf{A}$ into the product of a lower triangular matrix $\mathbf{L}$ and an upper triangular matrix $\mathbf{U}$. More precisely we have

$$\mathbf{PA} = \mathbf{LU},$$

where $\mathbf{P}$ is a permutation matrix which may be necessary or advisable to introduce for the above-mentioned reasons. We may try to understand, at least intuitively, where the above factorization comes from. The permutation matrix $\mathbf{P}$ corresponds to the pivoting operations; if pivoting is not required for a matrix, then this matrix can be neglected. The upper triangular matrix $\mathbf{U}$ corresponds to the end result of Gaussian elimination we just described. The lower triangular matrix $\mathbf{L}$ corresponds to the transformations we must carry out to obtain the equivalent system in upper triangular form. These transformations are linear combinations of rows, which can be obtained by multiplying the original matrix by suitable elementary matrices; the matrix **L** is linked to the product of these elementary matrices. This factorization is called *LU-decomposition*.

Example 3.11 LU-decomposition is obtained in MATLAB by calling the lu function with a matrix argument.

```
>> A = [1 4 -2 ; -3 9 8; 5 1 -6];
>> [L,U,P] = lu(A)
L =
    1.0000         0         0
   -0.6000    1.0000         0
    0.2000    0.3958    1.0000
U =

    5.0000    1.0000   -6.0000
         0    9.6000    4.4000
         0         0   -2.5417
P =
     0     0     1
     0     1     0
     1     0     0
```

With such a factorization, solving a system like $\mathbf{Ax} = \mathbf{b}$ is equivalent to solving the two systems

$$\mathbf{Ly} = \mathbf{Pb}$$
$$\mathbf{Ux} = \mathbf{y}$$

in cascade.

```
>> b = [1;2;3];
>> x = A\b
x =
    1.0820
    0.1967
    0.4344
>> x = U \ ( L \ (P*b))
x =
    1.0820
    0.1967
    0.4344
```

```
% TryLU.m
N=2000;
A=rand(100,100);

tic
for i=1:1000
    b=rand(100,1);
    x=A\b;
end
toc

tic
[L,U,P] = lu(A);
for i=1:1000
    b=rand(100,1);
    x=U\(L\(P*b));
end
toc
```

Fig. 3.2 Script to check the advantage of using LU decomposition.

LU-decomposition may be advantageous when it is necessary to solve a system repeatedly with different right-hand sides, as it occurs in the solution of certain PDEs by finite difference methods. In order to appreciate the point immediately, let us try a little experiment by running the MATLAB script in figure 3.2. In the example we generate a random[9] matrix of order $n = 2000$ and then solve 1000 systems with randomly generated right-hand sides. We may compare the CPU time with standard Gaussian elimination (cold start) and LU decomposition (warm start):

```
>> TryLU
Elapsed time is 0.904283 seconds.
Elapsed time is 0.096623 seconds.
```

Basically, with LU decomposition we obtain the same advantage we would have with matrix inversion, without all of its potential numerical difficulties.

LU-decomposition takes a special form when applied to symmetric positive definite matrices; such matrices occur in many optimization problems, and a typical example is a covariance matrix. If **A** is a symmetric positive definite matrix, it can be shown that there exists a unique upper triangular matrix

---

[9]The function **rand** generates a pseudo-random variable in the interval $(0,1)$. It will be used extensively for Monte Carlo simulation.

U such that $\mathbf{A} = \mathbf{U}'\mathbf{U}$; this is called *Cholesky factorization.*[10] Cholesky factorization may be a suitable alternative to the usual Gaussian elimination for special matrices.

Example 3.12 The Cholesky factorization is computed in MATLAB by the chol function. For instance, let us define a matrix and check that it is positive definite, by verifying that its eigenvalues are positive:

```
>> A = [ 3 1 4 ; 1 5 3 ; 4 3 7 ]
A =
     3     1     4
     1     5     3
     4     3     7
>> eig(A)
ans =
    0.3803
    3.5690
   11.0507
```

Given a known term **b**, we may factor **A** and solve the system.

```
>> b=(1:3)';
>> U=chol(A)
U =
    1.7321    0.5774    2.3094
         0    2.1602    0.7715
         0         0    1.0351
>> U \ (U' \ b)
ans =
   -1.0000
   -0.0000
    1.0000
```

In chapter 4 we will see that the Cholesky factorization is also useful when we have to simulate random variables with a multivariate normal distribution.

3.2.4 Tridiagonal matrices

In certain applications, the matrix of a system of linear equations has a very specific form. One such case is the tridiagonal matrix, which may occur in the solution of option pricing problems by PDEs. A tridiagonal matrix has

---

[10]In many texts, a lower triangular matrix **L** is considered, and the factorization is written as $\mathbf{A} = \mathbf{LL}'$. It is easy to see that the two definitions are actually equivalent. We will stick to this one, since the MATLAB function chol returns an upper triangular matrix.

the following form:

$$A = \begin{bmatrix} a_{11} & a_{12} & 0 & 0 & 0 & \cdots & 0 \\ a_{21} & a_{22} & a_{23} & 0 & 0 & \cdots & 0 \\ 0 & a_{32} & a_{33} & a_{34} & 0 & \cdots & 0 \\ \vdots & \vdots & \vdots & \vdots & \vdots & \ddots & \vdots \\ 0 & \cdots & \cdots & a_{n-2,n-3} & a_{n-2,n-2} & a_{n-2,n-1} & 0 \\ 0 & \cdots & \cdots & 0 & a_{n-1,n-2} & a_{n-2,n-1} & a_{n-1,n} \\ 0 & \cdots & \cdots & 0 & 0 & a_{n,n-1} & a_{nn} \end{bmatrix}.$$

This matrix has a banded form, and it is *sparse*; i.e., it has few non-zero entries. Without loss of generality, assume that $a_{i,j+1} \neq 0$. If $a_{j,j+1} = 0$, it is easy to see that the original system may be decomposed into two subsystems, since in such a case we have an upper block of lower triangular form. We may solve the system by a specially structured direct method. Consider the first equation:

$$a_{11}x_1 + a_{12}x_2 = b_1.$$

We may solve for x_2, in terms of x_1:

$$x_2 = c_2 + d_2 x_1,$$

where $c_2 = b_1/a_{12}$ and $d_2 = -a_{11}/a_{12}$. By the same token, we may obtain an expression of x_3 in terms of x_1. In fact, given the second equation

$$a_{21}x_1 + a_{22}x_2 + a_{23}x_3 = b_2,$$

we may express x_3 as a function of x_1 and x_2. But since we know x_2 as a function of x_1, we may get an expression of the form

$$x_3 = c_3 + d_3 x_1.$$

Going on the same way for all equations up to the $(n-1)$th one, we obtain expressions like $x_k = c_k + d_k x_1$, for all $k = 2, \ldots, n$. Finally, plugging the expressions for x_{n-1} and x_n into the last equation, we end up with

$$a_{n,n-1}x_{n-1} + a_{nn}x_n = a_{n,n-1}(c_{n-1} + d_{n-1}x_1) + a_{nn}(c_n + d_n x_1) = b_n,$$

which yields x_1, and, by substitution, all the other unknowns. The approach may be adapted in the case of similar banded matrices. It is also worth noting that memory savings may be obtained by storing only the non-zero matrix entries.

3.2.5 Iterative methods for solving systems of linear equations

In many situations we must solve a large system of linear equations, characterized by a sparse matrix. PDEs are a typical source of such systems, but there are others, such as computing the long-term probability distribution of some discrete-state, discrete-time stochastic systems (Markov chains). Storing a sparse matrix is a waste of memory, since many entries are zero; special techniques have been developed to avoid the problem. However, applying a direct method such as Gaussian elimination to a sparse matrix may destroy its characteristic. So we may try a different approach. One possibility is an iterative method, generating a sequence of vectors that converges to the solution desired. The process may be stopped when a reasonable accuracy has been achieved. Note that, unlike direct methods, the number of steps required by an iterative algorithm is not known a priori, and its behavior should be characterized in terms of convergence speed, along the lines illustrated in section 3.1.3. The first issue to consider is how to characterize the conditions under which an iterative method converges; in fact, the method could simply blow up due to instability, giving rise to an unbounded sequence.

Here we illustrate the basic iterative approaches described in any numerical analysis text. It is worth emphasizing that MATLAB has efficient capabilities to represent sparse matrices and provides the user with a rich set of iterative methods, which are much more sophisticated than the ones we describe here. Nevertheless, we believe that the background behind relatively simple iterative methods will be a useful reading, for at least a couple of reasons. On the one hand, they have been proposed in the literature on financial engineering to solve PDEs (see, e.g., [20, pp. 895–901] for a comparison of LU-decomposition and successive overrelaxation in option pricing). Second, in chapter 5 we investigate the numerical stability of finite difference methods for solving PDEs, using the same concepts we use here to study the convergence of iterative methods.

Iterative schemes are one possible approach when the fixed point of an operator is needed. Consider a generic operator $\mathbf{G}(\cdot)$ and assume that you want to find a fixed point of $\mathbf{G}$, i.e., a point satisfying the equation

$$\mathbf{x} = \mathbf{G}(\mathbf{x}).$$

A possible approach is to generate a sequence of approximations of the solution, according to the iteration scheme

$$\mathbf{x}^{(k+1)} = \mathbf{G}(\mathbf{x}^{(k)}), \tag{3.3}$$

starting from some initial approximation $\mathbf{x}^{(0)}$. This approach, called *fixed-point iteration*, may be used for both linear and non-linear equations, and for many other problems as well. Now the question is if and when this scheme will converge to a fixed point of $\mathbf{G}$. The general answer lies in the *contraction mapping* concept, which is widely applied in many diverse settings. To keep it simple, let us investigate the idea in the case of the familiar system of linear equations $\mathbf{A}\mathbf{x} = \mathbf{b}$, which can be rewritten as

$$\mathbf{x} = (\mathbf{A} + \mathbf{I})\mathbf{x} - \mathbf{b} = \hat{\mathbf{A}}\mathbf{x} - \mathbf{b}.$$

We want to find a fixed point of the operator $\mathbf{G}(x) = \hat{\mathbf{A}}\mathbf{x} - \mathbf{b}$, and we could consider the iterative approach (3.3). Would such a scheme converge? To begin with, consider starting from a first guess $\mathbf{x}^{(0)}$, and trace the first iteration steps:

$$\begin{aligned}
\mathbf{x}^{(1)} &= \hat{\mathbf{A}}\mathbf{x}^{(0)} - \mathbf{b} \\
\mathbf{x}^{(2)} &= \hat{\mathbf{A}}\mathbf{x}^{(1)} - \mathbf{b} = \hat{\mathbf{A}}^2 \mathbf{x}^{(0)} - \hat{\mathbf{A}}\mathbf{b} - \mathbf{b} \\
\mathbf{x}^{(3)} &= \hat{\mathbf{A}}^3 \mathbf{x}^{(0)} - \hat{\mathbf{A}}^2 \mathbf{b} - \hat{\mathbf{A}}\mathbf{b} - \mathbf{b}
\end{aligned}$$

...

Intuition suggests that if the elements of the matrix $\hat{\mathbf{A}}^n$ grow without bound as $n \to \infty$, the iteration scheme will diverge. Indeed, it can be shown that convergence will occur only if all the eigenvalues of $\hat{\mathbf{A}}$ have an absolute value less than 1 (see below). Since this may well not be the case for an arbitrary system of equations, it is better to take a slightly different approach and split the matrix $\mathbf{A}$ as follows:

$$\mathbf{A} = \mathbf{D} + \mathbf{C},$$

which yields an equivalent system

$$\mathbf{D}\mathbf{x} = -\mathbf{C}\mathbf{x} + \mathbf{b}.$$

Then we may apply the iteration scheme

$$\begin{aligned}
\mathbf{d}^{(k)} &= -\mathbf{C}\mathbf{x}^{(k)} + \mathbf{b} \\
\mathbf{D}\mathbf{x}^{(k+1)} &= \mathbf{d}^{(k)}
\end{aligned} \qquad (3.4)$$

in order to generate a sequence of approximations $\mathbf{x}^{(k)}$. In some sense, this is a generalization of the previous fixed-point approach, but the flexibility in choosing $\mathbf{D}$ may be exploited to improve convergence. To investigate the convergence issue further, we may write, as before,

$$\mathbf{x}^{(k+1)} = -\mathbf{D}^{-1}\mathbf{C}\mathbf{x}^{(k)} + \mathbf{D}^{-1}\mathbf{b}.$$

Letting $\mathbf{B} = -\mathbf{D}^{-1}\mathbf{C} = \mathbf{I} - \mathbf{D}^{-1}\mathbf{A}$, we may check how the absolute error $\mathbf{e}^{(k)} = \mathbf{x}^* - \mathbf{x}^{(k)}$ evolves, where $\mathbf{x}^*$ is the correct solution:

$$\mathbf{e}^{(k+1)} = \mathbf{x}^* - \mathbf{x}^{(k+1)} = (\mathbf{B}\mathbf{x}^* + \mathbf{D}^{-1}\mathbf{b}) - (\mathbf{B}\mathbf{x}^{(k)} + \mathbf{D}^{-1}\mathbf{b}) = \mathbf{B}(\mathbf{x}^* - \mathbf{x}^{(k)}) = \mathbf{B}\mathbf{e}^{(k)},$$

from which it is easy to see that

$$\lim_{k \to \infty} \mathbf{e}^{(k)} = \lim_{k \to \infty} \mathbf{B}^k \mathbf{e}^{(0)}.$$

It can be shown that

$$\lim_{k \to \infty} \mathbf{B}^k = 0$$

if and only if the spectral radius of **B** is strictly less than 1, i.e., if all of its eigenvalues have an absolute value less than 1. This implies that the approach will converge if and only if

$$\rho(\mathbf{I} - \mathbf{D}^{-1}\mathbf{A}) < 1.$$

To verify this condition, we should compute the eigenvalues of a possibly large matrix (actually, only the largest one in absolute value is needed). We may avoid this trouble by recalling that

$$\rho(\mathbf{B}) \leq \|\mathbf{B}\|$$

for any matrix norm compatible with a vector norm. Hence, we may settle the convergence question, in the sense of characterizing sufficient but not necessary conditions for convergence, by considering easily computable matrix norms such as $\|\mathbf{B}\|_1$ or $\|\mathbf{B}\|_\infty$. From a practical point of view, the whole approach makes sense only if solving the linear equation (3.4) is easy. By a proper choice of **D**, we obtain the methods described in the following.

Jacobi method A particularly convenient choice for **D** is a diagonal matrix:

$$\mathbf{D} = \begin{pmatrix} a_{11} & 0 & 0 & \cdots & 0 \\ 0 & a_{22} & 0 & \cdots & 0 \\ 0 & 0 & a_{33} & \cdots & 0 \\ \vdots & \vdots & \vdots & \ddots & \vdots \\ 0 & 0 & 0 & \cdots & a_{nn} \end{pmatrix},$$

which is easily inverted provided that $a_{ii} \neq 0$; this condition may be obtained by proper row/column permutations if **A** is non-singular. Choosing L_∞ norm, we obtain a sufficient condition for convergence:

$$\|\mathbf{I} - \mathbf{D}^{-1}\mathbf{A}\|_\infty = \max_{1 \leq i \leq n} \sum_{\substack{j=1 \\ j \neq i}}^n \left|\frac{a_{ij}}{a_{ii}}\right| < 1,$$

which actually boils down to diagonal dominance, i.e.,

$$|a_{ii}| > \sum_{\substack{j=1 \\ j \neq i}}^n |a_{ij}| \quad \forall i.$$

To implement the method, we must rewrite the initial equations as

$$x_i = \frac{1}{a_{ii}} \left(b_i - \sum_{\substack{j=1 \\ j \neq i}}^n a_{ij} x_j \right), \quad i = 1, \ldots, n,$$

```
function [x,i] = Jacobi(A,b,x0,eps,MaxIter)
dA = diag(A);   % get elements on the diagonal of A
C = A - diag(dA);
Dinv = diag(1./dA);
B = - Dinv * C;
b1 = Dinv * b;
oldx = x0;
for i=1:MaxIter
    x = B * oldx + b1;
    if norm(x-oldx) < eps*norm(oldx) break; end
    oldx = x;
end
```

Fig. 3.3 Implementation of the Jacobi iterative method.

which leads immediately to the iteration scheme

$$x_i^{(k+1)} = \frac{1}{a_{ii}}\left(b_i - \sum_{\substack{j=1\\j\neq i}}^{n} a_{ij} x_j^{(k)}\right), \quad i = 1,\ldots,n.$$

The iterations should be stopped when a satisfactory precision has been achieved. One possible condition to check is related to relative error. Having specified a tolerance parameter ϵ, we could stop the algorithm when

$$\|\mathbf{x}^{(k+1)} - \mathbf{x}^{(k)}\| < \epsilon \|\mathbf{x}^{(k)}\|.$$

Example 3.13 Jacobi method is easily coded in MATLAB, as illustrated in figure 3.3. Input arguments are matrix **A** and vector **b** of course, an initial approximation $\mathbf{x}_0$, convergence parameter ϵ, and maximum number of iterations. The implementation is based on vector and matrices as preferred in MATLAB. Note the twofold use of the `diag` function; given a matrix, it yields the vector of its elements on the diagonal ; given a vector, it builds a matrix with the elements of the vector on the diagonal.

To check `jacobi`, we may use the script of figure 3.4. Note that the first matrix is diagonally dominant; the second one is too, but to a lesser extent; the third one is not diagonally dominant. In the script, we compare the solution we get from the iterative method with the "correct" one obtained by Gaussian elimination; iterations are stopped after at most 10,000 steps. Please also note the use of the format string in `fprintf` (see online help). This is the output of the script.

```
Case of matrix
```

```
% ScriptJacobi
A1 = [3 1 1 0; 1 5 -1 2; 1 0 3 1; 0 1 1 4];
A2 = [2.5 1 1 0; 1 4.1 -1 2; 1 0 2.1 1; 0 1 1 2.1];
A3 = [2 1 1 0; 1 3.5 -1 2; 1 0 2.1 1; 0 1 1 2.1];
b = [1 4 -2 1]';

exact1 = A1\b;
[x1,i1] = Jacobi(A1,b,zeros(4,1),1e-08,10000);
fprintf(1, 'Case of matrix\n');
disp(A1);
fprintf(1, 'Terminated after %d iterations\n', i1);
fprintf(1, '   Exact       Jacobi\n');
fprintf(1, ' % -10.5g % -10.5g \n', [exact1' ; x1']);

exact2 = A2\b;
[x2,i2] = Jacobi(A2,b,zeros(4,1),1e-08,10000);
fprintf(1, '\nCase of matrix\n');
disp(A2);
fprintf(1, 'Terminated after %d iterations\n', i2);
fprintf(1, '   Exact       Jacobi\n');
fprintf(1, ' % -10.5g % -10.5g \n', [exact2' ; x2']);

exact3 = A3\b;
[x3,i3] = Jacobi(A3,b,zeros(4,1),1e-08,10000);
fprintf(1, '\nCase of matrix\n');
disp(A3);
fprintf(1, 'Terminated after %d iterations\n', i3);
fprintf(1, '   Exact       Jacobi\n');
fprintf(1, ' % -10.5g % -10.5g \n', [exact3' ; x3']);
```

Fig. 3.4 Script to check jacobi.m.

```
             3     1     1     0
             1     5    -1     2
             1     0     3     1
             0     1     1     4
Terminated after 41 iterations
   Exact          Jacobi
   0.55556        0.55556
   0.32407        0.32407
  -0.99074       -0.99074
   0.41667        0.41667

Case of matrix
     2.5000    1.0000    1.0000         0
     1.0000    4.1000   -1.0000    2.0000
     1.0000         0    2.1000    1.0000
          0    1.0000    1.0000    2.1000
Terminated after 207 iterations
   Exact          Jacobi
   3.1996         3.1996
  -2.7091        -2.7091
  -4.2898        -4.2898
   3.809          3.809

Case of matrix
     2.0000    1.0000    1.0000         0
     1.0000    3.5000   -1.0000    2.0000
     1.0000         0    2.1000    1.0000
          0    1.0000    1.0000    2.1000
Terminated after 10000 iterations
   Exact          Jacobi
  -42.808         1.6603e+027
   47.769        -1.8057e+027
   38.846        -1.5345e+027
  -40.769         1.5812e+027
```

We see that convergence is faster in the first case than in the second, and that divergence occurs in the third case. This is no surprise, if we check the degree of diagonal dominance, but we should note that lack of diagonal dominance does not necessarily imply divergence. The reader is urged to check the spectral radius of matrix **B** in the three cases:

$$\rho(\mathbf{B}_1) = 0.6489, \qquad \rho(\mathbf{B}_2) = 0.9257, \qquad \rho(\mathbf{B}_3) = 1.0059.$$

It may also be interesting to check the speed of convergence by plotting the norm of relative error with respect to the true solution. To this aim we modify

```
function [x,i] = JacobiBIS(A,b,x0,eps,MaxIter)
TrueSol = A\b;
aux = norm(TrueSol);
Error = zeros(MaxIter,1);
dA = diag(A);  % get elements on the diagonal of A
C = A - diag(dA);
Dinv = diag(1./dA);
B = - Dinv * C;
b1 = Dinv * b;
oldx = x0;
for i=1:MaxIter
   x = B * oldx + b1;
   Error(i) = norm(x-TrueSol)/aux;
   if norm(x-oldx) < eps*norm(oldx) break; end
   oldx = x;
end
plot(1:i,Error(1:i))
```

Fig. 3.5 Modifying Jacobi to plot residual.

```
% ScriptJacobiBIS
A1 = [3 1 1 0; 1 5 -1 2; 1 0 3 1; 0 1 1 4];
A2 = [2.5 1 1 0; 1 4.1 -1 2; 1 0 2.1 1; 0 1 1 2.1];
A3 = [2 1 1 0; 1 3.5 -1 2; 1 0 2.1 1; 0 1 1 2.1];
b = [1 4 -2 1]';
hold on
[x1,i1] = JacobiBIS(A1,b,zeros(4,1),1e-08,10000);
pause(3);
[x2,i2] = JacobiBIS(A2,b,zeros(4,1),1e-08,10000);
pause(3);
[x3,i3] = JacobiBIS(A3,b,zeros(4,1),1e-08,10000);
pause(3);
axis([1 100 0 2])
```

Fig. 3.6 Script to tun JacobiBIS.

jacobi and the relative script as shown in figures 3.5 and 3.6. The resulting plot is displayed in figure 3.7.

We see how important the spectral radius of matrix **B** is. In fact, later we discuss methods aimed at shifting its eigenvalues to speed up convergence.

☐

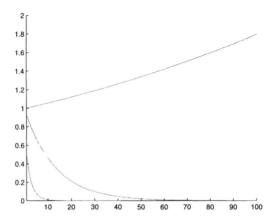

Fig. 3.7 Error in Jacobi method.

Gauss–Seidel method The Gauss–Seidel method is a variant of the Jacobi method. The idea is to use the updated values of $x_i^{(k+1)}$ immediately, as soon as they are computed. The iteration scheme is therefore

$$x_i^{(k+1)} = \frac{b_i - \sum_{j=1}^{i-1} a_{ij} x_j^{(k+1)} - \sum_{j=i+1}^{n} a_{ij} x_j^{(k)}}{a_{ii}}, \quad i = 1, \ldots, n. \qquad (3.5)$$

To analyze convergence of this method, we may note that this corresponds to choosing as $\mathbf{D}$ the lower triangle of $\mathbf{A}$:

$$\mathbf{D} = \begin{pmatrix} a_{11} & 0 & 0 & \cdots & 0 \\ a_{21} & a_{22} & 0 & \cdots & 0 \\ a_{31} & a_{32} & a_{33} & \cdots & 0 \\ \vdots & \vdots & \vdots & \ddots & \vdots \\ a_{n1} & a_{n2} & a_{n3} & \cdots & a_{nn} \end{pmatrix}.$$

Then it can be shown that diagonal dominance is again a sufficient condition for convergence:

$$\|\mathbf{I} - \mathbf{D}^{-1}\mathbf{A}\|_\infty \leq \max_{1 \leq i \leq n} \sum_{\substack{j=1 \\ j \neq i}}^{n} \left|\frac{a_{ij}}{a_{ii}}\right| < 1.$$

Speeding up convergence: successive overrelaxation Consider the iteration scheme

$$\mathbf{x}^{(k+1)} = \mathbf{B}\mathbf{x}^{(k)} + \mathbf{d}.$$

Since we move from the current point $\mathbf{x}^{(k)}$ to the updated point $\mathbf{x}^{(k+1)}$, we may think of it as the addition of a displacement to the old approximation:

$$\mathbf{x}^{(k+1)} = \mathbf{x}^{(k)} + \mathbf{r}^{(k)}.$$

Even though this method will converge if $\rho(\mathbf{B}) < 1$, convergence will be slow if the spectral radius of $\mathbf{B}$ is close to 1 (see example 3.13). We could try to speed up convergence by modifying the iteration:

$$\mathbf{x}^{(k+1)} = \mathbf{x}^{(k)} + \omega \mathbf{r}^{(k)} = \omega \mathbf{x}^{(k+1)} + (1 - \omega)\mathbf{x}^{(k)}.$$

Intuitively, if $\mathbf{r}^{(k)}$ is a good direction, we might think of accelerating the movement by setting $\omega > 1$. We must make sure that a poor choice of ω does not lead to instability. On the other hand, if the starting iteration is itself unstable, we might think that the difficulty stems from moving "too much" along the directions $\mathbf{r}^{(k)}$, which leads to oscillations and instability. In this case, we might think of dampening the oscillations with a suitable modification of the iteration scheme. To pursue this dampening, we may form a convex combination[11] of the new and the old point as follows:

$$\begin{aligned}\hat{\mathbf{x}}^{(k+1)} &= \omega \mathbf{x}^{(k+1)} + (1-\omega)\mathbf{x}^{(k)} \\ &= \omega(\mathbf{B}\mathbf{x}^{(k)} + \mathbf{d}) + (1-\omega)\mathbf{x}^{(k)} = \mathbf{B}_\omega \mathbf{x}^{(k)} + \omega \mathbf{d}.\end{aligned} \quad (3.6)$$

This is actually a convex combination if $\omega \in (0,1)$. It is worth noting that it looks like common exponential smoothing methods for time series analysis, where the aim is just to dampen oscillations in the estimates. The iterative scheme is stable if $\rho(\mathbf{B}_\omega) < 1$. Moreover, by a suitable choice of ω, the spectral radius will be reduced, with a corresponding improvement in convergence speed.

The reasoning above suggests that we may try to pursue modifications of the iterative approaches we have just described. For instance, we may try the idea on the Gauss–Seidel scheme. We may replace (3.5) by the following iteration:

$$z_i^{(k+1)} = \frac{1}{a_{ii}}\left[b_i - \sum_{j=1}^{i-1} a_{ij}x_j^{(k+1)} - \sum_{j=i+1}^{n} a_{ij}x_j^{(k)}\right]$$

$$x_i^{(k+1)} = \omega z_i^{(k+1)} + (1-\omega)x_i^{(k)}$$

In order to analyze the effect of this modification, let us rewrite the Gauss–Seidel scheme in a compact form, based on the following decomposition of $\mathbf{A}$:

$$\mathbf{A} = \mathbf{L} + \mathbf{D} + \mathbf{U},$$

---

[11] A convex combination of two points $\mathbf{x}_1$ and $\mathbf{x}_2$ is just a particular linear combination with nonnegative weights, such that their sum is 1: $\lambda \mathbf{x}_1 + (1-\lambda)\mathbf{x}_2$ for $\lambda \in [0,1]$.

where

$$L = \begin{bmatrix} 0 & 0 & 0 & \cdots & 0 & 0 \\ a_{21} & 0 & 0 & \cdots & 0 & 0 \\ a_{31} & a_{32} & 0 & \cdots & 0 & 0 \\ \vdots & \vdots & \vdots & \ddots & \vdots & \vdots \\ a_{n-1,1} & a_{n-1,2} & a_{n-1,3} & \cdots & 0 & 0 \\ a_{n1} & a_{n2} & a_{n3} & \cdots & a_{n,n-1} & 0 \end{bmatrix}$$

$$D = \begin{bmatrix} a_{11} & 0 & 0 & \cdots & 0 & 0 \\ 0 & a_{22} & 0 & \cdots & 0 & 0 \\ 0 & 0 & a_{33} & \cdots & 0 & 0 \\ \vdots & \vdots & \vdots & \ddots & \vdots & \vdots \\ 0 & 0 & 0 & \cdots & a_{n-1,n-1} & 0 \\ 0 & 0 & 0 & \cdots & 0 & a_{nn} \end{bmatrix}$$

$$U = \begin{bmatrix} 0 & a_{12} & a_{13} & \cdots & a_{1,n-1} & a_{1n} \\ 0 & 0 & a_{23} & \cdots & a_{2,n-1} & a_{2n} \\ 0 & 0 & 0 & \cdots & a_{3,n-1} & a_{3n} \\ \vdots & \vdots & \vdots & \ddots & \vdots & \vdots \\ 0 & 0 & 0 & \cdots & 0 & a_{n-1,n} \\ 0 & 0 & 0 & \cdots & 0 & 0 \end{bmatrix}.$$

With this notation, the modified Gauss–Seidel scheme may be rewritten in matrix form as

$$z^{(k+1)} = D^{-1}(b - Lx^{(k+1)} - Ux^{(k)})$$
$$x^{(k+1)} = \omega z^{(k+1)} + (1-\omega)x^{(k)}.$$

Eliminating $z^{(k+1)}$ and rearranging yields

$$(I + \omega D^{-1}L)x^{(k+1)} = [(1-\omega)I - \omega D^{-1}U]x^{(k)} + \omega D^{-1}b,$$

which will be stable if

$$\rho\left((I + \omega D^{-1}L)^{-1}[(1-\omega)I - \omega D^{-1}U]\right) < 1.$$

This method is called SOR (Successive OverRelaxation) and by proper selection of the parameter ω, we may reduce the spectral radius of the matrix, thus improving convergence.

Example 3.14 Figure 3.8 shows a possible implementation of successive overrelaxation, based on the Gauss–Seidel scheme. We may try to see the effect on convergence on the second matrix of example 3.13, which took 207 steps to converge with the Jacobi method. We do so by plotting the number of iterations needed for convergence as a function of different values of ω in the interval $[0, 2]$, which is obtained by running the script of figure 3.9. We

```
function [x,k] = SORGaussSeidel(A,b,x0,omega,eps,MaxIter)
oldx = x0;
x = x0;
N = length(x0);
omega1 = 1-omega;
for k=1:MaxIter
    for i=1:N
        z = (b(i) - sum(A(i,(1:i-1))*x(1:(i-1))) ...
            - sum(A(i,(i+1):N)*x((i+1):N))) / A(i,i);
        x(i) = omega*z + omega1*oldx(i);
    end
    if norm(x-oldx) < eps*norm(oldx) break; end
    oldx = x;
end
```

Fig. 3.8 Implementation of SOR modification of Gauss–Seidel method.

```
% ScriptSOR
A2 = [2.5 1 1 0; 1 4.1 -1 2; 1 0 2.1 1; 0 1 1 2.1];
b = [1 4 -2 1]';
omega = 0:0.1:2;
N = length(omega);
NumIterations = zeros(N,1);
for i=1:N
    [x,k] = SORGaussSeidel(A2,b,zeros(4,1),omega(i),1e-08,1000);
    NumIterations(i) = k;
end
plot(omega,NumIterations)
grid on
```

Fig. 3.9 Script to check SOR modification of Gauss–Seidel method.

get the plot in figure 3.10. This shows the impact of ω on speed of convergence. Actually, when the number of iterations exceeds the limit, we have divergence, since by playing with the relaxation parameter a stable case may result in instability and vice versa. With $\omega = 1$, we have the standard Gauss–Seidel approach, which requires 117 iterations; the best result, 49 iterations, is obtained with $\omega = 1.4$.   ☐

This example shows that finding the right value of the relaxation parameter is far from trivial, and in fact it is subject of quite some literature. For specific applications, there are strategies to estimate a good value for ω. By the way,

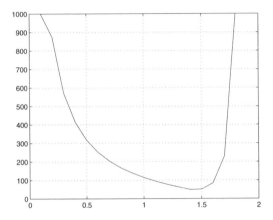

Fig. 3.10 Number of iterations in modified Gauss–Seidel as a function of the relaxation parameter ω.

the careful reader may wonder why in 3.10 we considered values of ω in the range $[0, 2]$. In fact, it can be proved that this acceleration method cannot converge for values of ω outside this interval. Finally, looking at equation (3.6) we may also guess why the method is actually called under-relaxation when $\omega < 1$, and overrelaxation when $\omega > 1$.

The conjugate gradient method   In MATLAB you will not find either Jacobi or Gauss–Seidel functions, as they are just the basic iterative methods to solve systems of linear equations. Some functions are related to an apparently weird approach to solving such systems, i.e., the solution of an optimization problem. In fact, solving the system $\mathbf{Ax} = \mathbf{b}$ is equivalent to solving the optimization problem:

$$\min_{\mathbf{x}} \| \mathbf{Ax} - \mathbf{b} \|^2,$$

where we are using Euclidean norm. Clearly, the objective function cannot be negative, and it will be zero for the solution of that system of equation (assuming it is unique). We may make the objective function more explicit:

$$\begin{aligned} \| \mathbf{Ax} - \mathbf{b} \|^2 &= (\mathbf{Ax} - \mathbf{b})'(\mathbf{Ax} - \mathbf{x}) = (\mathbf{x}'\mathbf{A}' - \mathbf{b}')(\mathbf{Ax} - \mathbf{b}) \\ &= \mathbf{x}'\mathbf{A}'\mathbf{Ax} - 2\mathbf{b}'\mathbf{Ax} + \mathbf{b}'\mathbf{b}, \end{aligned}$$

where the last term is actually irrelevant, as it is constant. We will see in chapter 6 that this is a quadratic programming problem (much like risk minimization in mean-variance portfolio optimization), and it can be solved by a number of ways. The most general approach, as we will see, is based on the gradient of the objective function, which yields a search direction to maximize or minimize its value.

In general, there is no advantage in using this approach, but for the case of a symmetric positive definite matrix, it can be shown that solving the system of equations is equivalent to the following problem:

$$\min \frac{1}{2}\mathbf{x}'\mathbf{A}\mathbf{x} - \mathbf{b}'\mathbf{x}.$$

The conjugate gradient method is based on a peculiar set of search directions, such that in theory the method would converge in a number of steps given by the order of the matrix. Hence, the method could be classified as a direct method. In practice, due to roundoff errors, this property does not hold and the method is considered as iterative. With recent improvements, conjugate gradient methods have become quite competitive for problems with specific structure. Such a structure occurs quite often in the numerical solution of PDEs.

3.3 FUNCTION APPROXIMATION AND INTERPOLATION

There are several reasons why we need the ability to approximate a function.

- Sometimes, we know an expression of the function, but it is impossible or expensive to evaluate. A typical example is the standard normal distribution function

$$N(x) = \frac{1}{\sqrt{2\pi}} \int_{-\infty}^{x} e^{-y^2/2} \, dy.$$

  which occurs in the Black–Scholes pricing formula.

- More generally, we may be able to evaluate the function itself, but we need something different, like the integral of the function. An approximation of the original function may be easier to integrate.

- Finally, there are situations in which the function is known or computed only at a discrete set of points (nodes), and we would like to find a suitable function which takes the same value (or a close one) at those nodes but can be evaluated outside this set.

In some cases, it is enough to find a *local* approximation, in the neighborhood of a given point x_0, in which case a Taylor expansion would suffice:

$$f(x) \approx f(x_0) + f'(x_0)(x - x_0) + \frac{1}{2}f''(x_0)(x - x_0)^2 + \cdots.$$

We have seen such an idea in the duration–convexity approximation used for bond portfolio immunization and the delta–gamma approximation used with derivatives (see examples 2.10 and 2.24 on pages 59 and 113, respectively).

In this section, however, we are interested in an approximation valid over an extended range of values of the independent variable.

Another criterion to classify approximation methods is based on the generality of the approach. In the case of the cumulative function for the normal distribution, we may look for some *ad hoc* approximation. In other cases we look for more general strategies based on classes of approximating functions. A possible choice for the class of approximating functions is represented by the class of polynomials of given degree m; let $P_m(x; \boldsymbol{\alpha})$ denote such a polynomial, with coefficients represented by the vector $\boldsymbol{\alpha}$. One reason behind this choice is that polynomials are continuous functions, as well as their derivatives, which lend themselves to easy differentiation and integration. One possible metric to select the best approximation is the least squares approximation, whereby we try to minimize the average square deviation of the approximating function from f on a set of selected points x_i, for which we know the value $f(x_i)$. The approximation problem can be stated as

$$\min_{\boldsymbol{\alpha}} \sum_{i=1}^{n} [f(x_i) - P_m(x_i; \boldsymbol{\alpha})]^2 .$$

Different objective functions could be used, basically corresponding to different ways of measuring the norm of the vector of the approximation errors. Another typical choice is the "min-max" metric, which is based on the $\|\cdot\|_\infty$ norm:

$$\min_{\boldsymbol{\alpha}} \max_{i=1,\ldots,n} |f(x_i) - P_m(x_i; \boldsymbol{\alpha})| .$$

Sometimes, it is very useful to take a slightly more explicit view of function approximation. What we usually try to find, given a function $f(\mathbf{x})$, is a suitable approximation expressed as a linear combination of a set of basis functions. If we consider a set of m basis functions $\phi_j(\mathbf{x})$, $j = 1, \ldots, m$, we want something like

$$f(\mathbf{x}) \approx \hat{f}(\mathbf{x}) = \sum_{j=1}^{m} c_j \phi_j(\mathbf{x}).$$

The basis functions may be polynomials, but there are alternatives. Finding the approximation means finding the m coefficients c_j in the linear combination. In function approximation by least squares, we have a set of n nodes at which we know the value of the function, and $n > m$. In this case, we have too few degrees of freedom, and we cannot enforce an exact match. In other words, we would like to find the approximation by solving a set of linear equations like

$$\sum_{j=1}^{m} c_j \phi_j(x_i) = f(x_i), \qquad i = 1, \ldots, n$$

or, in compact form,

$$\boldsymbol{\Phi} \mathbf{c} = \mathbf{y}, \tag{3.7}$$

where $y_i = f(x_i)$ and $\phi_{ij} = \phi_j(x_i)$. Unfortunately, if $n > m$ the system is overdetermined and it cannot be solved. What we can do is finding the least squares approximation, which requires the minimization of the sum of squared residuals e_i, where

$$e_i = f(x_i) - \sum_{j=1}^{m} c_j \phi_j(x_i), \quad i = 1, \ldots, n.$$

Using relatively straightforward calculus, we can show that the least squares solution is

$$\mathbf{c} = (\mathbf{\Phi}'\mathbf{\Phi})^{-1}\mathbf{\Phi}'\mathbf{y}.$$

If, however, the number of nodes and the number of basis function is the same, $m = n$, then we may be able to find an exact match of the function values at nodes. We find the solution by enforcing the interpolation conditions:

$$\sum_{j=1}^{m} c_j \phi_j(x_i) = f(x_i), \quad i = 1, \ldots, n.$$

This process is called function **interpolation** and, within this framework, it leads to the solution of a system of linear equations. The following example will illustrate the difference between approximation and interpolation.

Example 3.15 Say that we want to approximate/interpolate an increasing concave function, such as $\log(x)$. We are given a set of five nodes, which may be plotted as follows:

```
>> xdata = [1 5 10 30 50];
>> ydata = log(xdata);
>> plot(xdata,ydata,'o')
>> hold on
```

resulting in the plot of figure 3.11 We may try fitting a second-order polynomial, $ax^2 + bx + c$. Note that this *may* correspond to selecting basis functions:

$$\phi_1(x) = 1, \quad \phi_2(x) = x, \quad \phi_3(x) = x^2.$$

This choice is referred to as the *monomial* basis, but a different set of polynomials could be used. Polynomial fitting, in the least squares sense, can be accomplished by the MATLAB `polyfit` function:

```
>> p = polyfit(xdata,ydata,2)
p =
    -0.0022    0.1802    0.3544
>> xvet=1:0.1:50;
>> plot(xvet,polyval(p,xvet))
```

This snapshot produces the plot in figure 3.12. The approximating polynomial does not really pass through the data point, but this is expected, as the

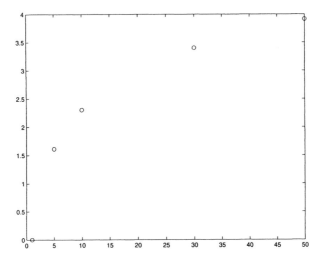

Fig. 3.11 Data points (nodes) for example 3.15.

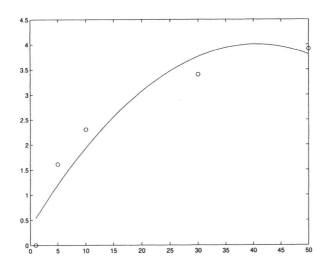

Fig. 3.12 Fitting a second-order polynomial in example 3.15.

number of nodes is larger than the set of coefficients in the polynomial. The trouble may be that, even if the fit is good, the approximating function is not monotonically increasing. If the logarithm is actually a utility function, we would require an increasing approximation which shows non-satiation. Since using a second-order polynomial is not that satisfactory, we could try increasing the order of the polynomial. We have five data points, and a fourth-order polynomial may result in exact polynomial interpolation. Note that the order of the polynomial is one less the number of nodes. To remember this, think

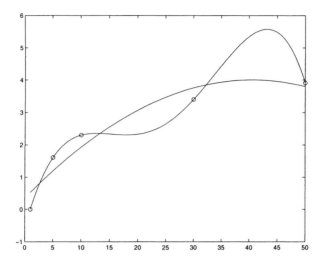

Fig. 3.13 Interpolation in example 3.15.

that there is one line (polynomial of order one) passing through two points. This is also easily accomplished in MATLAB:

```
>> p = polyfit(xdata,ydata,4)
p =
   -0.0000    0.0017   -0.0529    0.6705   -0.6192
>> plot(xvet,polyval(p,xvet))
```

and we get the plot in figure 3.13. Now we do pass through the data points, which is nice, but there are spurious oscillations and the approximating function is neither concave nor increasing, which is certainly bad for a utility function. In finance, we could have similar trouble when we try to define a term structure of interest rates fitted on the basis of a limited set of bond prices. Hence we see, that polynomial approximation and interpolation is not that trivial. ▫

Function interpolation and approximation is a vast sub-field of numerical analysis. In the next sections we will just cover the essentials: an example of ad hoc methods is given in section 3.3.1; straightforward polynomial interpolation is the topic of section 3.3.2; cubic splines are introduced in section 3.3.3; section 3.3.4 deals with least squares approximation at a more general level. We should also mention that the methods we illustrate here can be extended to multivariate cases.

3.3.1 Ad hoc approximation

In this section we consider an example of ad hoc approximation by a rational function. While polynomials enjoy nice characteristics, sometimes approxi-

```
function z = mynormcdf(x)
c = [ 0.31938153 , -0.356563782 , 1.781477937 , ...
      -1.821255978 , 1.330274429 ];
gamma = 0.2316419;
vx = abs(x);
k = 1./(1+gamma.*vx);
n = exp(-vx.^2./2)./sqrt(2*pi);
matk = ones(5,1) * k;
matexp = (ones(length(x),1)*(1:5))';
matv = matk.^matexp;
z = 1 - n.*(c*matv);
i = find(x < 0);
z(i) = 1-z(i);
```

Fig. 3.14 MATLAB code to approximate the cumulative normal distribution.

mations involving rational functions fit more nicely. For instance, there are various approximation formulas that can be used to evaluate the standard normal distribution function $N(x)$. One is the following[12]:

$$N(x) = \begin{cases} 1 - N'(x)(a_1 k + a_2 k^2 + a_3 k^3 + a_4 k^4 + a_5 k^5) & \text{if } x \geq 0 \\ 1 - N(-x) & \text{if } x < 0, \end{cases}$$

where

$$N'(x) = \frac{1}{\sqrt{2\pi}} e^{-x^2/2}, \qquad k = \frac{1}{1+\gamma x}$$

$$\gamma = 0.2316419, \qquad a_1 = 0.31938153$$

$$a_2 = -0.356563782, \qquad a_3 = 1.781477937$$

$$a_4 = -1.821255978, \qquad a_5 = 1.330274429.$$

The MATLAB code for this function is shown in figure 3.14; it is a little involved, as we have made sure it can operate on vector arguments (as it should be the case with good MATLAB functions). This is not really the formula used in the equivalent MATLAB function normcdf, but we may compare the two approximations:

```
>> normcdf([-1.5 -1 -0.5 0.5 1 1.5])
ans =
```

---

[12]This formula is proposed in [9, p. 248]. It is based on approximation 7.1.26 of the error function in [1], which in turn refers to [8]. If you have some archaeological instinct, you may go further back in time.

```
    0.0668    0.1587    0.3085    0.6915    0.8413    0.9332
>> mynormcdf([-1.5 -1 -0.5 0.5 1 1.5])
ans =
    0.0668    0.1587    0.3085    0.6915    0.8413    0.9332
```

3.3.2 Elementary polynomial interpolation

We consider here elementary interpolation by polynomials of sufficient degree. Let us consider a set of support points (x_i, y_i), $i = 0, 1, \ldots, n$, where $y_i = f(x_i)$ and $x_i \neq x_j$ for $i \neq j$. It is easy to find a polynomial of degree (at most) n such that $P_n(x_i) = y_i$ for any i. We may rely on the *Lagrange polynomials* $L_i(x)$, defined as

$$L_i(x) = \prod_{\substack{j=0 \\ j \neq i}}^{n} \frac{x - x_j}{x_i - x_j}. \tag{3.8}$$

Note that these are polynomials of degree n and that

$$L_i(x_k) = \begin{cases} 1 & \text{if } i = k \\ 0 & \text{otherwise.} \end{cases}$$

Now an interpolating polynomial can be easily written as

$$P_n(x) = \sum_{i=0}^{n} y_i L_i(x).$$

In practice, no one should use this form for computational purposes, and some tricks are needed for the sake of computational efficiency, but the idea is hopefully clear.

Example 3.16 We consider here the interpolation of a set of ten data points. We may try interpolating them by a polynomial of degree 9:

```
>> x=1:10;
>> y = [8 2.5 -2 0 5 2 4 7 4.5 2];
>> plot(x,y,'o')
>> hold on
>> x2=1:0.05:10;
>> p=polyfit(x,y,9);
Warning: Polynomial is badly conditioned. Remove repeated
         data points or try centering and scaling as described
         in HELP POLYFIT.
>> plot(x2,polyval(p,x2))
```

We get some warning from MATLAB, which we disregard for a moment. The result is shown in figure 3.15. We may see that the polynomial passes through the data set but, unfortunately, we also see that the interpolating polynomial

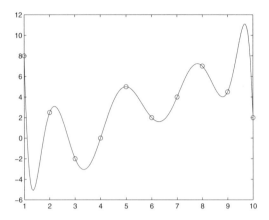

Fig. 3.15 Interpolating a given data set by a polynomial of degree 9.

has some undesirable oscillation behavior near the end points of the interval. This is not surprising: a polynomial of degree n may have up to n zeros, which means it may have up to $n-1$ local minima and maxima and oscillations are to be expected. □

The oscillation of high-degree interpolating polynomials is a typical difficulty, and there are a few ways to try overcoming it. One obvious way is to use more sophisticated functions, for both approximation and interpolation. But actually there is still another basic mistake we are doing in the last example: we did a poor choice in selecting nodes. In selecting nodes over an interval $[a, b]$, the natural choice is taking evenly spaced ones:

$$x_i = a + \frac{i-1}{n-1}(b-a), \qquad i = 1, 2, \ldots, n.$$

This choice may have nasty effects in itself. It turns out that a better choice is given by Chebyshev nodes:

$$x_i = \frac{a+b}{2} + \frac{b-a}{2} \cos\left(\frac{n-i+0.5}{n}\pi\right), \qquad i = 1, \ldots, n.$$

An investigation of why this seemingly odd choice is an improvement over a naive placement of nodes goes beyond the scope of this book, but we will illustrate the effect with a typical example.

Example 3.17 We consider polynomial interpolation for a well-known function, called Runge function:

$$\frac{1}{1+25x^2}$$

```
% RungeScript.m
% define inline function
runge = inline('1./(1+25*x.^2)');
% use equispaced nodes
EquiNodes = -5:5;
peq = polyfit(EquiNodes,runge(EquiNodes),10);
x=-5:0.01:5;
figure
plot(x,runge(x));
hold on
plot(x,polyval(p10,x));
% use Chebyshev nodes
ChebNodes = 5*cos(pi*(11 - (1:11) + 0.5)/11);
pcheb = polyfit(ChebNodes,runge(ChebNodes),10);
figure
plot(x,runge(x));
hold on
plot(x,polyval(pcheb,x));
```

Fig. 3.16 MATLAB script for example 3.17.

over the interval $[-5, 5]$. As we mentioned, a seemingly obvious and natural choice is to place equally spaced interpolation nodes, for instance $x_i = -5, -4, -3, \ldots, 4, 5$. These are eleven nodes, and we may try interpolating by a polynomial of degree ten.

Straightforward interpolation is accomplished by the MATLAB script in figure 3.16. Selecting equally spaced nodes results in the first plot, depicted in figure 3.17. We see the usual oscillation near the end points, but in this case the behavior looks really pathological. The reader is invited to verify that increasing the order of approximation only makes things worse. If we use Chebyshev nodes, which is done in the second half of the script, we get the result in figure 3.18. While the result is not yet satisfactory, at least it looks a bit less pathological.
▯

Even though choosing Chebyshev nodes helps in the last example, there is still something wrong with the quality of the approximation we get by interpolating with one high-degree polynomial. Using the right nodes, we may try increasing the order of the polynomial, but there is an easier way out: using piecewise polynomial functions. A look at figure 3.18 suggests that there are regions in which the function is essentially zero, and we should use a different approximation there. Using piecewise polynomial interpolation is pursued in the next section on splines.

We close the section here by noting that we have still another issue when using simple-minded polynomial interpolation. Consider again the basis func-

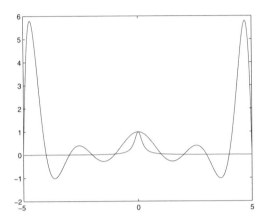

Fig. 3.17 Polynomial interpolation for Runge function: equally spaced nodes.

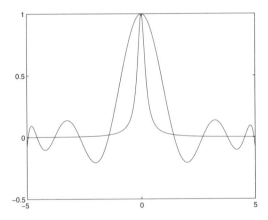

Fig. 3.18 Polynomial interpolation for Runge function: Chebyshev nodes.

tion framework. If we want polynomial approximation or interpolation, the monomial basis $(1, x, \ldots, x^{n-1})$ is the natural choice when selecting basis functions. However, this may lead to a badly conditioned matrix Φ in equation (3.7), along with a few numerical difficulties. In fact, several alternative families of polynomials have been proposed to avoid them. Since we mentioned Chebyshev nodes, we should at least mention in passing Chebyshev polynomials, which are recursively defined as follows:

$$
\begin{aligned}
T_0(x) &= 1 \\
T_1(x) &= x \\
T_2(x) &= 2x^2 - 1 \\
T_3(x) &= 4x^3 - 3x \\
&\ldots \\
T_n(x) &= 2xT_{j-1}(x) - T_{j-2}(x).
\end{aligned} \tag{3.9}
$$

3.3.3 Interpolation by cubic splines

One possible way to avoid oscillating polynomials in function interpolation is resorting to low-degree polynomials, interpolating the data points piecewise. The simplest idea is to use piecewise linear interpolation. Given the $N + 1$ nodes (or knots) (x_i, y_i), we may use N first-degree polynomials $S_i(x)$, each one valid on the interval (x_i, x_{i+1}). An obvious requirement is that the resulting function is continuous, i.e., $S_i(x_{i+1}) = S_{i+1}(x_{i+1})$. Recalling the Lagrange polynomials defined in equation (3.8), we have

$$S_i(x) = y_i \frac{x - x_{i+1}}{x_i - x_{i+1}} + y_{i+1} \frac{x - x_i}{x_{i+1} - x_i} \quad x \in [x_i, x_{i+1}].$$

This type of interpolation is called *linear spline*. Whereas the interpolating function is continuous, its derivative is not, which may have undesirable consequences. If the data we are interpolating are prices of an asset as a function of an underlying factor, non-differentiability prevents the estimation of sensitivities. If we are approximating a function which must then be optimized, as is the case with the value function in dynamic programming, non-differentiability is clearly a complication.

We may enforce the continuity of the derivatives of the spline by increasing the degree of the polynomials. The most common spline is obtained by "joining" N third-degree polynomials $S_i(x)$, with coefficients $s_{i0}, s_{i1}, s_{i2}, s_{i3}$, which must satisfy the following requirements:

$$
\begin{aligned}
S(x) &= S_i(x) = s_{i0} + s_{i1}(x - x_i) + s_{i2}(x - x_i)^2 + s_{i3}(x - x_i)^3 \\
&\quad x \in [x_i, x_{i+1}], \quad i = 0, 1, \ldots, N - 1 \\
S(x_i) &= y_i, \quad i = 0, 1, \ldots, N
\end{aligned}
$$

$$S_i(x_{i+1}) = S_{i+1}(x_{i+1}), \quad i = 0, 1, \ldots, N-2$$
$$S'_i(x_{i+1}) = S'_{i+1}(x_{i+1}), \quad i = 0, 1, \ldots, N-2$$
$$S''_i(x_{i+1}) = S''_{i+1}(x_{i+1}) \quad i = 0, 1, \ldots, N-2.$$

The resulting spline $S(x)$ is called a *cubic spline*. The condition above require continuity for the spline itself and for its first and second derivatives. To specify a spline, we must give $4N$ coefficients. Passage through the support points gives $N+1$ conditions; the continuity of the spline and the two derivatives enforces $3(N-1)$ conditions, yielding a total of $4N-2$ conditions. Hence, we have two degrees of freedom which may be eliminated by enforcing further requirements. Usually, they involve some conditions at, or near, the end points x_0 and x_N. Among the most common conditions, we recall the following ones:

- $S''(x_0) = S''(x_N) = 0$, which leads to *natural splines*.

- $S'(x_0) = f'(x_0)$ and $S'(x_N) = f'(x_N)$, which may be used if we have a precise idea of the behavior of $f(x)$ near the end points.

- The *not-a-knot* condition, which is obtained by requiring that the third-order derivative $S'''(x)$ be continuous in x_1 and x_{N-1}. This implies that $S(x)$ would be a spline for knots $x_0, x_2, x_3, \ldots, x_{N-2}, x_N$, but it would interpolate through x_1 and x_{N-1} too (hence the name).

We should note that these conditions are symmetric with respect to the end points of the interval; actually we could make different choices for the two end points. It is also interesting to note that we have no degree of freedom in linear splines; in the case of splines of degree 2, we would have one degree of freedom, with a corresponding asymmetry in end points. Despite the appealing name, natural splines are usually avoided. Their importance stems from the following theorem, which we state without proof.[13]

THEOREM 3.3 *Let f'' be continuous in (a, b) and let $a = x_0 < x_1 < \cdots < x_N = b$. If S is the natural cubic spline interpolating f on the knots x_i, then*

$$\int_a^b [S''(x)]^2 \, dx \leq \int_a^b [f''(x)]^2 \, dx.$$

The importance of this theorem can be understood by recalling that the curvature of the curve described by the equation $y = f(x)$ is given by

$$|f''(x)| \cdot \{1 + f'(x)^2\}^{-3/2}.$$

If f' is sufficiently small, we see that $|f''(x)|$ approximates the curvature; hence, the natural spline is, in some sense, an approximation of minimal

---

[13] See, e.g., [13, pp. 380–381].

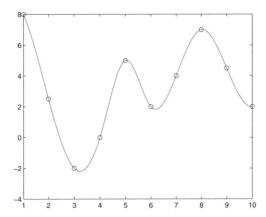

Fig. 3.19 Interpolating a given data set by a cubic spline.

curvature over the interval (a, b). When nothing is known about the function, the not-a-knot condition is the recommendable choice; in fact, this is the default option in MATLAB.

To find the unknown coefficients, we have to set up a system of linear equations; the details are a bit tedious, and since they are implemented in a ready-to-use MATLAB function, they are omitted. Yet it is interesting to note that for most choices of free conditions, the resulting system has a tridiagonal form like that discussed in section 3.2.4; furthermore, it is symmetric and diagonally dominant, hence it is particularly easy to solve.

Splines are so important that an entire MATLAB toolbox is devoted to them. In the base MATLAB system, you have two functions that may be used for cubic spline interpolation. One is `interp1`, provided that you call it with the parameter 'spline'; the other one is `spline`.

Example 3.18 Let us compare the interpolation we obtain for the cases we have already discussed in examples 3.16 and 3.17. Running the following script, we get the result in figure 3.19:

```
x=1:10;
y = [8 2.5 -2 0 5 2 4 7 4.5 2];
plot(x,y,'o')
hold on
x2=1:0.05:10;
y2=interp1(x,y,x2,'spline');
plot(x,y,'o',x2,y2);
```

We see that spurious oscillations are avoided. The same result is obtained by calling `spline`, which also returns a spline object; this object may be used for later evaluations by the function `ppval`:

```
% RungeSpline.m
% define inline function
runge = inline('1./(1+25*x.^2)');
% use 11 equispaced nodes
EquiNodes11 = -5:5;
ppeq11 = spline(EquiNodes11,runge(EquiNodes11));
x=-5:0.01:5;
subplot(3,1,1)
plot(x,runge(x));
hold on
plot(x,ppval(ppeq11,x));
axis([-5 5 -0.15 1])
title('11 equispaced points');
% use 20 equispaced nodes
EquiNodes20 = linspace(-5,5,20);
ppeq20 = spline(EquiNodes20,runge(EquiNodes20));
subplot(3,1,2)
plot(x,runge(x));
hold on
plot(x,ppval(ppeq20,x));
axis([-5 5 -0.15 1])
title('20 equispaced points');
% use 21 equispaced nodes
EquiNodes21 = linspace(-5,5,21);
ppeq21 = spline(EquiNodes21,runge(EquiNodes21));
subplot(3,1,3)
plot(x,runge(x));
hold on
plot(x,ppval(ppeq21,x));
axis([-5 5 -0.15 1])
title('21 equispaced points');
```

Fig. 3.20 MATLAB script to interpolate Runge function by cubic splines.

```
x=1:10;
y = [8 2.5 -2 0 5 2 4 7 4.5 2];
plot(x,y,'o')
hold on
pp=spline(x,y);
x2=1:0.05:10;
y2 = ppval(pp,x2);
plot(x,y,'o',x2,y2);
```

We may also check the result with the Runge function. Running the script of figure 3.20 we get the plots in figure 3.21. We may notice that using 21

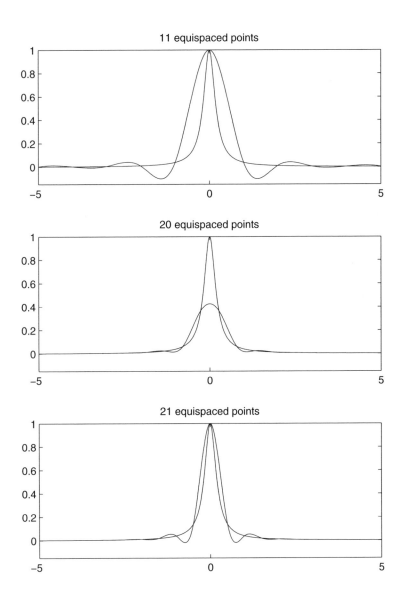

Fig. 3.21 Interpolating Runge function by a cubic spline.

points rather than 11 improves the approximation, whereas an even number of points result in a very poor match near the maximum. The approximation is still not satisfactory: the reader is urged to try placing nodes in points $-5, -3, 3, 5$ and distributing 17 nodes on the interval $[-2, 2]$. ☐

As we have pointed out, in MATLAB the default way to define the two degrees of freedom in cubic splines is the not-a-knot condition. But if you provide the function spline with an y vector with two more components than the vector x, the first and last components are used to enforce a value for the spline slopes at the extreme points of the interval.

Cubic splines are only the basic type of spline; many more have been proposed. A typical application in finance is in estimating term structures of interest rates given a limited set of market data related to bond prices (see, e.g., [3], [4], and the references therein). In economics, *shape-preserving* splines are sometimes used, which make sure that the resulting spline has certain qualitative features which are essential from an economical point of view.

3.3.4 Theory of function approximation by least squares

This section is somewhat more theoretical, and basically aims at providing a more general and abstract framework for function approximation. The basic concept we use here is a generalization of orthogonality between vectors. We should start with a general formulation of the best approximation problem. We are given a normed linear space E and a subspace G of E. By "normed" we mean that the objects in that space have an associated norm (e.g., the vector norms we have discussed in section 3.2.1); by "linear" we mean that by taking any linear combination of objects in G or E, we get another object in that set.

Given a norm, we may define distances between arbitrary objects in the space. The distance between two elements $f, g \in E$ is simply given by $\|f - g\|$. More generally, the distance of $f \in E$ from the subspace G is defined by

$$\text{dist}(f, G) = \inf_{g \in G} \|f - g\|.$$

An interesting specific case occurs when we have an inner-product space, whereby norm is based on the inner product defined on the space:

$$\|f\| = \sqrt{<f, f>}.$$

Typical examples of inner products are

$$<x, y> = \sum_{i=1}^{n} x_i y_i, \qquad (3.10)$$

for $x, y \in \mathbb{R}^n$, and

$$<f, g> = \int_a^b f(x) g(x)\, dx$$

for $f, g \in C(a, b)$, i.e., the space of continuous functions on the interval (a, b). We say that two elements $f, g \in E$ are **orthogonal** (denoted by $f \perp g$) if

$$< f, g > = 0.$$

We say that a finite or infinite sequence of elements $f_1, f_2, f_3, \ldots \in E$ is an orthogonal system if

$$< f_i, f_j > = 0 \qquad \forall i \neq j.$$

Furthermore, if all elements in the subset have unit norm, we say that the system is **orthonormal**.

Example 3.19 The following polynomials:

$$p_0(x) = 1$$
$$p_1(x) = x$$
$$p_2(x) = x^2 - \frac{1}{3}$$
$$p_3(x) = x^3 - \frac{3}{5}x$$
$$p_4(x) = x^4 - \frac{6}{7}x^2 + \frac{3}{35}$$

form an orthogonal system, on interval $[-1, 1]$, if the inner product

$$< f, g > = \int_{-1}^{1} f(x) g(x) \, dx.$$

They are the first polynomials in the family of Legendre polynomials. Similarly, the Chebyshev polynomials defined in (3.9) form an orthogonal system with respect to the inner product:

$$< f, g > = \int_{-1}^{1} f(x) g(x) \frac{dx}{\sqrt{1 - x^2}}.$$

Actually, there are general strategies to build orthogonal systems, which will be outlined in section 4.1.2.

We should also mention that orthogonal systems of random variables can also be built. The idea, said very roughly in financial terms, is to decompose risk (a random variable) into the sum of uncorrelated sources of risk, each one carrying a piece of information in such a way that redundancy is avoided and a simple representation of risk is obtained.

The fundamental result of approximation in a normed space is that, if the space is equipped by an inner product, there is an equivalence between the two conditions:

1. g is a best approximation to f in G.

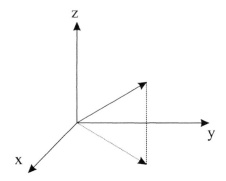

Fig. 3.22 Orthogonal projection.

2. The residual is orthogonal to the subspace: $f - g \perp G$.

This is again a generalization of the familiar geometric concept of orthogonal projection in Euclidean spaces (see figure 3.22). If we have a vector in the (x, y, z) space, and we want to find the closest vector on the (x, y) plane, we do an orthogonal projection. The following example shows how this equivalence can be exploited.

Example 3.20 Consider the space $E = C(0, 1)$ of continuous functions over interval $(0, 1)$, and assume that we want to find an optimal approximation (in the least-squares sense) in the subspace G consisting of polynomials of degree n. We may build the linear subspace G by using monomials $g_j(x) = x^j$, $j = 0, 1, \ldots, n$, as the basis. Thus, $g(x) = \sum_{j=0}^{n} a_j g_j(x) = \sum_{j=0}^{n} a_j x^j$. We want to minimize the deviation

$$\int_0^1 \left[\sum_{j=0}^{n} a_j x^j - f(x) \right]^2 dx.$$

If $f - g$ is orthogonal to G, then we must require

$$<g - f, g_i> = 0, \qquad i = 0, \ldots, n,$$

or, in other words,

$$\sum_{j=0}^{n} a_j <g_j(x), g_i(x)> = <f, g_i(x)>, \qquad i = 0, \ldots, n.$$

In our case, this yields a set of linear equations:

$$\sum_{j=0}^{n} a_j \int_0^1 x^j x^i \, dx = \int_0^1 f(x) x^i \, dx, \qquad i = 0, \ldots, n.$$

These equations are collectively called the *normal equations*. Unfortunately, the matrix of coefficients includes definite integrals evaluating to

$$\left.\frac{x^{i+j+1}}{i+j+1}\right|_0^1 = \frac{1}{i+j+1}.$$

But this is the (dreaded) Hilbert matrix that we already met in example 1.3 on page 18. ☐

The example shows that a simple-minded approach may lead to ill-conditioned numerical problems. Proper selection of the basis functions is fundamental from the numerical point of view, and this is why families of orthogonal polynomials are often used.

We have so far considered the *continuous* least squares problem, in order to motivate the introduction of orthogonal polynomials. Typically, in numerical applications, we have to solve a discrete problem in which a set of n data points (x_i, y_i), $i = 1, \ldots, n$, is given, where $y_i = f(x_i)$, and we look for an approximation in terms of a linear combination of m basis functions (e.g., polynomials). Using the Euclidean norm, as we have already seen, we get the ordinary least squares problem:

$$\min \|\mathbf{e}\|_2^2 = \sum_{i=1}^n \left| f(x_i) - \sum_{j=1}^m c_j \phi_j(x_i) \right|^2.$$

In this case the normal equations (or ordinary calculus) yield

$$\mathbf{c} = (\mathbf{\Phi}'\mathbf{\Phi})^{-1}\mathbf{\Phi}'\mathbf{y}.$$

In this case too, solving the normal equations may be easier with a proper selection of basis functions. In chapter 10 we will see an application of linear regression with polynomials to pricing American options by Monte Carlo simulation.

3.4 SOLVING NON-LINEAR EQUATIONS

Solving non-linear equations is a common task in finance; the most elementary example is the computation of the internal rate of return (see example 2.8 on page 47), which calls for finding the roots of a polynomial. A polynomial equation is a particular case of general non-linear equations, and it is a very lucky case, in the sense that we are typically able to find all of the roots of the equation by specific methods. For instance, if we consider

$$x^3 + 3x^2 - 2x^2 + 4 = 0,$$

we may use the MATLAB `roots` function and get

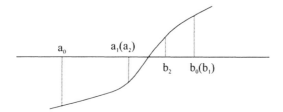

Fig. 3.23 Example of the bisection method.

```
>> roots([1 3 -2 4])
ans =
   -3.8026
    0.4013 + 0.9439i
    0.4013 - 0.9439i
```

In general we must settle for *one* root near some prespecified point.

You might wish to find a solution of an equation in a single variable, such as

$$f(x) = 0$$

or a system of equations in several variables, such as

$$\mathbf{F}(\mathbf{x}) = \mathbf{0}.$$

MATLAB offers different functionalities to this purpose. We first outline the basic features of numerical methods for non-linear equations, limiting the treatment to bisection and Newton methods.

3.4.1 Bisection method

The bisection method is the simplest method for solving the scalar equation

$$f(x) = 0$$

without requiring anything more than the ability to evaluate, or estimate, the function f at a given point. This is an important feature, since in some cases we do not even have an analytical expression for the function f, and therefore we are not able to apply more sophisticated methods such as Newton's method, which calls for computation of the derivative of f. Suppose that we know two points a, b ($a < b$) such that $f(a) < 0$ and $f(b) > 0$. Then, if the function is continuous, it is obvious that it must cross the zero axis somewhere in the interval $[a, b]$ (see figure 3.23). The same observation holds if the signs of the function in a and b are reversed. So $[a, b]$ is an interval encapsulating a root of the equation. Then we may try to reduce this interval by checking the sign of f in the midpoint of the interval, i.e.,

$$c = \frac{a+b}{2}.$$

If $f(c) = 0$, possibly within some prespecified tolerance, we are done. If $f(c) < 0$, we may conclude that a zero must be located somewhere in the interval $[c, b]$; otherwise, the interval to check is $[a, c]$. Going on this way, we build a sequence of smaller and smaller intervals bracketing the zero. Formally, you generate a non-decreasing sequence a_n and a non-increasing sequence b_n such that:

$$|r - c_n| \leq \frac{b_0 - a_0}{2^{n+1}},$$

where r is the (unknown) root and $c_n = (b_n + a_n)/2$. It can be shown that this method is characterized by a linear convergence rate.

The method, as usual, will not really find the exact root (in general), but only a suitable approximation. Furthermore, we should define some termination criteria to stop the algorithm. Possible choices are

- $b_n - a_n < \delta$
- $|f(c_n)| < \epsilon$
- maximum number of iterations

There is no best criterion and for a robust algorithm we must use *all* of them. Actually, the second one may depend on the chosen units of measure: by scaling the equation, this criterion may be met by any point. It is advisable to restate the criterion in relative terms.

Example 3.21 Consider a typical problem in Microeconomics. We want to find the price p such that supply $S(p)$ of some item equals demand $D(p)$. What we are looking for is a zero of the excess demand function $f(p) = D(p) - S(p)$. Asking for $|f(p)| < \epsilon$ is a bit arbitrary, as we have said. A better termination test could be $|D(p) - S(p)| < \delta D(p)$, i.e., demand minus supply is small with respect to demand. This is an example of "relative" rather than "absolute" condition. □

A possible difficulty of bisection is that you need an interval with a sign change to start. Library routines such as `fzero` may relieve the task, since they require a starting interval or one starting value, in which case they are supposed to locate a root near there; the search for an interval with a change in sign is carried out automatically. The following example shows what may go wrong with bisection.

Example 3.22 Consider the non-linear equation

$$\frac{1}{x} = 0.$$

Using MATLAB requires the definition of a function handle:

```
>> fzero(@(x) 1/x, 3)
```

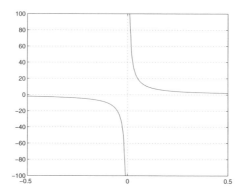

Fig. 3.24 Bisection cannot be applied to a discontinuous function.

```
ans =
-2.7776e-016
```

We get a very small number, virtually zero. But this is not really a root:

```
>> 1/ans
ans =
-3.6003e+015
```

In this case we get a "false" zero. Of course it is our fault: we are applying bisection to a discontinuous function (see figure 3.24). But what bisection sees is a function with a change in sign and a shrinking interval which eventually satisfies the first termination criterion, but not the second one.

In other cases (e.g., $x^2 = 0$) you do not get any root by bisection:

```
>> fzero(@(x) x^2, 3)
Exiting fzero: aborting search for an interval containing a
    sign change because NaN or Inf function value encountered
    during search.
(Function value at -1.8203e+154 is Inf.)
Check function or try again with a different starting value.
ans =
   NaN
```

The problem here is that we have a root where the graph is tangent to the x-axis and the initialization function is clearly not able to find an interval with a change in sign. □

Despite all of its weaknesses, the bisection method has the remarkable characteristic that it requires nothing more than the ability to evaluate, or estimate, the function f at a given point. To appreciate this, think of a function defined as a complicated expected value, or a function defined implicitly by an optimization problem:

$$f(x) = \mathrm{E}_\omega\left[F(x,\omega)\right] \quad \text{or} \quad g(x) = \min_{\mathbf{y}\in S} G(x,\mathbf{y}).$$

In both cases, getting more information on f and g (e.g., the value of the derivative, if it exists), may be no easy task. Moreover, bisection does not require the differentiability of the function. On the other hand, it can only be applied to problems in one unknown variable.

3.4.2 Newton's method

Unlike bisection, Newton's method exploits more knowledge of the function f; in particular, it requires computing the first-order derivative of the function f. The method can be applied to solving a system of non-linear equations, but let us first consider Newton's method for the scalar equation

$$f(x) = 0$$

and assume that $f \in C^2$, i.e., is sufficiently well-behaved in terms of continuity and differentiability. Consider a point $x^{(0)}$, which is not a solution of the equation since $f(x^{(0)}) \neq 0$. We would like to move by a step Δx, such that the new point $x = x^{(0)} + \Delta x$ solves the equation, i.e.,

$$f(x^{(0)} + \Delta x) = 0.$$

To obtain the displacement Δx, we may consider the Taylor expansion:

$$f(x^{(0)} + \Delta x) \approx f(x^{(0)}) + f'(x^{(0)})\,\Delta x.$$

Solving this equation for Δx, we get

$$\Delta x = -\frac{f(x^{(0)})}{f'(x^{(0)})}.$$

Since the Taylor expansion is truncated, we will not find a root of the equation in one step, but we may use the idea to define a sequence of points:

$$x^{(k+1)} = x^{(k)} - \frac{f(x^{(k)})}{f'(x^{(k)})}.$$

Geometrically, the method uses the tangent of f in $x^{(k)}$ to improve the estimate of the solution, as shown in figure 3.25. Like any method, Newton's method has strengths and weaknesses:

- Convergence, unlike bisection, is quadratic, which is good news.

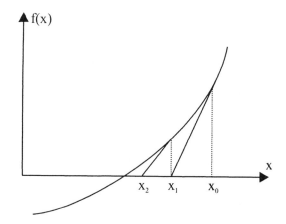

Fig. 3.25 Geometrical illustration of Newton's method.

- The bad news is that convergence is only local: This means that unless you start near the root, the method may fail; homotopy continuation methods (section 3.4.5) are a possible approach to ease this difficulty.

- Many things may go wrong, and stalling may result; in practice, many adjustments are needed to get a robust implementation of this methods.[14]

As an example of application of bisection and Newton's method, we consider next the computation of implicit volatility.

Example 3.23 As we have pointed out in section 2.6.5, sometimes Black–Scholes formula is used in an apparently weird way to find the value of volatility such that the theoretical price predicted by the formula matched the observed price. This is the *implied* volatility. This might be useful in order to estimate volatility as perceived by the market participants rather than using historical data; indeed, this approach has been advocated for VaR calculations.

This is easily accomplished in MATLAB. Consider a call option with strike price $54, expiring in five months, on a stock whose current price is $50, volatility is 30%, when the risk-free interest rate is 7%. Its price is obtained as follows:

```
>> c=blsprice(50, 54, 0.07, 5/12, 0.3)
c =
     2.8466
```

---

[14]For a full treatment of Newton's method, including MATLAB code, see [12].

Now let's go the other way around, and check which volatility would yield this price. We may define an anonymous function handle and find a zero using fzero:

```
>> fzero(@(x) blsprice(50, 54, 0.07, 5/12, x) - 2.8466, 1)
ans =
    0.3000
```

Alternatively, we could use an M-file to define the function.

Since in the Black–Scholes formula we have the option price in analytical terms, one might wonder if it is better to use Newton's method rather than simpler methods such as bisection. This requires computing the derivative of the non-linear function, but this effort could pay off in terms of efficiency. In fact, the Financial toolbox includes a function, blsimpv, which computes the implied volatility of a European call by Newton's method. Its performance may be compared with that of fzero.

```
>> tic, blsimpv(50,54,0.07,5/12,2.8466), toc
ans =
    0.3000
Elapsed time is 0.030920 seconds.
>> tic, fzero(@(x) blsprice(50, 54, 0.07, 5/12, x)-2.8466,1), toc
ans =
    0.3000
Elapsed time is 0.039830 seconds.
```

You see that there is a (small) advantage in using Newton's method. ☐

A significant advantage of Newton's method is that its is immediately generalized to a vector equation such as

$$\mathbf{F}(\mathbf{x}) = \mathbf{0},$$

where $\mathbf{F} = [f_1 \; f_2 \; \cdots \; f_n]'$. Given an approximation $\mathbf{x}^{(k)} = [x_1^{(k)} \; x_2^{(k)} \; \cdots \; x_n^{(k)}]'$ of the root $\mathbf{x}^* = [x_1^* \; x_2^* \; \cdots \; x_n^*]'$, we may write

$$f_1(\mathbf{x}^{(k)}) + (x_1^* - x_1^{(k)}) \left(\frac{\partial f_1}{\partial x_1}\right)_{\mathbf{x}=\mathbf{x}^{(k)}} + \cdots + (x_n^* - x_n^{(k)}) \left(\frac{\partial f_1}{\partial x_n}\right)_{\mathbf{x}=\mathbf{x}^{(k)}} \approx 0$$

$$f_2(\mathbf{x}^{(k)}) + (x_1^* - x_1^{(k)}) \left(\frac{\partial f_2}{\partial x_1}\right)_{\mathbf{x}=\mathbf{x}^{(k)}} + \cdots + (x_n^* - x_n^{(k)}) \left(\frac{\partial f_2}{\partial x_n}\right)_{\mathbf{x}=\mathbf{x}^{(k)}} \approx 0$$

$$\cdots$$

$$f_n(\mathbf{x}^{(k)}) + (x_1^* - x_1^{(k)}) \left(\frac{\partial f_n}{\partial x_1}\right)_{\mathbf{x}=\mathbf{x}^{(k)}} + \cdots + (x_n^* - x_n^{(k)}) \left(\frac{\partial f_n}{\partial x_n}\right)_{\mathbf{x}=\mathbf{x}^{(k)}} \approx 0,$$

which is simply a system of linear equations in which the matrix coefficients form the Jacobian matrix

$$\mathbf{J}^{(k)} = \mathbf{J}(\mathbf{x}^{(k)}) = \begin{bmatrix} \left(\frac{\partial f_1}{\partial x_1}\right)_{\mathbf{x}=\mathbf{x}^{(k)}} & \left(\frac{\partial f_1}{\partial x_2}\right)_{\mathbf{x}=\mathbf{x}^{(k)}} & \cdots & \left(\frac{\partial f_1}{\partial x_n}\right)_{\mathbf{x}=\mathbf{x}^{(k)}} \\ \left(\frac{\partial f_2}{\partial x_1}\right)_{\mathbf{x}=\mathbf{x}^{(k)}} & \left(\frac{\partial f_2}{\partial x_2}\right)_{\mathbf{x}=\mathbf{x}^{(k)}} & \cdots & \left(\frac{\partial f_2}{\partial x_n}\right)_{\mathbf{x}=\mathbf{x}^{(k)}} \\ \vdots & \vdots & \ddots & \vdots \\ \left(\frac{\partial f_n}{\partial x_1}\right)_{\mathbf{x}=\mathbf{x}^{(k)}} & \left(\frac{\partial f_n}{\partial x_2}\right)_{\mathbf{x}=\mathbf{x}^{(k)}} & \cdots & \left(\frac{\partial f_n}{\partial x_n}\right)_{\mathbf{x}=\mathbf{x}^{(k)}} \end{bmatrix}.$$

A sequence of solution estimates is built by solving the linear systems

$$\mathbf{J}^{(k)} \Delta \mathbf{x}^{(k)} = -\mathbf{F}(\mathbf{x}^{(k)})$$

and setting

$$\mathbf{x}^{(k+1)} = \mathbf{x}^{(k)} + \Delta \mathbf{x}^{(k)}.$$

A disadvantage of this approach is that it requires computation of the Jacobian matrix at each step. Coding that may be difficult and error-prone. Hence numerical approximations of the Jacobian are often used, leading to quasi-Newton methods.

3.4.3 Optimization-based solution of non-linear equations

Newton's method and its variants are a possible strategy to solve systems of non-linear equations. However, there are alternative approaches based on optimization. We have already established the connection between optimization and equation solving by the conjugate gradient method in section 3.2.5. When tackling a system of linear equations, like the one we have discussed in the previous section, we may consider the following reformulation:

$$\min \| \mathbf{F}(\mathbf{x}) \|_2^2 = \sum_{i=1}^{n} f_i^2(\mathbf{x}).$$

The idea is illustrated graphically in figure 3.26. Since the squared norm cannot be negative, if we find a minimizer such that the function value is zero, then the minimizer is a root of the equation. This is the approach taken in the MATLAB `fsolve` function; this function, unlike `fzero`, aims at solving systems of linear equations and is part of the Optimization Toolbox, not the MATLAB core. The figure also explains why, in general, finding the whole set of roots is a tough issue, corresponding to a non-convex optimization problem, possibly featuring several minima. The root we find will depend on the starting point. Furthermore, some numerical care is needed as shown in the following example.

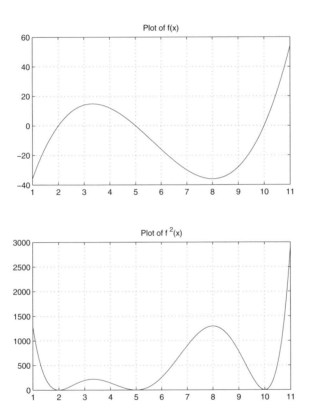

Fig. 3.26 Solving non-linear equations by optimization methods.

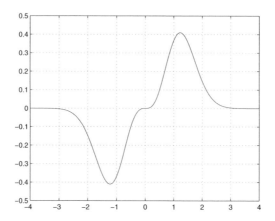

Fig. 3.27 Function for example 3.24.

Example 3.24 To solve the equation

$$x^3 e^{-x^2} = 0$$

we may use `fsolve` as follows. First we define the function (and we plot it, obtaining the graph illustrated in figure 3.27):

```
>> f = @(x) (x.^3).*exp(-x.^2);
>> vx=-4:0.05:4;
>> plot(vx,f(vx))
```

Then we may easily apply `fsolve`, providing a starting point:

```
>> fsolve(f,1)
Optimization terminated:
first-order optimality is less than options.TolFun.
ans =
     0
>> fsolve(f,2)
Optimization terminated:
first-order optimality is less than options.TolFun.
ans =
    3.4891
```

We see that the root we get depends on the starting point, which is expected. Unfortunately, the second point is not an actual root of the equation. Looking at the graph of the function, we may see that for $x \to \pm\infty$ the function *tends* to zero. This implies that we get a numerical "false" zero when the value of the function is smaller than a prescribed tolerance. □

Example 3.25 To illustrate the advantage of quasi-Newton methods, we consider here a classical example in Microeconomics, i.e., the computation of a Cournot equilibrium for a duopoly. For the unfamiliar reader, the problem is finding the two production outputs for two firms, in such a way that no firm would find advantageous to deviate (unilaterally) from that output. The problem each firm faces is that increasing output may increase revenue (the firm sells more) but it may also decrease prices (because of larger availability). Hence, we should look for production quantities maximizing net profit.

The two firms have cost functions:

$$C_i(q_i) = \frac{1}{2} c_i q_i^2, \qquad i = 1, 2,$$

which display increasing marginal cost. We assume the inverse demand function (for the whole market):

$$P(q) = q^{-1/\eta}.$$

This function yields the market price, given the joint supply $q = q_1 + q_2$. The profit for firm i is revenue minus cost:

$$\pi_i(q_1, q_2) = P(q_1 + q_2) q_i - C_i(q_i), \qquad i = 1, 2.$$

To find the Cournot equilibrium, we should enforce the optimality condition of profit for firm 1, as a function of its output q_1, and of profit for firm 2, as a function of q_2. The stationarity condition[15] yields the following set of two non-linear equations:

$$f_i(q) = (q_1 + q_2)^{-1/\eta} - (1/\eta)(q_1 + q_2)^{-1/\eta - 1} q_i - c_i q_i = 0 \qquad i = 1, 2.$$

We also need the Jacobian matrix, and to improve readability it is better to rewrite the function above as

$$f_i(q) = q^e + e q^{e-1} q_i - c_i q_i,$$

where $e = -1/\eta$. Then straightforward calculations yield

$$\frac{\partial f_i}{\partial q_i} = 2 e q^{e-1} + e(e-1) q^{e-2} q_i - c_i$$

$$\frac{\partial f_i}{\partial q_j} = e q^{e-1} + e(e-1) q^{e-2} q_i, \qquad i \neq j.$$

Assume $\eta = 1.6$, $c_1 = 0.6$, and $c_2 = 0.8$. To solve the problem by Newton's method, we need a function computing both the function itself and the Jacobian. This is accomplished by the code displayed in figure 3.28, which also includes a script to call the function.

---

[15] More on this in chapter 6.

```
function [fval,fjac] = cournotJac(q,c,eta)
e = -1/eta;
qtot = sum(q);
fval = qtot^e + e*qtot^(e-1)*q - c.*q;
fjac = zeros(2,2);
fjac(1,1) = 2*e*qtot^(e-1) + e*(e-1)*qtot^(e-2)*q(1) - c(1);
fjac(1,2) = e*qtot^(e-1) + e*(e-1)*qtot^(e-2)*q(2);
fjac(2,1) = e*qtot^(e-1) + e*(e-1)*qtot^(e-2)*q(1);
fjac(2,2) = 2*e*qtot^(e-1) + e*(e-1)*qtot^(e-2)*q(2) - c(2);
```

```
% CournotJacScript
c = [0.6; 0.8];
eta = 1.6;
q0 = [1; 1];
options = optimset('Jacobian', 'on', 'DerivativeCheck', 'on');
[q,fval,exitflag,output] = fsolve(@(q)cournotJac(q,c,eta), q0, options);
fprintf(1,' q1 = %f\n q2 = %f\n', q(1), q(2));
fprintf(1,' number of iterations = %d\n', output.iterations);
```

Fig. 3.28 Code and script for Cournot duopoly.

With optimset we tell MATLAB that we are going to provide the Jacobian, and we ask to check derivatives against a finite difference approximation. Running the script, we get

```
>> CournotJacScript
Maximum discrepancy between derivatives  = 3.12648e-009
Optimization terminated:
first-order optimality is less than options.TolFun.
 q1 = 0.839568
 q2 = 0.688796
 number of iterations = 5
```

It is interesting to note what happens if we introduce an error in the computation of the Jacobian. For instance, if the last line in cournotJac is changed to

```
fjac(2,2) = e*qtot^(e-1) + e*(e-1)*qtot^(e-2)*q(2) - c(2);
```

we get an error message:

```
>> CournotJacScript
Maximum discrepancy between derivatives  = 0.202631
Warning: Derivatives do not match within tolerance
```

```
function [fval,fjac] = cournotNoJac(q,c,eta)
e = -1/eta;
qtot = sum(q);
fval = qtot^e + e*qtot^(e-1)*q - c.*q;
```

```
% CournotNoJacScript
c = [0.6; 0.8];
eta = 1.6;
q0 = [1; 1];
[q, fval, exitflag, output] = fsolve(@(q) cournotNoJac(q,c,eta), q0);
fprintf(1,' q1 = %f\n q2 = %f\n', q(1), q(2));
fprintf(1,' number of iterations = %d\n', output.iterations);
```

Fig. 3.29 Code and script for Cournot duopoly using quasi-Newton method.

```
Derivative from finite difference calculation:
    -0.8406    -0.0380
    -0.0380    -1.0406
User-supplied derivative, @(q) cournotJac(q,c,eta):
    -0.8406    -0.0380
    -0.0380    -0.8380
Difference:
     0.0000     0.0000
     0.0000     0.2026
Strike any key to continue or Ctrl-C to abort
```

To avoid this kind of potential trouble, we may rely on numerical approximations of derivatives. This is easily accomplished by writing a function which does not compute the Jacobian, and by calling `fsolve` with default options. This is accomplished by the function and script in figure 3.29, which is definitely less prone to errors. Running the script, we get

```
>> CournotNoJacScript
Optimization terminated:
first-order optimality is less than options.TolFun.
 q1 = 0.839568
 q2 = 0.688796
 number of iterations = 3
```

We get the same solution, and what looks surprising is that less iterations are reported. Intuitively, we would expect less iterations by providing more information in the form of the Jacobian. However, we are not really using

Newton's method for non-linear equations, and intuition may fail. In fact, the performance of an algorithm depends on many features: `fsolve` is based on a choice of three optimization methods and several options may be selected influencing the number and speed of iterations. □

3.4.4 Putting two things together: solving a functional equation by a collocation method

Assume we have to solve a functional equation of the form

$$g(x, f(x)) = 0 \quad \forall x \in [a, b],$$

where g is given and f is the unknown function. Note that since we want to find a function defined over a real interval, this is an infinite-dimensional problem. The first step to deal with such a problem numerically is to find a suitable way to discretize it. One possibility would be to select a discrete subset of n points x_i in the interval and solve a system of non-linear equations:

$$g(x_i, y_i) = 0, \quad i = 1, \ldots, n,$$

where the unknown is really $y_i = f(x_i)$. Then we may use interpolation to "complete" the function on the whole interval.

However there is a more elegant alternative, known as the collocation method. The idea is still to fix a set of n points, called collocation nodes, and to approximate f by a linear combination of n basis functions:

$$f(x) \approx \sum_{i=1}^{n} c_i \phi_i(x).$$

Then our problem boils down to finding the coefficients c_i by solving a system of non-linear equations:

$$g\left(x_i, \sum_{j=1}^{n} c_j \phi_j(x_i)\right) = 0, \quad i = 1, \ldots, n.$$

We will meet other functional equations in the form of partial differential equations or recursive equations associated to dynamic programming. The collocation method is at the heart of the finite element method for solving PDEs and of some computational approaches to solve stochastic optimization problems by dynamic programming.

3.4.5 Homotopy continuation methods

Since Newton's method is not globally convergent, a good initial guess may be necessary. To overcome this difficulty, and enhance global convergence, we may embed the problem within a parameterized family of problems. Assume that we want to solve the equation $f(x;t) = 0$ for a specific value t^* of the parameter t. If we know that for $t = t^0$ we have a solution x^0, then we may generate a sequence of problems corresponding to parameters $t^0, t^1, t^2, \ldots$, using x^{i-1} as the initial guess for problem i. More generally, if we know a solution of the equation $g(x) = 0$, in order to solve $f(x) = 0$ we may define

$$h(t, x) = tf(x) + (1-t)g(x) \quad (3.11)$$

and "move" t from 0 to 1. In practice we are "deforming" an easy problem into a hard one. This idea may be formalized by a *homotopy*. Given two functions $f, g : X \longrightarrow Y$, a homotopy between f and g is a continuous map

$$h : [0, 1] \times X \longrightarrow Y$$

such that $h(0, x) = g(x)$ and $h(1, x) = f(x)$ Equation (3.11) is the linear homotopy. Newton's homotopy is

$$h(t, x) = tf(x) + (1-t)[f(x) - f(x_0)] = f(x) + (t-1)f(x_0),$$

where x_0 is the solution for $t = 0$.

We have a parameterized family of problems, such that a path of solutions $x(t)$ results. Strictly speaking, this makes sense if $h(t, x) = 0$ has one root for each $t \in [0, 1]$. Assuming this property holds, we must come up with a way to follow the path of solutions, leading to the one we are interested in. In the following example, based on [13, pp. 140–141], we give an idea of a path following strategy.

Example 3.26 Assuming differentiability of the involved functions, we may differentiate the equation

$$h(t, x(t)) = 0$$

and get

$$\frac{\partial h}{\partial t}(t, x(t)) + \frac{\partial h}{\partial x}(t, x(t)) \cdot x'(t) = 0.$$

This yields the following differential equation

$$x'(t) = -[h_x(t, x(t))]^{-1} h_t(t, x(t)),$$

where we have eased the notation by using h_x and h_t to denote partial derivatives. We could integrate this equation, with initial condition $x(0)$, to get the solution $x(1)$.

As a numerical example, consider the following problem, where $X = Y = \mathbb{R}^2$:

$$\mathbf{F}(\mathbf{x}) = \begin{bmatrix} x_1^2 - 3x_2^2 + 3 \\ x_1 x_2 + 6 \end{bmatrix} = \mathbf{0}.$$

Using Newton's homotopy with $\mathbf{x}^0 = (1, 1)$, we have

$$h_x = \frac{\partial \mathbf{F}}{\partial \mathbf{x}} = \begin{bmatrix} \partial f_1/\partial x_1 & \partial f_1/\partial x_2 \\ \partial f_2/\partial x_1 & \partial f_2/\partial x_2 \end{bmatrix} = \begin{bmatrix} 2x_1 & -6x_2 \\ x_2 & x_1 \end{bmatrix}$$

$$h_t = \mathbf{F}(\mathbf{x}^0) = \begin{bmatrix} f_1(\mathbf{x}^0) \\ f_2(\mathbf{x}^0) \end{bmatrix} = \begin{bmatrix} 1 \\ 7 \end{bmatrix}.$$

We may invert h_x:

$$h_x^{-1} = \frac{1}{\Delta} \begin{bmatrix} x_1 & 6x_2 \\ -x_2 & 2x_1 \end{bmatrix}, \qquad \Delta = 2x_1^2 + 6x_2^2.$$

Finally, we get the ordinary differential equations:

$$\begin{bmatrix} x_1' \\ x_2' \end{bmatrix} = -\frac{1}{\Delta} \begin{bmatrix} x_1 & 6x_2 \\ -x_2 & 2x_1 \end{bmatrix} \begin{bmatrix} 1 \\ 7 \end{bmatrix} = -\frac{1}{\Delta} \begin{bmatrix} x_1 + 42x_2 \\ x_2 + 14x_1 \end{bmatrix}.$$

By numerical integration, we get $\mathbf{x}(1) = (-2.961, 1.978)$. Now we are in a neighborhood of the solution of the original equation; to polish the solution, we may take a few iterations of Newton's method, which yields the solution $(-3, 2)$. □

We have included the example above to illustrate the overall idea, but there is a rich set of path following approaches. The same idea can be applied to optimization problems; in fact, we will meet path following again, since it is the foundation of advanced optimization methods such as interior point methods for linear programming (section 6.4.4). The homotopy continuation method is quite sophisticated and powerful; for advanced applications to economics, see [7] and [10].

For further reading

In the literature

- The literature on numerical methods is quite extensive. One classical reference is [18]. Other references are [2], [13], and [17].

- An interesting book on numerical methods from an economist's point of view is [10].

- Splines are dealt with in depth in [5]. They are a widespread tool, both in engineering (e.g., in computer-aided design) and in economics. For a recent application in financial economics, see [11].

- A classical source for special function evaluation is [1].

- Approximation theory is the subject of [15] and [19].

- If you would like a "cookbook" collection of algorithms, [16] is a well-known reference providing many C-language codes implementing numerical methods (a Fortran version is available, too).

- Several numerical analysis books have been written based on MATLAB; see, e.g., [6] and [14].

On the Web

- http://www.netlib.org is a web site offering many pointers to numerical analysis material.

- http://www.mathworks.com/support/books lists several MATLAB-based books, including basic numerical analysis texts.

REFERENCES

1. M. Abramowitz and I.A. Stegun, editors. *Handbook of Mathematical Functions.* Dover Publications, New York, 1972.

2. K.E. Atkinson. *An Introduction to Numerical Analysis (2nd ed.).* Wiley, Chichester, West Sussex, England, 1989.

3. L. Barzanti and C. Corradi. A Note on Interest Rate Term Structure Estimation Using Tension Splines. *Insurance Mathematics and Economics*, 22:139–143, 1998.

4. J.F. Carriere. Nonparametric Confidence Intervals of Instantaneous Forward Rates. *Insurance Mathematics and Economics*, 26:193–202, 2000.

5. C. de Boor. *A Practical Guide to Splines.* Springer-Verlag, New York, 1978.

6. L.V. Fausett. *Applied Numerical Analysis Using MATLAB.* Prentice Hall, Upper Saddle River, NJ, 1999.

7. C.B. Garcia and W.I. Zangwill. *Pathways to Solutions, Fixed Points, and Equilibria.* Prentice Hall, Englewood Cliffs, NJ, 1981.

8. C. Hastings. *Approximations for Digital Computers.* Princeton University Press, Princeton, NJ, 1955.

9. J.C. Hull. *Options, Futures, and Other Derivatives (5th ed.).* Prentice Hall, Upper Saddle River, NJ, 2003.

10. K.L. Judd. *Numerical Methods in Economics*. MIT Press, Cambridge, MA, 1998.

11. K.L. Judd, F. Kubler, and K. Schmedders. Computing Equilibria in Infinite-Horizon Finance Economies: The Case of One Asset. *Journal of Economic Dynamics and Control*, 24:1047–1078, 2000.

12. C.T. Kelley. *Solving Nonlinear Equations with Newton's Method*. SIAM, Philadelphia, PA, 2003.

13. D. Kincaid and W. Cheney. *Numerical Analysis: Mathematics of Scientific Computing*. Brooks/Cole Publishing Company, Pacific Grove, CA, 1991.

14. J.H. Mathews and K.D. Fink. *Numerical Methods Using MATLAB (3rd ed.)*. Prentice Hall, Upper Saddle River, NJ, 1999.

15. M.J.D. Powell. *Approximation Theory and Methods*. Cambridge University Press, Cambridge, 1981.

16. W.H. Press, S.A. Teukolsky, W.T. Vetterling, and B.P. Flannery. *Numerical Recipes in C (2nd ed.)*. Cambridge University Press, Cambridge, 1992.

17. H.R. Schwarz. *Numerical Analysis: A Comprehensive Introduction*. Wiley, Chichester, West Sussex, England, 1989.

18. J. Stoer and R. Burlisch. *Introduction to Numerical Analysis*. Springer-Verlag, New York, 1980.

19. G.A. Watson. *Approximation Theory and Numerical Methods*. Wiley, Chichester, West Sussex, England, 1980.

20. P. Wilmott. *Quantitative Finance (vols. I and II)*. Wiley, Chichester, West Sussex, England, 2000.

4
Numerical Integration: Deterministic and Monte Carlo Methods

Numerical integration is a standard topic in numerical analysis. We have preferred to dedicate a specific chapter to it because of its importance in computational finance. Furthermore, we include topics such as Monte Carlo integration which are not always covered in standard textbooks on numerical analysis. Usually, the term Monte Carlo *simulation* is used, which is somewhat more appealing, but it is important to cast this approach within a numerical integration framework in order to pave the way to quasi-Monte Carlo methods. Classical approaches to numerical integration based on quadrature formulas are deterministic, just as quasi-Monte Carlo methods. Monte Carlo methods are based on random sampling, at least conceptually, and so some connection with statistics is expected.

We have seen that option pricing requires computing an expected value under a risk-neutral measure, but an expected value is actually an integral. The expected value of a function $g(\cdot)$ of a random variable X with probability density $f_X(x)$ is

$$\mathrm{E}[g(X)] = \int_{-\infty}^{+\infty} g(x) f_X(x)\, dx.$$

In one-dimensional cases, we may find an analytical solution, like in the Black–Scholes case, but this is difficult in general. If the random variable X is a scalar, classical deterministic methods work quite well, but when expectation is taken with respect to a random vector and we must integrate over a high-dimensional space, random sampling may be necessary. Random sampling is a natural way to simulate dynamics affected by uncertainty, such as prices modeled by stochastic differential equations. Natural applications, apart from option pricing, are portfolio optimization, risk management, and estimation of Value at Risk.

It is worth noting that numerical integration may be implicitly used to estimate probabilities. If A is an event which may occur or not depending on a random variable X, then

$$P(A) = \int_{-\infty}^{+\infty} \mathcal{I}_A(x) f_X(x)\, dx,$$

where $\mathcal{I}_A(x)$ is the indicator function for event A (taking the value 1 if A occurs when $X = x$, 0 otherwise). When A is a rare event, clever strategies are needed to get an accurate estimate with a reasonable computational effort.

Finally, there are situations in which we define a function by an integral. A typical case is the expected value of a function depending on a control variable (modeling our decisions) and a random variable (modeling what we cannot control):

$$H(z) = E_X[g(X, z)] = \int_{-\infty}^{+\infty} g(x, z) f_X(x)\, dx.$$

This is quite common in stochastic optimization and dynamic programming, whereby we want to find a maximizer (or minimizer) of $H(z)$, and this calls for a suitable approximation of H by discretization of the continuous distribution. In other words, we want to generate a discrete set of scenarios yielding a reasonable approximation of the underlying uncertainty. Numerical methods such as Gaussian quadrature are helpful here. Indeed, all numerical integration methods require some form of discretization, or sampling, via regular grids or other mechanisms. We should also note that we may also be interested in the derivative of $H(z)$, not only for optimization purposes, but also to evaluate sensitivities. A familiar case is computing the Greeks of an option.

We start the chapter with a very brief overview of classical deterministic quadrature in section 4.1. We will just present very basic approaches in order to point out the conceptual basis of quadrature functions available in MATLAB. We will also deal with Gaussian quadrature because of its importance in computational dynamic programming.

Then we introduce Monte Carlo integration in section 4.2. Monte Carlo simulation is based on random number generation; actually, we must speak of *pseudorandom* numbers, since nothing is random on a computer. How this is accomplished is described in section 4.3.

If we feed random numbers into a simulation procedure, the output will be a sequence of random numbers. Given this output, we use statistical techniques to build an estimate of a quantity of interest. We would like to evaluate the reliability of this estimate in some way, e.g., by a confidence interval, or the other way around, we would like to carry out the simulation experiments in such a way that the estimation error is controlled. Section 4.4 deals with the issue of setting the number of simulation experiments (replications) properly. Intuitively, the more replications we run, the more reliable our estimates will be. Unfortunately, reaching a suitable precision might require a prohibitive number of experiments. Improving the quality of the estimates without incurring huge CPU times calls for proper variance reduction techniques, which are the subject of section 4.5. Using pseudorandom numbers on a computer and then applying statistical techniques may raise some philosophical issues; after all, the sequences of numbers we use are deterministic. It can be argued that the success of Monte Carlo simulation simply shows that there are some deterministic sequences that work well and that there could be others that work even better. Pursuing this idea leads to quasi-Monte Carlo simulation, which is dealt with in section 4.6.

A final consideration is that simulation may be used to evaluate the consequences of a certain policy, but it cannot generate the policy itself. To this end, we should use the optimization methods which will be described in chapter 6. Unfortunately, most of those techniques require an analytical model that may be too complex or not available at all, which is the very reason why we resort to simulation so often. Possible ways to couple simulation and optimization techniques are described in section 6.6.

In order to better illustrate the material we will use simple examples from elementary integration and pricing of vanilla options. We should bear in mind that for those vanilla options analytical formulas are available, and that our examples are just illustrative. We will consider practically relevant cases in chapter 8.

4.1 DETERMINISTIC QUADRATURE

Consider the problem of approximating the value of a definite integral like

$$I[f] = \int_a^b f(x) \, dx$$

over a bounded interval $[a, b]$ for a function f of a single variable. Since the integration is a linear operator, it is natural to look for an approximation preserving this property. Using a finite number of values of f over a set of nodes x_j such that

$$a = x_0 < x_1 < \cdots < x_N = b,$$

we may define a quadrature formula such as

$$Q[f] = \sum_{j=0}^{n} w_j f(x_j).$$

A quadrature formula is characterized by the weights w_j and by the nodes x_j. To be precise, a quadrature formula like the one we are describing is called a *closed* formula, since evaluation of the function in the extreme points of the interval is used. Sometimes, open formulas are used when the function is not well-behaved near a or b, or when we are integrating on an infinite interval.

Any quadrature formula is characterized by a truncation error:

$$E = I[f] - Q[f].$$

A reasonable requirement is that the error should be zero for sufficiently simple functions such as polynomials. We may define the order of a certain quadrature formula as the maximum degree m such that the truncation error is zero for all the polynomials of degree m or less. In other words, if the original function is substituted by an interpolating polynomial, we should not commit any error in integrating the polynomial. It is quite common to see expressions for the truncation error like

$$E = \gamma f^{(k)}(\xi),$$

where γ is some constant depending, among other things, on a and b, ξ is some unknown point in the interval (a, b), and k is the order of some derivative. Since the derivative of order k is zero for a polynomial of degree not exceeding $k - 1$, there is clear link between k and the order of the quadrature formula. If the function f is smooth enough, we may hope that high order translates to high accuracy.

4.1.1 Classical interpolatory formulas

One way to derive quadrature formula is to consider equally spaced nodes:

$$x_j = a + jh, \qquad j = 0, 1, 2, \ldots, n,$$

where $h = (b - a)/n$; also let $f_j = f(x_j)$. We have seen in function interpolation that this choice need not be the best one, but it is a natural starting point. Selecting equally spaced nodes yields the set of Newton–Cotes quadrature formulas.

Given those $n + 1$ nodes, we may consider the interpolating polynomial $P_n(x)$ using Lagrange polynomials of degree n:

$$f(x) \approx P_n(x) = \sum_{j=0}^{n} f_j L_j(x).$$

Then we may compute the correct weights as follows:

$$\begin{aligned}
\int_a^b f(x)\,dx &\approx \int_a^b P_n(x)\,dx = \int_a^b \left[\sum_{j=0}^{n} f_j L_j(x)\right] dx \\
&= \sum_{j=0}^{n} f_j \left[\int_a^b L_j(x)\,dx\right] = \sum_{j=0}^{n} w_j f_j.
\end{aligned}$$

DETERMINISTIC QUADRATURE

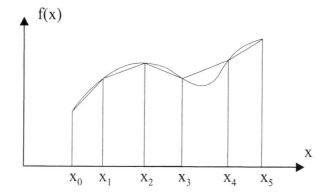

Fig. 4.1 Example of the trapezoidal quadrature formula.

Consider the case of two nodes only, $x_0 = a$ and $x_1 = b$. Here we are just interpolating f by a straight line:

$$P_1(x) = f_0 \frac{x - x_1}{x_0 - x_1} + f_1 \frac{x - x_0}{x_1 - x_0}.$$

A straightforward calculation yields

$$\int_{x_0}^{x_1} P_1(x)\,dx = h \frac{f_1 + f_0}{2}.$$

Actually, what we are saying is that we may approximate the area below the function using trapezoidal elements, as depicted in figure 4.1, and the formula above gives the area of one element. Applying the idea to more subintervals, we get the trapezoidal rule:

$$Q[f] = h \left[\frac{1}{2} f_0 + \sum_{j=1}^{n-1} f_j + \frac{1}{2} f_n \right].$$

Given any quadrature formula for an interval, we may get a *composite* formula by applying the same pattern to small subintervals of a large one.

A quadrature formula based on $n + 1$ nodes is by construction exact for polynomials of degree $\leq n$. We may go the other way around, and build a formula by requiring a certain order. Consider the case

$$\int_0^1 f(x)\,dx \approx w_0 f(0) + w_1 f(0.5) + w_2 f(1),$$

and say we want a formula that is exact for polynomials of degree ≤ 2. Having fixed the nodes, we may find the weights by solving the system of linear equations:

$$1 = \int_0^1 dx = w_0 + w_1 + w_2$$

$$\frac{1}{2} = \int_0^1 x\, dx = \frac{1}{2}w_1 + w_2$$

$$\frac{1}{3} = \int_0^1 x^2\, dx = \frac{1}{4}w_1 + w_2,$$

which yields $w_0 = 1/6$, $w_1 = 2/3$, $w_2 = 1/6$. Applying the same idea on the interval $[a, b]$, we get Simpson's rule:

$$\int_a^b f(x)\, dx \approx \frac{b-a}{6}\left[f(a) + 4f\left(\frac{b+a}{2}\right) + f(b)\right].$$

It fairly easy to see that, somewhat surprisingly, this formula is actually exact for polynomials of degree ≤ 3. In fact, we have

$$\int_a^b x^3 dx = \frac{b^4 - a^4}{4}.$$

Applying Simpson's rule we have, by some straightforward algebra:

$$\frac{b-a}{6}\left[a^3 + 4\left(\frac{b+a}{2}\right)^3 + b^3\right]$$

$$= \frac{b-a}{6}\left[a^3 + \frac{1}{2}\left(a^3 + 3a^2b + 3ab^2 + b^3\right) + b^3\right] = \frac{b^4 - a^4}{4}$$

Simpson's rule may be applied to subintervals of (a, b) in order to get a composite formula.

It is important to see the connection between the approach we have just pursued and the idea of moment matching in probability. We may discretize a continuous probability distribution in such a way that the discrete distribution matches moments of the continuous distribution, e.g., expected value and variance. This idea is used to approximate stochastic processes, such as geometric Brownian motion, by binomial and trinomial lattices and it will be pursued in chapter 7. Now, what we have seen is that for given nodes we may find suitable weights to obtain a quadrature formula with desired order. We have also seen that in function interpolation equispaced nodes need not be the best choice. Generalizing the idea we should wonder if there is a way to find weights and nodes jointly, in order to obtain a quadrature formula of maximal order. This idea leads to Gaussian quadrature formulas.

4.1.2 Gaussian quadrature

In Newton–Cotes formulas we fix nodes and try to find suitable weights so that the order of the formula is as large as possible. The rationale behind

Gaussian quadrature is that if we do not fix nodes a priori, we essentially double the degrees of freedom, in such a way that the order can be more or less doubled. Furthermore, Gaussian quadrature formulas are developed with respect to a non-negative weight function $w(x)$. We look for a quadrature formula like

$$\int_a^b w(x) f(x) \, dx \approx \sum_{i=1}^n w_i f(x_i), \qquad (4.1)$$

which is exact when f is a polynomial. Note that in this section, unlike the previous one, it is convenient to consider n nodes x_i, $i = 1, \ldots, n$. The weight function $w(x)$ can be used to encapsulate undesired singularities of the integrand function. In our setting, $w(x)$ will be interpreted as a probability density. In fact we will only outline the development of Gauss–Hermite quadrature, where $w(x) = e^{-x^2}$, and there is a clear connection with computing the expected value of a function of a normal random variable.

Let Y be a random variable with normal distribution $\mathcal{N}(\mu, \sigma^2)$. Then

$$\mathrm{E}[f(Y)] = \int_{-\infty}^{+\infty} \frac{1}{\sqrt{2\pi}\sigma} e^{-\frac{1}{2}\left(\frac{y-\mu}{\sigma}\right)^2} f(y) \, dy.$$

In order to use weights and nodes from a Gauss–Hermite formula, we need the following change of variable

$$-x^2 = -\frac{1}{2}\left(\frac{y-\mu}{\sigma}\right)^2 \;\Rightarrow\; y = \sqrt{2}\sigma x + \mu \;\Rightarrow\; \frac{dy}{\sqrt{2}\sigma} = dx.$$

Hence

$$\mathrm{E}[f(Y)] \approx \frac{1}{\sqrt{\pi}} \sum_{i=1}^n w_i f(\sqrt{2}\sigma x_i + \mu).$$

Now, how should we select the nodes and weights in (4.1) in order to get a quadrature formula with maximum order? We should choose as nodes the n roots of a polynomial of order n, selected within a family of orthogonal polynomials with respect to the inner product (see also section 3.3.4):

$$<f, g> = \int_a^b w(x) f(x) g(x) \, dx.$$

It can be shown that a polynomial of degree k within that family has k distinct real roots. Furthermore, these roots are interleaved, in the sense that each of the $k-1$ roots of the polynomial of degree $k-1$ lies in an interval defined by a pair of consecutive roots of the polynomial of degree k. Using this choice of nodes, along with a proper choice of weights, yields a quadrature formula with order $2n - 1$. To see this, consider a polynomial $q \in \Pi_n$, i.e., a polynomial of degree n, which is orthogonal to all polynomials in Π_{n-1}:

$$\int_a^b w(x) q(x) p(x) \, dx = 0 \qquad \forall p \in \Pi_{n-1}.$$

Any polynomial $f \in \Pi_{2n-1}$ can be divided by q, obtaining a quotient p and a remainder r:
$$f = qp + r,$$
where $p, r \in \Pi_{n-1}$. Now let us integrate wf by a quadrature formula on n nodes x_i, $i = 1, \ldots, n$, which are the zeros of q:

$$\int_a^b w(x)f(x)\,dx$$
$$= \int_a^b w(x)p(x)q(x)\,dx + \int_a^b w(x)r(x)\,dx \quad \text{(division)}$$
$$= 0 + \int_a^b w(x)r(x)\,dx \quad \text{(q is orthogonal to p)}$$
$$= \sum_{i=1}^n w_i r(x_i) \quad \text{(quadrature is exact for $r \in \Pi_{n-1}$)}$$
$$= \sum_{i=1}^n w_i f(x_i) \quad \text{(x_i is a zero of q)}.$$

A family of orthogonal polynomials $p_j(x)$ may be built by the following procedure:

$$p_{-1}(x) = 0$$
$$p_0(x) = 1$$
$$p_{j+1}(x) = (x - a_j)p_j(x) - b_j p_{j-1}(x), \quad j = 0, 1, 2, 3, \ldots,$$

where

$$a_j = \frac{<xp_j, p_j>}{<p_j, p_j>}, \quad j = 0, 1, 2, \ldots$$
$$b_j = \frac{<p_j, p_j>}{<p_{j-1}, p_{j-1}>}, \quad j = 1, 2, \ldots.$$

Here coefficient b_0 is arbitrary and it can be set to 0. At each step, the procedure generates a new polynomial of degree one plus the degree of the previous polynomial. In the end, we have a family of orthogonal polynomials, one for each degree. Actually there are different choices of normalizations yielding different families of polynomials.

In the Gauss–Hermite case, whereby $w(x) = e^{-x^2}$, applying the procedure above results in the following recursive procedure yielding a sequence of Hermite polynomials H_j:

$$H_{j+1} = 2xH_j - 2jH_{j-1}.$$

It is worth noting that this procedure is not quite numerically viable, as it implicitly computes factorials which tend to overflow for large n. This is why a different normalization can be used, yielding a family of orthonormal polynomials[1]:

$$\tilde{H}_{-1} = 0$$
$$\tilde{H}_0 = \frac{1}{\pi^{1/4}} \qquad (4.2)$$
$$\tilde{H}_{j+1} = x\sqrt{\frac{2}{j+1}}\,\tilde{H}_j - \sqrt{\frac{j}{j+1}}\,\tilde{H}_{j-1} \qquad j = 0, 1, 2, 3, \ldots.$$

In order to select weights, one possibility would be to require exact integration for the first n polynomials in the family, including the polynomial of degree 0. Since $p_0(x) \equiv 1$, this means that the (weighted) integrals of $p_j(x)$, $j = 1, \ldots, n-1$ should be zero, since they are all orthogonal to $p_0(x)$. These conditions yield the following system of linear equations:

$$\begin{bmatrix} p_0(x_1) & \cdots & p_0(x_n) \\ p_1(x_1) & \cdots & p_1(x_n) \\ \vdots & \ddots & \vdots \\ p_{n-1}(x_1) & \cdots & p_{n-1}(x_n) \end{bmatrix} \begin{bmatrix} w_1 \\ w_2 \\ \vdots \\ w_n \end{bmatrix} = \begin{bmatrix} \int_a^b w(x)\,dx \\ 0 \\ \vdots \\ 0 \end{bmatrix}$$

It can be shown that a possibly more convenient way of getting weights is by using the following recursion:

$$w_j = \frac{<p_{n-1}, p_{n-1}>}{p_{n-1}(x_j)p'_n(x_j)},$$

where $p'_n(x_j)$ is the derivative of the polynomial. In the Gauss–Hermite case, using the orthonormal set of polynomials, this boils down to:

$$w_j = \frac{2}{\left(\tilde{H}'_n(x_j)\right)^2}$$

where the derivative of polynomial j is

$$\tilde{H}'_j = \sqrt{2j}\,\tilde{H}_{j-1}.$$

MATLAB code to implement Gauss–Hermite quadrature is displayed in figure 4.2. Polynomials are stored in vectors; HPoly1, HPoly2, and HPoly3 play the roles of polynomials $\tilde{H}_{j-1}$, $\tilde{H}_j$, and $\tilde{H}_{j+1}$ in recursion (4.2), respectively. In the for loop, we must pay attention to i, since there is the typical shift in index values because of the MATLAB convention (array indexing starts from 1). On exit from the loop, HPoly3 contains $\tilde{H}_n$ and HPoly1 contains $\tilde{H}_{n-1}$. In computing roots, we use the standard roots function. This need not be

---

[1]See, e.g., [13, pp. 150–154].

```
function [x,w] = GaussHermite(mu,sigma2,N)
HPoly1 = [ 1/pi^0.25 ];
HPoly2 = [sqrt(2) / pi^0.25, 0];
for j=1:N-1
    HPoly3 = [sqrt(2/(j+1))*HPoly2 , 0] - [0, 0, sqrt(j/(j+1)*HPoly1];
    HPoly1 = HPoly2;
    HPoly2 = HPoly3;
end
x1 = roots(HPoly3);
w1 = zeros(N,1);
for i=1:N
    w1(i) = 1/(N)/(polyval(HPoly1, x1(i)))^2;
end
[x, index] = sort(x1*sqrt(2*sigma2)+mu);
w = w1(index)/sqrt(pi);
```

Fig. 4.2 Code to implement Gauss–Hermite quadrature.

the best approach, as using the interleaving property one can compute roots for each polynomial in the sequence by the Newton's method, using previous roots for initialization.[2] The last two lines are used to sort nodes in increasing order, and the `index` vector is used to sort weights accordingly.

It is interesting to check the weights and nodes we get from this function. For instance, let us consider a normal random variable with $\mu = 10$ and $\sigma^2 = 20$, and let us apply a quadrature formula based on five nodes:

```
>> [x,w] = GaussHermite(10,20,5)
x =
    -2.7768
     3.9375
    10.0000
    16.0625
    22.7768
w =
     0.0113
     0.2221
     0.5333
     0.2221
     0.0113
>> sum(w)
```

---

[2]This is the approach taken in [13]. A MATLAB implementation, which generalizes to multidimensional integration, can be found in the Computational Economics Toolbox described in [10].

```
% GHScript.m
N = [5, 10, 15, 20];
mu = 4;
sigma2 = 4;
TrueValue = exp(mu+0.5*sigma2);
for i=1:length(N)
    [x,w] = GaussHermite(mu,sigma2,N(i));
    ApproxValue = dot(w,exp(x));
    fprintf(1,'N=%2d True=%g Approx=%g PercError=%g \n', N(i), ...
        TrueValue, ApproxValue, abs(TrueValue-ApproxValue)/TrueValue);
end
```

Fig. 4.3 Script to check Gauss–Hermite quadrature.

ans =
    1.0000

Nodes, as expected, are symmetrically centered around the expected value; furthermore, the sum of weights is 1, which is only convenient, since this should be a discretization of a continuous distribution. As a complete example, we may deal with the case of integrating an exponential function. From the properties of the lognormal distribution (see section B.2.1) we know that if $X \sim \mathcal{N}(\mu, \sigma^2)$, then

$$E[e^X] = e^{\mu+\sigma^2/2}$$

A script to check this is displayed in figure 4.3. Running the script, we may see that remarkable precision is obtained with a fairly modest number of nodes:

```
>> GHScript
N= 5 True=403.429 Approx=398.657 PercError=0.0118287
N=10 True=403.429 Approx=403.429 PercError=5.53771e-007
N=15 True=403.429 Approx=403.429 PercError=1.90343e-012
N=20 True=403.429 Approx=403.429 PercError=3.95931e-014
```

Actually, the number of nodes needed to obtain a suitable accuracy depends on variance. The reader is urged to write a function pricing a vanilla European call option using Gauss–Hermite quadrature and to compare the result with blsprice.

4.1.3 Extensions and product rules

The interpolatory rules of section 4.1.1 are extended in many ways, which we just outline here. To begin with, nodes should be added dynamically until a prespecified accuracy is obtained. This can done according to clever strategies in order to avoid unnecessary function re-evaluations. This leads to recursive quadrature formulas and to Romberg integration. Furthermore, the choice of the nodes may be improved by adapting it to the function characteristics; more nodes are needed where there is more variation, and less are needed where the function is more "constant"; this leads to adaptive quadrature formulas. All these improvements are exploited in scientific libraries, including MATLAB functions.

Product rules are used when we want to extend quadrature formulas to multidimensional integration. Suppose we want to compute an integral on the unit hypercube

$$\int_{[0,1]^d} f(\mathbf{x})\, d\mathbf{x},$$

where $[0,1]^d = [0,1] \times [0,1] \times \cdots \times [0,1]$, and that we have weights and nodes for, say, a Newton–Cotes quadrature formula along each dimension; more precisely, for dimension k, $k = 1, \ldots, d$, we have weights w_i^k and nodes x_i^k, $i = 1, \ldots, m_k$. A product rule approximates the integral above as

$$\sum_{i_1=1}^{m_1} \sum_{i_2=1}^{m_2} \cdots \sum_{i_d=1}^{m_d} w_{i_1}^1 w_{i_2}^2 \cdots w_{i_d}^d f\left(x_{i_1}^1, x_{i_2}^2, \ldots, x_{i_d}^d\right).$$

A product rule builds nodes taking the Cartesian product of node sets along each dimension. It is easy to see that this regular grid is going to be impractical for large d, and this motivates Monte Carlo integration based on random sampling.

4.1.4 Numerical integration in MATLAB

There are a few MATLAB functions to compute one-dimensional integrals. They are based on refinements of basic schemes, such as adaptive extensions of Simpson's rule.

Example 4.1 Consider the integral

$$I = \int_0^{2\pi} e^{-x} \sin(10x)\, dx.$$

Integration by parts yields

$$I = -\frac{1}{101} e^{-x} \left[\sin(10x) + 10\cos(10x)\right]\Big|_0^{2\pi} \approx 0.0988.$$

Using the quad function, we get

```
>> f=@(x) exp(-x).*sin(10*x)
f =
    @(x) exp(-x).*sin(10*x)
```

```
>> quad(f,0,2*pi)
ans =
    0.0987
```

Precision may be improved by specifying a tolerance parameter:

```
>> quad(f,0,2*pi, 10e-6)
ans =
    0.0987
>> quad(f,0,2*pi, 10e-8)
ans =
    0.0988
```

We may also adopt alternative strategies, based on adaptive Lobatto quadrature:

```
>> quadl(f,0,2*pi)
ans =
    0.0988
```

☐

MATLAB also provides us with functions for multidimensional integration. In the bidimensional case, `dblquad` can be used, whereas `triplequad` is used for triple integrals. Actually, the latter is a relatively recent addition and was not available in earlier MATLAB versions. You can see that we cannot go beyond three dimensions. This is due to the intrinsic difficulty of using regular grids when we integrate in several dimensions. The typical way to avoid this difficulty is resorting to random sampling.

4.2 MONTE CARLO INTEGRATION

The definite integral of a function is a number, and computing that number is a deterministic problem involving no randomness. Nevertheless, we may cast the problem within a stochastic framework by interpreting the integral as an expected value. Consider an integral on the unit interval $[0, 1]$:

$$I = \int_0^1 g(x)\,dx.$$

We may think of this integral as the expected value $\mathrm{E}[g(U)]$, where U is a uniform random variable on the interval $(0, 1)$, i.e., $U \sim (0, 1)$. We may estimate the expected value (a number) by a sample mean (a random variable). What we have to do is generating a sequence $\{U_i\}$ of *independent* random samples from the uniform distribution and then evaluate the sample mean:

$$\hat{I}_m = \frac{1}{m} \sum_{i=1}^m g(U_i).$$

The strong law of large numbers implies that, with probability 1,

$$\lim_{m \to \infty} \hat{I}_m = I.$$

Random sampling, which is where "Monte Carlo" comes from, is not really possible with a computer, but we can generate a sequence of *pseudo*-random numbers using generators provided by most programming languages and environments.

Example 4.2 Consider the trivial case

$$I = \int_0^1 e^x \, dx = e - 1 \approx 1.7183.$$

To generate uniformly distributed random numbers, we may use the MATLAB rand function; a call like rand(m,n) yields a $m \times n$ matrix of uniform random numbers. Please note that the parameters m and n have nothing to do with the distribution, which is $\mathcal{U}(0,1)$ anyway. We can see the reliability of our estimates as a function of the sample size m:

```
>> rand('state', 0)
>> mean(exp(rand(1,10)))
ans =
    1.8318
>> mean(exp(rand(1,10)))
ans =
    2.0358
>> mean(exp(rand(1,10)))
ans =
    1.3703
>> mean(exp(rand(1,1000000)))
ans =
    1.7189
>> mean(exp(rand(1,1000000)))
ans =
    1.7178
>> mean(exp(rand(1,1000000)))
ans =
    1.7174
```

In order to understand the role of the command rand('state',0), we should consider how "random" numbers are generated on a computer. For now it is enough to say that the command resets the generator so that the experiment can be replicated obtaining the same results. We see that the estimate is not quite reliable for $m = 10$, whereas variance of the estimator is much lower when $m = 1{,}000{,}000$, and the result is close to the correct number. Needless to say, we do not know the exact result in practice, and we should wonder how to qualify the reliability of the estimate, and how to improve it. □

For one-dimensional integration, Monte Carlo is hardly competitive with deterministic quadrature, but when computing a multidimensional integral it may be the only viable option. In general, if we have an integral like

$$I = \int_{\mathcal{A}} \phi(\mathbf{x}) \, d\mathbf{x}, \tag{4.3}$$

where $\mathcal{A} \subset \mathbb{R}^n$, we may estimate I by randomly sampling a sequence of points $\mathbf{x}^i \in \mathcal{A}$, $i = 1, \ldots, m$, and building the estimator

$$\hat{I}_m = \frac{\text{vol}(\mathcal{A})}{m} \sum_{i=1}^{m} \phi(\mathbf{x}_i), \tag{4.4}$$

where vol($\mathcal{A}$) denotes the volume of the region $\mathcal{A}$. To understand the formula, we should think that the ratio $(1/m) \sum_{i=1}^{m} \phi(\mathbf{x}_i)$ estimates the average value of the function, which must be multiplied by the volume of the integration region in order to get the integral.

We will see that in practice we need only to integrate over the unit hypercube, i.e.,

$$\mathcal{A} = [0,1] \times [0,1] \times \cdots \times [0,1],$$

hence vol($\mathcal{A}$) = 1. Considering the unit hypercube looks restrictive. In general, we have a vector random variable

$$\mathbf{X} = \begin{bmatrix} X_1 \\ X_2 \\ \vdots \\ X_n \end{bmatrix},$$

with joint density function $f(x_1, \ldots, x_n)$, and we use Monte Carlo integration to estimate the expected value of an arbitrary function of $\mathbf{X}$:

$$\mathrm{E}[g(\mathbf{X})] = \int \int \cdots \int g(x_1, \ldots, x_n) f(x_1, \ldots, x_n) \, dx_1 \cdots dx_n.$$

MATLAB provides us with many functions to generate random variables, but we will see that the primary input is always a stream of uniform random numbers $U \sim \mathcal{U}(0,1)$. These generators are actually part of the Statistics Toolbox, but the core MATLAB environment also offers a function (**randn**) to sample the standard normal distribution. Using this function, we may use Monte Carlo integration to price a vanilla call option.

Example 4.3 We know that the price of a European style option is the expected value, under the risk-neutral measure, of the discounted payoff of the option:

$$f = e^{-rT} \hat{\mathrm{E}}[f_T],$$

```
% BlsMC1.m
function Price = BlsMC1(S0,K,r,T,sigma,NRepl)
nuT = (r - 0.5*sigma^2)*T;
siT = sigma * sqrt(T);
DiscPayoff = exp(-r*T)*max(0, S0*exp(nuT+siT*randn(NRepl,1))-K);
Price = mean(DiscPayoff);
```

Fig. 4.4 Code to price a vanilla European call by Monte Carlo simulation.

where f_T is the payoff at the maturity date T and a constant risk-free rate r is assumed. The notation $\hat{E}[\cdot]$ is used to emphasize that expectation is taken with respect to the risk-neutral measure. If we assume geometric Brownian motion, this means that the drift μ for the asset price must be replaced by the risk-free rate r (see section 2.6). Depending on the nature of the option at hand, we may need to generate the full sample paths, or simply the terminal asset price. Path generation will be dealt with in chapter 8, but a vanilla call option requires just sampling the payoff $\max\{0, S(T) - K\}$, where $S(T)$ is the price of the underlying asset at maturity and K is the strike price. From example 2.20 on page 98, we know that we may easily accomplish this by generating a standard normal random variable $\epsilon \sim \mathcal{N}(0,1)$:

$$f_T = \max\{0, S(0)e^{(r-\sigma^2/2)T+\sigma\sqrt{T}\epsilon} - K\}.$$

A MATLAB function to price the call option is displayed in figure 4.4. The first five input parameters are self-explanatory and are those required by the blsprice function implementing Black–Scholes formula. The last parameter NRepl is the number of replications, i.e., samples we want to take. We may check the impact of this parameter:

```
>> S0=50;
>> K=60;
>> r=0.05;
>> T=1;
>> sigma=0.2;
>> randn('state', 0)
>> BlsMC1(S0,K,r,T,sigma,1000)
ans =
    1.2562
>> BlsMC1(S0,K,r,T,sigma,1000)
ans =
    1.8783
>> BlsMC1(S0,K,r,T,sigma,1000)
ans =
    1.7864
```

```
>> BlsMC1(S0,K,r,T,sigma,1000000)
ans =
    1.6295
>> BlsMC1(S0,K,r,T,sigma,1000000)
ans =
    1.6164
>> BlsMC1(S0,K,r,T,sigma,1000000)
ans =
    1.6141
```

As before, we reset first the state of the randn generator, so that the experiment can be repeated by the reader. With only 1000 samples, we see quite some variability in the estimate, which starts looking reasonable when the number of samples is increased considerably. Clearly, we cannot yield just a point estimate: we should also compute some confidence interval for the estimate. Possibly, we should understand how many samples are needed in order to attain a given precision. Another point is that too many samples are needed; things may be worse with higher volatility, and with complex path-dependent options we cannot afford taking a huge number of samples. Hence we need clever ways to reduce the variance of the estimator. ☐

Needless to say, the example above is presented for illustrative purposes, as there is no need to resort to Monte Carlo simulation to price a vanilla European-style call option. What we need is a numerical approximation of the integrals involved in the cumulative distribution function for standard normal random variables. Nevertheless, we will see that pricing "easy" options by simulation may be useful in variance reduction by control variates.

4.3 GENERATING PSEUDORANDOM VARIATES

The usual way to generate pseudorandom variates, i.e., samples from a given probability distribution, starts from the generation of pseudorandom numbers, which are simply variates from the uniform distribution on the interval (0,1). Then, suitable transformations are applied in order to obtain the desired distribution. We discuss briefly the most common transformations: the inverse transform method, the acceptance–rejection approach, and ad hoc strategies such as those used to generate standard normal variates. The MATLAB Statistics toolbox provides the user with a rich library of random variate generators, so that the user need not herself program the procedures we describe in the following. Nevertheless, we believe it is important to have at least a grasp of what is done, in order to properly apply variance reduction procedures to improve the estimates.

```
function [USeq, ZSeq] = LCG(a,c,m,seed,N)
ZSeq = zeros(N,1);
USeq = zeros(N,1);
for i=1:N
    seed = mod(a*seed+c, m);
    ZSeq(i) = seed;
    USeq(i) = seed/m;
end
```

Fig. 4.5 Code to generate random numbers by a linear congruential generator.

4.3.1 Generating pseudorandom numbers

The standard textbook method to generate $\mathcal{U}(0,1)$ variates, is based on linear congruential generators (LCGs). A LCG generates a sequence of non-negative integer numbers Z_i as follows; given an integer number Z_{i-1}, we generate the next number in the sequence by computing

$$Z_i = (aZ_{i-1} + c) \bmod m,$$

where a (the multiplier), c (the shift), and m (the modulus) are properly chosen parameters and mod denotes the remainder of integer division (e.g., 15 mod 6 = 3). Then, to generate a uniform variate on the unit interval, we return the number (Z_i/m).

Example 4.4 In figure 4.5 we display MATLAB code to implement a LCG. Running the code with some choice of the parameters a, c, and m yields

```
>> a=5;
>> c=3;
>> m=16;
>> seed=7;
>> N=20;
>> [USeq, ZSeq] = LCG(a,c,m,seed,N);
>> fprintf(1,'%2d %2d %6.4f \n', [(1:N)', ZSeq, USeq]')
 1  6 0.3750
 2  1 0.0625
 3  8 0.5000
 4 11 0.6875
 5 10 0.6250
 6  5 0.3125
 7 12 0.7500
 8 15 0.9375
 9 14 0.8750
10  9 0.5625
```

```
11  0  0.0000
12  3  0.1875
13  2  0.1250
14 13  0.8125
15  4  0.2500
16  7  0.4375
17  6  0.3750
18  1  0.0625
19  8  0.5000
20 11  0.6875
```
                                                                    ☐

It is clear that there is nothing random in the sequence generated by a LCG. To begin with, it must start from an initial number Z_0; this is called the *seed* of the sequence. Starting the sequence from the same seed will always yield the same sequence. Indeed, any time you start MATLAB and type rand, you get the same number; if you keep typing rand, you see a sequence of numbers that look random and uniformly distributed. However, this sequence is always the same, since starting MATLAB sets the seed to a precise value. This may seem rather dull, and using a command like

rand('seed',sum(100*clock)),

which sets the seed of the random generator to a number depending on the current clock value, may seem a brilliant idea. In practice this is not a good idea at all; on the one hand, it makes debugging difficult; on the other one, the variance reduction techniques we describe in the following may call for the ability to control the seeds.[3]

A few remarks are in order. A first observation is that with a LCG we actually generate rational numbers rather than real ones; this is not a serious problem, provided that m is large enough. But there is another reason to choose a large value for m; the generator is periodic. In fact, we may generate at most m distinct integer numbers Z_i, in the range from 0 to $m-1$, and whenever we repeat a previously generated number, the sequence repeats itself (which is not very random at all). We may see from the previous example that we get back to the initial seed $Z_0 = 7$ after 16 steps. This is not too bad, as 16 is the maximum possible period, for $m = 16$. We do much worse if we select $a = 11$, $c = 5$, and $m = 16$. In this case, starting from $Z_0 = 3$, we get the following sequence of integer numbers Z_i:

$$6, \quad 7, \quad 2, \quad 11, \quad 14, \quad 15, \quad 10, \quad 3$$

which has half the maximal period. Since the maximum possible period is m, we should make it very large in order to have a large period. The proper

---
[3] Actually, this need is evident when we have a complex simulation with multiple sources of uncertainty.

choice of a and c ensures that the period is maximized and that the sequence looks random. A sequence like

$$U_i = \frac{i}{m}, \quad i = 0, 1, \ldots, m-1,$$

which is obtained if $a = c = 1$, has a maximum period and is, in some sense, uniformly distributed on the interval (0,1), but it is far from satisfactory. The point is that the samples should also look independent; to be more precise, they should be able to trick statistical testing procedures into "believing" that they are a sequence of independent samples from the uniform distribution. This is why designing a good random number generator is not easy; luckily, when you purchase good numerical software, someone has already solved the issue for you.

Example 4.5 Consider the generator $Z_i = (aZ_{i-1}) \bmod m$ with $a = 2^{16} + 3$ and $m = 2^{31}$. It is fairly easy to show that for the sequence $U_i = Z_i/m$ the expression

$$U_{i+2} - 6U_{i+1} + 9U_i$$

takes integer values.[4] In fact, given Z_i (integer) we have

$$Z_{i+1} = aZ_i \bmod m = aZ_i - k_1 m$$

for some integer k_1. We also have

$$\begin{aligned} Z_{i+2} &= aZ_{i+1} \bmod m = a\left(aZ_i \bmod m\right) \bmod m = a\left(aZ_i - k_1 m\right) - k_2 m \\ &= a^2 Z_i - (ak_1 + k_2)m = a^2 Z_i \bmod m \end{aligned}$$

for some integer k_2. This implies

$$\begin{aligned} Z_{i+2} - 6Z_{i+1} + 9Z_i &= (2^{16} + 3)^2 Z_i \bmod m - 6(2^{16} + 3) Z_i \bmod m + 9 Z_i \\ &= \left[(2^{16} + 3)^2 Z_i - 6(2^{16} + 3) Z_i + 9 Z_i\right] - km \\ &= (2^{32} + 6 \cdot 2^{16} + 9 - 6 \cdot 2^{16} - 18 + 9) Z_i - km \\ &= 2^{32} Z_i - km. \end{aligned}$$

Therefore

$$U_{i+2} - 6U_{i+1} + 9U_i = \frac{2^{32} Z_i - k2^{31}}{2^{31}}$$

is integer. This means that points of the form (U_i, U_{i+1}, U_{i+2}) lie on a limited number of hyperplanes. □

The type of phenomenon illustrated in the example results in a lattice structure of LCGs. This concept may also be illustrated by the MATLAB script in

---

[4]The examples below are taken from [14, pp. 22–25].

```
% RipleyLCG.m
m = 2048;
a = 65;
c = 1;
seed = 0;
U = LCG(a,c,m,seed, 2048);
subplot(2,1,1)
plot(U(1:m-1), U(2:m), '.');
subplot(2,1,2)
plot(U(1:511), U(2:512), '.');
%
a=1365;
c=1;
U = LCG(a,c,m,seed, 2048);
figure
plot(U(1:m-1), U(2:m), '.');
```

Fig. 4.6 Script to illustrate the lattice structure of LCGs.

figure 4.6, which yields the plot displayed in figures 4.7 and 4.8. The top part of figure 4.6 shows a fairly good filling of the unit square by points of the form (U_i, U_{i+1}), for the choice $a = 65$, $c = 1$, and $m = 2048$. This may suggest that the distribution is uniform and that consecutive samples behave as if they were "statistically independent". However, the second part of the figure shows that the first part of the sequence follows some definite pattern. This is even worse in the second case, where $a = 1365$, whose pattern is displayed in figure 4.8. We see that selecting parameters for LCGs is not trivial, and many commercially used generators in the past were indeed flawed.

The examples above show that LCGs may have several limitations. Indeed, LCGs were state of the art in the past. In fact, they were used in the release 4 of MATLAB. Now a different approach is taken; we will not enter in any detail, but it suffices to say that the new generator is based on a state vector with 35 components (see [11] for more information). By issuing a command like rand(state',0), we tell MATLAB to reset this state vector to the configuration which is loaded when MATLAB is started. Another important point is that when generating normal variates, MATLAB uses the randn function; this function generates standard normal variates, and it has a separate state from the uniform generator. The state mechanism for randn is similar to that of rand; the important point to keep in mind is that they are separate, and resetting the state for the uniform generator is no use when you are generating normal variates (which is a common task when pricing options).

230 NUMERICAL INTEGRATION: DETERMINISTIC AND MONTE CARLO METHODS

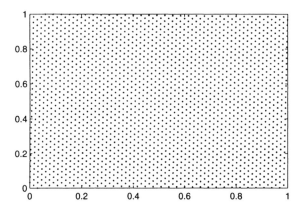

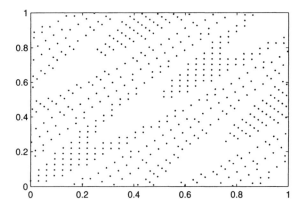

Fig. 4.7 Plots obtained by running the `RipleyLCG` script.

4.3.2 Inverse transform method

Suppose we are given the distribution function $F(x) = P\{X \leq x\}$, and that we want to generate random variates according to F. If we are able to invert F easily, we may apply the following inverse transform method:

1. We draw a random number $U \sim U(0, 1)$.

2. We return $X = F^{-1}(U)$.

It is easy to see that the random variate X generated by this method is actually characterized by the distribution function F:

$$P\{X \leq x\} = P\{F^{-1}(U) \leq x\} = P\{U \leq F(x)\} = F(x),$$

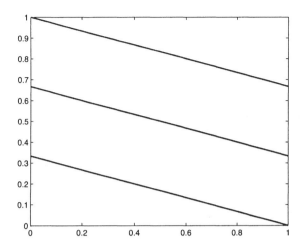

Fig. 4.8 Plot obtained by running the RipleyLCG script.

where we have used the monotonicity of F and the fact that U is uniformly distributed.

Example 4.6 A typical distribution which can be simulated easily by the inverse transform method is the exponential distribution. If $X \sim \exp(\mu)$, where $1/\mu$ is the expected value of X, its distribution function is

$$F(x) = 1 - e^{-\mu x}.$$

Direct application of the inverse transform yields

$$x = -\frac{1}{\mu}\ln(1 - U).$$

Since the distributions of U and $(1 - U)$ are actually the same, it is customary to generate exponential variates by drawing a random number U and by returning $-\ln(U)/\mu$. We may check that this is indeed the method used in the Statistics toolbox to simulate exponential random variables through the exprnd function:

```
>> rand('state',0)
>> exprnd(1)
ans =
    0.0512
>> rand('state',0)
>> -log(rand)
ans =
    0.0512
```

Generating exponential random variables is useful when you have to simulate a Poisson process, which is a possible model for shocks in asset prices or credit rating. □

The inverse transform method is quite simple, and it may also be applied when no theoretical distribution model is available and all you have is a set of empirical data. You just have to build a sensible distribution function based on your data set (see, e.g., [9]); one way to build a distribution function in this case is linear interpolation, and inverting a piecewise linear function is easily accomplished. However, we may not apply the inverse transform method when F is not invertible, as it happens with discrete distributions (in this case the distribution function is piecewise constant, with jumps where probability mass is concentrated). Nevertheless, we may adapt the method. Consider a discrete empirical distribution with a finite support:

$$P\{X = x_j\} = p_j, \qquad j = 1, 2, \ldots, n.$$

Then we should generate a uniform random variate U and return X as

$$X = \begin{cases} x_1 & \text{if } U < p_1 \\ x_2 & \text{if } p_1 \leq U < p_1 + p_2 \\ \vdots \\ x_j & \text{if } \sum_{k=1}^{j-1} p_k \leq U < \sum_{k=1}^{j} p_k \\ \vdots \end{cases}$$

It may be instructive to see how this code may be implemented in a simple way (not the most efficient one, however). Suppose we have a distribution defined by probabilities

$$0.1 \quad 0.2 \quad 0.4 \quad 0.2 \quad 0.1$$

over values $1, 2, 3, 4, 5$. First we define cumulative probabilities:

$$0.1 \quad 0.3 \quad 0.7 \quad 0.9 \quad 1.0,$$

then we draw a uniform random number, say $U = 0.82$. For each cumulative probability P we may check if $U > P$, yielding a vector

$$1 \quad 1 \quad 1 \quad 0 \quad 0,$$

where 1 corresponds to "true" and 0 to "false." To select the right value to return, we must sum the ones in this vector (the total is 3 here) and add 1; in this case we should return the value 4. Using MATLAB, this may be accomplished by working on vectors; code is displayed in figure 4.9 (howmany is the number of samples we want). For the example we are considering, we may check the function by plotting a histogram:

```
>> rand('state',0)
```

```
function samples = EmpiricalDrnd(values, probs, howmany)
% get cumulative probabilities
cumprobs = cumsum(probs);
N = length(probs);
samples = zeros(howmany,1);
for k=1:howmany
    loc=sum(rand*cumprobs(N) > cumprobs) + 1;
    samples(k)=values(loc);
end
```

Fig. 4.9 Sampling from an empirical discrete distribution.

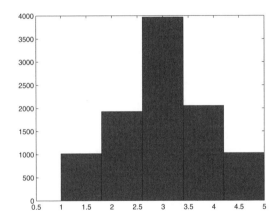

Fig. 4.10 Histogram produced by calling `EmpiricalDrnd`.

```
>> values=1:5;
>> probs=[0.1 0.2 0.4 0.2 0.1];
>> samples=EmpiricalDrnd(values,probs,10000);
>> hist(samples,5)
```

The resulting histogram is displayed in figure 4.10.

For many relevant distributions, the distribution function is invertible, but this is not easily accomplished. In such a case, one possibility is to resort to the acceptance–rejection method.

4.3.3 Acceptance–rejection method

Suppose we must generate random variates according to a probability density $f(x)$, and that the difficulty in inverting the corresponding distribution func-

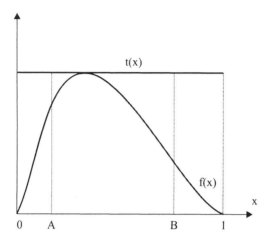

Fig. 4.11 Graphical example of the acceptance–rejection method.

tion makes the inverse transform method unattractive. Assume that we know a function $t(x)$ such that

$$t(x) \geq f(x) \qquad \forall x \in I,$$

where I is the support of f. The function $t(x)$ is not a probability density, but the related function $r(x) = t(x)/c$ is, provided that we select

$$c = \int_I t(x)\, dx.$$

If the distribution $r(x)$ is easy to simulate, it can be shown that the following acceptance–rejection method generates a random variate X distributed according to the density f:

1. Generate $Y \sim r$.

2. Generate $U \sim U(0, 1)$, independent of Y.

3. If $U \leq f(Y)/t(Y)$, return $X = Y$; otherwise, repeat the procedure.

If the support I is bounded, a natural choice for $r(x)$ is simply the uniform distribution on I, and we may choose

$$t(x) = \max_{x \in I} f(x).$$

We will not prove the correctness of the method, but an intuitive grasp can be gained from figure 4.11. In the figure, the support of $f(x)$ is the unit interval. A typical distribution that looks like f is the *beta distribution*:

$$f(x) = \frac{x^{\alpha_1 - 1}(1 - x)^{\alpha_2 - 1}}{\mathrm{B}(\alpha_1, \alpha_2)}, \qquad x \in [0, 1],$$

provided that the parameters satisfy $\alpha_1, \alpha_2 > 1$ (the beta distribution does not require this condition, but its appearance would be different from figure 4.11). The *beta function* is defined as

$$B(\alpha_1, \alpha_2) = \int_0^1 x^{\alpha_1-1}(1-x)^{\alpha_2-1}\, dx.$$

The Y variables are generated according to the uniform distribution and will spread evenly over the unit interval. Consider point A; since $f(A)$ is close to $t(A)$, A is likely to be accepted, as the ratio $f(A)/t(A)$ is close to 1. When we consider point B, where the value of the density f is small, we see that the ratio $f(B)/t(B)$ is small; hence, B is unlikely to be accepted, which is what we would expect. It can also be shown that the average number of iterations to terminate the procedure with an accepted value is c.

Example 4.7 Consider the density

$$f(x) = 30(x^2 - 2x^3 + x^4), \qquad x \in [0,1].$$

The reader is urged to verify that this is indeed a density (actually, it is the beta density with $\alpha_1 = \alpha_2 = 3$). If we apply the inverse transform method, we have to invert a fifth-degree polynomial at each generation, which suggests use of the acceptance–rejection method. By ordinary calculus we see that

$$\max_{x \in [0,1]} f(x) = 30/16$$

for $x^* = 0.5$. Using the uniform density as the easy density r, we get the following algorithm:

1. Draw two independent and uniformly distributed random variables U_1 and U_2.

2. If $U_2 \le 16(U_1^2 - 2U_1^3 + U_1^4)$, accept $X = U_1$; otherwise, reject and go back to step 1.

The average number of iterations to generate one random variate is $30/16$.

4.3.4 Generating normal variates by the polar approach

The inverse transform and acceptance–rejection methods are general purpose, but they are not always applicable. In the case of normal variables inverting the cumulative distribution function is no easy task, nor may we easily find a majorant function for the normal density, since its support is not finite. Actually, efficient approximations have been developed for the inverse distribution function for normal random variables. In MATLAB, a function call like

```
x = norminv(p,mu,sigma)
```

returns the quantile for probability p of a variable with expected value mu and standard deviation sigma This can be used to generate samples from the standard normal distribution, but it may not be the most efficient way:

```
>> tic, Z = norminv(rand(1000000,1));, toc
Elapsed time is 1.279080 seconds.
>> tic, Z = randn(1000000,1);, toc
Elapsed time is 0.048054 seconds.
```

Here, function `randn` uses a recent *ad hoc* method for the generation of normal variates. We outline here the basics of the classical polar approach, which may be outdated but is a nice example of ad hoc method.

Recall first that if $X \sim N(0,1)$, then $\mu + \sigma X \sim N(\mu, \sigma^2)$; hence we just need a method for generating standard normal variables. One old-fashioned possibility, which is still suggested in some textbooks, is to exploit the central limit theorem and to generate and sum a suitable number of uniform variates. Although this approach would work in the limit, computational efficiency would restrict the number of uniform variates that we use. The result is that we obtain a variate which could be of sufficient quality in noncritical simulations in which we are interested in average values, but is of debatable quality when we are interested in critical behavior in the tail of the distribution (as is the case in Value at Risk computations).

An alternative method is the Box–Muller approach. Consider two independent variables $X, Y \sim N(0,1)$, and let (R, θ) be the polar coordinates of the point of Cartesian coordinates (X, Y) in the plane, so that

$$d = R^2 = X^2 + Y^2 \qquad \theta = \tan^{-1} Y/X$$

The joint density of X and Y is

$$f(x,y) = \frac{1}{\sqrt{2\pi}} e^{-x^2/2} \frac{1}{\sqrt{2\pi}} e^{-y^2/2} = \frac{1}{2\pi} e^{-(x^2+y^2)/2} = \frac{1}{2\pi} e^{-d/2}.$$

The last expression looks like a product of an exponential density for d and a uniform distribution; the term $1/2\pi$ may be interpreted as the uniform distribution for the angle $\theta \in (0, 2\pi)$. However, we are missing some constant term in order to obtain the exponential density. To express the density in terms of (d, θ), we should properly take the Jacobian of the transformation from (x, y) to (d, θ) into account.[5] Some calculations yield

$$J = \begin{vmatrix} \dfrac{\partial d}{\partial x} & \dfrac{\partial d}{\partial y} \\ \dfrac{\partial \theta}{\partial x} & \dfrac{\partial \theta}{\partial y} \end{vmatrix} = 2,$$

---

[5] See, e.g., [16] for details.

and the correct density in the alternative coordinates is

$$f(d, \theta) = \frac{1}{2}\frac{1}{2\pi}e^{-d/2}.$$

Hence, we may generate R^2 as an exponential variable with mean 2 and θ as a uniformly distributed angle, and then transform back into Cartesian coordinates in order to obtain two independent standard normal variates. The Box–Muller algorithm may be implemented as follows:

1. Generate two independent uniform variates $U_1, U_2 \sim U(0, 1)$.

2. Set $R^2 = -2 \log U_1$ and $\theta = 2\pi U_2$.

3. Set $X = R\cos\theta$, $Y = R\sin\theta$.

In practice, this algorithm may be improved by avoiding the costly evaluation of trigonometric functions and integrating the Box–Muller approach with the rejection approach. The idea results in the following polar rejection method:

1. Generate two independent uniform variates $U_1, U_2 \sim U(0, 1)$.

2. Set $V_1 = 2U_1 - 1$, $V_2 = 2U_2 - 1$, $S = V_1^2 + V_2^2$.

3. If $S > 1$, return to step 1; otherwise, return the independent standard normal variates:

$$X = \sqrt{\frac{-2\ln S}{S}}V_1, \qquad Y = \sqrt{\frac{-2\ln S}{S}}V_2.$$

We refer the reader to [15, section 5.3] for a justification of the polar rejection method.

Example 4.8 We have seen that LCGs may exhibit a lattice structure. Since the Box–Muller transformation is non-linear, one might wonder if the composition of these two features may yield weird effects. We may check this in a somewhat peculiar case (see [14]), using the MATLAB script in figure 4.12. The script generates 2046 uniform random numbers for a sequence with modulus $m = 2048$; we discard the last pair, because the generator has maximum period and reverts back to the seed, which is 0 and causes trouble with the logarithm. Vectors U1 and U2 contain odd- and even-numbered random numbers in the sequence. The first part of the resulting plot, displayed in figure 4.13, shows poor coverage of the plane. The second part shows that swapping the pairs of random numbers may have a significant effect, whereas with truly random numbers the swap should be irrelevant. Of course, using better LCGs, or better random number generators prevents pathological behavior like this. However, it may be sometimes preferable to use the inverse transform method.

□

```
% Ripley2.m
m = 2048;
a = 1229;
c = 1;
N = m-2;
seed = 0;
U = LCG(a,c,m,seed,N);
U1 = U(1:2:N-1);
U2 = U(2:2:N);
X=sqrt(-2*log(U1)).* cos(2*pi*U2);
Y=sqrt(-2*log(U1)).* sin(2*pi*U2);
figure
subplot(2,1,1)
plot(X,Y,'.');
X=sqrt(-2*log(U2)).* cos(2*pi*U1);
Y=sqrt(-2*log(U2)).* sin(2*pi*U1);
subplot(2,1,2)
plot(X,Y,'.');
```

Fig. 4.12 Script to check Box–Muller approach.

In many financial applications one has to generate variates according to a multivariate normal distribution with (vector) expected value $\boldsymbol{\mu}$ and covariance matrix $\boldsymbol{\Sigma}$. This task may be accomplished by obtaining the Cholesky factor for $\boldsymbol{\Sigma}$, i.e., an upper triangular matrix $\mathbf{U}$ such that $\boldsymbol{\Sigma} = \mathbf{U}^T\mathbf{U}$ (see section 3.2.3). Then we may apply the following algorithm:

1. Generate n independent standard normal variates $Z_1, \ldots, Z_n \sim N(0,1)$.

2. Return $X = \boldsymbol{\mu} + \mathbf{U}^T\mathbf{Z}$, where $\mathbf{Z} = [Z_1, \ldots, Z_n]^T$.

Example 4.9 A rough code to simulate multivariate normal variables is illustrated in figure 4.14. The code builds a matrix whose columns correspond to the different variables, and the rows correspond to the different realizations of them. Assume that we have the following parameters:

```
>> Sigma = [4 1 -2 ; 1 3 1 ; -2 1 5];
>> mu = [ 8 ; 6 ; 10];
>> eig(Sigma)
ans =
    1.2855
    4.1433
    6.5712
```

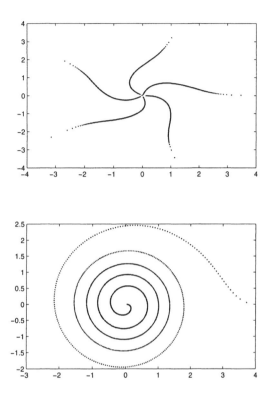

Fig. 4.13 Effect of swapping random numbers in the Box–Muller transformation.

Note that we make sure that the matrix Σ is positive definite, as it should be, by checking its eigenvalues. Now we may generate a few samples and verify the results.

```
>> rand('state',0);
>> Z = MultiNormrnd(mu,Sigma,10000);
>> mean(Z)
ans =
    8.0266    6.0234    9.9703
>> cov(Z)
ans =
    4.0159    1.0193   -1.9671
    1.0193    3.0011    1.0171
   -1.9671    1.0171    5.0060
```

We leave to the reader the exercise of improving the code, by checking that the vector and matrix sizes of the input arguments agree, by checking that the matrix Sigma is a positive definite symmetric matrix, and by avoiding the

```
function Z = MultiNormrnd(mu,sigma,howmany)
n = length(mu);
Z = zeros(howmany,n);
mu = mu(:); % make sure it's a column vector
U = chol(sigma);
for i=1:howmany
   Z(i,:) = mu' + randn(1,n) * U;
end
```

Fig. 4.14 Code to simulate multivariate normal variables.

for loop. Then have a look at the function mvnrnd, included in the Statistics toolbox, which does just this job. □

4.4 SETTING THE NUMBER OF REPLICATIONS

Carrying out a Monte Carlo simulation entails the generation of samples of the quantity of interest and then an estimation of the relevant parameters. One would expect that the larger the number of samples, or replications, the better the quality of the estimates will be. From appendix B we recall that given a sequence of *independent* (and we stress the independence) samples X_i, drawn from the same underlying distribution, we may build the sample mean:

$$\bar{X}(n) = \frac{1}{n}\sum_{i=1}^{n} X_i,$$

which is an unbiased estimator of the parameter $\mu = \mathrm{E}[X_i]$, and the sample variance:

$$S^2(n) = \frac{1}{n-1}\sum_{i=1}^{n} \left[X_i - \bar{X}(n)\right]^2.$$

We may try to quantify the quality of our estimator by considering the expected value of squared error of estimate:

$$\mathrm{E}[(\bar{X}(n)-\mu)^2] = \mathrm{Var}[\bar{X}(n)] = \frac{\sigma^2}{n},$$

where σ^2 may be estimated by the sample variance. Clearly, increasing the number n of replications improves the estimate; but how can we reasonably set the value of n?

Recall that the confidence interval at level $(1-\alpha)$ may be computed as

$$\bar{X}(n) \pm z_{1-\alpha/2}\sqrt{S^2(n)/n}, \tag{4.5}$$

where $z_{1-\alpha/2}$ is the quantile of the standard normal distribution corresponding to probability $1 - \alpha$. Strictly speaking, this is just an approximation, which will be a good one provided that n is large enough, both because $\bar{X}(n)$ will be approximately normal (central limit theorem) and because the quantile $t_{n-1,1-\alpha/2}$ from the t distribution with $n - 1$ degrees of freedom tends to $z_{1-\alpha}$.

Suppose you are interested in controlling the *absolute* error in such a way that, with probability $(1 - \alpha)$,

$$|\bar{X}(n) - \mu| \leq \beta,$$

where β is the maximum acceptable tolerance. But the confidence interval (4.5) is just built in such a way that

$$P\left\{\bar{X}(n) - H \leq \mu \leq \bar{X}(n) + H\right\} \approx 1 - \alpha,$$

where we denote the half-length of the interval by

$$H = z_{1-\alpha/2}\sqrt{S^2(n)/n}$$

of the confidence interval. This implies that, with probability $1 - \alpha$, we have

$$\bar{X}(n) - \mu \leq H, \quad \mu - \bar{X}(n) \leq H \quad \Rightarrow \quad |\bar{X}(n) - \mu| \leq H.$$

Hence, linking H to β, we should simply run replications until H is less than or equal to the tolerance β, and the number n must satisfy

$$z_{1-\alpha/2}\sqrt{S^2(n)/n} \leq \beta. \tag{4.6}$$

Actually, we are chasing our tail a bit here, since we cannot estimate the sample variance $S^2(n)$ until the number n has been set. One way out is to run a suitable number, say $k = 30$, of pilot replications, in order to come up with an estimate $S^2(k)$. Then we may apply (4.6) using $S^2(k)$ to determine n. After running the n replications, it is advisable to check that equation (4.6) holds with the new estimate $S^2(n)$. Alternatively, we may simply add replications, updating the sample variance, until the criterion is met; however, with this approach we do not control the amount of computation we are willing to spend.

If you are interested in controlling the relative error, so that

$$\frac{|\bar{X}(n) - \mu|}{|\mu|} \leq \gamma$$

holds with probability $(1 - \alpha)$, things are a little more involved. The difficulty is that we may run replications until the half-length H satisfies

$$\frac{H}{|\bar{X}(n)|} \leq \gamma,$$

```
function [Price, CI] = BlsMC2(S0,K,r,T,sigma,NRepl)
nuT = (r - 0.5*sigma^2)*T;
siT = sigma * sqrt(T);
DiscPayoff = exp(-r*T)*max(0, S0*exp(nuT+siT*randn(NRepl,1))-K);
[Price, VarPrice, CI] = normfit(DiscPayoff);
```

Fig. 4.15 Revised code to price a vanilla European call by Monte Carlo simulation.

but in this inequality we are using the known quantity $\bar{X}(n)$ rather than the unknown parameter μ. Nevertheless, if the inequality above holds, we may write

$$
\begin{aligned}
1-\alpha &\approx P\left\{\frac{|\bar{X}(n)-\mu|}{|\bar{X}(n)|} \leq \frac{H}{|\bar{X}(n)|}\right\} \\
&\leq P\{|\bar{X}(n)-\mu| \leq \gamma|\bar{X}(n)|\} \\
&= P\{|\bar{X}(n)-\mu| \leq \gamma|\bar{X}(n)-\mu+\mu|\} \\
&\leq P\{|\bar{X}(n)-\mu| \leq \gamma|\bar{X}(n)-\mu|+\gamma|\mu|\} \quad (4.7)\\
&= P\left\{\frac{|\bar{X}(n)-\mu|}{|\mu|} \leq \frac{\gamma}{1-\gamma}\right\},
\end{aligned}
$$

where inequality (4.7) follows from the triangle inequality and the last equation is obtained by a slight rearrangement. Therefore, we see that if we proceed without care, the actual relative error we get is bounded by $\gamma/(1-\gamma)$, which is larger than the desired bound γ; so, we should choose n such that the following criterion is met:

$$\frac{z_{1-\alpha/2}\sqrt{S^2(n)/n}}{|\bar{X}(n)|} \leq \gamma', \quad (4.8)$$

where

$$\gamma' = \frac{\gamma}{1+\gamma} < \gamma.$$

Again, we should run some pilot replications in order to get a first estimate of the sample variance $S^2(n)$.

Confidence intervals in MATLAB may be computed using the `normfit` function. This function is part of the Statistics Toolbox and it assumes that we are fitting a normal distribution based on samples from the normal distribution, which is not exactly what we have in mind; nevertheless, the way it computes confidence intervals fits our purpose. By default, `normfit` returns a 95% confidence interval, and different values may be specified, as usual in MATLAB, by passing an optional parameter.

Example 4.10 We may extend the code for pricing a vanilla call in order to compute confidence intervals on prices, as shown in figure 4.15. Note that in the last line we must collect three output arguments from `normfit`; the second one is sample variance, which is discarded. We can play a bit with BlsMC2 in order to get a feeling for how many replications are needed to get a fairly accurate estimate:

```
>> randn('state', 0)
>> S0=50;
>> K=55;
>> r=0.05;
>> T=5/12;
>> sigma=0.2;
>> Call = blsprice(S0,K,r,T,sigma)
Call =
    1.1718
>> [CallMC, CI] = BlsMC2(S0,K,r,T,sigma,50000)
CallMC =
    1.1953
CI =
    1.1704
    1.2201
>> (CI(2)-CI(1))/CallMC
ans =
    0.0416
```

We may notice that with 50000 samples the estimate is not quite satisfactory; however the true value is within the confidence interval, even though close to the left end-point. Of course, in a practically relevant case, we could only notice that the confidence interval is fairly wide. It may take a very large number of replications to get a reliable estimate:

```
>> [CallMC, CI] = BlsMC2(S0,K,r,T,sigma,1000000)
CallMC =
    1.1749
CI =
    1.1694
    1.1804
>> (CI(2)-CI(1))/CallMC
ans =
    0.0094
```

□

From equation (4.5) we see that the rate of improvement of the quality of our estimate, i.e., the rate of decrease of the error, is something like $O(1/\sqrt{n})$. In practice, this means that the more samples we get the better, but the rate of improvement is slower and slower as we keep adding samples. Thus a brute-force Monte Carlo simulation may take quite some amount of computation to yield an acceptable estimate. One way to overcome this issue is to adopt a clever sampling strategy in order to reduce the variance σ^2 of our samples; the other one is to adopt a quasi-Monte Carlo approach.

4.5 VARIANCE REDUCTION TECHNIQUES

We have seen in section 4.4 that one way to improve the accuracy of an estimate is to increase the number of replications n, since $\text{Var}(\bar{X}(n)) = \text{Var}(X_i)/n$. However, this brute-force approach may require an excessive computational effort. An alternative is to work on the numerator of this fraction and to reduce the variance of the samples X_i directly. This may be accomplished in different ways, more or less complicated, and more or less rewarding as well.

4.5.1 Antithetic sampling

A first approach that is easy to apply and does not require deep knowledge of what we are simulating is antithetic sampling. In plain Monte Carlo, we generate a sequence of independent samples. However, inducing some correlation in a clever way may be helpful. Consider the idea of generating a sequence of paired replications $(X_1^{(i)}, X_2^{(i)})$, $i = 1, \ldots, n$:

$$\begin{array}{cccc} X_1^{(1)} & X_1^{(2)} & \ldots & X_1^{(n)} \\ X_2^{(1)} & X_2^{(2)} & \ldots & X_2^{(n)}. \end{array}$$

These samples are "horizontally" independent, in the sense that $X_j^{(i_1)}$ and $X_k^{(i_2)}$ are independent however we choose $j, k = 1, 2$, provided $i_1 \neq i_2$. Thus the pair-averaged samples $X^{(i)} = (X_1^{(i)} + X_2^{(i)})/2$ are independent, and we may build a confidence interval based on them. However, we do not require "vertical" independence, since for a fixed i, $X_1^{(i)}$ and $X_2^{(i)}$ may be dependent. If we build the sample mean $\bar{X}(n)$ based on the samples $X^{(i)}$,

$$\begin{aligned} \text{Var}[\bar{X}(n)] &= \frac{\text{Var}(X^{(i)})}{n} \\ &= \frac{\text{Var}(X_1^{(i)}) + \text{Var}(X_2^{(i)}) + 2\,\text{Cov}(X_1^{(i)}, X_2^{(i)})}{4n} \\ &= \frac{\text{Var}(X)}{2n}\left(1 + \rho(X_1, X_2)\right). \end{aligned}$$

We see that, in order to reduce the variance of the sample mean, we should take negatively correlated replications within each pair. Each sample $X_{1,2}^{(i)}$ is obtained by generating random variates according to one of the methods we have described before; but all of these methods exploit a stream of uniformly distributed random numbers. Hence, to induce a negative correlation, we may use a random number sequence $\{U_k\}$ for the first replication in each pair, and then $\{1 - U_k\}$ in the second one. Since the input streams are negatively correlated, we hope that the output streams will be, too.

Example 4.11 Let us repeat example 4.2, where we used Monte Carlo integration to estimate

$$I = \int_0^1 e^x \, dx = e - 1 \approx 1.7183.$$

With only 100 samples, we do not get a reliable estimate:

```
>> randn('state',0)
>> X=exp(rand(100,1));
>> [I,dummy,CI] = normfit(X);
>> I
I =
    1.7631
>> (CI(2)-CI(1))/I
ans =
    0.1089
```

Antithetic sampling is easily accomplished here. We must store random numbers and take their complements to one. In order to have a fair comparison, we consider 50 antithetic pairs, which means 100 function samples as before:

```
>> randn('state',0)
>> U1=rand(50,1);
>> U2=1-U1;
>> X=0.5*(exp(U1)+exp(U2));
>> [I,dummy,CI] = normfit(X);
>> I
I =
    1.7021
>> (CI(2)-CI(1))/I
ans =
    0.0200
```

Now the confidence interval is much smaller and, despite the limited number of samples, the estimate is fairly reliable. □

The antithetic sampling method looks quite easy to apply and, in the example above, it works pretty well. May we always expect a similar pattern? Of course not. To begin with, if we integrate the exponential function over [0, 1] there is a strong positive correlation between U and e^U because the function is almost linear there. We should not expect impressive results in more complex cases. Moreover, the following counterexample shows that the method may actually backfire, resulting in an increase in the variance.

Example 4.12 Consider the function $h(x)$, defined as

$$h(x) = \begin{cases} 0, & x < 0 \\ 2x, & 0 \leq x \leq 0.5 \\ 2 - 2x, & 0.5 \leq x \leq 1 \\ 0, & x > 1 \end{cases}$$

and suppose that we want to take a Monte Carlo approach to estimate

$$\int_0^1 h(x)\,dx.$$

The function we want to integrate is obviously a triangle with both basis and height equal to 1; note that, unlike the exponential function of example 4.11, this is not a monotone function with respect to x. It is easy to compute the integral as the area of a triangle:

$$\int_0^1 h(x)\,dx \Rightarrow \mathrm{E}[h(U)] = \int_0^1 h(u) \cdot 1\,du = 1/2.$$

Now let
$$X_I = \frac{h(U_1) + h(U_2)}{2},$$
where U_1 and U_2 are independent uniform variates, be the usual sample based on independent sampling, and let

$$X_A = \frac{h(U) + h(1-U)}{2}$$

be the pair-averaged sample built by antithetic sampling. We may compare the two variances:

$$\mathrm{Var}(X_I) = \frac{\mathrm{Var}[h(U)]}{2}$$

$$\mathrm{Var}(X_A) = \frac{\mathrm{Var}[h(U)]}{2} + \frac{\mathrm{Cov}[h(U), h(1-U)]}{2}.$$

The difference between the two variances is

$$\begin{aligned}\Delta &= \mathrm{Var}(X_A) - \mathrm{Var}(X_I) = \frac{\mathrm{Cov}[h(U), h(1-U)]}{2} \\ &= \frac{1}{2}\{\mathrm{E}[h(U)h(1-U)] - \mathrm{E}[h(U)]\mathrm{E}[h(1-U)]\}.\end{aligned}$$

But in this case, due to the shape of h, we have

$$\mathrm{E}[h(U)] = \mathrm{E}[h(1-U)] = 1/2$$

and

$$\begin{aligned}\mathrm{E}[h(U)h(1-U)] &= \int_0^{1/2} 2u \cdot (2 - 2(1-u))\,du + \int_{1/2}^1 2(1-u) \cdot (2-2u)\,du \\ &= \int_0^{1/2} 4u^2\,du + \int_{1/2}^1 (2-2u)^2\,du = 1/3.\end{aligned}$$

Therefore, $\mathrm{Cov}[h(U), h(1-U)] = 1/3 - 1/4 = 1/12$ and $\Delta = 1/24 > 0$, and antithetic sampling actually increases variance in this case.

```
function [Price, CI] = BlsMCAV(S0,K,r,T,sigma,NRepl)
nuT = (r - 0.5*sigma^2)*T;
siT = sigma * sqrt(T);
Veps = randn(NRepl,1);
Payoff1 = max( 0 , S0*exp(nuT+siT*Veps) - K);
Payoff2 = max( 0 , S0*exp(nuT+siT*(-Veps)) - K);
DiscPayoff = exp(-r*T) * 0.5 * (Payoff1+Payoff2);
[Price, VarPrice, CI] = normfit(DiscPayoff);
```

Fig. 4.16 Using antithetic variates to price a vanilla European call by Monte Carlo simulation.

Indeed, there is a trivial explanation. The two antithetic samples have the same value $h(U) = h(1-U)$, so that $\text{Cov}[h(U), h(1-U)] = \text{Cov}[h(U), h(U)] = \text{Var}[h(U)]$. In this (pathological) case, the variance of the single sample is doubled by applying antithetic sampling. □

What is wrong with example 4.12? The variance of the antithetic pair is actually increased due to the non-monotonicity of $h(x)$. In fact, while it is true that the random numbers $\{U_i\}$ and $\{1 - U_i\}$ are negatively correlated, there is no guarantee that the same holds for $X_i^{(1)}$ and $X_i^{(2)}$ in general. To be sure that the negative correlation in the input random numbers yields a negative correlation in the output samples, we must require a monotonic relationship between them. The exponential function is a monotonic function, but the triangle function of the second example is not. We should also pay attention to how random variates are generated. The inverse transform method is based on the distribution function, which is a monotonic function; hence, there is a monotonic relationship between the input random numbers and the random variates generated. This is not necessarily the case with the acceptance–rejection method or the Box–Muller method. Luckily, when we need normal variates, we may simply generate a sequence Z_i, where $Z_i \sim N(0,1)$, and use the sequence $-Z_i$ for the antithetic samples. This idea is best illustrated by applying antithetic sampling to option pricing in the simplest setting.

We may easily incorporate antithetic sampling in our function BlsMC2 to price a European-style call option. MATLAB code is shown in figure 4.16. We simply generate a stream of standard normal variates and use the same sequence, with a change in sign, in the antithetic run. Each pair of antithetic samples is averaged and used as an estimator. Note that the last input parameter, NPairs, is the number of antithetic pairs, rather than samples; this must be taken into account when checking the variance reduction with respect to crude Monte Carlo:

```
>> randn('state',0)
```

248   NUMERICAL INTEGRATION: DETERMINISTIC AND MONTE CARLO METHODS

```
>> [Price, CI] = BlsMC2(50,50,0.05,1,0.4,200000)
Price =
    9.0843
CI =
    9.0154
    9.1532
>> (CI(2)-CI(1))/Price
ans =
    0.0152
>> randn('state',0)
>> [Price, CI] = BlsMCAV(50,50,0.05,1,0.4,100000)
Price =
    9.0553
CI =
    8.9987
    9.1118
>> (CI(2)-CI(1))/Price
ans =
    0.0125
```

We see that some improvement is obtained, but it is not that impressive, in this case. Clearly, one run for one example does not allow to draw any conclusion, but it is a fact that antithetic sampling is a simple technical trick which does not exploit too much knowledge.

In the case of a vanilla call option, the monotonicity condition required by antithetic sampling is met: the higher the sample from the standard normal distribution, the higher the terminal price of the underlying, and the higher the payoff. With non-monotonic payoffs, this need not be the case. We may illustrate this by using a payoff which is similar to the triangle function of example 4.12. The butterfly spread[6] is a trading strategy involving options on the same underlying asset, with the same maturity, but with different strike prices. The payoff from this combination is illustrated in figure 4.17. It can be obtained by buying one call option with strike price K_1, one call option with strike price K_3 ($K_1 < K_3$), and by selling two call options with a strike K_2 halfway between the other two. Since the butterfly spread is simply a combination of European calls, an option with that payoff may be directly priced by using Black–Scholes formula.

Since the payoff is clearly non-monotonic, and we know the "correct" price, it is interesting to check whether antithetic sampling works in this case. A crude Monte Carlo approach leads to the code in figure 4.18. The function MCButterfly receives the usual input arguments, plus the three strikes. Note the use of vectors In1 and In2 to collect the indexes corresponding to replications in which the terminal asset price falls in the increasing region of the

---

[6]See, e.g., [6, chapter 8] for more option trading strategies.

VARIANCE REDUCTION TECHNIQUES 249

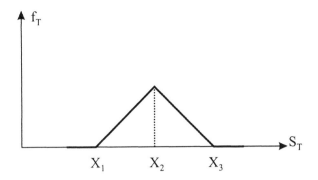

Fig. 4.17 Payoff from a butterfly spread.

```
function [P, CI] = MCButterfly(S0,r,T,sigma,NRepl,K1,K2,K3)
nuT = (r-0.5*sigma^2)*T;
siT = sigma*sqrt(T);
Veps = randn(NRepl,1);
Stocks = S0*exp(nuT + siT*Veps);
In1 = find((Stocks > K1) & (Stocks < K2));
In2 = find((Stocks >= K2) & (Stocks < K3));
Payoff = exp(-r*T)*[(Stocks(In1)-K1); (K3-Stocks(In2)); ...
         zeros(NRepl - length(In1) - length(In2),1)];
[P, V, CI] = normfit(Payoff);
```

Fig. 4.18 Crude Monte Carlo code to price a butterfly spread combination.

payoff ($K_1 < S_T < K_2$) or in the decreasing region ($K_2 \le S_T < K_3$); outside those regions the payoff is zero. The two vectors are used to avoid `for` loops.

The function `MCAVButterfly` of figure 4.19 is a modification based on antithetic sampling. The vector `Veps` contains the samples from the standard normal distribution, which are changed in sign to obtain the antithetic stock price samples `Stocks2`. Note that in this case we must preserve the order of the samples so as to pair the corresponding payoffs properly; this is why the code looks a bit more involved, and it uses `find` in order to spot samples falling in the interval of zero, increasing, or decreasing payoff.

It is common to choose K_2 close to the current stock price S_0, as this strategy is based on the bet that the stock price will not move too much. Let us check the results in such a case. Using `blsprice` we may get the theoretical result.

```
>> S0 = 60;
>> K1 = 55;
>> K2 = 60;
```

```
function [P, CI] = MCAVButterfly(S0,r,T,sigma,NPairs,K1,K2,K3)
nuT = (r-0.5*sigma^2)*T;
siT = sigma*sqrt(T);
Veps = randn(NPairs,1);
Stocks1 = S0*exp(nuT + siT*Veps);
Stocks2 = S0*exp(nuT - siT*Veps);
Payoff1 = zeros(NPairs,1);
Payoff2 = zeros(NPairs,1);
In = find((Stocks1 > K1) & (Stocks1 < K2));
Payoff1(In) = (Stocks1(In) - K1);
In = find((Stocks1 >= K2) & (Stocks1 < K3));
Payoff1(In) = (K3 - Stocks1(In));
In = find((Stocks2 > K1) & (Stocks2 < K2));
Payoff2(In) = (Stocks2(In) - K1);
In = find((Stocks2 >= K2) & (Stocks2 < K3));
Payoff2(In) = (K3 - Stocks2(In));
Payoff = 0.5 * exp(-r*T) * (Payoff1 + Payoff2);
[P, V, CI] = normfit(Payoff);
```

Fig. 4.19 Using antithetic sampling to price a butterfly spread combination.

```
>> K3 = 65;
>> T = 5/12;
>> r = 0.1;
>> sigma = 0.4;
>> calls = blsprice(S0, [K1, K2, K3], r, T, sigma);
>> Price = calls(1) - 2*calls(2) + calls(3)
Price =
    0.6124
```

Next, we may compare the two Monte Carlo methods:

```
>> randn('state',0)
[P, CI] = MCButterfly(S0,r,T,sigma,100000,K1,K2,K3);
>> P
P =
    0.6095
>> (CI(2)-CI(1))/P
ans =
    0.0256
>> randn('state',0)
>> [P, CI] = MCAVButterfly(S0,r,T,sigma,50000,K1,K2,K3);
>> P
P =
    0.6090
>> (CI(2)-CI(1))/P
```

```
ans =
    0.0355
```

We may see that variance is actually increased in this case. This does not mean that you will always have an increase in variance, as this depends on the input data (try changing the strikes to see this). Anyway, since one run does not tell us much, a better comparison may be carried out by checking the standard error of estimate with respect to the exact result using multiple runs:

```
>> randn('state',0)
>> V1 = zeros(100,1);
>> for i=1:100, V1(i)=MCButterfly(S0,r,T,sigma,100000,K1,K2,K3);, end
>> V2 = zeros(100,1);
>> for i=1:100, V2(i)=MCAVButterfly(S0,r,T,sigma,50000,K1,K2,K3);, end
>> sqrt(mean((V1 - Price).^2))
ans =
    0.0040
>> sqrt(mean((V2 - Price).^2))
ans =
    0.0055
```

Indeed, we see that the standard error of estimate is increased by antithetic sampling.

4.5.2 Common random numbers

The common random numbers (CRN) technique is very similar to antithetic sampling, but it is applied in a different situation. Suppose that we use Monte Carlo simulation to estimate a value depending on a parameter α. In formulas, we are trying to estimate something like

$$h(\alpha) = E_\omega[f(\alpha;\omega)],$$

where we have emphasized randomness through the variable ω. We could also be interested in evaluating the sensitivity of this value on the parameter α:

$$\frac{dh(\alpha)}{d\alpha}.$$

This would be of interest when dealing with option sensitivities beyond the Black–Scholes model. Clearly, we cannot compute the derivative analytically; otherwise, we wouldn't use simulation to evaluate h in the first place. So the simplest idea would be using simulation to estimate the value of the finite difference,

$$\frac{h(\alpha + \delta\alpha) - h(\alpha)}{\delta\alpha},$$

for a small value of the increment $\delta\alpha$. However, what we can really do is to generate samples of the difference

$$\frac{f(\alpha + \delta\alpha; \omega) - f(\alpha; \omega)}{\delta\alpha}$$

and to estimate its expected value. Unfortunately, when the increment $\delta\alpha$ is small, it is difficult to tell if the difference we obtain from the simulation is due to random noise or to variation in the parameter. A similar problem arises when we want to compare two portfolio management policies on a set of scenarios; in this case, too, what we need is an estimate of the expected value of the difference between two random variables.

Let us abstract a little and consider the difference of two random variables

$$Z = X_1 - X_2,$$

where, in general, $E[X_1] \neq E[X_2]$, since they come from simulating two different systems, possibly differing only in the value of a single parameter. By Monte Carlo simulation we get a sequence of independent samples

$$Z_j = X_{1,j} - X_{2,j}$$

and use statistical techniques to build a confidence interval for $E[X_1 - X_2]$. To improve our estimate, it would be useful to reduce the variance of the samples Z_j:

$$\text{Var}(X_{1j} - X_{2j}) = \text{Var}(X_{1j}) + \text{Var}(X_{2j}) - 2\,\text{Cov}(X_{1j}, X_{2j}).$$

To achieve this, we may try inducing some positive correlation between X_{1j} and X_{2j}. This can be obtained by using the same stream of random numbers in simulating both X_1 and X_2. The technique works much like antithetic sampling, and the same monotonicity assumption is required to ensure that the technique does not backfire. We will see an application of these concepts in section 8.5, where we apply Monte Carlo sampling to estimate option price sensitivities.

4.5.3 Control variates

Antithetic sampling and common random numbers are two almost foolproof techniques that, provided the monotonicity assumption is valid, do not require much knowledge about the systems we are simulating. Better results might be obtained by exploiting some more knowledge. Suppose that we want to estimate $\theta = E[X]$, and that there is another random variable Y, with a *known* expected value ν, which is somehow correlated with X. Such a case occurs when we use Monte Carlo simulation to price an option for which an analytical formula is not known: θ is the unknown price of the option, and ν is the price of a corresponding vanilla option.

The variable Y is called the *control variate*. Additional knowledge about Y may be exploited by adopting the controlled estimator

$$X_C = X + c(Y - \nu),$$

where c is a parameter we must choose. Intuitively, when we run a simulation and we observe that our estimates are such that

$$\hat{E}[Y] > \nu,$$

we may argue that the estimate $\hat{E}[X]$ should be increased or reduced accordingly, depending on the sign of the correlation between X and Y. Indeed, we may see that

$$E[X_C] = \theta$$
$$\mathrm{Var}(X_C) = \mathrm{Var}(X) + c^2 \mathrm{Var}(Y) + 2c\, \mathrm{Cov}(X, Y).$$

The first formula says that the controlled estimator is, for any choice of the control parameter c, an unbiased estimator of θ. The second formula suggests that by a suitable choice of c, we could reduce the variance of the estimator. We could even minimize the variance by choosing the optimal value for c:

$$c^* = -\frac{\mathrm{Cov}(X, Y)}{\mathrm{Var}(Y)},$$

in which case we get

$$\frac{\mathrm{Var}(X_C^*)}{\mathrm{Var}(X)} = 1 - \rho_{XY}^2,$$

where ρ_{XY} is the correlation between X and Y. Note that the sign of c depends on the sign of this correlation. For instance, if $\mathrm{Cov}(X, Y) > 0$, then $c < 0$. This implies that if $\hat{E}[Y] > \nu$, we should reduce $\hat{E}[X]$, which does make sense, because if our sample values for Y are larger than the average, the sample values for X are probably too.

In practice, the optimal value of c must be estimated, since $\mathrm{Cov}(X, Y)$ and possibly $\mathrm{Var}(Y)$ are not known. This may be accomplished by a set of pilot replications. It would be tempting to use these replications both for selecting c^* and to estimate θ; however, in doing so you induce some bias in the estimate of θ, since in this case c^* is a random variable depending on X itself. So, unless suitable statistical techniques are used, which are beyond the scope of this book, the pilot replications should be discarded.

The control variates approach may be generalized to as many control variates as we want, with a possible improvement in the quality of the estimates. Of course, this requires more knowledge about the system we are simulating and more effort in setting the control parameters. We may illustrate the approach using again the vanilla call option. In this case the stock price is a natural control variate, as both its expected value and the variance at the

```
function [Price, CI] = BlsMCCV(S0,K,r,T,sigma,NRepl,NPilot)
nuT = (r - 0.5*sigma^2)*T;
siT = sigma * sqrt(T);
% compute parameters first
StockVals = S0*exp(nuT+siT*randn(NPilot,1));
OptionVals = exp(-r*T) * max( 0 , StockVals - K);
MatCov = cov(StockVals, OptionVals);
VarY = S0^2 * exp(2*r*T) * (exp(T * sigma^2) - 1);
c = - MatCov(1,2) / VarY;
ExpY = S0 * exp(r*T);
%
NewStockVals = S0*exp(nuT+siT*randn(NRepl,1));
NewOptionVals = exp(-r*T) * max( 0 , NewStockVals - K);
ControlVars = NewOptionVals + c * (NewStockVals - ExpY);
[Price, VarPrice, CI] = normfit(ControlVars);
```

Fig. 4.20 Using control variates to price a vanilla European call by Monte Carlo simulation.

expiration of the option are known. To apply the method, we must compute an estimation of the covariance between the option value and the underlying asset price. The MATLAB code is illustrated in figure 4.20. The BlsMCCV function requires as an additional input parameter the number NPilot of pilot replications we want to run to estimate the covariance. Note that the first set of pilot replications is discarded to avoid biasing the estimator.

```
>> randn('state',0)
>> [P,CI] = BlsMC2(50,52,0.1,5/12,0.4,200000);
>> P
P =
    5.2328
>> (CI(2)-CI(1))/P
ans =
    0.0149
>> randn('state',0)
>> [P,CI] = BlsMCCV(50,52,0.1,5/12,0.4,195000,5000);
>> P
P =
    5.2008
>> (CI(2)-CI(1))/P
ans =
    0.0066
```

From these runs it would seem that there is some reduction in variance by using control variates. We should prepare a script in order to systematically check gain in efficiency. This is left as an exercise for the reader.

4.5.4 Variance reduction by conditioning

Computing expected values by conditioning is a common technique in probability theory. When we want to compute (or estimate) E[X], it is sometimes useful to condition with respect to another random variable Y, as the following formula holds:

$$E[X] = E[E[X \mid Y]]. \tag{4.9}$$

Variances may be computed by conditioning, too. We recall the conditional variance formula [see also equation (B.2) in appendix B]

$$\text{Var}(X) = E[\text{Var}(X \mid Y)] + \text{Var}(E[X \mid Y]).$$

We do not use the conditional variance formula directly in this book. However, since all the involved quantities are non-negative, we immediately see that the formula implies two consequences:

1. $\text{Var}(X) \geq E[\text{Var}(X \mid Y)]$.
2. $\text{Var}(X) \geq \text{Var}(E[X \mid Y])$.

Using the first inequality to reduce the variance of an estimator leads to variance reduction by stratification, which is discussed in the next section. The second one leads to variance reduction by conditioning.

Using conditioning is useful when our aim is to estimate $\theta = E[X]$ and there is another random variable Y such that the value of $E[X \mid Y = y]$ is known. From equation (4.9) we see that $E[X \mid Y]$ is also an unbiased estimator for θ, and the conditional variance formula implies that it may be a better one. In practice, to apply variance reduction by conditioning, we simulate Y rather than X. Unlike antithetic sampling, variance reduction by conditioning requires some careful thinking and is strongly problem dependent.

As an example of conditioning, we consider the problem of pricing an "as-you-like-it" option (also known as chooser option). The option is European-style and has maturity T_2. At time $T_1 < T_2$ you may choose if the option is a call or a put; the strike price K is fixed at time $t = 0$. Clearly, at time T_1 we should compare the values of the two options and choose the more valuable one. This can be done by using Black–Scholes formula to evaluate the price of call and put options with initial underlying price $S(T_1)$ and time to maturity $T_2 - T_1$. This means that, conditional on $S(T_1)$, we may get an exact estimate of the expected payoff at time T_2, under the risk-neutral probability. However, it is extremely instructive to write a pure Monte Carlo code, in which we only use sampling to get estimates.

In this case, this is not that trivial, as we must take a decision at time T_1; this is similar to the early exercise decision we must take with American options. To get a feeling for the issues involved, let us consider figure 4.21. Starting from the initial node, with price S_0, we generate four samples of price $S(T_1)$, and for each of these, we sample three prices $S(T_2)$. We have

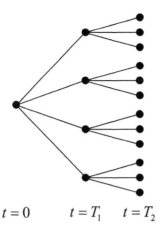

Fig. 4.21 Scenario tree for the as you like it option.

$4 \times 3 = 12$ scenarios, but they are tree-structured. We need this structure, because the decision at time T_1 (we like the put or the call) must be the same for all scenarios departing from each node at time T_1. Without this structure, our decisions would be based on perfect foresight about the future price at time T_2. This non-anticipativity concept is fundamental in dynamic stochastic optimization and in pricing American options.

A crude Monte Carlo code to price the option is displayed in figure 4.22. Here NRepl 1 is the number of samples (replications) at time T_1 and NRepl2 is the number of samples at time T_2, for each node at time T_1; hence, the overall number of scenarios is the product of NRepl1 and NRepl2. The vector DiscountedPayoffs has size corresponding to the overall number of scenarios. For each node at T_1, which is generated as usual with geometric Brownian motion, we generate nodes at time T_2, and we compare the estimates of expected payoff if we take the option as a call and if we take it as a put. Then we select *one* of the two alternatives and we fill a block (of size NRepl2) in the vector of discounted payoffs. Then we compute average and confidence intervals as usual. Later, we discuss if this is really correct.

Clearly, we are doing much more work than necessary in the crude Monte Carlo code. Conditional on a price $S(T_1)$, we know how to estimate expected payoff from each of the two choices, as this is given (apart from a discount factor) by the Black–Scholes formula. A code exploiting such a knowledge is displayed in figure 4.23. The code is actually much simpler: for each node at time $S(T_1)$ we take the larger value between the price of a call and the price of a put with initial price $S(T_1)$ and time to maturity $T_2 - T_1$, and we discount this value back from T_1 to time $t = 0$.

A script to compare crude and conditional Monte Carlo is given in figure 4.24. Running the script, we get

```
function [Price, CI] = AYLIMC(S0,K,r,T1,T2,sigma,NRepl1,NRepl2)
% compute auxiliary quantities outside the loop
DeltaT = T2-T1;
muT1 = (r-sigma^2/2)*T1;
muT2 = (r-sigma^2/2)*(T2-T1);
siT1 = sigma*sqrt(T1);
siT2 = sigma*sqrt(T2-T1);
% vector to contain payoffs
DiscountedPayoffs = zeros(NRepl1*NRepl2, 1);
% sample at time T1
Samples1 = randn(NRepl1,1);
PriceT1 = S0*exp(muT1 + siT1*Samples1);
for k=1:NRepl1
    Samples2 = randn(NRepl2,1);
    PriceT2 = PriceT1(k)*exp(muT2 + siT2*Samples2);
    ValueCall = exp(-r*DeltaT)*mean(max(PriceT2-K, 0));
    ValuePut = exp(-r*DeltaT)*mean(max(K-PriceT2, 0));
    if ValueCall > ValuePut
        DiscountedPayoffs(1+(k-1)*NRepl2:k*NRepl2) = ...
            exp(-r*T2)*max(PriceT2-K, 0);
    else
        DiscountedPayoffs(1+(k-1)*NRepl2:k*NRepl2) = ...
            exp(-r*T2)*max(K-PriceT2, 0);
    end
end
[Price, dummy, CI] = normfit(DiscountedPayoffs);
```

Fig. 4.22 Crude Monte Carlo code to price an as-you-like-it option.

```
function [Price, CI] = AYLIMCCond(S0,K,r,T1,T2,sigma,NRepl)
muT1 = (r-sigma^2/2)*T1;
siT1 = sigma*sqrt(T1);
Samples = randn(NRepl,1);
PriceT1 = S0*exp(muT1 + siT1*Samples);
[calls, puts] = blsprice(PriceT1,K,r,T2-T1,sigma);
Values = exp(-r*T1)*max(calls, puts);
[Price, dummy, CI] = normfit(Values);
```

Fig. 4.23 Using conditioning to price an as-you-like-it option.

```
% AYLIScript.m
S0 = 50;
K = 50;
r = 0.05;
T1 = 2/12;
T2 = 7/12;
sigma = 0.4;
NRepl1 = 100;
NRepl2 = 100;
[Call, Put] = blsprice(S0,K,r,T2,sigma);
randn('state',0);
[Price, CI] = AYLIMC(S0,K,r,T1,T2,sigma,NRepl1,NRepl2);
rand('state',0);
[PriceCond, CICond] = AYLIMCCond(S0,K,r,T1,T2,sigma,NRepl1*NRepl2);
fprintf(1,'Call = %f   Put = %f\n', Call, Put);
fprintf(1,'MC ->      Price = %f    CI = (%f, %f) \n', ...
    Price, CI(1), CI(2));
fprintf(1,'             Price = %6.4f%%\n', ...
    100*(CI(2)-CI(1))/Price);
fprintf(1,'MC+Cond -> Price = %f    CI = (%f, %f) \n', ...
    PriceCond, CICond(1), CICond(2));
fprintf(1,'             Price = %6.4f%%\n', ...
    100*(CICond(2)-CICond(1))/PriceCond);
```

Fig. 4.24 Script to compare pricing methods for an as-you-like-it option.

```
>> AYLIScript
Call = 6.728749 Put = 5.291478
MC ->      Price = 8.698173    CI = (8.489842, 8.906504)
           Ratio = 4.7902%
MC+Cond -> Price = 9.298894    CI = (9.218362, 9.379426)
           Ratio = 1.7321%
```

A few things should be noticed:

1. The value of the as you like it option is larger than the value of the call and the put options; deferring the choice has a significant value.

2. Conditioning seems to reduce variance, using the same number of scenarios in the two cases.

3. The value obtained by conditional Monte Carlo is larger.

The last point is quite relevant. Using conditional Monte Carlo, we do not only reduce variance; we take truly optimal decisions, whereas in crude Monte Carlo we may take the wrong choice at time T_1 because we are comparing estimates of the expected payoff. This may happen even if we estimate the payoffs with the same samples of price at time T_2 (which is essentially variance reduction by common random numbers). Hence, we have a *bias*. The estimator with crude Monte Carlo is biased low, since we are getting less money from a suboptimal strategy. And the bias does not disappear by increasing the number of replications. We urge the reader to run the script setting both NRepl1 and NRepl2 to 1000, which results in the following output:

```
>> AYLIScript
Call = 6.728749 Put = 5.291478
MC ->      Price = 8.930494    CI = (8.909643, 8.951345)
           Ratio = 0.4670%
MC+Cond -> Price = 9.259405    CI = (9.251437, 9.267372)
           Ratio = 0.1721%
```

We see that the bias is still there. This must be taken into account when using Monte Carlo methods to price American options (see chapter 10). If we use suboptimal exercise strategies, than we get a lower bound on the option price. It is also worth noting that this pricing problem is essentially a one-dimensional integration problem which may be solved more efficiently by other techniques.

To close this section, we should ask ourselves if the procedure we have followed is really correct. We have computed a confidence interval using the standard procedure, which assumes that samples are independent, but is this actually the case? Consider an intermediate node in our scenario tree, at time T_1, and its successor nodes at time T_2. Are the payoffs we receive in these successor nodes independent? Arguably, they are not, since we have used *all* of them to decide which option type we like at time T_1. The problem is that we are mixing two issues. The first one is *learning* an optimal decision rule by sampling; the second one is estimating the payoff we receive with that rule. A sound procedure would require two separate sampling phases. Doing so, we would be sure that payoffs are independent in the second sampling phase, and that the estimate we get is low-biased (since we are probably using a sub-optimal decision rule). We will meet such issues again in section 10.4, where we consider pricing American options by Monte Carlo sampling.

4.5.5 Stratified sampling

Suppose, as usual, that we are interested in estimating $E[X]$ and that X is somehow dependent on the value of another variable random Y, which may take a finite set of values y_j with known probability. Thus, Y has a discrete probability distribution with a known probability mass function:

$$P\{Y = y_j\} = p_j, \qquad j = 1, \ldots, m.$$

Using conditioning, we see that

$$E[X] = \sum_{j=1}^{m} E[X \mid Y = y_j] p_j.$$

So, we may use simulation to estimate the values $E[X \mid Y = y_j]$, for $j = 1, \ldots, m$, and use the formula above to put the results together. The conditional variance formula implies that this may yield a variance reduction with respect to crude sampling. The approach may look like variance reduction by conditioning. The key difference is that here we select a value for Y and then we sample X, conditioned on the event $Y = y_j$; this event is a *stratum*. In variance reduction by conditioning, you actually sample Y, not X. The following example justifies why such sampling is called *stratified*.

Example 4.13 As a simple example of stratification, consider using simulation to compute

$$\theta = \int_0^1 h(x) \, dx = E[h(U)].$$

In crude Monte Carlo simulation you would simply draw n uniform random numbers $U_i \sim U(0,1)$ and compute the sample mean

$$\frac{1}{n} \sum_{i=1}^{n} h(U_i).$$

An improved estimator over crude sampling may be obtained by partitioning the integration interval $(0, 1)$ into m subintervals $((j-1)/m, j/m)$, $j = 1, \ldots, m$. Each event $Y = y_j$ corresponds to a random number falling in the jth subinterval; in this case we have $p_j = 1/m$. For each stratum $j = 1, \ldots, m$ we may generate n_j random numbers $U_k \sim U(0,1)$ to estimate

$$\hat{\theta}_j = \frac{1}{n_j} \sum_{k=1}^{n_j} h\left(\frac{U_k + j - 1}{m}\right).$$

Then we build the overall estimator:

$$\hat{\theta} = \sum_{j=1}^{m} \hat{\theta}_j p_j. \qquad \square$$

How should we determine the number of samples n_j to be allocated to each stratum? A uniform allocation in example 4.13 makes sure that we sample uniformly over the integration interval $(0,1)$, but this need not be the optimal solution. Consider the variance of the estimator $\hat{\theta}$, and denote by X_j the random variable sampled in each stratum. If the strata are independently sampled, we have

$$\mathrm{Var}(\hat{\theta}) = \sum_{j=1}^{m} p_j^2 \, \mathrm{Var}(\hat{\theta}_j) = \sum_{j=1}^{m} \frac{p_j^2}{n_j} \, \mathrm{Var}(X_j).$$

To minimize the overall variance, we should allocate more samples to the strata where $\mathrm{Var}(X_j)$ is larger. So we could run a set of pilot replications to estimate $\mathrm{Var}(X_j)$ by sample variances S_j^2 and then obtain the fraction of samples to be allocated to each stratum by solving a non-linear programming problem:

$$\min \quad \sum_{j=1}^{m} \frac{p_j^2 S_j^2}{n_j}$$

$$\text{s.t.} \quad \sum_{j=1}^{m} n_j = n$$

$$n_j \geq 0.$$

4.5.6 Importance sampling

Unlike other variance reduction methods, importance sampling is based on the idea of "distorting" the underlying probability measure. It may be particularly useful when simulating rare events or sampling from the tails of a distribution. Consider the problem of estimating

$$\theta = \mathrm{E}[h(\mathbf{X})] = \int h(\mathbf{x}) f(\mathbf{x}) \, d\mathbf{x},$$

where $\mathbf{X}$ is a random vector with joint density $f(\mathbf{x})$. If we know another density g such that $f(\mathbf{x}) = 0$ whenever $g(\mathbf{x}) = 0$, we may write

$$\theta = \int \frac{h(\mathbf{x}) f(\mathbf{x})}{g(\mathbf{x})} g(\mathbf{x}) \, d\mathbf{x} = \mathrm{E}_g\left[\frac{h(\mathbf{X}) f(\mathbf{X})}{g(\mathbf{X})}\right], \qquad (4.10)$$

where the notation E_g is used to stress the fact that the last expected value is taken with respect to another measure. The ratio $f(\mathbf{x})/g(\mathbf{x})$ is used to correct the change in probability measure, and it is typically called the likelihood ratio: when using random sampling, this ratio will be a random variable.[7] That changing the underlying probability measure may be useful should not be a surprise for people interested in finance; risk-neutral valuation does just that. However, it is not so obvious why this should be helpful in reducing variance. Indeed, the method may backfire if g is not chosen with care. Intuitively, we may argue that when looking for rare but important events, as is the case in estimating Value at Risk, we should distort the probability measure in order to sample from the critical region, provided that we compensate for this bias. This is exactly what is done in equation (4.10).

To gain more insight into how density g should be chosen, let us introduce the notation

$$\theta = E_f[h(\mathbf{X})]$$

and assume for simplicity that $h(\mathbf{x}) \geq 0$. As we have pointed out above, there are two possible ways of estimating θ:

$$\begin{aligned} E_f[h(\mathbf{X})] &= \int h(\mathbf{x})f(\mathbf{x})\,d\mathbf{x} = \int \frac{h(\mathbf{x})f(\mathbf{x})}{g(\mathbf{x})} g(\mathbf{x})\,d\mathbf{x} \\ &= \int h^*(\mathbf{x})g(\mathbf{x})\,d\mathbf{x} = E_g[h^*(\mathbf{X})], \end{aligned}$$

where $h^*(\mathbf{X}) = h(\mathbf{x})f(\mathbf{x})/g(\mathbf{x})$. Note that the condition on the support of f and g is needed in order to avoid any trouble with the case $g(\mathbf{x}) = 0$ in the definition of h^*; we may think of integrating only on the support.

The two estimators have the same expectation, but what about the variance? Using the well-known properties of the variance, we obtain

$$\begin{aligned} \mathrm{Var}_f[h(\mathbf{X})] &= \int h^2(\mathbf{x})f(\mathbf{x})\,d\mathbf{x} - \theta^2 \\ \mathrm{Var}_g[h^*(\mathbf{X})] &= \int h^2(\mathbf{x})\frac{f(\mathbf{x})}{g(\mathbf{x})}f(\mathbf{x})\,d\mathbf{x} - \theta^2. \end{aligned}$$

From the second equation, it is easy to see that the choice

$$g(\mathbf{x}) = \frac{h(\mathbf{x})f(\mathbf{x})}{\theta}$$

leads to the ideal condition $\mathrm{Var}_g[h^*(\mathbf{X})] \equiv 0$. Unfortunately, this is indeed "ideal," as using this density requires knowledge of θ; still, we may at least try to use approximations of the ideal density (see the example below). Note

---

[7] Readers with a background in stochastic calculus would probably use the term "Radon–Nikodym derivative."

```
function out=estpi(m)
z=sqrt(1-rand(1,m).^2);
out = 4*sum(z)/m;
```

Fig. 4.25 Trivial code to estimate π.

also that the condition $h(\mathbf{x}) \geq 0$ is needed in order to ensure that this is a density; see, e.g., [17, p. 122] to see how to deal with a generic function h.

In general, the difference between the two variances is

$$\Delta \text{Var} = \text{Var}_f[h(\mathbf{X})] - \text{Var}_g[h^*(\mathbf{X})] = \int h^2(\mathbf{x}) \left[1 - \frac{f(\mathbf{x})}{g(\mathbf{x})}\right] f(\mathbf{x}) \, d\mathbf{x}.$$

From this expression we see that, in order to ensure that we do reduce variance, we should select a new density g such that

$$\begin{cases} g(\mathbf{x}) > f(\mathbf{x}) & \text{when the term } h^2(\mathbf{x})f(\mathbf{x}) \text{ is large,} \\ g(\mathbf{x}) < f(\mathbf{x}) & \text{when the term } h^2(\mathbf{x})f(\mathbf{x}) \text{ is small.} \end{cases}$$

The name "importance sampling" derives from this observation.

Example 4.14 We may use a trivial integration example to illustrate the idea. Let us consider a way to compute π. We know that[8]

$$\theta \equiv \int_0^1 \sqrt{1-x^2} \, dx = \frac{\pi}{4},$$

since this is simply the area of a quarter of a unit circle; hence, estimating the value of this integral is a possible way to obtain an estimate of π. A trivial code to do this is shown in figure 4.25, where the input parameter m is the number of points we want to sample. From the snapshot below we see that with 1000 samples, the estimates are not so reliable.

```
>> rand('state',0)
>> estpi(1000)
ans =
    3.1378
>> estpi(1000)
ans =
    3.1311
>> estpi(1000)
ans =
```

---

[8]This example is based on [2].

```
       3.0971
>> estpi(1000)
ans =
       3.1529
```

So, let us try to improve our estimates by using importance sampling. A possible idea to approximate the ideal probability distribution is to divide the integration interval $[0, 1]$ into L equally spaced subintervals of width $1/L$. The extreme points of the kth subinterval ($k = 1, \ldots, L$) are $(k-1)/L$ and k/L, and the midpoint of this subinterval is $s_k = (k-1)/L + 1/(2L)$. A rough estimate of the integral is obtained by computing

$$\frac{\sum_{k=1}^{L} h(s_k)}{L} = \tilde{\theta} \approx \theta.$$

Then, an approximation of the ideal density $g(x)$, we could use something like

$$\tilde{g}(x) \equiv \frac{h(x)f(x)}{\tilde{\theta}} = \frac{h(x)L}{\sum_{k=1}^{L} h(s_k)},$$

since $f(x) = 1$ (uniform distribution). Unfortunately, this need not be a density integrating to one over the unit interval. In order to avoid this difficulty and to simplify sampling, we may define a probability of sampling from a subinterval and use a uniform density within each subinterval. To this aim, consider the quantities

$$q_k = \frac{h(s_k)}{\sum_{j=1}^{L} h(s_j)}, \qquad k = 1, \ldots, L.$$

Clearly, $\sum_k q_k = 1$ and $q_k \geq 0$, since our function h is non-negative; hence, the numbers q_k may be interpreted as probabilities. In our case, they may be used as the probabilities of selecting a sample point from the kth subinterval. To summarize, and to cast the problem within the general framework, we have

$$\begin{aligned} h(x) &= \sqrt{1-x^2} \\ f(x) &= 1 \\ g(x) &= Lq_k, \qquad (k-1)/L \leq x < k/L. \end{aligned}$$

Here, $g(x)$ is a piecewise constant density; the L factor multiplying the q_k in $g(x)$ is just needed to obtain the uniform density over an interval of length $1/L$. The resulting code is illustrated in figure 4.26, where m is the number of sampled points and L is the number of subintervals. The code is fairly simple, and sub-intervals are selected as described in the last part of section 4.3.2, on page 233, where we have seen how to sample discrete empirical distributions by the function EmpiricalDrnd.

```
>> rand('state',0)
```

```
function z=estpiIS(m,L)
% define left end-points of sub-intervals
s= (0:(1/L):(1-1/L)) + 1/(2*L);
hvals = sqrt(1 - s.^2);
% get cumulative probabilities
cs=cumsum(hvals);
for j=1:m
   % locate sub-interval
   loc=sum(rand*cs(L) > cs) +1;
   % sample uniformly within sub-interval
   x=(loc-1)/L + rand/L;
   p=hvals(loc)/cs(L);
   est(j) = sqrt(1 - x.^2)/(p*L);
end
z = 4*sum(est)/m;
```

Fig. 4.26 Importance sampling-based code to estimate π.

```
>> estpiIS(1000,10)
ans =
    3.1491
>> estpiIS(1000,10)
ans =
    3.1434
>> estpiIS(1000,10)
ans =
    3.1311
>> estpiIS(1000,100)
ans =
    3.1403
>> estpiIS(1000,100)
ans =
    3.1416
>> estpiIS(1000,100)
ans =
    3.1411
```

We see that the improved code, although not a very sensible way to compute π, yields a remarkable reduction in variance.

The approach we have just taken looks suspiciously like stratified sampling. Actually, there is a subtle difference. In stratified sampling we define a set of strata, which correspond to events of known probability; here we have not used strata with known probability, as we have used sampling to estimate the probabilities q_k.

Importance sampling is often used when small probabilities are involved. Consider, for instance, a random vector $\mathbf{X}$ with joint density f, and suppose that we want to estimate

$$\theta = \mathrm{E}[h(\mathbf{X}) \mid \mathbf{X} \in \mathcal{A}],$$

where $\{\mathbf{X} \in \mathcal{A}\}$ is a rare event with a small, but unknown probability $P\{\mathbf{X} \in \mathcal{A}\}$. Such an event could be the occurrence of a loss larger than the Value at Risk. The conditional density is

$$f(\mathbf{x}|\mathbf{X} \in \mathcal{A}) = \frac{f(\mathbf{x})}{P\{\mathbf{X} \in \mathcal{A}\}}$$

for $\mathbf{x} \in \mathcal{A}$. Defining the indicator function $I_\mathcal{A}(\mathbf{X})$ as

$$I_\mathcal{A}(\mathbf{X}) = \begin{cases} 1 & \text{if } \mathbf{X} \in \mathcal{A} \\ 0 & \text{if } \mathbf{X} \notin \mathcal{A}, \end{cases}$$

we may rewrite θ as

$$\theta = \frac{\int_{\mathbf{x} \in \mathcal{A}} h(\mathbf{x}) f(\mathbf{x}) \, d\mathbf{x}}{P\{\mathbf{X} \in \mathcal{A}\}} = \frac{\mathrm{E}[h(\mathbf{X}) I_\mathcal{A}(\mathbf{X})]}{\mathrm{E}[I_\mathcal{A}(\mathbf{X})]}.$$

If we use crude Monte Carlo simulation, many samples will be wasted, as the event $\{\mathbf{X} \in \mathcal{A}\}$ will rarely occur. Now, assume that there is a density g such that this event is more likely under the corresponding probability measure. Then, we may generate the samples $\mathbf{X}_i$ according to g and estimate

$$\hat{\theta} = \frac{\sum_{i=1}^{k} h(\mathbf{X}_i) I_\mathcal{A}(\mathbf{X}_i) f(\mathbf{X}_i)/g(\mathbf{X}_i)}{\sum_{i=1}^{k} I_\mathcal{A}(\mathbf{X}_i) f(\mathbf{X}_i)/g(\mathbf{X}_i)}.$$

Importance sampling is certainly more difficult to apply than antithetic sampling or control variates: It requires more knowledge about what we are simulating, since we must be able to figure out a suitably distorted probability measure.

As an example, let us consider pricing a deep out-of-the-money vanilla call. If S_0 is the initial price of the underlying, we know that its expected value at maturity is, according to geometric Brownian motion under the risk-neutral measure, $S_0 e^{rT}$. If this expected value is small with respect to the strike price K, it is unlikely that the option will be in-the-money at maturity. If we apply crude Monte Carlo, many replications are wasted because the payoff will be zero in most of them. We should change the drift in order to increase the probability that the payoff is positive. It is easy to find a drift such that the expected value of S_T is the strike price:

$$S_0 e^{\mu T} = K \quad \Rightarrow \quad \mu = \frac{1}{T} \log\left(\frac{K}{S_0}\right).$$

QUASI-MONTE CARLO SIMULATION   267

While under the risk neutral measure we sample $S_T = S_0 e^Z$ by generating normal variates

$$Z \sim \mathcal{N}\left(\left(r - \frac{\sigma^2}{2}\right)T,\ \sigma\sqrt{T}\right),$$

we should sample by generating

$$Y \sim \mathcal{N}\left(\log\left(\frac{K}{S_0}\right) - \frac{\sigma^2 T}{2},\ \sigma\sqrt{T}\right),$$

which in turn requires generating standard normal variates ϵ and then using

$$Y = \log\left(\frac{K}{S_0}\right) - \frac{\sigma^2 T}{2} + \sigma\sqrt{T}\epsilon.$$

Now the tricky part is to compute the likelihood ratio. For the sake of clarity, assume that we sample Y from a normal distribution $\mathcal{N}(\beta, \xi)$ whereas the original distribution is $\mathcal{N}(\alpha, \xi)$. Then, the ratio of the two probability densities is

$$\frac{\frac{1}{\sqrt{2\pi}\xi}e^{-\frac{(Y-\alpha)^2}{2\xi^2}}}{\frac{1}{\sqrt{2\pi}\xi}e^{-\frac{(Y-\beta)^2}{2\xi^2}}} = e^{-\left[(Y-\alpha)^2 - (Y-\beta)^2\right]/2\xi^2} = e^{-\left[2(\alpha-\beta)Y - \alpha^2 + \beta^2\right]/2\xi^2}$$

Now it is easy to extend BlsMC2 to the function BlsMCIS displayed in figure 4.27. We may check the efficiency gain of importance sampling by running the script CheckBlsMCIS of figure 4.28. For a deep out-of-the-money option, we compute price with crude Monte Carlo and with importance sampling, and we compare the percentage error with respect to the exact price. We reset the random variate generator randn twice in order to use exactly the same stream of standard normal variates. Running the script, we get

```
>> CheckBlsMCIS
Average Percentage Error:
   MC    = 3.060%
   MC+IS = 1.155%
```

We should note that this improvement is not to be expected for at-the-money options.

4.6 QUASI-MONTE CARLO SIMULATION

In the preceding sections, we have considered the use of variance reduction techniques, which are based on the idea that random sampling is *really* random. However, the random numbers produced by a LCG or by more sophisticated algorithms are not random at all. Hence, one could take a philosophical

```
function [Price, CI] = BlsMCIS(S0,K,r,T,sigma,NRepl)
nuT = (r - 0.5*sigma^2)*T;
siT = sigma * sqrt(T);
ISnuT = log(K/S0) - 0.5*sigma^2*T;
Veps = randn(NRepl,1);
VY = ISnuT + siT*Veps;
ISRatios = exp( (2*(nuT - ISnuT)*VY - nuT^2 + ISnuT^2)/2/siT^2);
DiscPayoff = exp(-r*T)*max(0, (S0*exp(VY)-K));
[Price, VarPrice, CI] = normfit(DiscPayoff.*ISRatios);
```

Fig. 4.27 Importance sampling-based code to price an out-of-the-money vanilla call.

```
% CheckBlsMCIS.m
S0 = 50;
K = 80;
r = 0.05;
sigma = 0.4;
T = 5/12;
NRepl = 100000;
MCError = zeros(NRepl,1);
MCISError = zeros(NRepl,1);
TruePrice = blsprice(S0,K,r,sigma,T);
randn('state',0);
for k=1:100
    MCPrice = BlsMC2(S0,K,r,sigma,T,NRepl);
    MCError = abs(MCPrice - TruePrice)/TruePrice;
end
randn('state',0);
for k=1:100
    MCISPrice = BlsMCIS(S0,K,r,sigma,T,NRepl);
    MCISError = abs(MCISPrice - TruePrice)/TruePrice;
end
fprintf(1,'Average Percentage Error:\n');
fprintf(1,' MC    = %6.3f%%\n', 100*mean(MCError));
fprintf(1,' MC+IS = %6.3f%%\n', 100*mean(MCISError));
```

Fig. 4.28 Script to check effectiveness of importance sampling.

view and wonder about the very validity of variance reduction methods, and even the Monte Carlo approach itself. Taking a more pragmatic view, and considering the fact that Monte Carlo methods have proven their value over the years, we should conclude that this shows that there are some deterministic number sequences that work well in generating samples. So one could try to devise alternative deterministic sequences of numbers which are in some sense evenly distributed. This idea may be made more precise by defining the *discrepancy* of a sequence of numbers.

Assume that we want to generate a sequence of N "random" vectors $\mathbf{X}^1, \mathbf{X}^2, \ldots, \mathbf{X}^N$ in the m-dimensional hypercube $I^m = [0,1]^m \subset \mathbb{R}^m$. Now, given a sequence of such vectors, if they are well distributed, the number of points included in any subset G of I^m should be roughly proportional to its volume vol(G). Given a vector $\mathbf{X} = (x_1, x_2, \ldots, x_m)$, consider the rectangular subset $G_\mathbf{X}$ defined as

$$G_\mathbf{X} = [0, x_1) \times [0, x_2) \times \cdots \times [0, x_m),$$

which has a volume $x_1 x_2 \cdots x_m$. If we denote by $S_N(G)$ the function counting the number of points in the sequence, which are contained in a subset $G \subset I^m$, a possible definition of discrepancy is

$$D(\mathbf{x}^1, \ldots, \mathbf{x}^N) = \sup_{\mathbf{X} \in I^m} \mid S_N(G_\mathbf{X}) - N x_1 x_2 \cdots x_m \mid.$$

When computing a multidimensional integral on the unit hypercube, it is natural to look for low-discrepancy sequences; an alternative name for a low-discrepancy sequence is *quasirandom sequence*, which is why the term *quasi-Monte Carlo* is used. Actually, the quasirandom term is a bit misleading, as there is no randomness at all. Some theoretical results suggest that low-discrepancy sequences may perform better than pseudorandom sequences obtained through a LCG or its variations. The point is that from section 4.4 we know that the estimation error with Monte Carlo simulation is something like $O(1/\sqrt{N})$, where N is the number of samples. With certain low-discrepancy sequences, it can be shown that the error is something like $O(\ln N)^m/N$, where m is the dimension of the space in which we are integrating. We refer the reader to the comprehensive book [12] for a detailed and rigorous account on this subject. Different sequences have been proposed in the literature. In the following, we illustrate the basic ideas behind two low-discrepancy sequences, Halton and Sobol sequences, and their implementation. Low-discrepancy sequences are sequences in the unit interval $(0,1)$; from what we know about the generation of generally distributed random variates, we see that this is what we need to simulate according to any distribution we need.

4.6.1 Generating Halton low-discrepancy sequences

Halton low-discrepancy sequences are based on a simple recipe:

```
function h=Halton(n,b)
n0 = n;
h = 0;
f = 1/b;
while (n0 > 0)
   n1 = floor(n0/b);
   r = n0 - n1*b;
   h = h+f*r;
   f = f/b;
   n0=n1;
end
```

Fig. 4.29 MATLAB code to generate the nth element of a Halton sequence with a given base.

- Representing an integer number n in a base b, where b is a prime number:
$$n = (\cdots d_4 d_3 d_2 d_1 d_0)_b.$$

- Reflecting the digits and adding a radix point to obtain a number within the unit interval:
$$h = (0.d_0 d_1 d_2 d_3 d_4 \cdots)_b.$$

More formally, if we represent an integer number n as

$$n = \sum_{k=0}^{m} d_k b^k,$$

the nth number in the Halton sequence with base b is

$$h(n,b) = \sum_{k=0}^{m} d_k b^{-(k+1)}.$$

To be precise, what we get is known as Van der Corput sequence. Halton sequences are obtained in multiple dimensions when a Van der Corput generator is associated to each dimension, making sure different prime numbers are used for each base which is associated to each dimension. For the sake of simplicity we will only speak of Halton sequences.

Using the principles illustrated in section 3.1.1 on the binary representation of numbers on a computer, it is easy to generate the nth number in a Halton sequence with base b. The code is illustrated in figure 4.29. Let us generate the first 10 numbers in the sequence with base 2:

```
>> seq = zeros(10,1);
```

```
function Seq = GetHalton(HowMany, Base)
Seq = zeros(HowMany,1);
NumBits = 1+ceil(log(HowMany)/log(Base));
VetBase = Base.^(-(1:NumBits));
WorkVet = zeros(1,NumBits);
for i=1:HowMany
   % increment last bit and carry over if necessary
   j=1;
   ok = 0;
   while ok == 0
      WorkVet(j) = WorkVet(j)+1;
      if WorkVet(j) < Base
         ok = 1;
      else
         WorkVet(j) = 0;
         j = j+1;
      end
   end
   Seq(i) = dot(WorkVet,VetBase);
end
```

Fig. 4.30 MATLAB code to generate a Halton low-discrepancy sequence with a given base.

```
>> for i=1:10, seq(i) = Halton(i,2);, end
>> seq
seq =
    0.5000
    0.2500
    0.7500
    0.1250
    0.6250
    0.3750
    0.8750
    0.0625
    0.5625
    0.3125
```

We see how Halton sequences work; by reflecting and adding more bits, we fill the space between 0 and 1 with finer and finer intervals. A code to obtain a whole sequence is illustrated in figure 4.30; the input parameters are HowMany, i.e., how long the sequence should be, and the base Base. Rather than generating each number in the sequence one at a time, we generate the sequence $1, \ldots, n$ by incrementing the bit representation in base b, which is immediately converted into $H(n, b)$.

272 *NUMERICAL INTEGRATION: DETERMINISTIC AND MONTE CARLO METHODS*

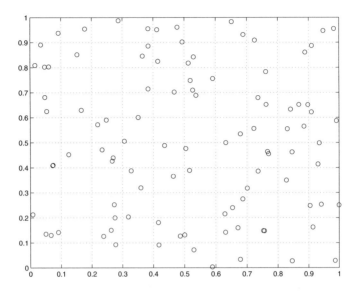

Fig. 4.31 Random sample in two dimensions.

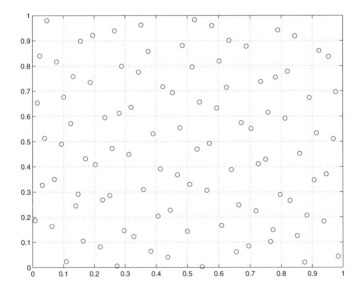

Fig. 4.32 Covering the bidimensional unit square with Halton sequences.

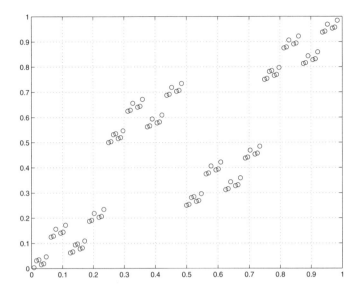

Fig. 4.33 Bad choice of bases in Halton sequences.

Example 4.15 It is instructive to compare how a pseudorandom sample covers the square $(0,1)\times(0,1)$ in two dimensions. Using the MATLAB random generator, we get the plot of figure 4.31:

```
>> plot(rand(100,1),rand(100,1),'o')
>> grid on
```

To do the same with Halton sequences we must use different bases, which should be prime numbers. Let us try with 2 and 7:

```
>> plot(GetHalton(100,2),GetHalton(100,7),'o')
>> grid on
```

The result is shown in figure 4.32. The judgment is a bit subjective here, but it could be argued that the covering of the Halton sequence is more even. On the other hand, using a non-prime number as the base, as in

```
>> plot(GetHalton(100,2), GetHalton(100,4), 'o')
>> grid on
```

may result in quite unsatisfactory patterns, such as the one shown in figure 4.33. □

Example 4.16 Let us explore the use of Halton low-discrepancy sequences in a bidimensional integration context. Suppose that we want to compute

$$\int_0^1 \int_0^1 e^{-xy} \left(\sin 6\pi x + \cos 8\pi y \right) \, dx \, dy.$$

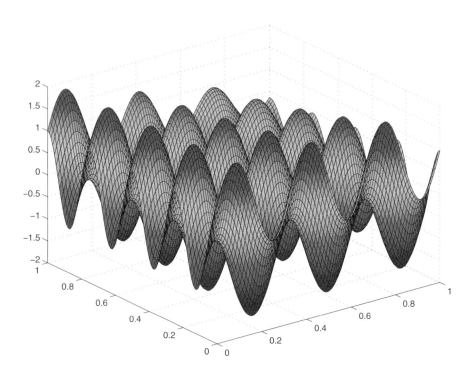

Fig. 4.34 Plot of the integrand function in example 4.16.

To begin with, let us set up a function in order to plot the integrand and to use the dblquad MATLAB function to get an estimate by traditional quadrature formulas.

```
>> f=@(x,y) exp(-x.*y).*(sin(6*pi*x)+cos(8*pi*y));
>> dblquad(f,0,1,0,1)
ans =
    0.0199
>> [X,Y] = meshgrid(0:0.01:1 , 0:0.01:1);
>> Z = f(X,Y);
>> surf(X,Y,Z)
```

Please note how the function is defined using the dot operator, in order to receive vector or matrix arguments and to compute the vector or matrix of the corresponding function values. The resulting surface is illustrated in figure 4.34. It is easy to see that Monte Carlo estimates with 10,000 sampled points are not reliable:

```
>> rand('state',0);
>> mean(f(rand(1,10000),rand(1,10000)))
ans =
    0.0276
>> mean(f(rand(1,10000),rand(1,10000)))
ans =
    0.0332
>> mean(f(rand(1,10000),rand(1,10000)))
ans =
    0.0098
```

So, we may try with Halton sequences, changing the bases and keeping the same number of samples:

```
>> seq2 = GetHalton(10000,2);
>> seq4 = GetHalton(10000,4);
>> seq5 = GetHalton(10000,5);
>> seq7 = GetHalton(10000,7);
>> mean(f(seq2,seq5))
ans =
    0.0200
>> mean(f(seq2,seq4))
ans =
    0.0224
>> mean(f(seq2,seq7))
ans =
    0.0199
>> mean(f(seq5,seq7))
ans =
    0.0198
```

276 NUMERICAL INTEGRATION: DETERMINISTIC AND MONTE CARLO METHODS

We see that, provided that we use prime numbers as the bases, the results are much more accurate. It is also instructive to compare the results for a small number of samples.

```
>> rand('state',0)
>> mean(f(rand(1,100),rand(1,100)))
ans =
    -0.0032
>> mean(f(rand(1,500),rand(1,500)))
ans =
    0.0197
>> mean(f(rand(1,1000),rand(1,1000)))
ans =
    0.0577
>> mean(f(rand(1,1500),rand(1,1500)))
ans =
    0.0461
>> mean(f(rand(1,2000),rand(1,2000)))
ans =
    0.0311

>> mean(f(seq2(1:100),seq7(1:100)))
ans =
    0.0267
>> mean(f(seq2(1:500),seq7(1:500)))
ans =
    0.0197
>> mean(f(seq2(1:1000),seq7(1:1000)))
ans =
    0.0210
>> mean(f(seq2(1:1500),seq7(1:1500)))
ans =
    0.0190
>> mean(f(seq2(1:2000),seq7(1:2000)))
ans =
    0.0197
```

The potential advantage of low-discrepancy sequences is evident even if the optimal choice of bases is an issue. ☐

Example 4.17 As a more practical exercise, we may try pricing the usual vanilla European call using a low-discrepancy sequence. We use here the simplest sequence, the Halton sequence. To generate normal variates, we may either use the Box–Muller method, which we described in section 4.3.4 or the inverse transform method. We *cannot* apply polar rejection, because when using low discrepancy sequences we must integrate over a space with a well-defined dimensionality. We must know exactly how many quasi-random numbers we need, whereas with rejection-based methods we cannot anticipate

```
function Price = BlsHaltonBM(S0,K,r,T,sigma,NPoints,Base1,Base2)
nuT = (r - 0.5*sigma^2)*T;
siT = sigma * sqrt(T);
% Use Box Muller to generate standard normals
H1 = GetHalton(ceil(NPoints/2),Base1);
H2 = GetHalton(ceil(NPoints/2),Base2);
VLog = sqrt(-2*log(H1));
Norm1 = VLog .* cos(2*pi*H2);
Norm2 = VLog .* sin(2*pi*H2);
Norm = [Norm1 ; Norm2];
%
DiscPayoff = exp(-r*T) * max( 0 , S0*exp(nuT+siT*Norm) - K);
Price = mean(DiscPayoff);
```

Fig. 4.35 Using Halton sequences and Box–Muller algorithm to price a vanilla European call.

that. This is an important remark to keep in mind when pricing complex options.

We recall the Box–Muller algorithm here for convenience. To generate two independent standard normal variates, we should first generate two independent random numbers U_1 and U_2, and then set

$$X = \sqrt{-2\ln U_1}\cos(2\pi U_2)$$
$$Y = \sqrt{-2\ln U_1}\sin(2\pi U_2).$$

Rather than generating pseudorandom numbers, we may use two Halton sequences with two prime numbers as bases. This is accomplished by the code displayed in figure 4.35.

An alternative approach is based on the inverse transform method. Given the potentially weird effects of the Box–Muller transformation, which we have illustrated in figure 4.12 on page 238, one could argue that this is a safer approach. The code is given in figure 4.36

Let us check first the use of Halton sequences with Box–Muller transformation first:

```
>> blsprice(50,52,0.1,5/12,0.4)
ans =
    5.1911
>> BlsHaltonBM(50,52,0.1,5/12,0.4,5000,2,7)
ans =
    5.1970
>> BlsHaltonBM(50,52,0.1,5/12,0.4,5000,11,7)
ans =
    5.2173
```

```
function Price = BlsHaltonINV(S0,K,r,T,sigma,NPoints,Base)
nuT = (r - 0.5*sigma^2)*T;
siT = sigma * sqrt(T);
% Use inverse transform to generate standard normals
H = GetHalton(NPoints,Base);
Veps = norminv(H);
%
DiscPayoff = exp(-r*T)*max(0,S0*exp(nuT+siT*Veps)-K);
Price = mean(DiscPayoff);
```

Fig. 4.36 Using Halton sequences and inverse transform to price a vanilla European call.

```
>> BlsHaltonBM(50,52,0.1,5/12,0.4,5000,2,4)
ans =
    6.2485
```

The first run shows the potential of low-discrepancy sequences; we get a good estimate of the option with a limited number of samples. It is instructive to see the variability of a Monte Carlo estimate with 5000 samples:

```
>> randn('state',0)
>> BlsMC2(50,52,0.1,5/12,0.4,5000)
ans =
    5.2549
>> BlsMC2(50,52,0.1,5/12,0.4,5000)
ans =
    5.1090
>> BlsMC2(50,52,0.1,5/12,0.4,5000)
ans =
    5.2777
```

From the second run with Halton sequences, we also see that the quality of the estimate may depend on the choice of the bases; the third run shows that using a non-prime number as a basis yields a very poor result.

Using the inverse transform, an interesting pattern emerges:

```
>> BlsHaltonINV(50,52,0.1,5/12,0.4,1000,2)
ans =
    5.1094
>> BlsHaltonINV(50,52,0.1,5/12,0.4,2000,2)
ans =
    5.1469
>> BlsHaltonINV(50,52,0.1,5/12,0.4,5000,2)
ans =
    5.1688
```

```
>> BlsHaltonINV(50,52,0.1,5/12,0.4,10000,2)
ans =
    5.1789
>> BlsHaltonINV(50,52,0.1,5/12,0.4,50000,2)
ans =
    5.1879
```

We see that prices *look* monotonically increasing with respect to the number of samples. This is not really the case, as a detailed plot of the price as a function of number of samples would show that there are oscillations, yet there is a tendency for the price to increase from below. We can try to find a reason for this trend: Using Halton sequence with base 2, we fill the unit interval with consecutive runs from a low extreme to a high extreme, according to the following scheme:

0.5

0.25   0.75

0.125   0.625   0.375   0.875

0.0625   0.5625   0.3125   0.8125   0.1875   0.6875   0.4375   0.9375

0.0313   ...

Each subsequence is delimited by the new lowest and the new highest point. We see that the current maximum found so far increases according to a regular pattern; and high values of these numbers correspond to large prices of the underlying asset, which are those contributing to the increase of the option price.

If we use 17 as the basis, we see longer monotonically increasing sequences:

```
>> GetHalton(17,17)
ans =
    0.0588
    0.1176
    0.1765
    0.2353
    0.2941
    0.3529
    0.4118
    0.4706
    0.5294
    0.5882
    0.6471
    0.7059
    0.7647
    0.8235
    0.8824
    0.9412
    0.0035
```

280 NUMERICAL INTEGRATION: DETERMINISTIC AND MONTE CARLO METHODS

Hence, it is not surprising that if we use a large prime number as the basis, the price we get is, in a sense, "more low-biased":

```
>> BlsHaltonINV(50,52,0.1,5/12,0.4,1000,499)
ans =
    5.1139
>> BlsHaltonINV(50,52,0.1,5/12,0.4,2000,499)
ans =
    5.1141
>> BlsHaltonINV(50,52,0.1,5/12,0.4,5000,499)
ans =
    5.1148
>> BlsHaltonINV(50,52,0.1,5/12,0.4,10000,499)
ans =
    5.1159
>> BlsHaltonINV(50,52,0.1,5/12,0.4,50000,499)
ans =
    5.1252
```

Using a large base, even if it is a prime number, has an even more detrimental effect if we use the Box–Muller transformation:

```
>> BlsHaltonBM(50,52,0.1,5/12,0.4,5000,59,83)
ans =
    5.3232
>> BlsHaltonBM(50,52,0.1,5/12,0.4,5000,101,103)
ans =
    6.0244
```

□

To understand why using large bases is a bad idea, we may plot the first 1000 points in the bidimensional sequence when 109 and 113 are used:

```
>> plot(GetHalton(1000,109), GetHalton(1000,113), 'o')
```

yields the plot displayed in figure 4.37. The result should be compared against figure 4.32.

Since pricing certain options is a high-dimensional problem, straightforward use of Halton sequences is not feasible, as this would require using large bases. As an alternative, Faure sequences have been proposed. The basic idea in Faure sequences is using only one base, a prime number which must be greater than problem dimensionality; coordinates are generated by suitable permutations of Van der Corput sequences. This net effect is using a smaller base than the largest one used by Halton sequences. Another alternative is represented by Sobol sequences, which are discussed in the next section. In Sobol sequences only the base 2 is used, which is good. In order to generate multidimensional sequences, the Van der Corput sequence with base 2 is permuted by a mechanism linked to polynomials in a binary arithmetic.

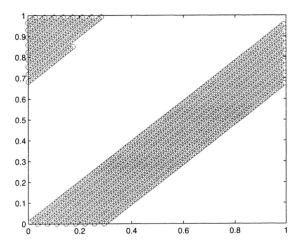

Fig. 4.37 Poor coverage of the unit square when large bases are used in Halton sequences.

4.6.2 Generating Sobol low-discrepancy sequences

In this section we would like at least to take a look at a more sophisticated alternative than Halton sequences, i.e., Sobol sequences. For the sake of clarity, it is better to consider the generation of a one-dimensional sequence x^n in the $[0,1]$ interval. A Sobol sequence is generated on the basis of a set of "direction numbers" $v_1, v_2, \ldots$; we will see shortly how direction numbers are selected, but for now just think of them as numbers which are less than 1. To get the nth number in the sequence, consider the binary representation of the integer n:

$$n = (\ldots b_3 b_2 b_1)_2.$$

The result is obtained by computing the bitwise exclusive or of the direction numbers v_i for which $b_i \neq 0$:

$$x^n = b_1 v_1 \oplus b_2 v_2 \oplus \cdots . \tag{4.11}$$

If direction numbers are chosen properly, a low-discrepancy sequence will be generated [18]. A direction number may be thought as a binary fraction:

$$v_i = (0.v_{i1} v_{i2} v_{i3} \ldots)_2,$$

or as

$$v_i = \frac{m_i}{2^i},$$

where $m_i < 2^i$ is an odd integer. To generate direction numbers, we exploit primitive polynomials over the field $\mathbb{Z}_2$, i.e., polynomials with binary coefficients:

$$P = x^d + a_1 x^{d-1} + \cdots + a_{d-1} x + 1, \qquad a_k \in \{0,1\}.$$

Irreducible polynomials are those polynomials which cannot be factored; primitive polynomials are a subset of the irreducible polynomials and are strongly linked to the theory of error-correcting codes, which is beyond the scope of the book. Some irreducible polynomials over the field $\mathbb{Z}_2$ are listed, e.g., in [13, chapter 7], to which the reader is referred for further information. Given a primitive polynomial of degree d, the procedure for generating direction numbers is based on the recurrence formula

$$v_i = a_1 v_{i-1} \oplus a_2 v_{i-2} \oplus \cdots \oplus a_{d-1} v_{i-d+1} \oplus v_{i-d} \oplus [v_{i-d}/2^d], \qquad i > d.$$

This is better implemented in integer arithmetic as

$$m_i = 2 a_1 m_{i-1} \oplus 2^2 a_2 m_{i-2} \oplus \cdots \oplus 2^{d-1} a_{d-1} m_{i-d+1} \oplus 2^d m_{i-d} \oplus m_{i-d}.$$

Some numbers $m_1, \ldots, m_d$ are needed to initialize the recursion. They may be chosen arbitrarily, provided that each m_i is odd and $m_i < 2^i$.

Example 4.18 As an example, let us build the set of direction numbers on the basis of the primitive polynomial

$$x^3 + x + 1.$$

The recursive scheme runs as follows:

$$m_1 = 4 m_{i-2} \oplus 8 m_{i-3} \oplus m_{i-3},$$

which may be initialized with $m_1 = 1$, $m_2 = 3$, $m_3 = 7$.[9] We may carry out the necessary computations step by step in MATLAB, using the bitxor function.

```
>> m = [1 3 7];
>> i=4;
>> m(i) = bitxor( 4 * m(i-2) , bitxor(8*m(i-3) , m(i-3)));
>> i=5;
>> m(i) = bitxor( 4 * m(i-2) , bitxor(8*m(i-3) , m(i-3)));
>> i=6;
>> m(i) = bitxor( 4 * m(i-2) , bitxor(8*m(i-3) , m(i-3)));
>> m
m =
      1     3     7     5     7    43
```

---

[9]The reasons why this may be a good choice are given in [3].

```
function [v, m] = GetDirNumbers(p,m0,n)
degree = length(p)-1;
p = p(2:degree);
m = [ m0 , zeros(1,n-degree) ];
for i= (degree+1):n
   m(i) = bitxor(m(i-degree), 2^degree * m(i-degree));
   for j=1:(degree-1)
      m(i) = bitxor(m(i), 2^j * p(j) * m(i-j));
   end
end
v=m./(2.^(1:length(m)));
```

Fig. 4.38 MATLAB code to generate direction numbers for Sobol sequences.

Given the integer numbers m_i, we may build the direction numbers v_i. To implement the generation of direction numbers, we may use a function like GetDirNumbers, which is given in figure 4.38. The function requires a primitive polynomial p, a vector of initial numbers m, and the number n of direction numbers we want to generate. On exit we obtain the direction numbers v and the integer numbers m.

```
>> p = [1 0 1 1];
>> m0 = [1 3 7];
>> [v,m]=GetDirNumbers(p,m0,6)
v =
    0.5000    0.7500    0.8750    0.3125    0.2188    0.6719
m =
    1    3    7    5    7    43
```

The code is not optimized; for instance, the first and last coefficients of the polynomial should be 1 by default, and no check is done on the congruence in size of the input vectors. □

After computing the direction numbers, we could generate a Sobol sequence according to equation (4.11). However, an improved method was proposed by Antonov and Saleev [1], who proved that the discrepancy is not changed by using the Gray code representation of n. Gray codes are discussed, e.g., in [13, chapter 20]; all we need to know is the following:

1. A Gray code is a function mapping an integer i to a corresponding binary representation $G(i)$; the function, for a given integer N, is one-to-one for $0 \leq i \leq 2^N - 1$.

2. A Gray code representation for the integer n is obtained from its binary representation by computing

$$\ldots g_3 g_2 g_1 = (\ldots b_3 b_2 b_1)_2 \oplus (\ldots b_4 b_3 b_2)_2.$$

3. The main feature of such a code is that the codes for consecutive numbers n and $n+1$ differ only in one position.

Example 4.19 Computing a Gray code is easily accomplished in MATLAB. For instance, we may define an inline function and compute the Gray codes for the numbers $i = 0, 1, \ldots, 15$ as follows:

```
>> gray = inline('bitxor(x,bitshift(x,-1))');
>> codes = zeros(16,4);
>> for i=1:16, codes(i,:)=bitget(gray(i-1), [4 3 2 1]);, end
>> codes
codes =
     0     0     0     0
     0     0     0     1
     0     0     1     1
     0     0     1     0
     0     1     1     0
     0     1     1     1
     0     1     0     1
     0     1     0     0
     1     1     0     0
     1     1     0     1
     1     1     1     1
     1     1     1     0
     1     0     1     0
     1     0     1     1
     1     0     0     1
     1     0     0     0
```

We have used the function `bitshift` to shift the binary representation of x one position to the right and the function `bitget` to get specific bits of the binary representation of a number. We see that indeed the Gray codes for consecutive numbers i and $i+1$ differ in one position; that position corresponds to the rightmost zero bit in the binary representation of i (adding leading zeros if necessary). □

Using the feature of Gray codes, we may streamline generation of a Sobol sequence. Given x^n, we have

$$x^{n+1} = x^n \oplus v_c,$$

where c is the index of the rightmost zero bit b_c in the binary representation of n.

Example 4.20 To implement the mechanism in MATLAB, we need a way to find the rightmost zero bit in the binary representation of a number. A function like the following one will do (provided that at most eight bits are used to represent x):

```
function SobSeq = GetSobol(GenNumbers, x0, HowMany)
Nbits = 20;
factor = 2^Nbits;
BitNumbers = GenNumbers * factor;
SobSeq = zeros(HowMany + 1, 1);
SobSeq(1) = fix(x0*factor);
for i=1:HowMany
   c = min(find( bitget(i-1,1:16) == 0));
   SobSeq(i+1) = bitxor(SobSeq(i), BitNumbers(c));
end
SobSeq = SobSeq / factor;
```

Fig. 4.39 MATLAB code to generate a Sobol sequence by the Antonov and Saleev approach.

```
rightbit = inline('min(find( bitget(x,1:8) == 0))')
```

Now we may put it all together. First, we generate the direction numbers. Then we initialize the sequence in some way, e.g., $x^0 = 0$, and apply the code of figure 4.39. The code is straightforward; the only point is that in theory we should compute the exclusive or on bits of a binary fraction; however, `bitxor` works on integer numbers only. This is why we shift everything to the left by `Nbits` position, which is accomplished multiplying by `factor` and dividing on exit from the function. Also, we truncate the initial number in order to make sure that we are "xoring" integer numbers.

```
>> p = [1 0 1 1];
>> m0 = [1 3 7];
>> [v,m]=GetDirNumbers(p,m0,6);
>> GetSobol(v,0,10)
ans =
         0
    0.5000
    0.2500
    0.7500
    0.1250
    0.6250
    0.3750
    0.8750
    0.6875
    0.1875
    0.9375
```

Using a different set of generating numbers and a different starting point, we generate different sequences.

```
>> p = [1 0 1 1 1 1];
>> m0 = [1 3 5 9 11];
>> [v,m]=GetDirNumbers(p,m0,8);
>> GetSobol(v,0.124,10)
ans =
    0.1240
    0.6240
    0.3740
    0.8740
    0.4990
    0.9990
    0.2490
    0.7490
    0.1865
    0.6865
    0.4365
```

Note that to generate longer sequences, more generating numbers are needed.

□

For further reading

In the literature

- For a general introduction to simulation, see [9] or [15], both of which have heavily influenced the presentation in this chapter; [14] is another classical reference.

- For a more theoretical treatment of Monte Carlo simulation and random number generation, see [4]. The random number generators used in MATLAB are described in [11].

- Low-discrepancy sequences are treated in [12], which is at a quite advanced level.

- An excellent and very readable introduction to Monte Carlo and quasi-Monte Carlo methods in finance is [5]. See also [7] for a discussion on selecting primitive polynomials for Sobol sequences. A table of primitive polynomials is also given in [13].

- See [8] for an early account on the use of low-discrepancy sequences within financial engineering.

On the Web

- For a list of resources on Monte Carlo and quasi-Monte Carlo simulation, see http://www.mcqmc.org.

- See also http://www.mat.sbg.ac.at/~schmidw/links.html.

REFERENCES

1. I.A. Antonov and V.M. Saleev. An Economic Method of Computing LP_τ Sequences. *USSR Computational Mathematics and Mathematical Physics*, 19:252–256, 1979.

2. I. Beichl and F. Sullivan. The Importance of Importance Sampling. *Computing in Science and Engineering*, 1:71–73, March-April 1999.

3. P. Bratley and B.L. Fox. Algorithm 659: Implementing Sobol's Quasirandom Sequence Generator. *ACM Transactions on Mathematical Software*, 14:88–100, 1988.

4. G.S. Fishman. *Monte Carlo: Concepts, Algorithms, and Applications*. Springer-Verlag, Berlin, 1996.

5. P. Glasserman. *Monte Carlo Methods in Financial Engineering*. Springer-Verlag, New York, NY, 2004.

6. J.C. Hull. *Options, Futures, and Other Derivatives (5th ed.)*. Prentice Hall, Upper Saddle River, NJ, 2003.

7. P. Jaeckel. *Monte Carlo Methods in Finance*. Wiley, Chichester, 2002.

8. C. Joy, P.P. Boyle, and K.S. Tan. Quasi-Monte Carlo Methods in Numerical Finance. *Management Science*, 42:926–938, 1996.

9. A.M. Law and W.D. Kelton. *Simulation Modeling and Analysis (3rd ed.)*. McGraw-Hill, New York, 1999.

10. M.J. Miranda and P.L. Fackler. *Applied Computational Economics and Finance*. MIT Press, Cambridge, MA, 2002.

11. C. Moler. Random Thoughts. *Matlab News & Notes*, pages 2–3, Fall 1995. This paper may be downloaded from The Mathworks' Web site at http://www.mathworks.com/company/newsletter/pdf/Cleve.pdf.

12. H. Niederreiter. *Random Number Generation and Quasi-Monte Carlo Methods*. Society for Industrial and Applied Mathematics, Philadelphia, PA, 1992.

13. W.H. Press, S.A. Teukolsky, W.T. Vetterling, and B.P. Flannery. *Numerical Recipes in C (2nd ed.)*. Cambridge University Press, Cambridge, 1992.

14. B.D. Ripley. *Stochastic Simulation.* Wiley, New York, 1987.

15. S. Ross. *Simulation.* Academic Press, San Diego, CA, 1997.

16. S. Ross. *Introduction to Probability Models (8th ed.).* Academic Press, San Diego, CA, 2002.

17. R.Y. Rubinstein. *Simulation and the Monte Carlo Method.* Wiley, Chichester, 1981.

18. I.M. Sobol. On the Distribution of Points in a Cube and the Approximate Evaluation of Integrals. *USSR Computational Mathematics and Mathematical Physics*, 7:86–112, 1967.

5
Finite Difference Methods for Partial Differential Equations

Partial differential equations (PDEs) play a major role in financial engineering. Since the seminal work leading to the Black–Scholes equation, which we introduced in section 2.6.2, PDEs have become an important tool in option valuation. It turns out that PDEs provide a powerful and consistent framework for pricing rather complex derivatives. Unfortunately, as analytical solutions like the Black and Scholes formula are not available in general, one must often resort to numerical methods.

The numerical solution of PDEs is a common tool in mathematical physics and engineering, and quite sophisticated methods have been developed. The complexity of the methods also depends on the specific type of PDE at hand. As expected, non-linear equations are generally more difficult than linear ones, but there is also a subtler dependence on numerical parameters, since a change in the value of a coefficient may drastically change the characteristics of an equation. In the financial engineering case, it happens that in many cases rather simple methods are enough to obtain a reasonably accurate solution. Indeed, we deal here only with relatively straightforward finite difference methods, which are based on the natural idea of approximating partial derivatives with difference quotients. Even so, the topic is not as trivial as one may think, since careless use of finite difference schemes may lead to unreasonable results. In fact, while some authors suggest the use of PDEs as the single most useful tool in derivatives pricing [9, p. 615], others suggest that they are quite vulnerable to numerical difficulties and, while acknowledging the role of finite difference methods, they suggest the use of lattice-based methods whenever possible (see, e.g., [2, p. 365]). Actually, this is a bit a matter of taste, and when confident with a method, one is able to squeeze the most out of it. Fortunately, when numerical difficulties occur in solving a PDE for a financial problem, often the answers we get from the algorithm are so blatantly senseless that we may easily spot the trouble; in other cases, however, unreliable answers may have nasty effects. In this chapter we also introduce concepts related to convergence, consistency, and stability in order to understand the basic issues connected with the numerical solution of PDEs. It should be stressed that PDEs are actually a difficult topic requiring advanced mathematical concepts for a rigorous treatment, and as usual we will rely mostly on relatively informal arguments and intuition.

We first classify PDEs in section 5.1. Then in section 5.2 we introduce different ways to approximate partial derivatives by finite differences, leading to different solution schemes which may turn out numerically stable or unstable. We devote a particular attention to the heat equation, which is the subject of section 5.3, since the Black–Scholes PDE is strongly linked to diffusion processes. We generalize to multiple spatial dimensions in section 5.4, where we consider the heat equation in two dimensions; the Alternating Direction Implicit approach is described. Finally, in section 5.5 we briefly point out a few theoretical concepts concerning the convergence of finite difference methods.

5.1 INTRODUCTION AND CLASSIFICATION OF PDEs

We introduced the Black–Scholes PDE in section 2.6.2 to find the theoretical price $f(S, t)$ of a derivative security depending on the price S of one underlying asset at time t. Using a stochastic differential equation to model the dynamics of the underlying asset price and using no arbitrage arguments, we have found that f must satisfy the PDE

$$\frac{\partial f}{\partial t} + \frac{1}{2}\sigma^2 S^2 \frac{\partial^2 f}{\partial S^2} + rS\frac{\partial f}{\partial S} - rf = 0, \tag{5.1}$$

where r is the risk-free interest rate and σ is the asset price volatility. Suitable boundary conditions must be added to find a specific solution corresponding to the option type we are considering. This equation has various features:

- It is second-order.

- It is linear.

- It is a parabolic equation.

All these features refer to how PDEs are classified; such a classification is relevant in that the choice of a numerical method to cope with a PDE generally depends on its characteristics.

In order to classify PDEs, let us abstract from the financial interpretation of the variables involved and refer to an unknown function $\phi(x, y)$, depending on variables x and y; for simplicity we deal with a function of two independent variables only, but the classification scheme may be applied in a more general setting. The *order* of a PDE is the highest order of the derivatives involved. For instance, a generic first-order equation has the form

$$a(x,y)\frac{\partial \phi}{\partial x} + b(x,y)\frac{\partial \phi}{\partial y} + c(x,y)\phi + d(x,y) = 0,$$

where a, b, c, d are given functions of the independent variables. This equation is first-order since only first-order derivatives are involved. Furthermore, it is linear, since the functions a, b, c, and d depend only on the independent variables x and y and not on ϕ itself. By the same token, the generic form of a linear second-order equation is

$$a\frac{\partial^2 \phi}{\partial x^2} + b\frac{\partial^2 \phi}{\partial x \partial y} + c\frac{\partial^2 \phi}{\partial y^2} + d\frac{\partial \phi}{\partial x} + e\frac{\partial \phi}{\partial y} + f\phi + g = 0,$$

where again all the given functions, from a to g, depend only on x and y. An example of a first-order non-linear equation is

$$\left(\frac{\partial \phi}{\partial x}\right)^2 + \left(\frac{\partial \phi}{\partial y}\right)^2 = 1. \qquad (5.2)$$

An example of a second-order non-linear equation is

$$a\left(x, y, \frac{\partial \phi}{\partial y}\right)\frac{\partial^2 \phi}{\partial x^2} + d(x, y, \phi)\frac{\partial \phi}{\partial x} + e(x, y)\frac{\partial \phi}{\partial y} + f(x, y)\phi = 0. \qquad (5.3)$$

Equation (5.3) is non-linear but in a different way than (5.2). In this equation, the coefficient a of the highest-order derivative depends only on the first-order derivative. We have a *quasilinear equation* whenever the highest-order derivatives occur linearly, with coefficients depending only on the independent variables, the unknown function ϕ, and its lower-order derivatives. For the sake of simplicity, in this introductory book we deal only with linear equations. It should be noted that while most of the models you will see in finance are linear, non-linear equations may be obtained when relaxing some of the assumptions behind the Black–Scholes model; for an example of a non-linear equation that arises when introducing transaction costs, see [9, chapter 21].

It is customary to classify quasilinear second-order equations depending on the sign of the expression $b^2 - 4ac$:

- If $b^2 - 4ac > 0$, the equation is hyperbolic.
- If $b^2 - 4ac = 0$, the equation is parabolic.
- If $b^2 - 4ac < 0$, the equation is elliptic.

It is easy to see that the discriminant term $b^2 - 4ac$ is formally similar to the analogous term we have in second-degree algebraic equations. Elliptic equations may arise in equilibrium models (where time is not involved). A typical example is the Laplace equation

$$\frac{\partial^2 \phi}{\partial x^2} + \frac{\partial^2 \phi}{\partial y^2} = 0.$$

Here we have $a = c = 1$ and $b = 0$, so that $b^2 - 4ac = -4 < 0$. The wave equation

$$\frac{\partial^2 \phi}{\partial t^2} - \rho^2 \frac{\partial^2 \phi}{\partial x^2} = 0,$$

where t is time, is a typical example of a hyperbolic equation, since the discriminant term is $4\rho^2 > 0$. The prototype parabolic equation is the *heat* (or *diffusion*) *equation*:

$$\frac{\partial \phi}{\partial t} = k \frac{\partial^2 \phi}{\partial x^2},$$

where t is time and ϕ is the temperature of a point with coordinate x on a line. In this case, $b^2 - 4ac = 0$. By a change of variables, the equation may be cast into a dimensionless form:

$$\frac{\partial \phi}{\partial t} = \frac{\partial^2 \phi}{\partial x^2}. \qquad (5.4)$$

Now consider the Black–Scholes equation; again $b = c = 0$, so the equation is parabolic. This does not happen by chance, since with a transformation of coordinates it can be shown that the Black–Scholes equation actually boils down to the heat equation.

An equation like (5.4) must be integrated with suitable conditions in order to pinpoint a meaningful solution. For instance, assume that $\phi(x,t)$ is the "temperature" at point $x \in [0,1]$ of a rod of length 1 at time t; the end points are kept at a constant temperature u_0, and the initial temperature of the rod is given over all of its length. Then we must add the initial condition

$$\phi(x, 0) = u(x), \qquad 0 \leq x \leq 1,$$

and the boundary conditions

$$\phi(0, t) = \phi(1, t) = u_0, \qquad t > 0.$$

Here the domain is bounded with respect to space and unbounded with respect to time. In financial problems, the initial condition is usually replaced by a terminal condition, as the option payoff is known at expiration; therefore, the time domain is bounded, whereas the domain with respect to the price of the underlying asset may be (in principle) unbounded. From a computational point of view, the domain must be limited in some sensible way. Boundary conditions are easy to spot for vanilla European options. With exotic options, enforcing boundary conditions may be more complicated, e.g., when the boundary conditions must themselves be approximated by some numerical scheme. In other cases, such as barrier options, the boundary conditions may actually result in a simplification of the problem. American options raise another issue; for each time before expiration, there is a critical value for the price of the underlying asset at which it is optimal to exercise the option (see figure 2.22 on page 118); depending on the option type (call or put), it will also be optimal to exercise the option for prices above and below the critical price.[1] So with American options we should cope with a *free boundary*, i.e., a boundary within the domain, which separates the exercise and no-exercise region. We deal with these issues in chapter 9.

A noteworthy feature of the heat equation is that any discontinuity in the initial conditions is somehow smoothed out, so that the solution for $t > 0$ is differentiable everywhere. On the contrary, in the wave equation, the irregularities are propagated along lines called characteristics.[2] Another feature of parabolic equations is that they are relatively easy to work with from the numerical point of view.

A final remark is that the form of the equation and the boundary conditions determine if a given problem involving a PDE is *well-posed*. A problem is well-posed if:

- There exists a solution.

- The solution is unique (at least within a certain class of functions of interest).

- The solution depends in a nice way on the problem data (i.e., a small perturbation in the problem data results in a small perturbation of the solution).

We will trust our intuition that the equations we write make sense and will assume implicitly that all our problems are well-posed.

5.2 NUMERICAL SOLUTION BY FINITE DIFFERENCE METHODS

Finite difference methods to solve PDEs are based on the simple idea of approximating each partial derivative by a difference quotient. This transforms the functional equation into a set of algebraic equations. As in many numerical algorithms, the starting point is a finite series approximation. Under suitable continuity and differentiability hypotheses, Taylor's theorem states

---

[1] Recall that a vanilla American call should be never exercised unless the stock pays dividends.
[2] In hyperbolic equations, two characteristic lines exist, and this is actually linked to the fact that the discriminant $b^2 - 4ac$ is positive, a property that is linked to the existence of two roots in algebraic second-order equations.

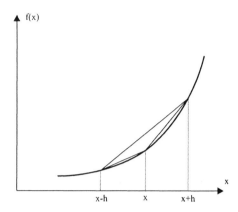

Fig. 5.1 Graphical illustration of forward, backward, and central approximations of a derivative.

that a function $f(x)$ may be represented as

$$f(x+h) = f(x) + hf'(x) + \frac{1}{2}h^2 f''(x) + \frac{1}{6}h^3 f'''(x) + \cdots. \qquad (5.5)$$

If we neglect the terms of order h^2 and higher, we get

$$f'(x) = \frac{f(x+h) - f(x)}{h} + O(h). \qquad (5.6)$$

This is the *forward* approximation for the derivative; indeed, the derivative is just defined as a limit of the difference quotient above as $h \to 0$. There are alternative ways to approximate first-order derivatives. By similar reasoning, we may write

$$f(x-h) = f(x) - hf'(x) + \frac{1}{2}h^2 f''(x) - \frac{1}{6}h^3 f'''(x) + \cdots, \qquad (5.7)$$

from which we obtain the *backward* approximation,

$$f'(x) = \frac{f(x) - f(x-h)}{h} + O(h). \qquad (5.8)$$

In both cases we get a truncation error of order $O(h)$. A better approximation can be obtained by subtracting equation (5.7) from equation (5.5) and rearranging:

$$f'(x) = \frac{f(x+h) - f(x-h)}{2h} + O(h^2). \qquad (5.9)$$

This is the *central* or *symmetric* approximation, and for small h it is a better approximation, since the truncation error is $O(h^2)$. Why this is the case may also be seen from figure 5.1. However, this does not imply that forward and backward approximations must be disregarded; they may be useful to come up with efficient numerical schemes, depending on the type of boundary conditions.

The reasoning may be extended to higher-order derivatives. To cope with the Black–Scholes equation, we must approximate second-order derivatives, too. This is obtained by adding equations (5.5) and (5.7), which yields

$$f(x+h) + f(x-h) = 2f(x) + h^2 f''(x) + O(h^4),$$

and rearranging yields

$$f''(x) = \frac{f(x+h) - 2f(x) + f(x-h)}{h^2} + O(h^2). \tag{5.10}$$

In order to apply the ideas above to a PDE involving a function $\phi(x, y)$, it is natural to set up a discrete grid of points of the form $(i\,\delta x, j\,\delta y)$, where δx and δy are discretization steps, and to look for the values of ϕ on this grid. It is customary to use the grid notation:

$$\phi_{ij} = \phi(i\,\delta x, j\,\delta y).$$

Depending on the type of equation and on how the derivatives are approximated, we obtain a set of algebraic equations which may be more or less easily solved. A possible difficulty is represented by boundary conditions. If the equation is defined over a rectangular domain in the (x, y) space, it is easy to set up a grid such that the boundary points are on the grid. Other cases might not be so easy, and a sensible way to approximate the boundary conditions must be devised. Nevertheless, we would expect that for $\delta x, \delta y \to 0$ the solution of this set of equations converges (in some sense) to the solution of the PDE. Actually, this is not granted at all, as different complications may arise.

5.2.1 Bad example of a finite difference scheme

Consider the following example of a first-order linear equation:[3]

$$\frac{\partial \phi}{\partial t} + c\frac{\partial \phi}{\partial x} = 0, \tag{5.11}$$

where $\phi = \phi(x, t)$, $c > 0$, and the initial condition

$$\phi(x, 0) = f(x) \qquad \forall x$$

is given. It is easy to verify that the solution is of the form

$$\phi(x, t) = f(x - ct);$$

---

[3] The example is taken from [1, chapter 2].

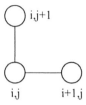

Fig. 5.2 Representing a finite difference scheme by a computational diagram.

in other words, the solution is simply a translation of $f(x)$ with velocity of propagation c. In fact, this type of equation is called the *transport equation*. A real transport equation typically involves a function $c(x)$ rather than a constant velocity c. We take for granted that the problem is well-posed, and we do not check the uniqueness of the solution (see [1, pp. 21–25] for a thorough discussion). Now let us ignore what we know about the solution and try a finite difference scheme based on forward approximations. Equation (5.11) may be approximated by

$$\frac{\phi(x, t + \delta t) - \phi(x, t)}{\delta t} + c \frac{\phi(x + \delta x, t) - \phi(x, t)}{\delta x} + O(\delta t) + O(\delta x) = 0,$$

which, neglecting the truncation error and using the grid notation $x = i\,\delta x$, $t = j\,\delta t$, yields

$$\frac{\phi_{i,j+1} - \phi_{ij}}{\delta t} + c \frac{\phi_{i+1,j} - \phi_{ij}}{\delta x} = 0, \qquad (5.12)$$

with the initial condition

$$\phi_{i0} = f(i\delta x) = f_i \qquad \forall i.$$

In practice, in order to solve the problem on a computer, we should restrict the domain in some way, enforcing some limits on i and j. For now, we simply assume that we are interested in the solution for $t > 0$, thus $j = 1, 2, 3, \ldots$. Now, how can we solve equation (5.12) in a systematic way? If we consider equation (5.12) for $j = 0$, we see that values $\phi_{i+1,0}$ and ϕ_{i0} are involved, and they are known from the initial conditions; the only unknown value is $\phi_{i,1}$, which may be obtained as an explicit function of known values. In fact, solving for the unknown value, we get

$$\phi_{i,j+1} = \left(1 + \frac{c}{\rho}\right)\phi_{ij} - \frac{c}{\rho}\phi_{i+1,j}, \qquad (5.13)$$

where $\rho = \delta x/\delta t$. This computational scheme can be represented by the computational diagram depicted in figure 5.2, and it is easy to understand and implement. Unfortunately, it need not converge to the solution of the equation. Consider the following initial condition:

$$f(x) = \begin{cases} 0, & x < -1, \\ x + 1, & -1 \le x \le 0, \\ 1, & x > 0, \end{cases} \qquad (5.14)$$

which implies
$$\phi_{i0} = f(i\delta x) = 1 \quad \forall i \geq 0.$$
Now, using the computational scheme (5.13), for $j = 0$ we have
$$\phi_{i,1} = \left(1 + \frac{c}{\rho}\right)\phi_{i0} - \frac{c}{\rho}\phi_{i+1,0} = 1 \quad \forall i \geq 0.$$

Repeating this argument for any time instant $(j = 2, 3, \ldots)$, it is easily seen that, however small we take the discretization steps,
$$\phi_{ij} = 1, \quad i, j \geq 0,$$
which is certainly not the correct solution. Some readers might wonder if this is due to some irregularity in the initial values. In fact, the derivative of $f(x)$ is discontinuous at certain points, but it is easy to see that using a smoothed version of this function would not change the issue. This example also shows that non-differentiable functions may look like acceptable solutions of a PDE, which is a bit odd since derivatives are not defined everywhere for such functions; a rigorous investigation of this question leads to the concept of weak solution of a PDE [1].

5.2.2 Instability in a finite difference scheme

The example illustrated in the previous section shows that a numerically reasonable scheme, with a truncation error that tends to zero as discretization steps get smaller and smaller, may fail to converge. From a mathematical point of view, there is a non trivial interplay between concepts such as consistency, stability, and convergence. A full investigation calls for a deep treatment, and we will just briefly outline the concepts in section 5.5. From a more intuitive point of view, the reason for the failure of the previous finite difference scheme is that it does not reflect the physical propagation process, where the initial condition is translated "to the right" with respect to space. Hence, we could try and fix the problem by adopting the computational scheme represented in figure 5.3, which is obtained by using a backward difference for the partial derivative with respect to x. This yields

$$\frac{\phi_{i,j+1} - \phi_{ij}}{\delta t} + c\frac{\phi_{ij} - \phi_{i-1,j}}{\delta x} = 0, \quad (5.15)$$

and solving for $\phi_{i,j+1}$, we get the scheme

$$\phi_{i,j+1} = \left(1 - \frac{c}{\rho}\right)\phi_{ij} + \frac{c}{\rho}\phi_{i-1,j}. \quad (5.16)$$

Note that here $\phi_{i,j+1}$ still depends on the data at the previous time instant but "to the left" with respect to space. Let us try this scheme with MATLAB.

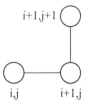

Fig. 5.3 Computational diagram of the modified scheme for the transport equation.

```
% f0transp.m
function y=f0transp(x)
if (x < -1)
   y=0;
elseif (x <= 0)
   y=x+1;
else
   y=1;
end
```

Fig. 5.4 Function to evaluate the initial values for the transport equation.

Example 5.1 In order to apply the computational scheme (5.16) with initial condition (5.14), we have to write a few M-files. In figure 5.4 we show code to evaluate the initial value at a given point x at $t = 0$. In figure 5.5 we see the MATLAB code for solving the equation. Note that we must truncate the domain between minimum and maximum x values, and with respect to time as well. We use a fixed value for the leftmost value in space, assuming that for smaller values of x the initial value is constant. Finally, the function TransportPlot illustrated in figure 5.6 is used to plot the numerical solution at different times: Four time subscripts are passed as an argument and the corresponding four plots are obtained. To begin with, we may solve the equation on the domain $-2 \leq x \leq 3$, $0 \leq t \leq 2$, with discretization steps $\delta x = 0.05$, $\delta t = 0.01$:

```
>> xmin = -2;
>> xmax = 3;
>> dx = 0.05;
>> tmax = 2;
>> dt = 0.01;
>> c = 1;
>> sol = transport(xmin, dx, xmax, dt, tmax, c, 'f0transp');
>> TransportPlot(xmin, dx, xmax, [1 51 101 201], sol)
```

We should note that, since array indexing in MATLAB starts from 1, the solution for $t = 2$ is in column 201 in the array. The solution, plotted in

```
% transport.m
function [solution, N, M] = transport(xmin, dx, xmax, dt, tmax, c, f0)
N = ceil((xmax - xmin) / dx);
xmax = xmin + N*dx;
M = ceil(tmax/dt);
k1 = 1 - dt*c/dx;
k2 = dt*c/dx;
solution = zeros(N+1,M+1);
vetx = xmin:dx:xmax;
for i=1:N+1
   solution(i,1) = feval(f0,vetx(i));
end
fixedvalue = solution(1,1);
% this is needed because of finite domain
for j=1:M
   solution(:,j+1) = k1*solution(:,j)+k2*[fixedvalue ; solution(1:N,j)];
end
```

Fig. 5.5 Code implementing the finite difference scheme for the transport equation.

```
% TransportPlot.m
function TransportPlot(xmin, dx, xmax, times, sol)
subplot(2,2,1)
plot(xmin:dx:xmax, sol(:,times(1)))
axis([xmin xmax -0.1 1.1])
subplot(2,2,2)
plot(xmin:dx:xmax, sol(:,times(2)))
axis([xmin xmax -0.1 1.1])
subplot(2,2,3)
plot(xmin:dx:xmax, sol(:,times(3)))
axis([xmin xmax -0.1 1.1])
subplot(2,2,4)
plot(xmin:dx:xmax, sol(:,times(4)))
axis([xmin xmax -0.1 1.1])
```

Fig. 5.6 Function for plotting the numerical solution of the transport equation.

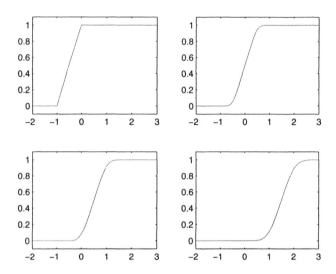

Fig. 5.7 Numerical solution of the transport equation for $\delta x = 0.05$, $\delta t = 0.01$; $t = 0$, $t = 0.5$, $t = 1$, and $t = 2$.

figure 5.7, gets progressively translated as we would expect, but it also looks progressively "smoothed." This could be due to a coarse discretization along the x axis. So we may try with $\delta x = 0.01$:

```
>> dx = 0.01;
>> sol = transport(xmin, dx, xmax, dt, tmax, c, 'f0transp');
>> TransportPlot(xmin, dx, xmax, [1 51 101 201], sol)
```

The solution is depicted in figure 5.8, and it looks much better. So, why don't we try a finer discretization, say $\delta x = 0.005$?

```
>> dx = 0.005;
>> sol = transport(xmin, dx, xmax, dt, tmax, c, 'f0transp');
>> TransportPlot(xmin, dx, xmax, [1 6 7 8], sol)
```

The solution we see in figure 5.9 is not really satisfactory. Something is definitely going wrong. □

As we may see, for certain settings of the discretization steps, the finite difference method is subject to numerical instability. By looking at equation (5.16), we may see that what we are doing is similar to a convex combination (i.e., an average) of two values; indeed, it will be a convex combination, provided that $c/\rho \geq 0$, which is the case as we assumed that $c > 0$, and $c/\rho \leq 1$, i.e.,

$$c\,\delta t \leq \delta x. \tag{5.17}$$

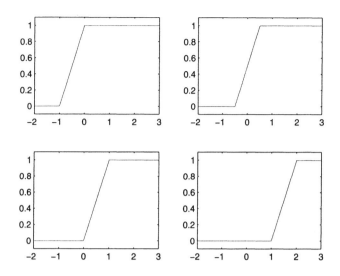

Fig. 5.8 Numerical solution of the transport equation for $\delta x = 0.01$, $\delta t = 0.01$; $t = 0$, $t = 0.5$, $t = 1$, and $t = 2$.

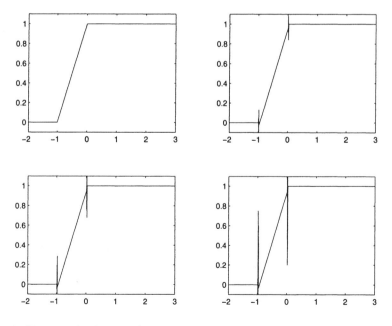

Fig. 5.9 Numerical solution of the transport equation for $\delta x = 0.005$, $\delta t = 0.01$; $t = 0$, $t = 0.05$, $t = 0.06$, and $t = 0.07$.

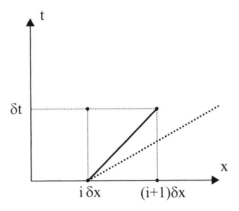

Fig. 5.10 Physical interpretation of the stability condition (5.17).

If this condition is not met, we have a negative coefficient in the linear combination (5.16); but if the initial data are positive, we would not expect negative quantities.

It is also possible to give a more physical interpretation of the stability condition (5.17) in terms of a domain of influence. Consider figure 5.10. Due to the structure of the numerical scheme (5.16), the value $\phi_{i+1,1}$ depends on the values ϕ_{i0} and $\phi_{i+1,0}$. The exact solution of the transport equation is such that the initial value at point $i\,\delta x$ should influence only the values on the characteristic[4] represented as a dotted line in figure 5.10. The slope of the characteristic line is $1/c$; the slope of the line joining the points $(i\,\delta x, 0)$ and $((i+1)\delta x, \delta t)$ is clearly $\delta t/\delta x$. In the figure this second line has a larger slope than the first one and the stability condition (5.17) is violated, since

$$\frac{\delta t}{\delta x} > \frac{1}{c}.$$

From a physical point of view this makes no sense, since in this case the numerical scheme is such that the initial value at point $i\,\delta x$ is influencing the value at a point *above* the characteristic line. In other words, the "speed" of the numerical scheme, $\delta x/\delta t$, should not be smaller than the transport speed c to ensure stability.

All of these considerations are nothing more than intuitive arguments. The instability problem may be analyzed rigorously in different ways. One approach, known as Von Neumann stability analysis, is related to Fourier analysis and is illustrated in the next example. Another approach, based on matrix theoretic arguments, will be illustrated in section 5.3, where we consider the heat equation. It should also be noted that in some cases a financial interpretation of instability may be given (see section 9.2.1).

---

[4]The characteristic is also a curve on which singularities in the solution may propagate.

Example 5.2 Consider again the transport equation, but with different initial values:
$$\phi(x,0) = f(x) = \epsilon \cos\left(\frac{\pi x}{\delta x}\right).$$
Since we know that the exact solution is $\phi(x,t) = f(x - ct)$, we see that the solution will be bounded everywhere, just like the initial values. Note also that after discretization we have a peculiar set of initial values on the grid:
$$\phi_{i,0} = \epsilon \cos\left(\frac{\pi i \, \delta x}{\delta x}\right) = \epsilon(-1)^i.$$

Going forward one layer of nodes in time, applying the scheme (5.16) yields
$$\begin{aligned}\phi_{i,1} &= \left(1 - \frac{c}{\rho}\right)\epsilon(-1)^i + \frac{c}{\rho}\epsilon(-1)^{i-1} = \left(1 - \frac{c}{\rho}\right)\epsilon(-1)^i - \frac{c}{\rho}\epsilon(-1)^i \\ &= \epsilon(-1)^i\left(1 - 2\frac{c}{\rho}\right).\end{aligned}$$

By the same token,
$$\begin{aligned}\phi_{i,2} &= \left(1 - \frac{c}{\rho}\right)\epsilon(-1)^i\left(1 - 2\frac{c}{\rho}\right) + \frac{c}{\rho}\epsilon(-1)^{i-1}\left(1 - 2\frac{c}{\rho}\right) \\ &= \epsilon(-1)^i\left(1 - 2\frac{c}{\rho}\right)^2,\end{aligned}$$

and in general we get
$$\phi_{ij} = \epsilon(-1)^i\left(1 - 2\frac{c}{\rho}\right)^j.$$

We see that the if the stability condition (5.17) is violated, i.e., if $c/\rho > 1$, we have
$$\left|1 - \frac{2c}{\rho}\right| > 1$$
and the initial data are amplified by a factor that goes to infinity for increasing values of j. □

5.3 EXPLICIT AND IMPLICIT METHODS FOR THE HEAT EQUATION

Let us consider the heat equation in dimensionless form:
$$\frac{\partial \phi}{\partial t} = \frac{\partial^2 \phi}{\partial x^2}.$$

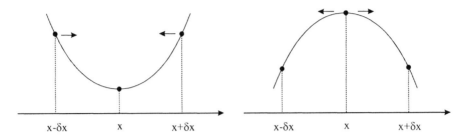

Fig. 5.11 Intuitive interpretation of the heat equation: heat flows from hot points to cold points.

With some work, the Black–Scholes equation can be transformed into this form, so it is worthwhile to investigate this equation in some detail. We also assume that the domain of interest is $x \in (0, 1)$ and $t \in (0, \infty)$; actually, in a practical scheme, we will also limit the domain with respect to time, $t \in (0, T)$. We have initial conditions for $t = 0$ and boundary conditions at $x = 0$ and $x = 1$ for any $t > 0$. We discretize with respect to x with a step δx, such that $N \delta x = 1$, and with respect to t with a step δt, such that $M \delta t = T$. Note that this results in a grid with $(N + 1) \times (M + 1)$ points.

Before proceeding with the treatment of standard methods for the heat equation, it may be useful to get an intuitive feeling for the physical sense of this equation. To this aim, let us consider figure 5.11. The figure on the left shows a temperature profile which is (at least) locally convex at point x. In this case, heat should diffuse from the warmer points $x - \delta x$ and $x + \delta x$ towards the center, and temperature in x should rise. In fact, the second-order derivative with respect to time is positive and the derivative with respect to time is positive as well. If the temperature profile is locally concave, in which case the second-order derivative is negative, heat should diffuse from the center to the left and to the right; temperature at point x should decrease, and its derivative with respect to time is negative.

In general, when we have a term like $\partial^2 \phi / \partial x^2$ in a PDE, it is called a *diffusion* term. In equation (5.11) we have seen that a term $\partial \phi / \partial x$ may be linked to transportation, or convection, phenomena. Indeed, an equation like

$$\frac{\partial \phi}{\partial t} + a \frac{\partial \phi}{\partial x} = b \frac{\partial^2 \phi}{\partial x^2} \qquad (5.18)$$

is called a *convection–diffusion* equation.

5.3.1 Solving the heat equation by an explicit method

A first possibility for coping with this equation is to approximate the derivative with respect to time by a forward approximation, and the second derivative

EXPLICIT AND IMPLICIT METHODS FOR THE HEAT EQUATION   305

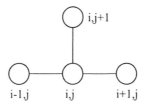

Fig. 5.12 Computational diagram of the explicit method for the heat equation.

by the approximation (5.10). This yields

$$\frac{\phi_{i,j+1} - \phi_{ij}}{\delta t} = \frac{\phi_{i+1,j} - 2\phi_{ij} + \phi_{i-1,j}}{(\delta x)^2}.$$

It is easy to see that we may rearrange this equation by solving for the unknown value $\phi_{i,j+1}$:

$$\phi_{i,j+1} = \rho \phi_{i-1,j} + (1 - 2\rho)\phi_{ij} + \rho \phi_{i+1,j}, \qquad (5.19)$$

where $\rho = \delta t/(\delta x)^2$.

Starting from the initial conditions ($j = 0$), we may solve the equation for increasing values of $j = 1, \ldots, M$. Note that for each j, i.e., for each layer in time, we must use equation (5.19) to find out $N-1$ values for $i = 1, \ldots, N-1$, as the remaining two are given by the boundary conditions. Since the unknown values are given by an explicit expression, this approach is called *explicit*. It can be represented by the computational diagram in figure 5.12.

Example 5.3 Consider the following initial data:

$$\phi(x,0) = f(x) = \begin{cases} 2x, & 0 \le x \le 0.5 \\ 2(1-x), & 0.5 \le x \le 1, \end{cases}$$

and boundary conditions

$$\phi(0,t) = \phi(1,t) = 0 \qquad \forall t.$$

The MATLAB code for solving the heat equation for this initial condition is shown in figure 5.13. Note that we store the results in a matrix; we could also store only two consecutive layers of points in time, but keeping the whole set of results makes plotting the solution easier. Let us solve the equation with $\delta x = 0.1$ and $\delta t = 0.001$, and plot the result for $t = 0, 10\delta t, 50\delta t, 100\delta t$.

```
>> dx = 0.1;
>> dt = 0.001;
>> tmax = dt*100;
>> sol=HeatExpl(dx, dt, tmax);
>> subplot(2,2,1);
```

```
% HeatExpl.m
function sol = HeatExpl(deltax, deltat, tmax)
N = round(1/deltax);
M = round(tmax/deltat);
sol = zeros(N+1,M+1);
rho = deltat / (deltax)^2;
rho2 = 1-2*rho;
vetx = 0:deltax:1;
for i=2:ceil((N+1)/2)
   sol(i,1) = 2*vetx(i);
   sol(N+2-i,1) = sol(i,1);
end
for j=1:M
   for i=2:N
      sol(i,j+1) = rho*sol(i-1,j) + ...
         rho2*sol(i,j) + rho*sol(i+1,j);
   end
end
```

Fig. 5.13 MATLAB code for solving the heat equation by the explicit method.

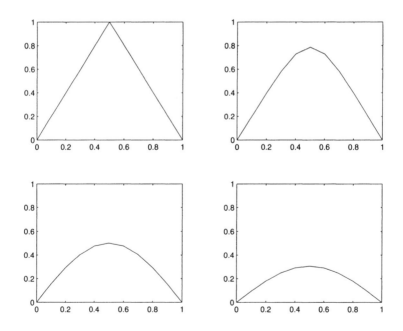

Fig. 5.14 Numerical solution of the heat equation with $\delta x = 0.1$ and $\delta t = 0.001$, by the explicit method, for $t = 0$, $t = 0.01$, $t = 0.05$, $t = 0.1$.

```
>> plot(0:dx:1,sol(:,1))
>> axis([0 1 0 1])
>> subplot(2,2,2);
>> plot(0:dx:1,sol(:,11))
>> axis([0 1 0 1])
>> subplot(2,2,3);
>> plot(0:dx:1,sol(:,51))
>> axis([0 1 0 1])
>> subplot(2,2,4);
>> plot(0:dx:1,sol(:,101))
>> axis([0 1 0 1])
```

The result, plotted in figure 5.14, looks reasonable, as the heat is progressively diffused and lost through the end points. At this point the reader may wish to refer back to figure 2.21, which depicts the value of a call option when the expiration date is approached. The only difference between figures 5.14 and 2.21 is that time goes forward for the heat equation, and it goes backward for the Black–Scholes equation; in fact, for an option we have a final condition rather than an initial one. Apart from this difference, the two solutions are qualitatively similar, as the boundary condition is a kinky function which is smoothed going forward or backward in time. This is a characteristic of parabolic equations, which smooth the irregularities of the boundary conditions out. On the contrary, these are propagated by hyperbolic equations and, as we have seen, by the transport equation.

However, we note that the discretization with respect to space is a bit coarse: we could increase precision by letting $\delta x = 0.01$. We can repeat the above set of MATLAB and plot the solution at time instants $t = \delta t, 2\delta t, 3\delta t, 4\delta t$. The result is shown in figure 5.15. We see that the solution does not make any sense; first, it assumes negative values, which should not be the case for intuitive physical reasons; then it shows an evident instability. The point is that here we have chosen discretization steps such that $\rho = 10$. In the following we show that for stability, the condition $0 < \rho \leq 0.5$ is required. □

How can we figure out a way to understand what condition should be required on the discretization steps to ensure numerical stability? In the case of the transport equation we have used one approach, based on Fourier analysis. Here we illustrate a matrix theoretic approach. The explicit method of equation (5.19), together with the boundary conditions

$$\phi_{0,j} = f_0(j\,\delta t) = f_{0j}, \qquad \phi_{1,j} = f_N(j\,\delta t) = f_{Nj}$$

can be represented in matrix terms as

$$\Phi_{j+1} = \mathbf{A}\Phi_j + \rho \mathbf{g}_j, \qquad j = 0, 1, 2, \ldots,$$

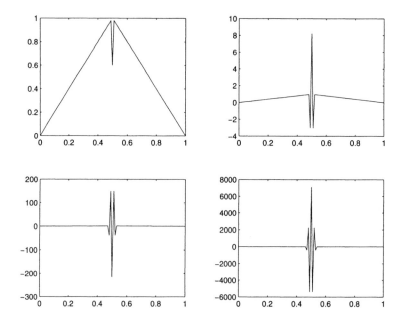

Fig. 5.15 Instability in the solution of the heat equation by an explicit method.

where

$$\mathbf{A} = \begin{bmatrix} 1-2\rho & \rho & 0 & \cdots & 0 & 0 \\ \rho & 1-2\rho & \rho & \cdots & 0 & 0 \\ 0 & \rho & 1-2\rho & \cdots & 0 & 0 \\ \vdots & \vdots & \vdots & \ddots & \vdots & \vdots \\ 0 & 0 & 0 & \cdots & \rho & 1-2\rho \end{bmatrix}$$

$$\Phi_j = \begin{bmatrix} \phi_{1,j} \\ \phi_{2,j} \\ \vdots \\ \phi_{N-1,j} \end{bmatrix}, \quad \mathbf{g}_j = \begin{bmatrix} f_{0,j} \\ 0 \\ \vdots \\ 0 \\ f_{N,j} \end{bmatrix}.$$

Note that $\mathbf{A} \in \mathbb{R}^{N-1,N-1}$ is a tridiagonal matrix. Recalling the convergence analysis that we carried out in section 3.2.5 for iterative algorithms, it is easy to see that the scheme will be stable when

$$\|\mathbf{A}\|_\infty \leq 1.$$

Now when $0 < \rho \leq 1/2$, then $1 - 2\rho \geq 0$ and

$$\|\mathbf{A}\|_\infty = \rho + (1-2\rho) + \rho = 1.$$

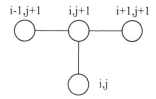

Fig. 5.16 Computational diagram of the implicit method for the heat equation.

But if $\rho > 1/2$, then $|1 - 2\rho| = 2\rho - 1$ and

$$\|\mathbf{A}\|_\infty = \rho + 2\rho - 1 + \rho = 4\rho - 1 > 1,$$

and stability cannot be guaranteed.

To get a more intuitive feeling for this stability condition, we see from figure 5.12 that the explicit scheme is based on a linear combination of three values in the previous time layer. Since the heat equation is a diffusion equation, we should take an average of these three values. But an average must be a convex combination, with positive weights; indeed, the stability condition makes the weight $1 - 2\rho$ positive. In a financial framework, similar interpretations can be found where weights are interpreted as risk-neutral probabilities, which must be positive as well.

To avoid instability, we may be forced to keep δt very small, since it must satisfy the condition $\delta t \leq 0.5(\delta x)^2$; if we want accuracy, we must take a small δx, which is smaller when squared, placing a severe restriction on δt. As this may require too much computational effort, an alternative approach may be pursued, based on implicit methods.

5.3.2 Solving the heat equation by a fully implicit method

If we use a forward approximation for the derivative with respect to time, we get an explicit method for the heat equation. We get a completely different scheme if we use a backward approximation:

$$\frac{\phi_{ij} - \phi_{i,j-1}}{\delta t} = \frac{\phi_{i+1,j} - 2\phi_{ij} + \phi_{i-1,j}}{(\delta x)^2}.$$

In this case we link one known value in time layer $j - 1$ to three unknown values in time layer j:

$$-\rho\phi_{i-1,j} + (1 + 2\rho)\phi_{ij} - \rho\phi_{i+1,j} = \phi_{i,j-1}, \quad (5.20)$$

where again $\rho = \delta t/(\delta x)^2$; see the computational diagram of figure 5.16. Thus, the unknown values are given implicitly, which is where the "implicit method" name comes from; a scheme like this is often referred to as *fully implicit*. We have to solve a system of linear equations for each time layer. Since boundary

```
% HeatImpl.m
function sol = HeatImpl(deltax, deltat, tmax)
N = round(1/deltax);
M = round(tmax/deltat);
sol = zeros(N+1,M+1);
rho = deltat / (deltax)^2;
B = diag((1+2*rho) * ones(N-1,1)) - ...
    diag(rho*ones(N-2,1),1) - diag(rho*ones(N-2,1),-1);
vetx = 0:deltax:1;
for i=2:ceil((N+1)/2)
    sol(i,1) = 2*vetx(i);
    sol(N+2-i,1) = sol(i,1);
end
for j=1:M
    sol(2:N,j+1) = B \ sol(2:N,j);
end
```

Fig. 5.17 MATLAB code for the implicit method.

conditions are given, we have $N-1$ equations in $N-1$ unknowns. In matrix terms, we have to solve a set of systems like

$$\mathbf{B}\Phi_{j+1} = \Phi_j + \rho\mathbf{g}_j, \quad j = 0, 1, 2, \ldots, \tag{5.21}$$

where $\mathbf{B} \in \mathbb{R}^{N-1,N-1}$ is a tridiagonal matrix,

$$\mathbf{B} = \begin{bmatrix} 1+2\rho & -\rho & 0 & \cdots & 0 & 0 \\ -\rho & 1+2\rho & -\rho & \cdots & 0 & 0 \\ 0 & -\rho & 1+2\rho & \cdots & 0 & 0 \\ \vdots & \vdots & \vdots & \ddots & \vdots & \vdots \\ 0 & 0 & 0 & \cdots & -\rho & 1+2\rho \end{bmatrix}.$$

Example 5.4 The MATLAB code for the implicit method to solve the heat equation is illustrated in figure 5.17 (here $\mathbf{g}_j = \mathbf{0}$). Note that we are not exploiting the fact that the matrix $\mathbf{B}$ is tridiagonal, as we simply leave to MATLAB the solution of the system of linear equations; the techniques described in section 3.2.4 could and should be used here. Furthermore, a matrix factorization like LU would be also useful, since the systems we are solving share the same matrix.

We may verify that the case $\delta x = 0.1$ and $\delta t = 0.001$ does not cause any trouble.

```
>> dx=0.01;
>> dt=0.001;
>> tmax=dt*100;
```

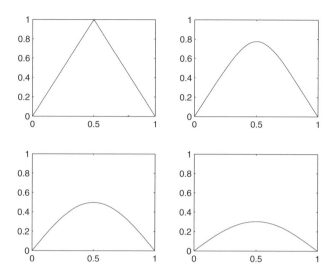

Fig. 5.18 Numerical solution of the heat equation with $\delta x = 0.1$ and $\delta t = 0.001$, by the implicit method, for $t = 0$, $t = 0.01$, $t = 0.05$, $t = 0.1$.

```
>> sol=HeatImpl(dx,dt,tmax);
>> subplot(2,2,1);
>> plot(0:dx:1,sol(:,1))
>> axis([0 1 0 1])
>> subplot(2,2,2);
>> plot(0:dx:1,sol(:,11))
>> axis([0 1 0 1])
>> subplot(2,2,3);
>> plot(0:dx:1,sol(:,51))
>> axis([0 1 0 1])
>> subplot(2,2,4);
>> plot(0:dx:1,sol(:,101))
>> axis([0 1 0 1])
```

The plots in figure 5.18 look less jagged than the plots of figure 5.14), because of the smaller discretization step with respect to space. In fact, we may prove that the implicit method is unconditionally stable. □

To prove that the implicit method of equation (5.21) is stable, we may rewrite the scheme as

$$\Phi_{j+1} = \mathbf{B}^{-1}(\Phi_j + \rho \mathbf{g}_j),$$

from which it is easy to see that stability depends on the spectral radius $\rho(\mathbf{B}^{-1})$. In this case, we may work directly on the spectral radius, rather than on a matrix norm. The scheme will be stable if the eigenvalues of $\mathbf{B}^{-1}$

are less than 1 in absolute value; to see that this is indeed the case, we may rewrite the matrix as follows:

$$\mathbf{B} = \mathbf{I} + \rho \mathbf{T},$$

where

$$\mathbf{T} = \begin{bmatrix} 2 & -1 & 0 & \cdots & 0 & 0 \\ -1 & 2 & -1 & \cdots & 0 & 0 \\ 0 & -1 & 2 & \cdots & 0 & 0 \\ \vdots & \vdots & \vdots & \ddots & \vdots & \vdots \\ 0 & 0 & 0 & \cdots & -1 & 2 \end{bmatrix}. \quad (5.22)$$

It can be shown that the eigenvalues of $\mathbf{T} \in \mathbb{R}^{N-1,N-1}$ are

$$\lambda_k = 4 \sin^2\left(\frac{k\pi}{2N}\right), \qquad k = 1, 2, \ldots, N-1.$$

We will not prove this claim, but we may have a quick informal check with MATLAB:

```
>> N=6;
>> T = diag(2*ones(N-1,1)) - diag(ones(N-2,1),1) - ...
   diag(ones(N-2,1),-1);
>> sort(eig(T))
ans =
    0.2679
    1.0000
    2.0000
    3.0000
    3.7321
>> sort(4*sin((1:N-1)*pi/(2*N)).^2)
ans =
    0.2679    1.0000    2.0000    3.0000    3.7321
```

Now we recall a couple of facts from matrix algebra, which are easily proved:

- If λ is an eigenvalue of the matrix $\mathbf{T}$, $1 + \rho\lambda$ is an eigenvalue of the matrix $\mathbf{I} + \rho\mathbf{T}$.

- If β is an eigenvalue of the matrix $\mathbf{B}$, β^{-1} is an eigenvalue of the matrix $\mathbf{B}^{-1}$.

Putting all together, we may conclude that the eigenvalues of $\mathbf{B}^{-1}$ are

$$\alpha_k = \frac{1}{1 + 4\rho \sin^2\left(\frac{k\pi}{2N}\right)} < 1, \qquad k = 1, 2, \ldots, N-1,$$

and the fully implicit scheme is unconditionally stable.

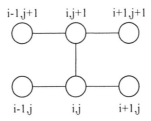

Fig. 5.19 Computational diagram of the Crank–Nicolson method for the heat equation.

5.3.3 Solving the heat equation by the Crank–Nicolson method

So far, we have seen methods involving three points on one time layer and one on a neighboring layer. It is natural to wonder if a better scheme may be obtained by considering three points on both layers. One way to do this is to consider the point $(x_i, t_{j+1/2}) = (x_i, t_j + \delta t/2)$, which is actually outside the grid, and to approximate the derivatives at that point using values in the six neighboring points on the grid. By using Taylor expansions, as we did in section 5.2, we may see that

$$\frac{\partial^2 \phi}{\partial x^2}(x_i, t_{j+1/2}) = \frac{1}{2}\left[\frac{\partial^2 \phi}{\partial x^2}(x_i, t_{j+1}) + \frac{\partial^2 \phi}{\partial x^2}(x_i, t_j)\right] + O(\delta x^2),$$

and a central difference approximation for the derivative with respect to time in $(x_i, t_{j+1/2})$ yields

$$\frac{\partial \phi}{\partial t}(x_i, t_{j+1/2}) = \frac{\phi(x_i, t_{j+1}) - \phi(x_i, t_j)}{\delta t} + O(\delta t^2).$$

Using these two approximations together with the usual ones, we get the Crank–Nicolson scheme:

$$-\rho\phi_{i-1,j+1} + 2(1+\rho)\phi_{i,j+1} - \rho\phi_{i+1,j+1} = \rho\phi_{i-1,j} + 2(1-\rho)\phi_{ij} + \rho\phi_{i+1,j}, \quad (5.23)$$

which is represented in figure 5.19. The fundamental feature of this scheme is that the error is both $O(\delta x^2)$ and $O(\delta t^2)$; this implies that less computational effort is required to obtain a satisfactory degree of accuracy in the numerical solution.

The Crank–Nicolson scheme may be analyzed in a more general framework. We may think of using a convex combination of two approximations of the second-order derivative in the finite difference scheme:

$$\frac{\phi_{i,j+1} - \phi_{ij}}{\delta t} = \frac{1}{(\delta x)^2}[\lambda(\phi_{i-1,j+1} - 2\phi_{i,j+1} + \phi_{i+1,j+1}) \\ + (1-\lambda)(\phi_{i-1,j} - 2\phi_{ij} + \phi_{i+1,j})] \quad (5.24)$$

for $0 \leq \lambda \leq 1$. Note that we get the explicit scheme by choosing $\lambda = 0$, the fully implicit scheme for $\lambda = 1$, and the Crank–Nicolson scheme for $\lambda = 1/2$.

To see that the Crank–Nicolson scheme is unconditionally stable, we may proceed just as with the first implicit scheme. We may rewrite equation (5.23) in matrix form:
$$\mathbf{C}\Phi_{j+1} = \mathbf{D}\Phi_j + \rho(\mathbf{g}_{j+1} + \mathbf{g}_j),$$
where

$$\mathbf{C} = \begin{bmatrix} 2(1+\rho) & -\rho & 0 & \cdots & 0 & 0 \\ -\rho & 2(1+\rho) & -\rho & \cdots & 0 & 0 \\ 0 & -\rho & 2(1+\rho) & \cdots & 0 & 0 \\ \vdots & \vdots & \vdots & \ddots & \vdots & \vdots \\ 0 & 0 & 0 & \cdots & -\rho & 2(1+\rho) \end{bmatrix}$$

$$\mathbf{D} = \begin{bmatrix} 2(1-\rho) & \rho & 0 & \cdots & 0 & 0 \\ \rho & 2(1-\rho) & \rho & \cdots & 0 & 0 \\ 0 & \rho & 2(1-\rho) & \cdots & 0 & 0 \\ \vdots & \vdots & \vdots & \ddots & \vdots & \vdots \\ 0 & 0 & 0 & \cdots & \rho & 2(1-\rho) \end{bmatrix}.$$

Then, using matrix the same matrix $\mathbf{T}$ of equation (5.22) again, we may see that the eigenvalues of $\mathbf{C}^{-1}\mathbf{D}$ are

$$\alpha_k = \frac{2 - 4\sin^2\left(\frac{k\pi}{2N}\right)}{2 + 4\sin^2\left(\frac{k\pi}{2N}\right)}, \quad k = 1, 2, \ldots, N-1.$$

As these eigenvalues are, in absolute value, less than 1, we see that the scheme is unconditionally stable.

5.4 SOLVING THE BIDIMENSIONAL HEAT EQUATION

Sometimes, PDEs arising in financial engineering involve two uncertain quantities. They may be the prices of two assets in a multidimensional option, or a price and an interest rate, or a price and a volatility. In these cases we have a more complex PDE to deal with. When the dimensionality of the equation goes beyond a certain limit, we must necessarily resort to Monte Carlo methods, but in two or three dimensions (plus time), finite difference schemes can be still applied. To get a feeling for the issues involved, we consider here the simplest generalization of the heat equation, i.e., the bidimensional heat equation

$$\frac{\partial \phi}{\partial t} = \frac{\partial^2 \phi}{\partial x^2} + \frac{\partial^2 \phi}{\partial y^2}, \tag{5.25}$$

where the unknown function $\phi(t, x, y)$ is the temperature of a point (x, y) in the plane at time t. We may extend the standard grid notation by introducing discretization steps δx, δy, and δt:

$$\phi(k\,\delta t, i\,\delta x, j\,\delta y) \Rightarrow \phi_{ij}^k,$$

where time index k is written as a superscript and should not be confused with a power. For the sake of simplicity we will assume that we are interested in the solution on the unit square

$$\{(x,y) \mid 0 \leq x \leq 1,\ 0 \leq y \leq 1\},$$

given initial and boundary conditions.

Just like in the one-dimensional case, we may use central differences for the second-order spatial derivatives. If we use the forward difference for the derivative with respect to time, we get the finite difference approximation:

$$\frac{\phi_{ij}^{k+1} - \phi_{ij}^k}{\delta t} = \frac{\phi_{i+1,j}^k - 2\phi_{ij}^k + \phi_{i-1,j}^k}{(\delta x)^2} + \frac{\phi_{i,j+1}^k - 2\phi_{ij}^k + \phi_{i,j-1}^k}{(\delta y)^2}.$$

This immediately leads to an explicit scheme:

$$\phi_{ij}^{k+1} = (1 - 2\rho_x - 2\rho_y)\phi_{ij}^k + \rho_x\left(\phi_{i+1,j}^k + \phi_{i-1,j}^k\right) + \rho_y\left(\phi_{i,j+1}^k + \phi_{i,j-1}^k\right), \quad (5.26)$$

where

$$\rho_x = \frac{\delta t}{(\delta x)^2}, \quad \rho_y = \frac{\delta t}{(\delta x)^2}$$

This method is relatively straightforward to implement, but it suffers from instability. It can be shown that a stability condition is:

$$\rho_x + \rho_y \leq \frac{1}{2}.$$

This condition may be interpreted intuitively as usual: it just makes sure that we are taking a convex combination of five neighboring values in the previous time layer to get the value ϕ_{ij}^{k+1}. This implies a rather severe condition on δt, just like the one-dimensional case. However, in this case an explicit algorithm is more time-consuming and requires more memory. In fact, now we must solve the equation by avoiding storage of a tridimensional array, whereas in the one-dimensional case we stored all the solution in one matrix. We alternate time layers, keeping track of two consecutive ones, and swapping them as time goes forward.

A code to implement this explicit method is shown in figure 5.20. A few comments are in order here.

- The input arguments are:
    - the three discretization steps (dt, dx, dy)
    - the time Tmax at which we want to stop the solution process

```
function U = Heat2D(dt, dx, dy, Tmax, Tsnap, value, bounds)
% make sure steps are consistent
Nx = round(1/dx);
dx = 1/Nx;
Ny = round(1/dy);
dy = 1/Ny;
Nt = round(Tmax/dt);
dt = Tmax/Nt;
rhox = dt/dx^2;
rhoy = dt/dy^2;
if   rhox + rhoy > 0.5
    fprintf(1,'Warning: bad selection of steps\n');
end
C1 = 1-2*rhox-2*rhoy;
Layers = zeros(2, 1+Nx, 1+Ny);
tpast = 1;
tnow = 2;
iTsnap = Tsnap/dt;
[X, Y] = meshgrid(0:dx:1, 0:dy:1);
% set up initial conditions and plot
Layers(tpast, (1+round(bounds(1)/dx)):(1+round(bounds(2)/dx)), ...
    (1+round(bounds(3)/dy)):(1+round(bounds(4)/dy))) = value;
U = shiftdim(Layers(tpast,:,:));
figure;
surf(X,Y,U);
title('t=0','Fontsize',12);
% Carry out iterations
for t=1:Nt
    for i=2:Nx
        for j=2:Ny
            Layers(tnow,i,j) = C1*Layers(tpast,i,j) + ...
                rhox*(Layers(tpast,i+1,j) + Layers(tpast,i-1,j)) + ...
                rhoy*(Layers(tpast,i,j+1) + Layers(tpast,i,j-1));
        end
    end
    if find(iTsnap == t)            % Plot if required
        U = shiftdim(Layers(tnow,:,:));
        figure;
        surf(X,Y,U);
        title(['t=', num2str(Tsnap(1)) ],'Fontsize',12);
        Tsnap(1) = [];
    end
    tnow = 1+mod(t+1,2);        % Swap layers
    tpast = 1+mod(t,2);
end
```

Fig. 5.20 Code to solve the bidimensional heat equation by an implicit method.

```
dt = 0.0001;
dx = 0.05;
dy = 0.05;
value = 10;
bounds = [0.7, 0.9, 0.1, 0.9];
Tmax = 0.1;
Tsnap = [0.01, 0.02, 0.03, 0.04, 0.05, 0.06];
U = Heat2D(dt, dx, dy, Tmax, Tsnap, value, bounds);
```

Fig. 5.21 Script to test `Heat2D`.

- a vector `Tsnap` of time instants at which we want to display a plot of the solution
- a value `value` and a four-dimensional vector `bounds` to store the initial conditions, which we assume of the form

$$\phi(x, y, 0) = \begin{cases} V & \text{for } 0 < b_1 \leq x \leq b_2 < 1 \text{ and } 0 < b_3 \leq y \leq b_4 < 1 \\ 0 & \text{otherwise.} \end{cases}$$

- In the first few lines we check consistency of discretization steps with the boundaries of the domain, changing discretization steps a bit if necessary. Then we precompute fixed quantities outside the main loop of the procedure (issuing a warning message if discretization steps may lead to instability).

- The solution is stored in two consecutive layers of size $(1 + N_x) \times (1 + N_y)$, which form the tridimensional array `Layers`. The two layers are alternated, as one is indexed by `tnow` and the other one by `tpast`; these two indexes are incremented modulo two at the end of the main loop (so that copying a matrix is not necessary).

- Plots are displayed in separate figures (with some heading) at time $t = 0$ and when required; to that purpose we must use `meshgrid` to set up matrices of coordinates in the plane, and `shiftdim` to transform one layer in the tridimensional array `Layers` to the bidimensional array `U`.

- Finally, things are made a bit more complicated by the fact that in mathematics we start subscripts from 0, whereas in MATLAB array indexing starts from 1.

Running the script of figure 5.21 we get a set of surfaces, three of which are displayed in figure 5.22.

The explicit method may prove time-consuming because of the restriction on the time step, and we would like to have stability guarantees typically associated with implicit methods. A fully implicit method is easily obtained

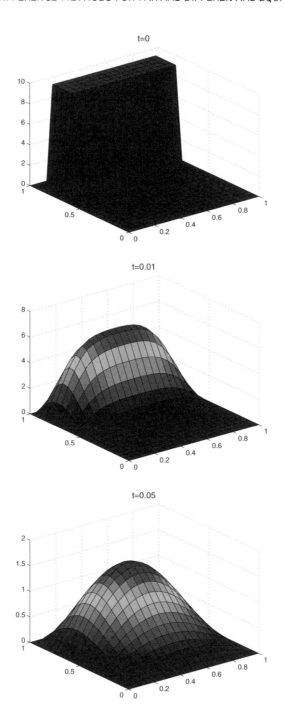

Fig. 5.22 Numerical solution of the bidimensional heat equation by an explicit method.

by taking a backward approximation of the derivative with respect to time, but in the bidimensional case we have a system of linear equations which may be time-consuming to solve, since there is no easy structure to exploit.

Alternative approaches have been proposed, including the Alternating Direction Implicit (ADI) method. There are several variations on this theme, and we will just describe the simplest one, due to Peaceman and Rachford. A sound motivation of the scheme would call for a detailed analysis of finite difference operators and their truncation errors, which together with a stability analysis would prove convergence. Since this is not trivial, we refer the reader to the references listed at the end of the chapter.[5] The intuitive idea is to introduce an intermediate time layer in the solution process, stepping from t to $t + \delta t/2$, and to use an approximation scheme which is implicit with respect to one of the two space dimensions, and explicit with respect to the other one. Then we step from $t + \delta t/2$ to $t + \delta t$, swapping the role of the two space dimensions. The net effect is to solve the bidimensional problem as a set of one-dimensional ones.

We can specify the method in detail by using first a difference scheme based on points (x, y, t) and $(x, y, t + \delta t/2)$:

$$\frac{\phi_{ij}^{k+\frac{1}{2}} - \phi_{ij}^k}{\delta t/2} = \frac{\phi_{i+1,j}^{k+\frac{1}{2}} - 2\phi_{ij}^{k+\frac{1}{2}} + \phi_{i-1,j}^{k+\frac{1}{2}}}{(\delta x)^2} + \frac{\phi_{i,j+1}^k - 2\phi_{ij}^k + \phi_{i,j-1}^k}{(\delta y)^2}. \quad (5.27)$$

Note that the scheme is implicit in x but explicit in y, since the second-order derivative for x is approximated by a central difference on time layer $k + 1/2$ rather than on time layer k. This may look arbitrary, but it introduces a truncation error which is comparable to other terms. Equation (5.27) can be rewritten as

$$\phi_{ij}^{k+\frac{1}{2}} - \phi_{ij}^k = \frac{\rho_x}{2}\left(\phi_{i+1,j}^{k+\frac{1}{2}} - 2\phi_{ij}^{k+\frac{1}{2}} + \phi_{i-1,j}^{k+\frac{1}{2}}\right) + \frac{\rho_y}{2}\left(\phi_{i,j+1}^k - 2\phi_{ij}^k + \phi_{i,j-1}^k\right),$$

which can be rearranged by separating what is known and what is not:

$$-\frac{\rho_x}{2}\phi_{i-1,j}^{k+\frac{1}{2}} + (1+\rho_x)\phi_{ij}^{k+\frac{1}{2}} - \frac{\rho_x}{2}\phi_{i+1,j}^{k+\frac{1}{2}} = \frac{\rho_y}{2}\phi_{i,j-1}^k + (1-\rho_y)\phi_{ij}^k + \frac{\rho_y}{2}\phi_{i,j+1}^k. \quad (5.28)$$

We should note that everything is known on the right-hand side, whereas on the left-hand side subscript j is fixed; hence we may solve one tridiagonal system for each j, i.e., for given y. Indeed, we see that a bidimensional problem is decomposed into a sequence of one-dimensional problems. By the same token, we can step forward to $k+1$, reversing the roles of i and j. The starting point is the finite difference scheme:

$$\frac{\phi_{ij}^{k+1} - \phi_{ij}^{k+\frac{1}{2}}}{\delta t/2} = \frac{\phi_{i+1,j}^{k+\frac{1}{2}} - 2\phi_{ij}^{k+\frac{1}{2}} + \phi_{i-1,j}^{k+\frac{1}{2}}}{(\delta x)^2} + \frac{\phi_{i,j+1}^{k+1} - 2\phi_{ij}^{k+1} + \phi_{i,j-1}^{k+1}}{(\delta y)^2}, \quad (5.29)$$

---

[5] In particular, we suggest section 7.3 of [7] or chapter 3 of [4].

which is explicit in x and implicit in y and can be rearranged to

$$-\frac{\rho_y}{2}\phi_{i,j-1}^{k+1} + (1+\rho_y)\phi_{ij}^{k+1} - \frac{\rho_y}{2}\phi_{i,j+1}^{k+1} = \frac{\rho_x}{2}\phi_{i-1,j}^{k+\frac{1}{2}} + (1-\rho_x)\phi_{ij}^{k+\frac{1}{2}} + \frac{\rho_x}{2}\phi_{i+1,j}^{k+\frac{1}{2}}. \tag{5.30}$$

In this case, we solve one tridiagonal system for each value of x in the time layer.

The idea is implemented in the MATLAB code displayed in figures 5.23 and 5.24. The remarks we have made for the implementation of the explicit method apply here too, with some additional issues:

- We use LU factorization of both matrices involved, since they are constant with respect to time, resulting in matrices L1, U1, L2, and U2; right-hand sides of systems are stored in vectors Rhs1 and Rhs2.

- The intermediate layer for time $t + \delta t/2$ is stored in the bidimensional array Auxlayer, which is the unknown in the first system, and makes up the right-hand side in the second one.

- In checking the code pay attention to the shift from mathematical subscripts to MATLAB array indexing.

The code may be easily tested by adapting the script of figure 5.21.

5.5 CONVERGENCE, CONSISTENCY, AND STABILITY

We have developed finite difference schemes, and we have informally noted that there is some truncation error that tends to zero as the discretization steps tend to zero. We would expect that this ensures the convergence of the solution to the difference equations to the solution of the differential equation. However, the counterexample of section 5.2.1 shows that the matter is not so trivial, since we should consider carefully the interplay of three concepts: convergence, stability, and consistency. The point is that the solution of the finite difference equations for discretization steps $\delta x, \delta t \to 0$ could converge to a function which is not the solution of the PDE. A rigorous analysis of these concepts and their relationships is beyond the scope of the book, but we would like to give at least a glimpse into these topics.

An initial value problem such as the familiar heat equation is defined over a space/time domain

$$\mathcal{D} \times (0 < t < \infty).$$

The problem can be cast in a more abstract way as

$$L\phi = f,$$

where L is a differential operator, f is a known function, and ϕ is the unknown function we seek to determine. When we set up a discrete grid $\mathcal{G}_\Delta$, we also

```
function U = Heat2DADI(dt, dx, dy, Tmax, Tsnap, value, bounds)
% make sure steps are consistent
Nx = round(1/dx);
dx = 1/Nx;
Ny = round(1/dy);
dy = 1/Ny;
Nt = round(Tmax/dt);
dt = Tmax/Nt;
rhox = dt/dx^2;
rhoy = dt/dy^2;

Layers = zeros(2, 1+Nx, 1+Ny);
Auxlayer = zeros(1+Nx, 1+Ny);
tpast = 1;
tnow = 2;
iTsnap = Tsnap/dt;
[X, Y] = meshgrid(0:dx:1, 0:dy:1);
% set up initial conditions
Layers(tpast, (1+round(bounds(1)/dx)):(1+round(bounds(2)/dx)), ...
    (1+round(bounds(3)/dy)):(1+round(bounds(4)/dy))) = value;
U = shiftdim(Layers(tpast,:,:));
figure;
surf(X,Y,U);
title('t=0','Fontsize',12);
% Prepare matrices and LU decomposition
Matrix1 = diag((1+rhox)*ones(Nx-1,1)) + ...
    diag(-rhox/2*ones(Nx-2,1),1) + ...
    diag(-rhox/2*ones(Nx-2,1),-1);
[L1, U1] = lu(Matrix1);
Matrix2 = diag((1+rhoy)*ones(Ny-1,1)) + ...
    diag(-rhoy/2*ones(Ny-2,1),1) + ...
    diag(-rhoy/2*ones(Ny-2,1),-1);
[L2, U2] = lu(Matrix2);
Rhs1 = zeros(Nx-1,1);
Rhs2 = zeros(Ny-1,1);
```

Fig. 5.23 Code to solve the bidimensional heat equation by an ADI method (continued in figure 5.24).

```
% Carry out iterations
for t=1:Nt
    % first half step
    for j=1:Ny-1
        % set up right hand side
        for i=1:Nx-1
            Rhs1(i) = rhoy/2*Layers(tpast,i+1,j) + ...
                (1-rhoy)*Layers(tpast,i+1,j+1) + ...
                rhoy/2*Layers(tpast,i+1,j+2);
        end
        % solve
        Auxlayer(2:Nx,j+1) = U1 \ (L1 \ Rhs1);
    end
    % second half step
    for i=1:Nx-1
        % set up right hand side
        for j=1:Ny-1
            Rhs2(j) = rhox/2*Auxlayer(i,j+1) + ...
                (1-rhox)*Auxlayer(i+1,j+1) + ...
                rhox/2*Auxlayer(i+2,j+1);
        end
        % solve
        Layers(tnow, i+1,2:Ny) = (U2 \ (L2 \ Rhs2))';
    end
    % plot if necessary
    if find(iTsnap == t)
        U = shiftdim(Layers(tnow,:,:));
        figure;
        surf(X,Y,U);
        title(['t=', num2str(Tsnap(1)) ],'Fontsize',12);
        Tsnap(1) = [];
    end
    % swap layers
    tnow = 1+mod(t+1,2);
    tpast = 1+mod(t,2);
end
```

Fig. 5.24 Code to solve the bidimensional heat equation by an ADI method (continued from figure 5.23).

discretize the operator L by an operator L_Δ. Given a function ψ and a point $(\mathbf{P}_i, t_j) \in \mathcal{G}_\Delta$, we may consider the truncation error

$$t_\psi(\mathbf{P}_i, t_j) = L\psi(\mathbf{P}_i, t_j) - L_\Delta \psi(\mathbf{P}_i, t_j).$$

If, when the grid is refined and the discretization steps tend to zero, this truncation error tends to zero,[6] the numerical scheme is said to be *consistent*. This essentially says that the finite difference representation we are using tends to the PDEs we are interested in.

The stability issue is concerned basically with whether or not the difference between the numerical solution and the exact solution remains bounded as time progresses. To be more specific, consider the heat equation of section 5.3. Let ϕ_{ij} be the solution of the finite difference scheme and $\phi(x,t)$ the correct solution of the PDE. We may investigate

- the behavior of $|\phi_{ij} - \phi(i\,\delta x, j\,\delta t)|$ as $j \to \infty$ for fixed discretization steps δx and δt,

- or the behavior of $|\phi_{ij} - \phi(i\,\delta x, j\,\delta t)|$ as $\delta x, \delta t \to 0$ for a fixed value of $j\,\delta t$.

The first issue is related to stability; the second issue is related to convergence. To ensure the convergence of the numerical solution to the exact solution, the consistency condition is not enough. However, it can be shown (Lax's equivalence theorem; see [5]) that for a well-posed linear initial value problem, stability is a necessary and sufficient condition for convergence of a consistent numerical scheme. As the following example shows, the numerical scheme of section 5.2.1 is not stable, and this is why it fails to converge.

Example 5.5 For the sake of convenience, let us recall the numerical scheme of section 5.2.1 for the transport equation with constant velocity c:

$$\phi_{i,j+1} = \left(1 + \frac{c}{\rho}\right)\phi_{ij} - \frac{c}{\rho}\phi_{i+1,j},$$

where $\rho = \delta x / \delta t$. We may apply the same Von Neumann analysis of stability that we applied in example 5.2. Leaving the details as an exercise, we may see that in this case

$$\phi_{ij} = \epsilon\,(-1)^i \left(1 + 2\frac{c}{\rho}\right)^j.$$

Since c and ρ are both positive, we see that ϕ_{ij} goes to infinity as $j \to \infty$. Hence, the scheme is unconditionally unstable and convergence is not ensured even if the discretization steps tend to zero. □

---

[6]This should be made more precise, as the space and time discretization steps could tend to zero in an arbitrary way, or with some relationship between them.

For further reading

- Partial differential equations are a large and complicated topic. For an introduction including both classical and advanced concepts, see, e.g., [3].[7]

- Another book covering PDEs in a relatively general setting is [1], which also includes many pieces of MATLAB code.

- A classical reference on finite difference methods for PDEs is [6]. See also [4] and [8].

- A recent addition to the literature on finite difference schemes is [7].

- Advanced issues, including the important Lax theorem, are covered in [5].

- To see extensive examples of PDEs in action to tackle financial engineering problems, see [9] or [10].

REFERENCES

1. J. Cooper. *Introduction to Partial Differential Equations with MATLAB.* Birkhäuser, Berlin, 1998.

2. D.G. Luenberger. *Investment Science.* Oxford University Press, New York, 1998.

3. R. McOwen. *Partial Differential Equations: Methods and Applications.* Prentice Hall, Upper Saddle River, NJ, 1996.

4. K.W. Morton and D.F. Mayers. *Numerical Solution of Partial Differential Equations (2nd ed.).* Cambridge University Press, Cambridge, 2005.

5. R.D. Richtmyer and K.W. Morton. *Difference Methods for Initial Value Problems (2nd ed.).* Wiley, New York, 1967. Reprinted in 1994 by Krieger, New York.

6. G.D. Smith. *Numerical Solution of Partial Differential Equations: Finite Difference Methods (3rd ed.).* Oxford University Press, Oxford, 1985.

7. J.C. Strickwerda. *Finite Difference Schemes and Partial Difference Equations.* SIAM, Philadelphia, PA, 2004.

---

[7]An errata sheet for this book is available at www.math.neu.edu/~mcowen/mathindex.html.

8. J.W. Thomas. *Numerical Partial Differential Equations: Finite Difference Methods.* Springer-Verlag, New York, 1995.

9. P. Wilmott. *Derivatives: The Theory and Practice of Financial Engineering.* Wiley, Chichester, West Sussex, England, 1999.

10. P. Wilmott. *Quantitative Finance (vols. I and II).* Wiley, Chichester, West Sussex, England, 2000.

6
Convex Optimization

Optimization methods play an important role in finance. As we have seen in chapter 2, optimization models may be used in portfolio management, in which case they are used as a decision support tool; sometimes, optimization methods are somewhat more instrumental and are used, e.g., to solve model calibration problems. Covering in depth all optimization methods that could be useful in solving finance-related problems would require a few books (tough ones, by the way). The aim of this chapter is much less ambitious. We want to provide the reader with a minimal background required to grasp what MATLAB offers in the Optimization toolbox; in particular, one should know what she's doing when choosing one among the various methods that are available to cope with the same type of problem.

To simplify things, we consider only basic optimization problems in this chapter. In particular, we assume they are convex and deterministic. Basic notions on convexity are summarized in supplement S6.1 at the end of this chapter. Basically, convexity ensures that a local optimum is a global one, and allows to find easy characterizations of optimal solutions, which pave the way to solution algorithms. Optimization models and methods in non-convex cases are dealt with in chapter 12. When data are uncertain, we should resort to stochastic optimization models, which is quite important in the context of dynamic decision making over time. There are two basic approaches to cope with dynamic decision making under uncertainty: dynamic programming and stochastic programming with recourse. Dynamic programming is described in chapter 10, where we also describe its role in pricing American options by Monte Carlo simulation; stochastic programming with recourse is covered in chapter 11. Actually, these two approaches have a lot in common, but apparently the first one is quite common in Economics, whereas the second one is more appreciated within the engineering community. We will try to explain why in later chapters.

We first provide a framework to classify optimization models in section 6.1. In fact, models may be classified along many directions, including constrained and unconstrained problems. Unconstrained optimization is covered in section 6.2. Methods for unconstrained optimization differ in their requirements; many are gradient-based, and require the ability of computing or approximating function derivatives; other methods are derivative-free, in the sense that they are just based on function evaluations.[1] Constrained optimization is dealt with in section 6.3, where we also introduce fundamental theoretical concepts like Kuhn–Tucker conditions and duality theory. A specific case of constrained optimization is linear programming, which is the topic of section 6.4; quite often, non-trivial problems may be expressed as linear programming models, and the ability to solve really huge optimization problems efficiently make linear programming a fundamental tool. We illustrate MATLAB functions all along the way with small toy examples, and we close with more significant examples in section 6.5.

Finally, we should bear in mind that optimization methods typically assume that we are able to capture the desirability of a solution by a function given in closed form. But analytical models may be too complex or not available at all, and we may be forced to resort to simulation tools for performance evaluation. The integration of simulation and optimization techniques is described in section 6.6.

6.1 CLASSIFICATION OF OPTIMIZATION PROBLEMS

There is a huge variety of optimization models that we meet in financial applications, which can be tackled by an equally vast array of methods. Hence, the starting point of this chapter should be a listing of the basic features by which an optimization model may be characterized.

6.1.1 Finite- vs. infinite-dimensional problems

In this chapter we are concerned with problems whose abstract form is

$$\begin{aligned}\min \quad & f(\mathbf{x}) \\ \text{s.t.} \quad & \mathbf{x} \in S \subseteq \mathbb{R}^n.\end{aligned} \quad (6.1)$$

---

[1] Derivative-free optimization methods are the core of a recently released MATLAB toolbox, called Genetic Algorithm and Direct Search. We outline genetic algorithms in section 12.4.

The objective function f is a scalar function quantifying the suitability of a solution $\mathbf{x}$, which is a vector of decision variables and must belong to a feasible set S, which is a subset of the set of vectors with n real components. Since the solution is expressed by a finite-dimensional vector, we speak of a *finite-dimensional problem*. There is no loss of generality in considering only minimization problems, since a maximization problem may be transformed into a minimization problem simply by changing the sign in the objective:

$$\max f(x) \;\Rightarrow\; -\min[-f(x)].$$

Indeed, all MATLAB functions in the Optimization toolbox assume a minimization problem. Solving an optimization problem like (6.1) means finding a point $\mathbf{x}^* \in S$ such that

$$f(\mathbf{x}^*) \leq f(\mathbf{x}) \quad \forall \mathbf{x} \in S. \tag{6.2}$$

The point $\mathbf{x}^*$ is said to be a *global optimum* (the terms *optimizer* or *minimizer* are also used to avoid confusion between the optimal point and the corresponding value of the objective function). Neither the existence nor the uniqueness of a global optimum should be taken for granted. To begin with, the problem may be unbounded, which is the case if there is a sequence of solutions $\mathbf{x}^{(k)} \in S$ such that

$$\lim_{k \to \infty} f(\mathbf{x}^{(k)}) = -\infty.$$

Furthermore, the problem may be infeasible, i.e., the feasible set S may be empty. Finally, the solution is not unique when condition (6.2) is satisfied by a set of alternative optima, which may be a discrete and finite set, or an infinite set. If the condition (6.2) holds only in a neighborhood of $\mathbf{x}^*$, we speak of a *local optimum*.

Example 6.1 A typical objective function that gives rise to local optima is a polynomial function; recall that the oscillatory behavior of high-order polynomials is the reason why they are not well-suited to function interpolation (see example 3.16 on page 179). We may check this with a simple MATLAB snapshot. Consider a polynomial like

$$f(x) = x^4 - 10.5x^3 + 39x^2 - 59.5x + 30$$

and use MATLAB to plot it.

```
>> g = @(x) polyval( [ 1 -10.5 39 -59.5 30], x);
>> xvet=1:0.05:4;
>> plot(xvet,g(xvet))
```

The plot produced is illustrated in figure 6.1, from which it is clear that there are two local minimizers. One MATLAB function to solve a minimization

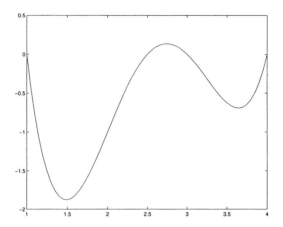

Fig. 6.1 Global and local optima for a polynomial.

problem is fminunc; the "unc" stands for *unconstrained*, since we are not enforcing any requirement on the decision variable. This function requires an argument which is the initial point of the search process.

```
>> [x,fval] = fminunc(g, 0)
Warning: Gradient must be provided for trust-region method;
   using line-search method instead.
x =
    1.4878
fval =
   -1.8757
>> [x,fval] = fminunc(g, 5)
Warning: Gradient must be provided for trust-region method;
   using line-search method instead.
x =
    3.6437
fval =
   -0.6935
```

We see that depending on the starting point, we get the global or the local minimizer. The MATLAB output has been cut a little, but we see some messages concerning trust regions and line search; the meaning of these terms is illustrated in the following (this is all this chapter is about, after all). A different situation occurs in the following case:

```
>> f = @(x) polyval( [ 1 -8 22 -24 1], x);
>> xvet=0:0.05:4;
>> plot(xvet,f(xvet))
```

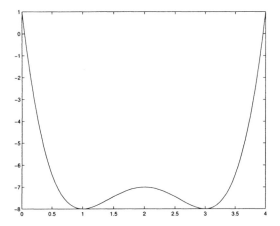

Fig. 6.2 Objective function with two global optima.

The plot is shown in figure 6.2. It may be seen that we have two alternative global minima. □

Example 6.2 It is easy to build problems which are, respectively:

1. Unbounded:

$$\begin{aligned}\max \quad & x_1^2 + x_2^2 \\ \text{s.t.} \quad & x_1 + x_2 \geq 4 \\ & x_1, x_2 \geq 0.\end{aligned}$$

2. Infeasible:

$$\begin{aligned}\max \quad & 2x_1 + 3x_2 \\ \text{s.t.} \quad & x_1 + x_2 \geq 4 \\ & 0 \leq x_1, x_2 \leq 1.\end{aligned}$$

3. Characterized by an infinite set of optima:

$$\begin{aligned}\max \quad & x_1 + x_2 \\ \text{s.t.} \quad & x_1 + x_2 \leq 4 \\ & x_1, x_2 \geq 0.\end{aligned}$$

The reader is urged to check this by drawing the feasible set and the level curves of the objective function.

Another important remark is that some problems may have no solution because they are posed the wrong way. Consider the innocent-looking example

$$\begin{aligned}\min \quad & x \\ \text{s.t.} \quad & x > 2.\end{aligned}$$

This problem has no solution, as the feasible set is open, and the apparently obvious solution $x = 2$ is not feasible. In fact, there is not a minimum but only an infimum. This is why in any optimization software you only get constraints such as $\geq$ or $\leq$, so that the feasible set is a closed region. □

So far we have assumed that the feasible set is a subset of the space on n-dimensional vectors with real components. In *infinite-dimensional problems* the solution is represented by an infinite collection of decision variables. This is the case when the solution we are seeking is a function of time over a continuous interval. Consider, for instance, a continuous-time dynamic system represented by the vector differential equation

$$\dot{\mathbf{x}}(t) = \mathbf{h}[\mathbf{x}(t), \mathbf{u}(t)],$$

where $\mathbf{x}$ is the vector of state variables and $\mathbf{u}$ is the vector of control inputs. An optimal control $\mathbf{u}(t)$, $t \in [0, T]$ for this system may be found by solving

$$\begin{aligned}
\min \quad & \int_0^T f[\mathbf{x}(t), \mathbf{u}(t)] \, dt + g[\mathbf{x}(T)] \\
\text{s.t.} \quad & \dot{\mathbf{x}}(t) = \mathbf{h}[\mathbf{x}(t), \mathbf{u}(t)] \quad \forall t \in [0, T] \\
& \mathbf{x}(0) = \mathbf{x}_0 \\
& \mathbf{u}(t) \in \Omega \quad \forall t \in [0, T],
\end{aligned}$$

where $[0, T]$ is the time horizon we are interested in, $\mathbf{x}_0$ is the (known) initial state of the system, and Ω is the set of admissible controls. The objective function includes both a *trajectory cost*, depending on both states and controls, and a *terminal cost*, depending on the terminal state $\mathbf{x}(T)$. It is also possible to specify some constraints on the terminal state.

There is a vast literature on optimal control models in finance. They are actually formulated within a stochastic setting (returns are random and modeled by stochastic differential equations as discussed in chapter 2) and solved by dynamic programming (see, e.g., [13]). Optimal control methods are an excellent tool to analyze relatively simple models and to derive valuable insights from a qualitative and theoretical point of view; however, it might be argued that, in general, complex and realistic problems are usually best formulated and solved as finite-dimensional models. This is an admittedly debatable point, as many would disagree, particularly when it comes to stochastic models for finance (see, e.g., [12] for an alternative view). Anyway, we do not deal with this class of models, essentially to keep the book to a reasonable size. It is worth noting that finite-dimensional models may be used to approximate infinite-dimensional problems by discretizing the continuous-time model. For instance, the infinite-dimensional problem above can be transformed into the finite-dimensional problem

$$\min \quad \sum_{k=1}^{K} f(\mathbf{x}_k, \mathbf{u}_k) + g(\mathbf{x}_K)$$

s.t. $\quad \mathbf{x}_k = \mathbf{h}(\mathbf{x}_{k-1}, \mathbf{u}_k) \quad k = 1, \ldots, K$
$\quad\quad\;\; \mathbf{u}_k \in \Omega \quad\quad\quad\quad k = 1, \ldots, K,$

where the time horizon has been discretized in time intervals of width δt and $\mathbf{x}_k = \mathbf{x}(k\,\delta t)$. Note that $\mathbf{x}_k$ is the state *at the end* of the kth period [i.e., the period between $(k-1)\delta t$ and $k\,\delta t$], whereas $\mathbf{u}_k$ is the control applied *during* the kth period.

6.1.2 Unconstrained vs. constrained problems

If $S \equiv \mathbb{R}^n$, we have an *unconstrained problem*; otherwise, we have a *constrained problem*. Needless to say, real-life problems are rarely unconstrained; yet methods for unconstrained optimization are the foundation for many constrained optimization methods. The set S is usually specified by enforcing the following types of constraints on the decision variables.

- Equality constraints:

$$h_i(\mathbf{x}) = 0, \quad i \in E,$$

  or in vector form:

$$\mathbf{h}(\mathbf{x}) = \mathbf{0}.$$

- Inequality constraints:

$$g_i(\mathbf{x}) \le 0, \quad i \in I.$$

  or in vector form:

$$\mathbf{g}(\mathbf{x}) \le \mathbf{0},$$

  having stipulated that a vector inequality is interpreted componentwise. The constraint $g_i(\mathbf{x}) \le 0$ is said to be active at the point $\hat{\mathbf{x}}$ if $g_i(\hat{\mathbf{x}}) = 0$, and inactive if $g_i(\hat{\mathbf{x}}) < 0$. A "greater than" constraint such as $g_k(\mathbf{x}) \ge 0$ can be rewritten immediately in the form $-g_k(\mathbf{x}) \le 0$. In MATLAB, inequality constraints are assumed in the "less than" form. Non-negativity restrictions such as $x \ge 0$, also denoted by $x \in \mathbb{R}_+$, may be thought of as inequality constraints. However, simple bounding constraints of the form $l \le x \le u$ are usually dealt with in a special way by optimization algorithms; hence, inequality constraints and bounds are passed separately to optimization procedures.

6.1.3 Convex vs. non-convex problems

Depending on the nature of the objective function f and of the feasible set S, problem (6.1) may or may not be easy. In particular, when there is only one local optimum which is also the global optimum, the problem should be expected to be relatively easy. The key concept here, and in most optimization theory as well, is convexity. Some background in convex analysis is given in supplement S6.1 at the end of the chapter.

Problem (6.1) is a *convex problem* if f is a convex function and S is a convex set. Problem (6.1) is a *concave problem* if f is a concave function and S is a convex set. Assuming that the optimization problem has a finite solution, the following properties can be proved.

PROPERTY 6.1 *In a convex problem a local optimum is also a global optimum.*

PROPERTY 6.2 *In a concave problem the global optimum lies on the boundary of the feasible region S.*

To get a feeling for the second property, the reader is urged to solve the following problem graphically:

$$\min \quad -(x-2)^2 + 3$$
$$\text{s.t.} \quad 1 \leq x \leq 4.$$

Ideally, we would like to come up with a set of necessary and sufficient conditions for global optimality. Regrettably, what we have, in general, are just either sufficient or necessary conditions for local or global optimality. However, when the problem is unconstrained and the function is convex, it is easy to find a convenient characterization of a global minimizer.

THEOREM 6.3 *If the function f is convex and differentiable on $\mathbb{R}^n$, the point $\mathbf{x}^*$ is a global minimizer of f if and only if it satisfies the stationarity condition:*

$$\nabla f(\mathbf{x}^*) = \mathbf{0}.$$

Proof. If f is convex and differentiable, then we have

$$f(\mathbf{x}) \geq f(\mathbf{x}_0) + \nabla f'(\mathbf{x}_0)(\mathbf{x} - \mathbf{x}_0) \qquad \forall \mathbf{x}, \mathbf{x}_0.$$

But if the function is stationary at point $\mathbf{x}^*$,

$$f(\mathbf{x}) \geq f(\mathbf{x}^*) + \nabla f'(\mathbf{x}^*)(\mathbf{x} - \mathbf{x}^*) = f(\mathbf{x}^*) + \mathbf{0}'(\mathbf{x} - \mathbf{x}^*) = f(\mathbf{x}^*) \qquad \forall \mathbf{x},$$

which simply says that $\mathbf{x}^*$ is a global optimum. $\square$

The stationarity condition is a *first-order condition*; for generic functions, *second-order conditions* involving the Hessian matrix are required to guarantee that a stationary point is actually a (local) minimizer. The stationarity condition is easily extended to the case of a convex non-differentiable function.

THEOREM 6.4 *If the function f is convex on $\mathbb{R}^n$, the point $\mathbf{x}^*$ is a global minimizer of f if and only if the subdifferential of f at $\mathbf{x}^*$ includes the zero vector:*

$$\mathbf{0} \in \partial f(\mathbf{x}^*).$$

Proof. As discussed in supplement S6.1, a convex function f is subdifferentiable at any point[2]; that is, at any point $\mathbf{x}_0$ there is a set of subgradients, which is called the *subdifferential*. A subgradient at $\mathbf{x}_0$ is a vector $\boldsymbol{\gamma}$ such that

$$f(\mathbf{x}) \geq f(\mathbf{x}_0) + \boldsymbol{\gamma}'(\mathbf{x} - \mathbf{x}_0) \qquad \forall \mathbf{x}.$$

It is easy to see that if $\mathbf{0}$ belongs to the subdifferential at $\mathbf{x}^*$, we have $f(\mathbf{x}) \geq f(\mathbf{x}^*)$ for any $\mathbf{x}$. It is worth noting that this theorem is a generalization of the previous one, as if the function is differentiable in $\mathbf{x}^*$, the subdifferential includes only the gradient, and this condition boils down to stationarity. □

It should be noted that a set $S = \{\mathbf{x} \in \mathbb{R}^n \mid g_i(\mathbf{x}) \leq 0, i \in I\}$ is convex if the functions g_i are convex. To see this for a single function $g(\mathbf{x})$, assume that $\mathbf{x}_1, \mathbf{x}_2 \in S$. Convexity of g implies that

$$g[\lambda \mathbf{x}_1 + (1-\lambda) \mathbf{x}_2] \leq \lambda g(\mathbf{x}_1) + (1-\lambda) g(\mathbf{x}_2) \leq 0 \qquad \forall \lambda \in [0,1].$$

Since the intersection of convex sets is a convex set, the result is valid for an arbitrary number of convex functions. The equality-constrained case is more critical. Since an inequality constraint $h_i(\mathbf{x}) = 0$ can be thought of as two inequalities,

$$h_i(\mathbf{x}) \leq 0, \qquad -h_i(\mathbf{x}) \leq 0,$$

we see that it will describe a convex set only if the function h_i is both convex and concave. This will be the case only if h_i is affine, i.e., it is of the form

$$\mathbf{a}_i' \mathbf{x} = b_i.$$

6.1.4 Linear vs. non-linear problems

A finite-dimensional problem is called a *linear programming* (LP) *problem* when both the constraints and the objective are expressed by affine functions. The general form of a linear programming problem is

$$\min \quad \sum_{j=1}^{n} c_j x_j$$

$$\text{s.t.} \quad \sum_{j=1}^{n} a_{ij} x_j = b_i \qquad \forall i \in E$$

$$\sum_{j=1}^{n} d_{ij} x_j \leq e_i \qquad \forall i \in I,$$

---

[2] Strictly speaking, this is true only for the *interior* of the domain over which the function is convex.

which can be written in matrix form as

$$\min \quad \mathbf{c'x}$$
$$\text{s.t.} \quad \mathbf{Ax = b}$$
$$\mathbf{Dx \le e}.$$

Linear programming problems have two important features; they are both convex and concave problems. Thus, a local optimum is also a global one, and it lies on the boundary of the feasible solution; actually, it turns out that the feasible set is a polyhedron and that there is an optimal solution which corresponds to one of its vertices.

Example 6.3 Here is an example of an LP problem:

$$\min \quad 2x_1 + 3x_2 + 3x_3$$
$$\text{s.t.} \quad x_1 + 2x_2 = 3$$
$$x_1 + x_3 \ge 3$$
$$x_1, x_2, x_3 \ge 0.$$
□

If either condition is not met, i.e., if the objective function or a constraint is expressed by a non-linear function, we have a non-linear programming problem.

Example 6.4 The following are examples of non-linear programming problems:

$$\min \quad 2x_1 + 3x_2 + 3x_3$$
$$\text{s.t.} \quad x_1 + x_2^2 = 3$$
$$x_1 + x_3 \ge 3$$
$$x_1, x_2, x_3 \ge 0.$$

$$\min \quad 2x_1 + 3x_2 x_3$$
$$\text{s.t.} \quad x_1 + 2x_2 = 3$$
$$x_1 x_3 \ge 3$$
$$x_1, x_2, x_3 \ge 0.$$

$$\min \quad 2x_1^2 + 3x_2^2 + 3x_1 x_3$$
$$\text{s.t.} \quad x_1 + 2x_2 = 3$$
$$x_1 + x_3 \ge 3$$
$$x_1, x_2, x_3 \ge 0.$$

The last problem is characterized by a quadratic objective function and by linear constraints. This kind of problem is called a *quadratic programming problem*. Quadratic programming problems are the simplest non-linear programming problems, provided that the objective function is convex. If the quadratic part of the objective is related to a covariance matrix, as it happens for mean-variance portfolio optimization, the objective function is convex, as the covariance matrix is positive semidefinite (see theorem 6.11 in supplement S6.1.1). □

6.1.5 Continuous vs. discrete problems

Linear and quadratic programming problems are rather easy to solve, as they are convex problems. In some decision problems, it is necessary to enforce integrality constraints on some decision variables:

$$\mathbf{x} \in \mathbb{Z}_+^n,$$

where $\mathbb{Z}_+ = \{0, 1, 2, \ldots\}$ is the set of non-negative integers (models involving negative integer variables are quite rare). If the integrality constraint applies to all of the decision variables, we have a *pure integer program*; otherwise, we have a *mixed-integer program*. Such a restriction makes the problem much harder, mainly because a discrete feasible region is not convex. While non-linear integer programming techniques are known, robust commercial tools are available only for mixed-integer linear programs.[3]

Quite often, an integrality restriction has the form $x \in \{0, 1\}$, which is used when we have to model all-or-nothing decisions. One such case is the knapsack problem we met in example 1.2 on page 15. We will illustrate several "modeling tricks" based on logical variables in section 12.1.1. We should mention that past versions of the Optimization toolbox were not able to cope with discrete optimization problems. At the time of writing, a function bintprog is available to solve pure binary problems, i.e., linear programming problems in which *all* the decision variables are restricted to the set $\{0, 1\}$. This is a limited functionality which could be improved in future versions to cope with general mixed-integer problems. Nevertheless, we should mention that large-scale mixed-integer problems are a hard nut to crack and that specialized state-of-the-art packages are required.

6.1.6 Deterministic vs. stochastic problems

All the model classes we have considered so far assume, on the one hand, that there is no uncertainty in the data and, on the other one, that a sensible analytical model can be built. In some cases, building an analytical model

---

[3]However, recently released versions of ILOG CPLEX are able to solve mixed-integer quadratic problems.

is out of the question, because of both the randomness and the complexity involved. As an example, consider a set of rules for portfolio rebalancing; say these rules depend on a set of parameters and that you would like to find the optimal value of these parameters. It may be the case that a thorough testing of the rules may be carried out only by running a set of simulated experiments. This means that a simulator acts as a black box mapping a vector of decision variables $\mathbf{x}$ into an estimate of an objective function $f(\mathbf{x}) = \mathrm{E}[U(\mathbf{x})]$, possibly related to an expected utility. In this case, you have to integrate stochastic simulation and optimization methods, as described in section 6.6.

In other cases, we may be able to build an optimization model, but uncertainty in the problem data prevents the application of standard optimization methods. It is fairly obvious that coping with uncertain data is a significant complication, but there is a subtler issue. When uncertainty is involved, we should consider how and when the "true" values of the problem data are discovered: In fact, time and information are likely to play a role, since decision making under uncertainty typically involves a dynamic process in which decisions are "adjusted" when more and more information is revealed. Dealing with this dynamic decision process calls for an appropriate framework which is discussed in chapters 10 and 11.

6.2 NUMERICAL METHODS FOR UNCONSTRAINED OPTIMIZATION

In principle, an unconstrained problem $\min_{\mathbf{x} \in \mathbb{R}^n} f(\mathbf{x})$ may be solved by looking for a stationary point. Some care is needed for the non-convex case, since second-order information should be checked; furthermore, what we get in general is a local optimizer; indeed, almost all the non-linear programming libraries commercially available are aimed at local optimization. The stationarity condition yields a set of non-linear equations which could be solved to spot candidate optima; in fact, there are a few links between unconstrained optimization and the numerical solution of non-linear equations.

In optimization, one avoids direct solution of the non-linear equations. The computational approaches are generally based on the generation of a sequence of points $\mathbf{x}^{(k)}$, converging to a local optimum $\mathbf{x}^*$. In order to drive the search process in the right direction, one should find, for each point $\mathbf{x}^{(k)}$ in the sequence, a descent direction, i.e., a vector $\mathbf{s}^{(k)} \in \mathbb{R}^n$ such that

$$f(\mathbf{x}^{(k)} + \alpha \mathbf{s}^{(k)}) < f(\mathbf{x}^{(k)})$$

for some $\alpha > 0$. If we consider the function $h(\alpha) = f(\mathbf{x} + \alpha \mathbf{s})$, a descent direction is characterized by

$$\left. \frac{dh}{d\alpha} \right|_{\alpha=0} = [\nabla f(\mathbf{x})]' \mathbf{s} < 0.$$

It may be convenient to consider true direction vectors, i.e., unit norm vectors such that $\|\mathbf{s}\| = 1$. A general iteration scheme is, after initialization with a starting guess $\mathbf{x}^{(0)}$:

1. Find a descent direction $\mathbf{s}^{(k)}$.
2. Find a step length $\alpha^{(k)} \in \mathbb{R}_+$.
3. Update $\mathbf{x}^{(k+1)} = \mathbf{x}^{(k)} + \alpha^{(k)} \mathbf{s}^{(k)}$.

The scheme is iterated until some convergence criterion is met. There are a variety of choices, which lead to different algorithms, some of which are briefly outlined in the following. It should be noted that this approach can be extended to deal with constrained optimization problems. An easy case is when we have to solve
$$\min_{\mathbf{x} \in \mathbb{R}^n_+} f(\mathbf{x}).$$
Here it is sufficient to slightly modify the updating rule as follows:
$$\mathbf{x}^{(k+1)} = \max\left\{\mathbf{0}, \mathbf{x}^{(k)} + \alpha^{(k)} \mathbf{s}^{(k)}\right\},$$
which should be interpreted componentwise; if some component becomes negative, set it at zero. This operation essentially amounts to *projecting* $\mathbf{x}^{(k+1)}$ onto the feasible set $\mathbb{R}^n_+$ (projection can be exploited for more general feasible sets, with computational difficulties depending on their nature).

6.2.1 Steepest descent method

One seemingly obvious choice for the descent direction is
$$\mathbf{s}^{(k)} = -\frac{\nabla f^{(k)}}{\|\nabla f^{(k)}\|},$$
which yields the steepest descent or gradient method. The step length α may be chosen by solving the one-dimensional problem
$$\min_{\alpha \geq 0} h(\alpha) = f(\mathbf{x}^{(k)} + \alpha \mathbf{s}^{(k)}).$$
This one-dimensional problem is easier than the original problem, as it is a scalar optimization problem. It can be solved by a variety of *line search methods*. One possibility, which works for convex functions, is using a quadratic fit. Assume that we have three points $0 \leq \alpha_1 < \alpha_2 < \alpha_3$, such that
$$h(\alpha_1) > h(\alpha_2), \qquad h(\alpha_2) < h(\alpha_3).$$
An initial set of points satisfying these conditions can be found by some search procedure. Now we may fit a quadratic curve passing through the three points;

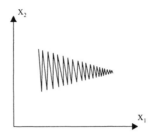

Fig. 6.3 Zig-zagging in the steepest descent procedure.

minimization of the quadratic curve is easily accomplished, under convexity assumption, by setting its derivative to zero. This yields another point, α^*. Assume that $\alpha^* > \alpha_2$. If $h(\alpha^*) \geq h(\alpha_2)$, we proceed with the new set of points $(\alpha_1, \alpha_2, \alpha^*)$; otherwise, we proceed with $(\alpha_2, \alpha^*, \alpha_3)$. Actually, there is a rich set of line search methods, involving, e.g., cubic interpolation and other tricks of the trade; some may be selected by setting MATLAB option parameters.

Despite its apparent appeal, the steepest descent method may suffer from poor convergence near the minimizer. In some cases, pathological behavior called "zig-zagging" is observed. The zig-zagging phenomenon is illustrated in figure 6.3.[4] Furthermore, roundoff errors may make the straightforward steepest descent method rather unreliable.

6.2.2 The subgradient method

It is obvious that the gradient method cannot be applied to a non-differentiable function. In supplement S6.1.1 we note that the subgradient is a generalization of the gradient concept to the case of non-smooth functions. Hence, assuming we can compute a subgradient $\gamma^{(k)}$ for a convex function f at any point $\mathbf{x}^{(k)}$, we may wonder if a scheme like

$$\mathbf{x}^{(k+1)} = \mathbf{x}^{(k)} - \alpha^{(k)} \gamma^{(k)}$$

could work. The answer is not easy, since there is no guarantee that by changing the sign of the subgradient we find a descent direction. However, if some condition is enforced on the step lengths $\alpha^{(k)}$, it can be shown that the subgradient method converges to the optimal solution. An intuitive justification runs as follows.

---

[4]The purpose of the figure is just to illustrate the phenomenon, as the angles between successive segments are not necessarily realistic. To really see zig-zagging, the reader is urged to try the Optimization toolbox demos. Just type **demo**, which opens a window in which you should select the Optimization toolbox. Then try the "minimization of the banana function" demo.

Consider a point $\mathbf{x}_0$ and let $\boldsymbol{\gamma}_0$ be a subgradient of f at $\mathbf{x}_0$. Then, by definition of a subgradient:

$$f(\mathbf{x}) \geq f(\mathbf{x}_0) + \boldsymbol{\gamma}_0'(\mathbf{x} - \mathbf{x}_0) \qquad \forall \mathbf{x}.$$

By applying this inequality to the optimal solution $\mathbf{x}^*$ and rearranging, we obtain

$$-\boldsymbol{\gamma}_0'(\mathbf{x}^* - \mathbf{x}_0) \geq f(\mathbf{x}_0) - f(\mathbf{x}^*) \geq 0.$$

Note that the vector $\mathbf{x}^* - \mathbf{x}_0$ is the direction along which we should move to reach the optimal solution from $\mathbf{x}_0$. The inequality above shows that this vector forms an angle less than 90 degrees with $-\boldsymbol{\gamma}_0$. Hence, the subgradient, changed in sign, need not be a descent direction, but at least it points to the "right" half-space, where the optimal solution lies.

6.2.3 Newton and the trust region methods

The convergence problems in the gradient method are essentially due to the fact that the gradient method uses a *first-order* local approximation of f ignoring curvature information. The situation could be improved by using a second-order approximation, for a displacement vector $\boldsymbol{\delta}$:

$$f(\mathbf{x} + \boldsymbol{\delta}) \approx f(\mathbf{x}) + [\nabla f(\mathbf{x})]'\boldsymbol{\delta} + \frac{1}{2}\boldsymbol{\delta}'\mathbf{H}(x)\boldsymbol{\delta},$$

where $\mathbf{H}$ is the Hessian matrix. If $\mathbf{H}$ is positive definite, the function is locally strictly convex and we may find a minimizer for the quadratic approximation by solving the system of linear equations

$$\mathbf{H}(\mathbf{x})\boldsymbol{\delta} = -\nabla f(\mathbf{x}).$$

This method is known as Newton's method (for optimization) and it has better convergence properties as well as higher computational costs. However, we are in trouble if the Hessian is not positive definite.

Another approach is to restrict the step α taken along the direction given by the gradient. The rationale is that the first-order approximation is valid only in a neighborhood of the current iterate $\mathbf{x}^{(k)}$. To find the displacement $\boldsymbol{\delta}$, we could consider the restricted minimization subproblem:

$$\min_{\boldsymbol{\delta}} \quad f(\mathbf{x}^{(k)}) + [\nabla f(\mathbf{x}^{(k)})]'\boldsymbol{\delta}$$
$$\text{s.t.} \quad \|\boldsymbol{\delta}\| \leq h^{(k)}.$$

Exploiting this idea leads to *trust region methods*, which are actually used in MATLAB for large-scale problems. The trust region is delimited by the parameter $h^{(k)}$, which controls the step length and should be adjusted dynamically. We may compare the predicted improvement in the objective function (according to the approximating function) with the actual improvement. A large difference suggests that the approximation is not reliable and that the step length should be reduced. Otherwise, the step length can be increased.

6.2.4 No-derivatives algorithms: quasi-Newton method and simplex search

One problem with Newton's method is that the Hessian matrix is required. Since providing the software with this information requires a good deal of error-prone work, alternative approaches have been developed in order to approximate this matrix based on function evaluations only. This leads to quasi-Newton methods, which we have already met in the case of non-linear equations (see example 3.25 on page 201). The same observation applies to providing the gradient of the objective function. As we have seen in chapter 5, one idea is to approximate the gradient by finite differences like

$$\frac{\partial f(\mathbf{x})}{\partial x_i}\bigg|_{\mathbf{x}=\hat{\mathbf{x}}} \approx \frac{f(\hat{\mathbf{x}} + h_i \mathbf{1}_i) - f(\hat{\mathbf{x}})}{h_i}$$

or

$$\frac{\partial f(\mathbf{x})}{\partial x_i}\bigg|_{\mathbf{x}=\hat{\mathbf{x}}} \approx \frac{f(\hat{\mathbf{x}} + h_i \mathbf{1}_i) - f(\hat{\mathbf{x}} - h_i \mathbf{1}_i)}{2h_i},$$

where $\mathbf{1}_i$ is the ith unit vector. By the same token, we may devise suitable approximations of the Hessian matrix.

In some circumstances, you would not be able to compute the gradient anyway; one case is when the objective function is not known, but it is implicitly computed by a simulation model; another one is when there are discontinuities in the objective function. In such cases, it is useful to adopt methods that rely only on function evaluations. One such approach is the simplex search method developed by Nelder and Mead.[5] The rationale behind the method is illustrated in figure 6.4 for a minimization problem in $\mathbb{R}^2$. A simplex in $\mathbb{R}^n$ is the convex hull of a set of $n+1$ affinely independent points $\mathbf{x}_1, \ldots, \mathbf{x}_{n+1}$.[6] In two dimensions, a simplex is simply a triangle, whereas in three dimensions it is a tetrahedron. The simplex search method works by building and transforming a set of $n+1$ points rather than generating a sequence of single points; the point with the worst value of the objective is spotted and replaced by another point. For instance, consider the three vertices of the triangle in figure 6.4 and assume that $f(\mathbf{x}_3)$ is the worst objective value; then it seems reasonable to move away from $\mathbf{x}_3$ by reflecting it through the center of the

---

[5]This method should not be confused with the celebrated simplex method for linear programming.
[6]Affine independence here means that the vectors $(\mathbf{x}_2 - \mathbf{x}_1), \ldots, (\mathbf{x}_{n+1} - \mathbf{x}_1)$ are linearly independent. For $n = 2$ this means that the three points do not lie on the same line. For $n = 3$ this means that the four points do not lie on the same plane.

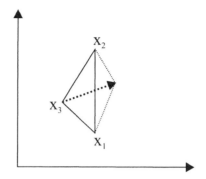

Fig. 6.4 Reflection of the worst value point in the Nelder–Mead simplex search procedure.

face formed by the other points. This is easily accomplished algebraically. Assume that $\mathbf{x}_{n+1}$ is the worst point; then we compute the centroid of the best n points as

$$\mathbf{c} = \frac{1}{n} \sum_{i=1}^{n} \mathbf{x}_i,$$

and we try a new point of the form

$$\mathbf{x}_r = \mathbf{c} + \alpha(\mathbf{c} - \mathbf{x}_{n+1}).$$

The reflection coefficient $\alpha > 0$ is adjusted depending on the circumstances. If $\mathbf{x}_r$ turns out to be even worse than $\mathbf{x}_{n+1}$, we may argue that the step was too long, and the simplex should be contracted. If $\mathbf{x}_r$ turns out to be the new best point, we have found a good direction and the simplex should be expanded. Different tricks have been devised in order to improve the convergence of the method.

6.2.5 Unconstrained optimization in MATLAB

Consider the unconstrained optimization problem

$$\min f(x_1, x_2) = (x_1 - 2)^4 + (x_1 - 2x_2)^2.$$

Clearly, $f(x_1, x_2) \geq 0$ and $f(2, 1) = 0$; hence $(2, 1)$ is a globally optimal solution. The gradient of f is given by

$$\nabla f(x_1, x_2) = \begin{bmatrix} 4(x_1 - 2)^3 + 2(x_1 - 2x_2) \\ -4(x_1 - 2x_2) \end{bmatrix}.$$

It is easy to see that $\nabla f(2, 1) = \mathbf{0}$.

In the Optimization toolbox we have two functions that can be used for unconstrained optimization:

344　CONVEX OPTIMIZATION

- `fminsearch`, which implements a variant of the simplex search method.
- `fminunc`, which actually implements a variety of methods, which are selected according to a set of options controlled by the user.

Both functions require an M-file, a function handle, or an inline function to evaluate the objective, and an initial estimate of the solution. An optional parameter may be used to set the desired options through the `optimset` function.

Let us first try the simplex search procedure, giving an initial estimate $x_0 = 0$:

```
>> f = @(x) (x(1) - 2)^4 + (x(1) - 2 * x(2))^2;
>> x=fminsearch(f,[0 0])
x =
    2.0000    1.0000
>> f(x)
ans =
  2.9563e-017
>>
```

Now we may try `fminunc`:

```
>> x=fminunc(f,[0 0])
Warning: Gradient must be provided for trust-region method;
  using line-search method instead.
Optimization terminated: relative infinity-norm of gradient less
  than options.TolFun.
x =
    1.9897    0.9948
>> f(x)
ans =
  1.1274e-008
```

The result is not exact really. The point is that the function is rather "flat" around the minimizer; in fact the objective function is close to zero in the solution reported by MATLAB. We could change the tolerance parameters in order to improve the solution, but this could make no sense in practice. We may also note that MATLAB complains about the lack of gradient information, so that it cannot apply a trust region method. This is not much trouble, as the gradient may be estimated numerically. However, we could ask MATLAB not to use the default "large-scale" algorithm, which is a trust region method, but a "medium-scale" algorithm.

```
>> options=optimset('largescale', 'off');
x=fminunc(f,[0 0], options)
Optimization terminated: relative infinity-norm of gradient less
  than options.TolFun.
x =
```

```
        1.9897    0.9948
```

Alternatively, we may provide a function to compute the gradient and tell MATLAB to use it within a large-scale algorithm, possibly with a stricter tolerance:

```
>> f = @(x) (x(1) - 2)^4 + (x(1) - 2 * x(2))^2;
>> gradf = @(x) [4*(x(1)-2)^3+2*(x(1)-2*x(2)) , -4*(x(1)-2*x(2))];
>> options=optimset('gradobj','on', 'largescale','on', 'tolfun',1e-13);
>> x=fminunc({f, gradf}, [0 0], options)
Optimization terminated: relative function value changing by less
  than OPTIONS.TolFun.
x =
    1.9997    0.9998
```

Computing a gradient analytically is clearly an error-prone activity. To help with this task, it is possible to ask MATLAB to compare the gradient we provide with a numerical estimate. All we have to do is to reset the options and to set the derivativecheck option on.[7] Here we may try this functionality, providing MATLAB with an incorrect expression for the gradient implemented in the function gradf1:

```
>> options = optimset;
>> options=optimset('gradobj', 'on', 'largescale', 'off', ...
'derivativecheck', 'on');
>> gradf1 = @(x) [6*(x(1)-2)^3+2*(x(1)-2*x(2)) , -4*(x(1)-2*x(2))];
>> x=fminunc({f, gradf1}, [0 0], options)
Maximum discrepancy between derivatives  = 16
Warning: Derivatives do not match within tolerance
Derivative from finite difference calculation:
  -32.0000
        0
User-supplied derivative,
  @(x) [6*(x(1)-2)^3+2*(x(1)-2*x(2)) , -4*(x(1)-2*x(2))]:
   -48
     0
Difference:
  -16.0000
        0
Strike any key to continue or Ctrl-C to abort
Optimization terminated:
  relative infinity-norm of gradient less than options.TolFun.
x =
    1.9841    0.9921
```

Indeed, we see that a warning is issued by the system, spotting a likely trouble with our analytical gradient.

---
[7] See also the code displayed in figure 3.28 on page 202.

6.3 METHODS FOR CONSTRAINED OPTIMIZATION

Consider a general constrained optimization problem, such as

$$\begin{align} \min \quad & f(\mathbf{x}) \\ \text{s.t.} \quad & h_i(\mathbf{x}) = 0 \quad i \in E \\ & g_i(\mathbf{x}) \leq 0 \quad i \in I. \end{align}$$

In this section we assume that all the involved functions have suitable differentiability properties. For a constrained problem, stationarity is not a necessary condition anymore, since the optimal solution may be a non-stationary point on the boundary of the feasible set (this means that there are descent directions, but they all lead outside the feasible region). One possible approach to cope with this difficulty is trying to transform the problem in such a way that stationarity condition may be applied again; this leads to the penalty function approach (section 6.3.1). Another idea is to develop optimality conditions which include some form of stationarity, plus some additional requirements; this leads to the Kuhn–Tucker conditions (section 6.3.2). Kuhn–Tucker conditions generalize the Lagrange multiplier method for equality-constrained problems, and they are linked to a body of optimization theory called duality theory (section 6.3.3), which leads both to theoretical insights and to practical algorithms. Another important observation is that a constrained problem is relatively easy when all the involved function are affine; indeed, linear programming is a very well developed branch of optimization theory (section 6.4). So it may be interesting to develop algorithms which somehow transform a non-linear problem to a linear problem. This may be accomplished easily if the constraints are linear and the objective function is convex; Kelley's cutting planes algorithm (section 6.3.4) is based on this idea, and it is the conceptual basis of some methods for stochastic problems. In general, it is reasonable to assume that a linearly constrained problem has some specific features that may be exploited in a computational algorithm. The active set method (section 6.3.5) is one such strategy; it also worth noting that, in the earlier versions of the Optimization toolbox, the active set method was the basis of the functions for both linear and quadratic programming.

Due to its introductory nature, this book has been written sacrificing the mathematical rigor. This is particularly true for this chapter, as optimization theory is a tough subject in which simplistic approaches may lead to disasters. Hence, the serious reader is urged not to take what we illustrate in the following as a foolproof set of recipes; it is a good starting point, but the references at the end of the chapter should be consulted for a more thorough treatment.

6.3.1 Penalty function approach

Penalty functions are based on the idea of relaxing constraints through the addition of a suitable term to the objective function. Consider a problem with equality constraints:

$$\min \quad f(\mathbf{x})$$
$$\text{s.t.} \quad h_i(\mathbf{x}) = 0, \quad i \in E.$$

It is possible to approximate this constrained problem by the unconstrained one

$$\min \Phi(\mathbf{x}, \sigma) = f(\mathbf{x}) + \sigma \sum_{i \in E} h_i^2(\mathbf{x}).$$

This function penalizes both positive and negative values of h_i. If σ is large enough, the optimization algorithm will, in some sense, first drive the solution toward the feasible region by minimizing the penalty term; then it will try to minimize the objective f. Actually, convergence difficulties will arise if we try solving the unconstrained problem with a large value of the penalty coefficient σ. So it is advisable to solve a sequence of unconstrained problems using the optimal solution of each subproblem as the initial solution of the next one:

1. Choose a sequence $\{\sigma^{(k)}\} \to \infty$.

2. Find the minimizer $\mathbf{x}^*(\sigma^{(k)})$ of $\Phi(\mathbf{x}, \sigma)$.

3. Stop if $h_i(\mathbf{x}^*)$ is sufficiently small for all i.

We can see this as an example of a continuation strategy (see section 3.4.5). The case of inequality constraints

$$\min \quad f(\mathbf{x})$$
$$\text{s.t.} \quad g_i(\mathbf{x}) \leq 0 \quad i \in I.$$

can be tackled by a similar approach. In this case, however, we must only penalize positive values of the constraint functions g_i. Using the notation $[y]^+ = \max\{y, 0\}$, we may use a penalty function like

$$f(\mathbf{x}) + \sigma \sum_{i \in I} \left[g_i^+(\mathbf{x})\right]^2$$

or

$$f(\mathbf{x}) + \sigma \sum_{i \in I} g_i^+(\mathbf{x})$$

for increasing values of σ. The first penalty function is differentiable, whereas the second one is not, as you may see in figure 6.5a; however, the second function may be advantageous from the numerical point of view, as there is no need to use too large values of the penalty coefficients. Indeed, one of the driving forces behind the development of non-smooth optimization algorithms was the use of exact penalty functions.

In both cases, we are actually using an *exterior penalty function*. The name stems from the fact that the feasible set is approached from outside for

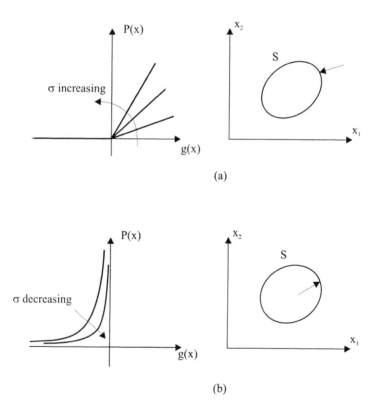

Fig. 6.5 Exterior (a) and interior (b) penalties.

increasing values of σ, as illustrated in figure 6.5a. If the optimal solution is on the boundary of the feasible set (which is usually the case, since some inequality constraints are active), a feasible solution is obtained only in the limit. In some cases, this is quite natural, as the constraints may be soft or "elastic" and express some desirable feature rather than a hard requirement. In other cases, we would like to be able to stop the algorithm whenever we want and still come up with a strictly feasible solution. To overcome this problem, an *interior penalty approach* can be pursued, by introducing a suitable *barrier function*. One example is

$$B(\mathbf{x}) = -\sum_{i \in I} \frac{1}{g_i(\mathbf{x})}.$$

The barrier function goes to infinity when $\mathbf{x}$ tends to the boundary of the feasible region from inside. Then an unconstrained problem,

$$\min f(\mathbf{x}) + \sigma B(\mathbf{x}),$$

is solved for decreasing values of σ, until the term $\sigma B(\mathbf{x})$ is small enough. As shown in figure 6.5b, in this case we approach the optimal solution on the boundary staying within the feasible region; this may be an advantage, provided that we have a way to start the iterations with a feasible point. From figure 6.5 it should also be clear that both exterior and interior penalty functions are numerically feasible ways of approximating the ideal penalty:

$$P_i(x) = \begin{cases} 0, & g_i(\mathbf{x}) \leq 0 \\ +\infty, & g_i(\mathbf{x}) > 0. \end{cases}$$

Example 6.5 Consider the problem

$$\min \quad (x - 1.5)^2 + (y - 0.5)^2$$
$$\text{s.t.} \quad x, y \leq 1,$$

whose optimal solution is clearly $x^* = 1$, $y^* = 0.5$. An interior penalty function could be

$$(x - 1.5)^2 + (y - 0.5)^2 + \frac{\sigma}{1-x} + \frac{\sigma}{1-y}.$$

Using MATLAB graphics, we may easily plot the level curves of the penalty function for different values of the parameter σ. We need to define a function and to use the functions meshgrid, to define the grid of points on which we want to evaluate the function, and contour, to plot a set of level curves.

```
>> f=@(sigma,x,y) (x-1.5).^2+(y-0.5).^2+sigma./(1-x)+sigma./(1-y);
>> [x y] = meshgrid(0.01 : 0.01 : 0.99);
>> subplot(2,2,1)
>> contour(f(0.1,x,y),30)
```

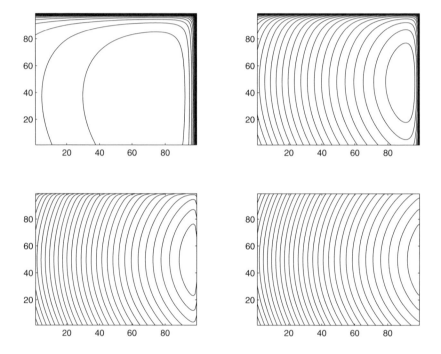

Fig. 6.6 Plots of the level curves for the interior penalty function of example 6.5 for $\sigma = 0.1$, $\sigma = 0.01$, $\sigma = 0.001$, $\sigma = 0.0001$.

```
>> subplot(2,2,2)
>> contour(f(0.01,x,y),30)
>> subplot(2,2,3)
>> contour(f(0.001,x,y),30)
>> subplot(2,2,4)
>> contour(f(0.0001,x,y),30)
```

The three plots are shown in figure 6.6. We see that the optimal solution of the unconstrained problem tends to the optimal solution of the original one from the inside. □

The penalty function approach is conceptually very simple, and some convergence properties can be proved. However, severe numerical difficulties may arise, for instance, when σ gets very large in the case of an exterior penalty. Nevertheless, penalty functions are most useful in providing a starting point for other, more sophisticated methods. They may be integrated with the Lagrangian methods described below, giving rise to the *augmented Lagrangian methods*, and they are one of the ingredients of the increasingly popular interior point methods for linear programming.

6.3.2 Kuhn–Tucker conditions

Consider a general constrained problem (P_{EI}):

$$\begin{aligned} \min \quad & f(\mathbf{x}) \\ \text{s.t.} \quad & h_i(\mathbf{x}) = 0, \quad i \in E \\ & g_i(\mathbf{x}) \le 0, \quad i \in I. \end{aligned}$$

The stationarity of f plays no role in proving optimality here, but the stationarity of a related function does. Consider the Lagrangian function

$$\mathcal{L}(\mathbf{x}, \boldsymbol{\lambda}, \boldsymbol{\mu}) = f(\mathbf{x}) + \sum_{i \in E} \lambda_i h_i(\mathbf{x}) + \sum_{i \in I} \mu_i g_i(\mathbf{x}). \qquad (6.3)$$

The stationarity of $\mathcal{L}$ does play a role in the following conditions.

THEOREM 6.5 (Kuhn–Tucker conditions) *Assume that the functions f, h_i, g_i in (P_{EI}) are continuously differentiable, and that $\mathbf{x}^*$ is feasible and satisfies a constraint qualification condition. Then a necessary condition for the local optimality of $\mathbf{x}^*$ is that there exist numbers λ_i^* $(i \in E)$ and $\mu_i^* \ge 0$ $(i \in I)$ such that*

$$\nabla f(\mathbf{x}^*) + \sum_{i \in E} \lambda_i^* \nabla h_i(\mathbf{x}^*) + \sum_{i \in I} \mu_i^* \nabla g_i(\mathbf{x}^*) = \mathbf{0}$$
$$\mu_i^* g_i(\mathbf{x}^*) = 0 \qquad \forall i \in I.$$

352 CONVEX OPTIMIZATION

The first condition is the stationarity of the Lagrangian function; if the set of inequality constraints is empty, these conditions boil down to the older Lagrange method to deal with equality-constrained problems. The numbers λ_i and μ_i are called *Lagrange multipliers*; note that the multipliers for inequality constraints are restricted in sign. For reasons that will be clear in the next section, the multipliers are also called *dual variables* (as opposed to the primal variables **x**). The Kuhn–Tucker conditions are, in a sense, rather weak, as they are only necessary conditions for local optimality, and they further require differentiability properties and some additional qualification condition on the constraints (to be clarified in example 6.7). They are, however, necessary and sufficient for global optimality in the convex case.

Example 6.6 As a first example, we may solve the optimization problem we have considered in example 2.3, on page 37. It is a non-linear programming problem with one equality constraint:

$$\max \quad x_1^\alpha x_2^{1-\alpha}$$
$$\text{s.t.} \quad p_1 x_1 + p_2 x_2 = W.$$

We introduce a Lagrange multiplier λ and build the Lagrangian function:

$$\mathcal{L}(x_1, x_2, \lambda) = x_1^\alpha x_2^{1-\alpha} + \lambda \left(p_1 x_1 + p_2 x_2 - W \right).$$

Since there is no inequality, we have just to write first-order optimality conditions:

$$\frac{\partial \mathcal{L}}{\partial x_1} = \alpha x_1^{\alpha-1} x_2^{1-\alpha} + \lambda p_1 = \alpha \left(\frac{x_1}{x_2}\right)^{\alpha-1} + \lambda p_1 = 0$$

$$\frac{\partial \mathcal{L}}{\partial x_2} = (1-\alpha) x_1^\alpha x_2^{-\alpha} + \lambda p_2 = (1-\alpha) \left(\frac{x_1}{x_2}\right)^{\alpha} + \lambda p_2 = 0$$

$$\frac{\partial \mathcal{L}}{\partial \lambda} = p_1 x_1 + p_2 x_2 - W = 0.$$

Dividing the first two equations term by term, after a rearrangement, we get

$$\frac{\alpha}{1-\alpha} \frac{x_2}{x_1} = \frac{p_1}{p_2} \quad \Rightarrow \quad (1-\alpha) p_1 x_1 - \alpha p_2 x_2 = 0. \tag{6.4}$$

From the budget equation we may get x_2:

$$x_2 = \frac{W - p_1 x_1}{p_2},$$

which may be substituted in (6.4):

$$(1-\alpha) p_1 x_1 - \alpha \left(W - p_1 x_1 \right) = 0,$$

which yields

$$x_1 = \alpha \frac{W}{p_1} \quad \Rightarrow \quad x_2 = (1-\alpha) \frac{W}{p_2}.$$

We see that consumption of each good is inversely proportional to its price, and it depends on the preference parameter α. □

We will not prove the Kuhn–Tucker conditions, as a rigorous proof is beyond the scope of the book; informally, they can be derived by characterizing a local optimum as a point such that an improvement in the objective function can only be obtained by going outside the feasible region. It is worth noting that the stationarity condition says that the gradient of the objective function can be expressed as a linear combination of the gradients of the objectives; this clarifies a little what we mean by *constraint qualification*; if the gradients of the constraints are not linearly independent at $\mathbf{x}^*$, it might be the case that we cannot use them as a basis to express ∇f. So it may happen that the Kuhn–Tucker conditions are not satisfied by a point that is actually a local minimizer.

Example 6.7 To understand the issue behind the constraint qualification condition, consider the problem:

$$\begin{aligned}
\min \quad & x_1 + x_2 \\
\text{s.t.} \quad & h_1(\mathbf{x}) = x_2 - x_1^3 = 0 \\
& h_2(\mathbf{x}) = x_2 = 0.
\end{aligned}$$

It is easy to see that the feasible set is the single point $(0, 0)$, which is the (trivial) optimal solution. If we try applying the Kuhn–Tucker conditions, we first build the Lagrangian function

$$\mathcal{L}(x_1, x_2, \lambda_1, \lambda_2) = x_1 + x_2 + \lambda_1(x_2 - x_1^3) + \lambda_2 x_2.$$

Writing the stationarity yields the system

$$\begin{aligned}
\frac{\partial \mathcal{L}}{\partial x_1} &= 1 - 3\lambda_1 x_1^2 = 0 \\
\frac{\partial \mathcal{L}}{\partial x_2} &= 1 + \lambda_1 + \lambda_2 = 0 \\
\frac{\partial \mathcal{L}}{\partial \lambda_1} &= x_2 - x_1^3 = 0 \\
\frac{\partial \mathcal{L}}{\partial \lambda_2} &= x_2 = 0,
\end{aligned}$$

which has no solution (the first equation requires that $x_1 \neq 0$, which is not compatible with the last two equations). This is due to the fact that the gradients of the two constraints are parallel at the origin:

$$\nabla h_1(0, 0) = \begin{bmatrix} -3x_1^2 \\ 1 \end{bmatrix}_{\mathbf{x}=0} = \begin{bmatrix} 0 \\ 1 \end{bmatrix}$$

$$\nabla h_2(0, 0) = \begin{bmatrix} 0 \\ 1 \end{bmatrix}_{\mathbf{x}=0} = \begin{bmatrix} 0 \\ 1 \end{bmatrix},$$

and they are not a basis able to express the gradient of f

$$\nabla f(0,0) = \begin{bmatrix} 1 \\ 1 \end{bmatrix}_{x=0} = \begin{bmatrix} 1 \\ 1 \end{bmatrix}.$$

□

Different constraint qualification conditions have been proposed in the literature. Sufficient conditions to avoid trouble are that the gradients of the active constraints are linearly independent, or that the constraints are all linear. We will not pursue this issue any further, but we recommend a book like [18] as a warning against easy cookbook recipes in optimization.

The Kuhn–Tucker theorem also includes a second set of conditions, which are known as *complementary slackness* conditions. They may be interpreted by noting that if a constraint is inactive at $\mathbf{x}^*$, i.e., if $g_i(\mathbf{x}^*) < 0$, the corresponding multiplier must be zero; by the same token, if the multiplier μ_i^* is strictly positive, the corresponding constraints must be active (which roughly means that it could be substituted by an equality constraint without changing the optimal solution). The complementary slackness conditions could be used, in principle, to find a feasible point and a set of multipliers satisfying the Kuhn–Tucker conditions.

Example 6.8 Consider the convex problem

$$\begin{aligned}
\min \quad & x_1^2 + x_2^2 \\
\text{s.t.} \quad & x_1 \geq 0 \\
& x_2 \geq 3 \\
& x_1 + x_2 = 4.
\end{aligned}$$

First write the Lagrangian function:

$$\mathcal{L}(\mathbf{x}, \boldsymbol{\mu}, \lambda) = x_1^2 + x_2^2 - \mu_1 x_1 - \mu_2(x_2 - 3) + \lambda(x_1 + x_2 - 4).$$

A set of numbers satisfying the Kuhn–Tucker conditions can be found by solving the following system:

$$\begin{aligned}
& 2x_1 - \mu_1 + \lambda = 0 \\
& 2x_2 - \mu_2 + \lambda = 0 \\
& x_1 \geq 0, \quad x_2 \geq 3 \\
& x_1 + x_2 = 4 \\
& \mu_1 x_1 = 0, \quad \mu_1 \geq 0 \\
& \mu_2(x_2 - 3) = 0, \quad \mu_2 \geq 0.
\end{aligned}$$

We may proceed with a case-by-case analysis exploiting the complementary slackness conditions. If a multiplier is strictly positive, the corresponding inequality is active, which helps us in finding the value of a decision variable.

Case 1 ($\mu_1 = \mu_2 = 0$). In this case, the inequality constraints are dropped from the Lagrangian function. From the stationarity conditions we obtain the system

$$2x_1 + \lambda = 0$$
$$2x_2 + \lambda = 0$$
$$x_1 + x_2 - 4 = 0.$$

This yields a solution $x_1 = x_2 = 2$, which violates the second inequality constraint.

Case 2 ($\mu_1, \mu_2 \neq 0$). The complementary slackness conditions immediately yield $x_1 = 0, x_2 = 3$, violating the equality constraint.

Case 3 ($\mu_1 \neq 0, \mu_2 = 0$). We obtain

$$x_1 = 0$$
$$x_2 = 4$$
$$\lambda = -2x_2 = -8$$
$$\mu_1 = \lambda = -8.$$

The Kuhn–Tucker conditions are not satisfied since the value of μ_1 is negative.

Case 4 ($\mu_1 = 0, \mu_2 \neq 0$). We obtain

$$x_2 = 3$$
$$x_1 = 1$$
$$\lambda = -2$$
$$\mu_2 = 4,$$

which satisfy all the necessary conditions.

Since this is a convex problem, we have obtained the global optimum. Note how non-zero multipliers correspond to the active constraints, whereas the inactive constraint $x_1 \geq 0$ is associated to a multiplier $\mu_1 = 0$. The same result can easily be obtained through MATLAB. The **quadprog** function deals with quadratic programming problems such as

$$\min \quad \frac{1}{2}\mathbf{x}'\mathbf{H}\mathbf{x} + \mathbf{f}'\mathbf{x}$$
$$\text{s.t.} \quad \mathbf{A}\mathbf{x} \leq \mathbf{b}$$
$$\mathbf{A}_{eq}\mathbf{x} = \mathbf{b}_{eq}$$
$$\mathbf{l} \leq \mathbf{x} \leq \mathbf{u}.$$

For our example, some entries of the problem are empty. Note also that simple bounds are treated apart in practice and that the quadratic term in the objective function must be written in a specific way, as it involves a 1/2 factor and it assumes a symmetric Hessian matrix **H**.

```
>> H = 2*eye(2);
>> f = [0 0];
>> Aeq = [1 1];
>> beq = 4;
>> lb = [0; 3];
>> options=optimset('LargeScale', 'off');
>> [x,fval,exitflag,output,lambda] = quadprog(H,f,[],[],Aeq,beq,...
         lb,[],[],options);
Optimization terminated.
>> x
x =
   1.0000
   3.0000
>> lambda.eqlin
ans =
  -2.0000
>> lambda.lower
ans =
        0
   4.0000
```

The output arguments include the optimal decision variables, the optimal value of the objective function, an exit flag containing information about the termination of the algorithm, additional output information, and the multipliers included in the structure `lambda`. The multipliers in our case are associated to the linear equalities and to the lower bounds on the decision variables. □

Clearly, the approach we have taken in the example is not practical. Some alternative way must be found to spot the optimal multipliers. This leads to duality theory, which is the topic of next section. Before proceeding, it is also useful to get an intuitive grasp of the meaning of the Lagrange multipliers.

Example 6.9 Consider the parameterized problem

$$\min \quad x_1^2 + x_2^2$$
$$\text{s.t.} \quad x_1 + x_2 = b.$$

The stationarity conditions on the Lagrangian function,

$$\mathcal{L}(x_1, x_2, \lambda) = x_1^2 + x_2^2 + \lambda(x_1 + x_2 - b),$$

immediately yield $x_1^* = x_2^* = b/2$ and $\lambda^* = -b$. Now, ask how slight changes in the parameter b will affect the optimal value $f^* = b^2/2$:

$$\frac{df^*}{db} = b = -\lambda^*.$$

This suggests that, neglecting the sign, the dual variables are linked to the sensitivity of the optimal value with respect to perturbations in the right hand side of the constraints. □

The intuition suggested by the example is correct, provided we assume that the derivative makes sense. Consider an equality-constrained problem and apply a small perturbation to the constraints

$$h_i(\mathbf{x}) = \epsilon_i, \qquad i \in E.$$

Applying the Lagrangian approach to the perturbed problem, we get a new solution $\mathbf{x}^*(\epsilon)$ and a new multiplier vector $\boldsymbol{\lambda}^*(\epsilon)$, both depending on ϵ. The Lagrangian function for the perturbed problem is

$$\mathcal{L}(\mathbf{x}, \boldsymbol{\lambda}, \epsilon) = f(\mathbf{x}) + \sum_{i \in E} \lambda_i (h_i(\mathbf{x}) - \epsilon_i). \tag{6.5}$$

Equality constraints must be satisfied by the optimal solution of the perturbed problem. Hence:

$$f^* = f(\mathbf{x}^*(\epsilon)) = \mathcal{L}(\mathbf{x}^*(\epsilon), \boldsymbol{\lambda}^*(\epsilon), \epsilon). \tag{6.6}$$

We can evaluate the derivative of the optimal value with respect to each component of ϵ,

$$\frac{df^*}{d\epsilon_i} = \frac{d\mathcal{L}}{d\epsilon_i} = \underbrace{[\nabla_{\mathbf{x}}\mathcal{L}]' \frac{\partial \mathbf{x}}{\partial \epsilon_i} + [\nabla_{\boldsymbol{\lambda}}\mathcal{L}]' \frac{\partial \boldsymbol{\lambda}}{\partial \epsilon_i}}_{=0} + \frac{\partial \mathcal{L}}{\partial \epsilon_i} = \frac{\partial \mathcal{L}}{\partial \epsilon_i} = -\lambda_i, \tag{6.7}$$

where we have used the stationarity condition of $\mathcal{L}$. As to inequality constraints, they are either inactive or active in $\mathbf{x}^*$: in the first case, they play no role for small enough perturbations; in the second one, they essentially act as equality constraints. It may be tempting to conclude that if a constraint is associated to a null multiplier, then it can be dropped without changing the optimal solution. The counterexample shown in figure 6.7 shows that this is not the case. Here we have a convex quadratic objective, to which the two concentric level curves are associated; the feasible region is the portion of the "bean" S below the constraint $g(\mathbf{x}) \leq 0$, which is actually an upper bound on x_2. The optimal solution is the point A, and the constraint $g(\mathbf{x}) \leq 0$ is inactive at that point; however, if we eliminate the constraint, the optimal solution is B (it remains true that A is a locally optimal solution). The issue here is that the overall problem is not convex.

6.3.3 Duality theory

In preceding sections we have shown that the stationarity of the Lagrangian function plays a crucial role in constrained optimization. Stationarity is linked

358 CONVEX OPTIMIZATION

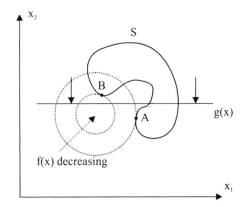

Fig. 6.7 Counterexample showing that a constraint may be relevant even if it has a null multiplier.

to an optimality condition for either minimization or maximization. It is rather intuitive that we should minimize the Lagrangian function with respect to the primal variables, but what about the dual variables? This is an important point if we want to devise a numerical way to find optimal values for both the primal and dual variables. In this section we show that interesting results are obtained by *maximizing* a dual function with respect to the dual variables, leading to duality theory.

Consider the inequality-constrained problem

$$(P) \quad \min \quad f(\mathbf{x})$$
$$\text{s.t.} \quad g_i(\mathbf{x}) \leq 0 \quad i \in I \quad (6.8)$$
$$\mathbf{x} \in S \subseteq \mathbb{R}^n.$$

This problem is called the *primal problem*. Note that the set S is any subset of $\mathbb{R}^n$, possibly a *discrete* one; furthermore, in this section we do not assume the differentiability nor the convexity of the objective function. The results we get are therefore extremely general.

Consider the Lagrangian function obtained by *dualizing* constraints (6.8):

$$\mathcal{L}(\mathbf{x}, \boldsymbol{\mu}) = f(\mathbf{x}) + \sum_{i \in I} \mu_i g_i(\mathbf{x}).$$

For a given multiplier vector $\boldsymbol{\mu}$, the minimization of the Lagrangian function with respect to $\mathbf{x} \in S$ is called the *relaxed problem*; the solution of the relaxed problem defines a function $w(\boldsymbol{\mu})$, called the *dual function*:

$$w(\boldsymbol{\mu}) = \min_{\mathbf{x} \in S} \mathcal{L}(\mathbf{x}, \boldsymbol{\mu}).$$

Consider the *dual problem*:

$$(D) \quad \max_{\boldsymbol{\mu} \geq \mathbf{0}} w(\boldsymbol{\mu}) = \max_{\boldsymbol{\mu} \geq \mathbf{0}} \left\{ \min_{\mathbf{x} \in S} \mathcal{L}(\mathbf{x}, \boldsymbol{\mu}) \right\}. \quad (6.9)$$

The following theorem holds.

THEOREM 6.6 (Weak duality theorem) *For any $\mu \geq 0$, the dual function is a lower bound for the optimum $f(\mathbf{x}^*)$ of the primal problem (P), i.e.,*

$$w(\mu) \leq f(\mathbf{x}^*) \qquad \forall \mu \geq 0.$$

Proof. Let us adopt the notation $\nu(P)$ to denote the optimal value of the objective function for an optimization problem P. Under the hypothesis $\mu \geq 0$, it is easy to see that

$$\nu(P) \geq \nu \begin{pmatrix} \min & f(\mathbf{x}) \\ \text{s.t.} & \mathbf{x} \in S \\ & \mu' \mathbf{g}(\mathbf{x}) \leq 0 \end{pmatrix} \qquad (6.10)$$

$$\geq \nu \begin{pmatrix} \min & f(\mathbf{x}) + \mu' \mathbf{g}(\mathbf{x}) \\ \text{s.t.} & \mathbf{x} \in S \\ & \mu' \mathbf{g}(\mathbf{x}) \leq 0 \end{pmatrix} \qquad (6.11)$$

$$\geq \nu \begin{pmatrix} \min & f(\mathbf{x}) + \mu' \mathbf{g}(\mathbf{x}) \\ \text{s.t.} & \mathbf{x} \in S \end{pmatrix}. \qquad (6.12)$$

Inequality (6.10) is justified by the fact that the points satisfying the set of constraints $g_i(\mathbf{x}) \leq 0$, for all i, also satisfy the aggregate constraint $\mu' \mathbf{g}(\mathbf{x}) \leq 0$ if $\mu \geq 0$, but not vice versa. In other words, the feasible set of the first problem is a subset of the feasible set of the second one. Clearly, when we relax the feasible set, the optimal value cannot increase. Inequality (6.11) holds since the third problem involves the same feasible set as the second problem, but we have added a non-positive term to the objective function. Finally, inequality (6.12) holds since the fourth problem is a relaxation of the third one (we delete a constraint).

□

We obtain a very general but weak relationship. Under suitable conditions (essentially convexity), a stronger property holds, known as *strong duality*:

$$\nu(D) = w(\mu^*) = f(\mathbf{x}^*) = \nu(P).$$

The convexity assumption does not hold, in particular, for the case of a discrete set; therefore, in general, duality yields only a lower bound for discrete optimization problems. The following theorem is useful in establishing when the dual problem yields an optimal solution of the primal problem.

THEOREM 6.7 *If there is a pair $(\mathbf{x}^*, \mu^*)$, where $\mathbf{x}^* \in S$ and $\mu^* \geq 0$, satisfying the following conditions:*

1. $f(\mathbf{x}^*) + (\mu^*)' \mathbf{g}(\mathbf{x}^*) = \min_{\mathbf{x} \in S}\{f(\mathbf{x}) + (\mu^*)' \mathbf{g}(\mathbf{x})\}$;

2. $(\mu^*)' \mathbf{g}(\mathbf{x}^*) = \mathbf{0}$;

3. $g(x^*) \leq 0$;

then x^* is a global optimum for the primal problem (P).

In other words, the optimal solution x^* of the relaxed problem for a multiplier vector μ^* is a global optimum for the primal problem if the pair (x^*, μ^*) is primal feasible, dual feasible, and it satisfies the complementary slackness conditions. Note that these are *sufficient* conditions for global optimality.

Weak duality also holds in the equality-constrained case. Consider the optimal solution x^* of the primal problem:

$$\begin{aligned} \min \quad & f(x) \\ \text{s.t.} \quad & h_i(x) = 0, \quad i \in E \\ & x \in S, \end{aligned}$$

and the optimal solution $\bar{x}$ of the relaxed problem:

$$\min_{x \in S} \left\{ f(x) + \lambda' h(x) \right\}.$$

For any multiplier vector λ (not restricted in sign), it is easy to see that

$$f(\bar{x}) + \lambda' h(\bar{x}) \leq f(x^*) + \lambda' h(x^*) = f(x^*).$$

Unfortunately, convexity does not hold easily for equality constraints. In fact, it holds only for linear equality constraints such as $a_i' x = b_i$. Hence, strong duality with equality constraints holds only in specific cases; a very important one is linear programming (see section 6.4.3).

Example 6.10 Consider the problem

$$\begin{aligned} \min \quad & x_1^2 + x_2^2 \\ \text{s.t.} \quad & x_1 + x_2 \geq 4 \\ & x_1, x_2 \geq 0. \end{aligned}$$

The optimal value is 8, corresponding to the optimal solution $(2, 2)$. Since this is a convex problem, we can apply strong duality. The dual function is

$$\begin{aligned} w(\mu) &= \min_{x_1, x_2} \{ x_1^2 + x_2^2 + \mu(-x_1 - x_2 + 4) \}; \text{ s.t. } x_1, x_2 \geq 0 \} \\ &= \min_{x_1} \{ x_1^2 - \mu x_1; \text{ s.t. } x_1 \geq 0 \} + \min_{x_2} \{ x_2^2 - \mu x_2; \text{ s.t. } x_2 \geq 0 \} + 4\mu. \end{aligned}$$

Since $\mu \geq 0$, the optima with respect to x_1, x_2 are obtained for

$$x_1^* = x_2^* = \frac{\mu}{2}.$$

Hence,

$$w(\mu) = -\frac{1}{2} \mu^2 + 4\mu.$$

The maximum of the dual function is reached for $\mu^* = 4$, and we have $w(4) = f^* = 8$. □

In example 6.10, we have found an explicit representation of the dual function. In general, the maximization of the dual function must be tackled by a numerical method. In practice, the following iterative procedure can be adopted (assuming the inequality-constrained case):

1. Assign an initial value $\boldsymbol{\mu}^{(0)} \geq \mathbf{0}$; set $k \leftarrow 0$.

2. Solve the relaxed problem with multipliers $\boldsymbol{\mu}^{(k)}$.

3. Given the solution $\hat{\mathbf{x}}^{(k)}$ of the relaxed problem, compute a search direction $\mathbf{s}^{(k)}$ and a step length $\alpha^{(k)}$, and update the multipliers (making sure they stay non-negative):

$$\boldsymbol{\mu}^{(k+1)} = \max\left\{\mathbf{0}, \boldsymbol{\mu}^{(k)} + \alpha^{(k)}\mathbf{s}^{(k)}\right\}.$$

Then set $k \leftarrow k+1$, and go to step 2.

In order to find a search direction, one would be tempted to compute a gradient of the dual function. Unfortunately, the dual function need not be everywhere differentiable, as we can see from the following example.

Example 6.11 Consider the discrete optimization problem

$$\begin{aligned} \min \quad & \mathbf{c}'\mathbf{x} \\ \text{s.t.} \quad & \mathbf{a}'\mathbf{x} \geq b & (6.13) \\ & \mathbf{x} \in S = \left\{\mathbf{x}^1, \mathbf{x}^2, \ldots, \mathbf{x}^m\right\}, & (6.14) \end{aligned}$$

where $\mathbf{c}, \mathbf{a}, \mathbf{x} \in \mathbb{R}^n$, $b \in \mathbb{R}$ and S is a *discrete* set. Dualizing constraint (6.13) with a multiplier $\mu \geq 0$, we obtain the dual function:

$$w(\mu) = \min_{j=1,\ldots,m} \left\{(b - \mathbf{a}'\mathbf{x}^j)\mu + \mathbf{c}'\mathbf{x}^j\right\}.$$

It is easy to see that the dual function is the lower envelope of a family of affine functions, as shown in figure 6.8. We have a non-differentiability point when the relaxed problem has multiple optimal solutions. □

From example 6.11 we may conclude that there is no differentiability guarantee for the dual function; however, the dual function for this case is concave. In fact, we may easily prove that the dual function is always concave.

THEOREM 6.8 *The dual function $w(\boldsymbol{\mu})$ is a concave function.*

Proof. We must show that for any multiplier vectors $\boldsymbol{\mu}_1$ and $\boldsymbol{\mu}_2$,

$$w[\lambda \boldsymbol{\mu}_1 + (1-\lambda)\boldsymbol{\mu}_2] \geq \lambda w(\boldsymbol{\mu}_1) + (1-\lambda)w(\boldsymbol{\mu}_2), \qquad \lambda \in [0,1].$$

362 CONVEX OPTIMIZATION

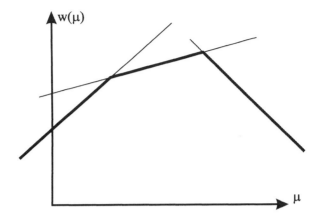

Fig. 6.8 Non-differentiable dual function.

Let us denote by $\hat{\mathbf{x}}_1$ and $\hat{\mathbf{x}}_2$ the optimal solutions of the relaxed subproblems with multipliers $\boldsymbol{\mu}_1$ and $\boldsymbol{\mu}_2$, respectively. We have

$$w(\boldsymbol{\mu}_1) = f(\hat{\mathbf{x}}_1) + \boldsymbol{\mu}_1'\mathbf{g}(\hat{\mathbf{x}}_1) \leq f(\mathbf{x}_\lambda) + \boldsymbol{\mu}_1'\mathbf{g}(\mathbf{x}_\lambda)$$
$$w(\boldsymbol{\mu}_2) = f(\hat{\mathbf{x}}_2) + \boldsymbol{\mu}_2'\mathbf{g}(\hat{\mathbf{x}}_2) \leq f(\mathbf{x}_\lambda) + \boldsymbol{\mu}_2'\mathbf{g}(\mathbf{x}_\lambda),$$

where $\mathbf{x}_\lambda$ is the optimal solution corresponding to the multiplier vector $\lambda\boldsymbol{\mu}_1 + (1-\lambda)\boldsymbol{\mu}_2$. The result is obtained by multiplying the first inequality by λ, the second one by $1-\lambda$, and summing. □

Since maximizing a concave function is equivalent to minimizing a convex function, this is a reassuring result. In fact, we may apply a subgradient algorithm (see section 6.2.2) provided that we are able to find a subgradient of the dual function for any value of the multipliers.

THEOREM 6.9 *Let $\hat{\mathbf{x}}$ be an optimal solution of the relaxed problem for a multiplier vector $\hat{\boldsymbol{\mu}}$. Then $\mathbf{g}(\hat{\mathbf{x}})$ is a subgradient of the dual function at $\hat{\boldsymbol{\mu}}$.*

Proof. To show that $\mathbf{g}(\hat{\mathbf{x}}) \in \partial w(\hat{\boldsymbol{\mu}})$, we must show that, for any $\boldsymbol{\mu}$, we have

$$w(\boldsymbol{\mu}) \leq w(\hat{\boldsymbol{\mu}}) + \mathbf{g}(\hat{\mathbf{x}})'(\boldsymbol{\mu} - \hat{\boldsymbol{\mu}}).$$

Here the inequality is reversed with respect to the definition of a subgradient for a convex function, since w is concave. We know that $\hat{\mathbf{x}}$ is the optimal solution of the relaxed problem for $\hat{\boldsymbol{\mu}}$:

$$w(\hat{\boldsymbol{\mu}}) = f(\hat{\mathbf{x}}) + \hat{\boldsymbol{\mu}}'\mathbf{g}(\hat{\mathbf{x}}) \qquad (6.15)$$

but not for a generic $\boldsymbol{\mu}$:

$$w(\boldsymbol{\mu}) = \min_{\mathbf{x} \in S} \left\{ f(\mathbf{x}) + \boldsymbol{\mu}'\mathbf{g}(\mathbf{x}) \right\} \leq f(\hat{\mathbf{x}}) + \boldsymbol{\mu}'\mathbf{g}(\hat{\mathbf{x}}). \qquad (6.16)$$

METHODS FOR CONSTRAINED OPTIMIZATION   363

Subtracting equation (6.15) from inequality (6.16), we get

$$w(\mu) - w(\hat{\mu}) \leq \mathbf{g}'(\hat{\mathbf{x}})(\mu - \hat{\mu}),$$

and the result follows.   □

Theorem 6.9 allows us to solve the dual problem (6.9) by a subgradient algorithm. A remarkable point is that we are able to optimize a function, even if it is not known in explicit form, provided we know how to find a subgradient; this applies to the dual function, which is implicitly defined by an optimization problem, and to the recourse function that we will meet in stochastic programming (chapter 11).

In order to maximize the dual function, a sequence of relaxed problems is solved, updating the dual variables as follows:

$$\mu^{(k+1)} = \max\left\{0, \mu^{(k)} + \alpha^{(k)}\mathbf{g}(\hat{\mathbf{x}}^{(k)})\right\}$$

where $\hat{\mathbf{x}}^{(k)}$ is the solution of the kth relaxed problem. Note that this solution need not be feasible for the original (primal) problem. Provided that strong duality holds, the method converges to the optimal solution of the original problem. When only weak duality applies, we obtain a lower bound on the optimal value of the primal problem (which may be valuable in itself), and probably a near-feasible solution, from which a feasible near-optimal solution may be obtained with some problem-dependent procedure. It should be noted that duality theory in itself does not generally yield numerically efficient algorithms directly. Nevertheless, it may be fruitfully exploited for specially structured problems; in fact, we have seen in example 6.10 that dualizing certain constraints may decompose an optimization problem into independent subproblems; certain model formulations lend themselves to a decomposition by dualization of the interaction constraints. Furthermore, duality theory is a fundamental theoretical tool paving the way for important algorithmic developments.

6.3.4 Kelley's cutting plane algorithm

In the last section, we have been able to maximize the dual function, even if it is not known in an explicit form. We have relied on the fact that the function was concave and we were able to evaluate the function and to find a subgradient at a given point. A similar idea leads to Kelley's cutting plane method for the minimization of a convex function. Assume that we have to solve a convex problem $\min_{\mathbf{x} \in S} f(\mathbf{x})$, where the objective function f is actually not known in analytical form. Suppose that, for a given point $\mathbf{x}^k$, we are not only able to compute the function value $f(\mathbf{x}^k) = \alpha_k$, but also a subgradient γ_k, which does exist if the function is convex on the set S. In other words, we are able to find an affine function such that

$$f(\mathbf{x}^k) = \alpha_k + \gamma_k'\mathbf{x}^k \qquad (6.17)$$

CONVEX OPTIMIZATION

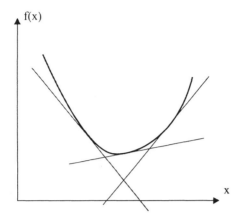

Fig. 6.9 Example of Kelley's cutting plane algorithm.

$$f(\mathbf{x}) \geq \alpha_k + \gamma'_k \mathbf{x} \quad \forall \mathbf{x} \in S. \tag{6.18}$$

The availability of such a support hyperplane suggests the possibility of approximating f from below, by the upper envelope of support hyperplanes, as illustrated in figure 6.9. The Kelley's cutting plane algorithm exploits this idea by building and improving a lower bounding function until some convergence criterion is met.

Step 0. Let $\mathbf{x}^1 \in S$ be an initial feasible solution; initialize the iteration counter $k \leftarrow 0$, the upper bound $u_0 = f(\mathbf{x}^1)$, the lower bound $l_0 = -\infty$, and the lower bounding function $\beta_0(\mathbf{x}) = -\infty$.

Step 1. Increment the iteration counter $k \leftarrow k+1$. Find a subgradient of f at $\mathbf{x}^k$, such that equation (6.17) and condition (6.18) hold.

Step 2. Update the upper bound

$$u_k = \min\{u_{k-1}, f(\mathbf{x}^k)\}$$

and the lower bounding function

$$\beta_k(\mathbf{x}) = \max\{\beta_{k-1}(\mathbf{x}), \alpha_k + \gamma'_k \mathbf{x}\}.$$

Step 3. Solve the problem

$$l_k = \min_{\mathbf{x} \in S} \beta_k(\mathbf{x}),$$

and let $\mathbf{x}^{k+1}$ be the optimal solution.

Step 4. If $u_k - l_k < \epsilon$, stop: $\mathbf{x}^{k+1}$ is a satisfactory approximation of the optimal solution; otherwise, go to step 1.

It is worth noting that, if the feasible set S is polyhedral, then all the subproblems we solve are LP problems. The Kelley's cutting plane algorithm is the conceptual basis of some algorithms for stochastic programming, such as the L-shaped decomposition for stochastic programming, which will be described in section 11.4.

6.3.5 Active set method

Although duality theory is a powerful tool, both in theory and practice, dual algorithms have the general drawback that a feasible solution is generally obtained only in the limit. A natural aim of many constrained optimization algorithms is to stay within the feasible region. This is particularly easy to accomplish if the problem is linearly constrained. Consider the problem

$$\min \quad f(\mathbf{x})$$
$$\text{s.t.} \quad \mathbf{A}\mathbf{x} = \mathbf{b},$$

where the matrix $\mathbf{A} \in \mathbb{R}^{m,n}$, $m < n$, is assumed of full row rank for simplicity. Given a feasible solution $\hat{\mathbf{x}}$, how can we characterize descent directions $\boldsymbol{\delta}$ such that the new solution $\hat{\mathbf{x}} + \alpha\boldsymbol{\delta}$ remains feasible for some $\alpha > 0$? Since both solutions must be feasible,

$$\mathbf{A}(\hat{\mathbf{x}} + \alpha\boldsymbol{\delta}) = \mathbf{b} + \alpha\mathbf{A}\boldsymbol{\delta} = \mathbf{b} \quad \Rightarrow \quad \mathbf{A}\boldsymbol{\delta} = \mathbf{0}.$$

Technically speaking, the vector $\boldsymbol{\delta}$ must lie in the null space of the matrix $\mathbf{A}$; since this is a linear space, there must be a basis for it. Let $\mathbf{Z} \in \mathbb{R}^{n,(n-m)}$ be a matrix whose columns are a basis for this space; then we have

$$\mathbf{A}\mathbf{Z} = \mathbf{0},$$

and the direction $\boldsymbol{\delta}$ is a linear combination of the columns of $\mathbf{Z}$:

$$\boldsymbol{\delta} = \mathbf{Z}\mathbf{d}.$$

The basis consists of $(n-m)$ vectors. To see why, consider that the m equality constraints eliminate m degrees of freedom for the n decision variables. Then we may move in some space with $(n-m)$ degrees of freedom. The first-order Taylor expansion for a perturbed point along the feasible direction is

$$f(\hat{\mathbf{x}} + \epsilon\mathbf{Z}\mathbf{d}) \approx f(\hat{\mathbf{x}}) + \epsilon\mathbf{d}'\mathbf{Z}'\nabla f(\hat{\mathbf{x}}).$$

A descent direction is obtained when $\mathbf{d}'\mathbf{Z}'\nabla f(\hat{\mathbf{x}}) < 0$; furthermore, the first-order necessary optimality condition is

$$\mathbf{Z}'\nabla f(\mathbf{x}^*) = \mathbf{0}. \tag{6.19}$$

The vector $\mathbf{Z}'\nabla f$ is called the *reduced gradient*, and we see that a stationarity condition must be required for this reduced gradient. By the way, the condition (6.19) implies that the gradient ∇f is a linear combination of the rows

of **A**. This means that
$$\nabla f(\mathbf{x}^*) = \mathbf{A}'\boldsymbol{\lambda},$$
which could also be obtained by the Lagrange multipliers approach.

Provided that we are able to find a suitable matrix **Z**, an algorithm is readily devised, as we must simply spot descent directions and select the step length α in order to reduce the objective function while keeping the iterates feasible. One possible choice of **Z** is obtained by exploiting *QR factorization*. This factorization, which is implemented by the `qr` function in MATLAB, allows us to write

$$\mathbf{A}' = \mathbf{Q} \begin{bmatrix} \mathbf{R} \\ \mathbf{0} \end{bmatrix} = \begin{bmatrix} \mathbf{Q}_1 & \mathbf{Q}_2 \end{bmatrix} \begin{bmatrix} \mathbf{R} \\ \mathbf{0} \end{bmatrix} = \mathbf{Q}_1\mathbf{R},$$

where $\mathbf{Q} \in \mathbb{R}^{n,n}$ is an orthogonal matrix (i.e., its columns are orthogonal vectors), and $\mathbf{R} \in \mathbb{R}^{n,n-m}$ is upper triangular. The choice $\mathbf{Z} = \mathbf{Q}_2$ satisfies our requirements, since the orthogonality of **Q** implies that

$$\mathbf{A} = \mathbf{R}'\mathbf{Q}_1' \quad \Rightarrow \quad \mathbf{A}\mathbf{Z} = \mathbf{R}'\mathbf{Q}_1'\mathbf{Q}_2 = \mathbf{0}.$$

Different choices of **Z** and different approaches in selecting the descent direction and the step length result in a variety of methods which are described in the literature. It should also be mentioned that second-order conditions should be checked if f is not convex.

The approach may be extended to linear inequalities. To cope with a problem like

$$\min \quad f(\mathbf{x})$$
$$\text{s.t.} \quad \mathbf{Ax} \leq \mathbf{b},$$

a possible idea is to restrict the attention to the active constraints, i.e., the constraints which are satisfied at equality. In principle, if we knew which constraints are active in the optimal solution, we could treat the problem like an equality-constrained problem. The active set strategy works on a pool of active constraints, trying to identify which constraints must be brought in and out of the active set. Roughly speaking, if we see that a relaxed constraint would get violated by a move along the feasible direction, it should be added to the set. Similarly, an inactive constraints can be dropped. The details of the method are not so easy, but it is enough to know the qualitative aspects of its working and that it is actually implemented and used in MATLAB functions for both quadratic and linear programming (see section 6.5.1 to appreciate this point).

6.4 LINEAR PROGRAMMING

A general LP problem can be expressed as

$$\min \quad \mathbf{c}'\mathbf{x}$$

$$\text{s.t.} \quad \mathbf{a}'_i\mathbf{x} = b_i, \quad i \in E$$
$$\mathbf{a}'_i\mathbf{x} \geq b_i, \quad i \in I,$$

where $\mathbf{c}, \mathbf{a}_i, \mathbf{x} \in \mathbb{R}^n$, $b_i \in \mathbb{R}$. When dealing with solution algorithms for LP problems, it is convenient to assume that the problem has a specific form. An LP problem is said to be in *canonical form* if it involves only inequality constraints, and all the decision variables are restricted in sign. A canonical form for a maximization problem is

$$\max \quad \mathbf{c}'\mathbf{x}$$
$$\text{s.t.} \quad \mathbf{A}\mathbf{x} \leq \mathbf{b}$$
$$\mathbf{x} \geq \mathbf{0},$$

where $\mathbf{c}, \mathbf{x} \in \mathbb{R}^n$, $\mathbf{b} \in \mathbb{R}^m$, $\mathbf{A} \in \mathbb{R}^{m,n}$. We denote the ith row (corresponding to the ith constraint) of $\mathbf{A}$ by $\mathbf{a}'_i$ and the jth column (corresponding to the jth variable) by $\mathbf{A}^j$. An LP problem is said to be in *standard form* if it involves only equality constraints:

$$\min \quad \mathbf{c}'\mathbf{x}$$
$$\text{s.t.} \quad \mathbf{A}\mathbf{x} = \mathbf{b}$$
$$\mathbf{x} \geq \mathbf{0},$$

with the same notation as in the case of the canonical form. Clearly, we must have $m < n$, so that the system of linear equations is underdetermined and there are multiple solutions.

The reader might think that the canonical and standard forms are somewhat restrictive; in fact, this is not true, since a generic LP problem can be reduced to either form using the following transformations:

- If a variable x_j is not restricted in sign, it can be rewritten as $x_j = x_j^+ - x_j^-$, where $x_j^+, x_j^- \geq 0$.
- An inequality constraint
$$\mathbf{a}'_i\mathbf{x} \geq b_i$$
can be transformed into an equality constraint by introducing a slack variable $s_i \geq 0$:
$$\mathbf{a}'_i\mathbf{x} - s_i = b_i.$$
- An equality constraint
$$\mathbf{a}'_i\mathbf{x} = b_i$$
can be transformed into two inequality constraints:
$$\mathbf{a}'_i\mathbf{x} \geq b_i, \quad -\mathbf{a}'_i\mathbf{x} \geq -b_i.$$

We know from supplement S6.1 that the feasible set of a LP problem is convex and polyhedral. Furthermore, the problem is both convex and concave. This implies that an optimal solution (if any exists) may be found on the boundary of the feasible set; more specifically, it will be a vertex of the feasible set. This is easy to see by expressing the feasible region S as the convex hull of its extreme points $\mathbf{X}^k$, $k = 1, \ldots, I$. Strictly speaking, if S is unbounded, we should also consider its extreme rays; however, if we assume that the optimal value is finite, there is no loss of generality by discarding the possibility of going to infinity along a ray. Denoting by C^k the cost of the extreme point $\mathbf{X}^k$, we may transform the LP problem

$$\min \quad \mathbf{c}'\mathbf{x}$$
$$\text{s.t.} \quad \mathbf{x} \in S$$

into the equivalent problem

$$\min \quad \sum_{k=1}^{I} \lambda_k C^k \mathbf{X}^k$$
$$\text{s.t.} \quad \sum_{i=1}^{I} \lambda_k = 1$$
$$\lambda_k \geq 0.$$

This problem has just one constraint, but a possibly huge number of variables; nevertheless, it is easy to see that an optimal solution can be found as the least cost extreme point.

If the problem is cast in standard form, the extreme points correspond to special solutions of the system of linear equations $\mathbf{Ax} = \mathbf{b}$; this is explained briefly in section 6.4.1 and is the basis of the simplex algorithm to which section 6.4.2 is devoted. Applying the duality principles to LP problems produces an interesting theory, outlined in section 6.4.3. The simplex algorithm is certainly the best known method for LP problems, but it is not the only method that you get in MATLAB. The Optimization toolbox provides the user with two options: for medium-scale problems, a version of the active set method is also implemented; for large-scale problems, an interior point method is available. Some ideas behind interior point methods are described in section 6.4.4. It is interesting to note that the simplex algorithm is not, in the worst case, a polynomial complexity algorithm, whereas polynomial complexity may be proved for interior point methods. In fact, interior point methods are faster on many problem instances, but not always.

6.4.1 Geometric and algebraic features of linear programming

Given an LP problem, one of the three following cases occurs:

1. The feasible set is empty, and the problem has no solution.

2. The optimal solution is, loosely speaking, "unbounded." This case may occur only if the feasible set is an unbounded polyhedron, and we may

keep improving the objective value by going to infinity along an extreme ray.

3. The problem has a finite optimal solution, corresponding to an extreme point of the feasible set; note that we have an infinite set of optimal solutions if the level curves of the objective function are parallel to a face of the polyhedron (see example 6.2).

Since there is a finite number of extreme points in a polyhedron, one way to solve an LP problem is to explore the set of extreme points of the feasible set without considering the interior points. Furthermore, a local minimizer will also be a global one; hence, if we find an extreme point such that no adjacent extreme point improves the objective function, then we have found the optimal solution.

To implement this idea, the geometrical intuition must be translated into algebraic terms. To this end, it is convenient to work on the standard LP form. To avoid unnecessary complications, let us assume that the matrix $\mathbf{A} \in \mathbb{R}^{m,n}$, $m < n$, has full row rank. This assumption is not necessary in practice, as redundant equations are easily spot and eliminated. It is useful to consider a solution of the system $\mathbf{A}\mathbf{x} = \mathbf{b}$ as a way to express the vector $\mathbf{b}$ as a linear combination of the columns of $\mathbf{A}$:

$$\mathbf{A}\mathbf{x} = \begin{bmatrix} \mathbf{A}^1 \mathbf{A}^2 \cdots \mathbf{A}^n \end{bmatrix} \begin{bmatrix} x_1 \\ x_2 \\ \vdots \\ x_n \end{bmatrix} = \sum_{j=1}^{n} x_j \mathbf{A}^j = \mathbf{b}.$$

This system has infinite solutions, but not all of them are feasible with respect to the requirement $\mathbf{x} \geq \mathbf{0}$. Furthermore, we would like to work on feasible solutions which are extreme points of the feasible set. This is easily accomplished by considering only solutions in which at most m components x_j are strictly positive, and the remaining $n - m$ variables are zero. Such solutions are called *basic solutions*; the name derives from the fact that the m column vectors associated with the m possibly non-null variables are sufficient to express the m-dimensional vector $\mathbf{b}$. Any basic solution is associated with a basis of $\mathbb{R}^m$ consisting of m columns of $\mathbf{A}$. The m variables corresponding to the columns selected are called *basic variables*; the others are called *non-basic variables*. A basic solution with non-negative components is called a *basic feasible solution*.

Example 6.12 Consider the following system of linear equations:

$$\begin{bmatrix} -1 & 1 & 1 & -1 & 0 \\ 0 & 1 & 0 & 4 & 0 \\ 0 & 0 & 2 & 2 & 1 \end{bmatrix} \begin{bmatrix} x_1 \\ x_2 \\ x_3 \\ x_4 \\ x_5 \end{bmatrix} = \begin{bmatrix} 1 \\ 3 \\ 1 \end{bmatrix}.$$

A basic solution is

$$x_1 = 2, \ x_2 = 3, \ x_3 = x_4 = 0, \ x_5 = 1,$$

which corresponds to the basis formed by the columns $\mathbf{A}^1, \mathbf{A}^2, \mathbf{A}^5$. This solution is also feasible. If we take the basis formed by $\mathbf{A}^2, \mathbf{A}^3, \mathbf{A}^5$, we obtain the basic solution

$$x_1 = 0, \ x_2 = 3, \ x_3 = -2, \ x_4 = 0, \ x_5 = 5,$$

which is not feasible since $x_3 < 0$. □

Basic feasible solutions are fundamental because it can be shown that they actually correspond to the extreme points of the feasible set. Furthermore, given a current extreme point, the adjacent extreme point may be obtained by exchanging one basic variable with a non-basic one; this means that we may move from a vertex to another one by driving one basic variable out of the basis and driving one non-basic variable into the basis.

6.4.2 Simplex method

The simplex method is an iterative algorithm; given a current extreme point (or basic feasible solution, or basis), it looks for an adjacent extreme point such that the objective function is improved, and it stops when no improving adjacent extreme point is found.

Assume that we have a basic feasible solution $\mathbf{x}$; we will consider later how to obtain an initial basic feasible solution. We can partition the vector $\mathbf{x}$ into two subvectors: the subvector $\mathbf{x}_B \in \mathbb{R}^m$ of the basic variables and the subvector $\mathbf{x}_N \in \mathbb{R}^{n-m}$ of the non-basic variables. Using a suitable permutation of the variable indexes, we may rewrite the system of linear equations

$$\mathbf{A}\mathbf{x} = \mathbf{b}$$

as

$$[\mathbf{A}_B \mathbf{A}_N] \begin{bmatrix} \mathbf{x}_B \\ \mathbf{x}_N \end{bmatrix} = \mathbf{A}_B \mathbf{x}_B + \mathbf{A}_N \mathbf{x}_N = \mathbf{b}, \quad (6.20)$$

where $\mathbf{A}_B \in \mathbb{R}^{m,m}$ is non-singular and $\mathbf{A}_N \in \mathbb{R}^{m,n-m}$. If $\mathbf{x}$ is basic feasible, it may be written as

$$\mathbf{x} = \begin{bmatrix} \mathbf{x}_B \\ \mathbf{x}_N \end{bmatrix} = \begin{bmatrix} \hat{\mathbf{b}} \\ 0 \end{bmatrix},$$

where

$$\hat{\mathbf{b}} = \mathbf{A}_B^{-1}\mathbf{b} \geq 0.$$

The objective function value corresponding to $\mathbf{x}$ is

$$\hat{f} = [\mathbf{c}'_B \ \mathbf{c}'_N] \begin{bmatrix} \hat{\mathbf{b}} \\ 0 \end{bmatrix} = \mathbf{c}'_B \hat{\mathbf{b}}. \quad (6.21)$$

Now we must find out if it is possible to improve the current solution by slightly changing the basis, i.e., by replacing one basic variable with a non-basic one. To assess the potential benefit of introducing a non-basic variable into the basis, we may eliminate the basic variables in equation (6.21). Using equation (6.20), we may express the basic variables as

$$\mathbf{x}_B = \mathbf{A}_B^{-1}(\mathbf{b} - \mathbf{A}_N \mathbf{x}_N) = \hat{\mathbf{b}} - \mathbf{A}_B^{-1} \mathbf{A}_N \mathbf{x}_N; \tag{6.22}$$

then we rewrite the objective function value

$$\mathbf{c}'\mathbf{x} = \mathbf{c}'_B \mathbf{x}_B + \mathbf{c}'_N \mathbf{x}_N = \mathbf{c}'_B (\hat{\mathbf{b}} - \mathbf{A}_B^{-1} \mathbf{A}_N \mathbf{x}_N) + \mathbf{c}'_N = \hat{f} + \hat{\mathbf{c}}'_N \mathbf{x}_N,$$

where

$$\hat{\mathbf{c}}'_N = \mathbf{c}'_N - \mathbf{c}'_B \mathbf{A}_B^{-1} \mathbf{A}_N. \tag{6.23}$$

The quantities $\hat{\mathbf{c}}_N$ are called *reduced costs*, as they measure the marginal variation of the objective function with respect to the non-basic variables. If $\hat{\mathbf{c}}_N \geq \mathbf{0}$, it is not possible to improve the objective function; in this case, bringing a non-basic variable into the basis at some positive value cannot reduce the overall cost. Therefore, the current basis is optimal if $\hat{\mathbf{c}}_N \geq \mathbf{0}$. If, on the contrary, there exists a $q \in N$ such that $\hat{c}_q < 0$, it is possible to improve the objective function by bringing x_q into the basis. A simple strategy is to choose q such that

$$\hat{c}_q = \min_{j \in N} \hat{c}_j.$$

This selection does not necessarily result in the best performance of the algorithm; we should consider not only the rate of change in the objective function, but also the value attained by the new basic variable. Furthermore, it may happen that the entering variable is stuck to zero and does not change the value of the objective. In such a case, there is danger of cycling on a set of bases; ways to overcome this difficulty are well explained in the literature.

When x_q is brought into the basis, a basic variable must "leave" in order to maintain $\mathbf{A}\mathbf{x} = \mathbf{b}$. To spot the leaving variable we can reason as follows. Given the current basis, we can use it to express both $\mathbf{b}$ and the column $\mathbf{A}^q$ corresponding to the entering variable:

$$\mathbf{b} = \sum_{i=1}^{m} x_{B(i)} \mathbf{A}^{B(i)} \tag{6.24}$$

$$\mathbf{A}^q = \sum_{i=1}^{m} d_i \mathbf{A}^{B(i)}, \tag{6.25}$$

where $B(i)$ is the index of the ith basic variable ($i = 1, \ldots, m$) and

$$\mathbf{d} = \mathbf{A}_B^{-1} \mathbf{A}^q.$$

If we multiply equation (6.25) by a number θ and subtract it from equation (6.24), we obtain

$$\mathbf{b} = \sum_{i=1}^{m} \left(x_{B(i)} - \theta d_i \right) \mathbf{A}^{B(i)} + \theta \mathbf{A}^q. \tag{6.26}$$

From equation (6.26) we see that θ is the value of the entering variable in the new solution, and that the value of the current basic is affected in a way depending on the sign of d_i. If $d_i \leq 0$, $x_{B(i)}$ remains non-negative when x_q increases. But if there is an index i such that $d_i > 0$, then we cannot increase x_q at will, since there is a limit value for which a currently basic variable becomes zero. This limit value is attained by the entering variable x_q, and the first current basic variable which gets zero leaves the basis

$$x_q = \min_{\substack{i=1,\ldots,m \\ d_i > 0}} \frac{\hat{b}_i}{d_i}.$$

If $\mathbf{d} \leq \mathbf{0}$, there is no limit on the increase of x_q, and the optimal solution is unbounded.

In order to start the iterations, a starting basis is needed. One possibility is to introduce a set of auxiliary artificial variables $\mathbf{z}$ in the constraints:

$$\mathbf{A}\mathbf{x} + \mathbf{z} = \mathbf{b}$$
$$\mathbf{x}, \mathbf{z} \geq \mathbf{0}.$$

Assume also that the equations have been rearranged in such a way that $\mathbf{b} \geq \mathbf{0}$; then a basic feasible solution is trivially $\mathbf{z} = \mathbf{b}$. Minimizing the inadmissibility form

$$\phi = \min \sum_{i=1}^{m} z_i$$

by the simplex method itself, we may find a basic feasible solution if $\phi = 0$; otherwise, the original problem is not feasible.

At this point, one should wonder what is the connection, if any exists, between the simplex method for LP problems and the simplex search we have hinted at in section 6.2.4. Actually, they are quite different approaches for different problems. The name of the simplex method comes from the fact that it works on a simplex in the reduced space of the non-basic variables. In this space, the origin corresponds to the current basic solution, as the non-basic variables are zero; the remaining extreme points of the simplex correspond to the adjacent bases. The simplex method checks, in the reduced space, if any of these extreme points improves the objective function.

6.4.3 Duality in linear programming

We dealt with duality in non-linear programming in section 6.3.3. Duality in LP can be developed without considering the more general non-linear case, but we prefer to put it in a more general framework. Note that, due to the convexity of LP problems, strong duality holds. Let us start with an LP problem (P_1) in the following canonical form:

$$(P_1) \quad \min \quad \mathbf{c'x}$$
$$\text{s.t.} \quad \mathbf{Ax} \geq \mathbf{b}.$$

If we dualize the inequality constraints with a vector $\boldsymbol{\mu} \in \mathbb{R}^m_+$ of dual variables, we get the dual problem

$$\max_{\boldsymbol{\mu} \geq 0} \min_{\mathbf{x}} \{\mathbf{c'x} + \boldsymbol{\mu}'(\mathbf{b} - \mathbf{Ax})\} = \max_{\boldsymbol{\mu} \geq 0} \left\{ \boldsymbol{\mu}'\mathbf{b} + \min_{\mathbf{x}} (\mathbf{c}' - \boldsymbol{\mu}'\mathbf{A})\mathbf{x} \right\}.$$

Since $\mathbf{x}$ is unrestricted in sign, the inner minimization problem has a finite value if and only if

$$\mathbf{c}' - \boldsymbol{\mu}'\mathbf{A} = \mathbf{0};$$

otherwise, each component of $\mathbf{x}$ is set to $\pm\infty$, depending on the sign of the corresponding cost coefficient, and this results in a value $-\infty$ for the dual function. Since we want to maximize the dual function, we may enforce the condition above, and the dual problem (D_1) turns out to be

$$(D_1) \quad \max \quad \boldsymbol{\mu}'\mathbf{b}$$
$$\text{s.t.} \quad \mathbf{A}'\boldsymbol{\mu} = \mathbf{c}$$
$$\boldsymbol{\mu} \geq \mathbf{0}.$$

The dual problem is still an LP problem, resulting from exchanging $\mathbf{b}$ with $\mathbf{c}$ and by transposing $\mathbf{A}$. The duality relationship between (P_1) and (D_1) can be interpreted the other way round, too:

$$\begin{pmatrix} \max & \mathbf{x'c} \\ \text{s.t.} & \mathbf{Ax} = \mathbf{b} \\ & \mathbf{x} \geq \mathbf{0} \end{pmatrix} \iff \begin{pmatrix} \min & \mathbf{b'}\boldsymbol{\nu} \\ \text{s.t.} & \mathbf{A'}\boldsymbol{\nu} \geq \mathbf{c} \end{pmatrix}.$$

Given an LP problem (P_2) in standard form,

$$(P_2) \quad \min \quad \mathbf{c'x}$$
$$\text{s.t.} \quad \mathbf{Ax} = \mathbf{b}$$
$$\mathbf{x} \geq \mathbf{0},$$

we can use the relationship above to find its dual:

$$\begin{pmatrix} \min & \mathbf{c'x} \\ \text{s.t.} & \mathbf{Ax} = \mathbf{b} \\ & \mathbf{x} \geq \mathbf{0} \end{pmatrix} \iff \begin{pmatrix} \max & \mathbf{x'}(-\mathbf{c}) \\ \text{s.t.} & (\mathbf{A}')'\mathbf{x} = \mathbf{b} \\ & \mathbf{x} \geq \mathbf{0} \end{pmatrix}$$

$$\iff \begin{pmatrix} \min & \mathbf{b'}\boldsymbol{\nu} \\ \text{s.t.} & \mathbf{A'}\boldsymbol{\nu} \geq -\mathbf{c} \end{pmatrix} \iff \begin{pmatrix} \min & -\mathbf{b'}\boldsymbol{\mu} \\ \text{s.t.} & -\mathbf{A'}\boldsymbol{\mu} \geq -\mathbf{c} \end{pmatrix}$$

$$\iff \begin{pmatrix} \max & \mathbf{b'}\boldsymbol{\mu} \\ \text{s.t.} & \mathbf{A'}\boldsymbol{\mu} \leq \mathbf{c} \end{pmatrix},$$

Table 6.1 Duality Relationships

Primal	Dual
min $\mathbf{c}'\mathbf{x}$	max $\boldsymbol{\mu}'\mathbf{b}$
$\mathbf{a}_i'\mathbf{x} = b_i$	μ_i unrestricted
$\mathbf{a}_i'\mathbf{x} \geq b_i$	$\mu_i \geq 0$
$x_j \geq 0$	$\boldsymbol{\mu}'\mathbf{A}^j \leq c_j$
x_j unrestricted	$\boldsymbol{\mu}'\mathbf{A}^j = c_j$

where we have introduced $\boldsymbol{\mu} = -\boldsymbol{\nu}$; we obtain the dual (D_2) of problem (P_2).

Note the similarities and the differences between the two dual pairs. The dual variables are restricted in sign when the constraints of the primal problem are inequalities, and are unrestricted in sign in the other case (this is coherent with the Kuhn–Tucker conditions). When the variables are restricted in sign in the primal, we have inequality constraints in the dual, whereas in the case of unrestricted variables we have equality constraints in the dual. In table 6.1 we summarize the "recipe" for building the dual of a generic LP.

Given a primal–dual pair of LP problems, the following cases may occur:

- Both problems have a finite optimal solution, in which case the two objectives have the same value at the optimum.

- Both problems are infeasible.

- One problem is unbounded, in which case the other one is infeasible.

As a final remark, it is important to note that the dual feasibility constraint $\mathbf{A}'\boldsymbol{\mu} \leq \mathbf{c}$ for the dual of the problem in standard form can be read as the non-negativity condition on the reduced costs by equating $\boldsymbol{\mu}' = \mathbf{c}'\mathbf{A}_B^{-1}$. Recall the sufficient conditions (6.7) for global optimality. They correspond to

- Primal feasibility

- Dual feasibility

- Complementary slackness

In fact, the simplex method works by maintaining primal feasibility and complementary slackness, and it iterates until dual feasibility is obtained. Switching roles between primal and dual problems, it is possible to devise a dual simplex method which works toward primal feasibility. This is sometimes advantageous over the primal simplex approach. However, there is still a third possibility: We can keep a pair of primal and dual feasible solutions and work to obtain complementary slackness. This approach leads to primal–dual algorithms, and it is exploited in the interior point method described in the next section.

6.4.4 Interior point methods

The simplex method works only on the extreme points of the feasible set. As the name suggests, interior point methods move on a path that lies within the feasible set. There are several variants of interior point algorithms; we describe just the basics of a rather simple approach, which may be called the *primal–dual barrier method*, as it exploits the correspondence between a primal and dual problems, and an interior penalty function. It is convenient to start with the LP problem written in canonical form for a maximization problem[8]:

$$\begin{aligned} \max \quad & \mathbf{c}'\mathbf{x} \\ \text{s.t.} \quad & \mathbf{A}\mathbf{x} \leq \mathbf{b} \\ & \mathbf{x} \geq \mathbf{0}, \end{aligned}$$

which may be converted to the standard form by adding slack variables $\mathbf{w}$:

$$\begin{aligned} \max \quad & \mathbf{c}'\mathbf{x} \\ \text{s.t.} \quad & \mathbf{A}\mathbf{x} + \mathbf{w} = \mathbf{b} \\ & \mathbf{x}, \mathbf{w} \geq \mathbf{0}. \end{aligned}$$

Now suppose that we do not know anything about the simplex method. We could try applying what we know from the general theory of constrained optimization; one idea would be getting rid of the non-negativity restriction by a suitable penalty function and then apply the method of Lagrange multipliers. Using an interior penalty function based on a logarithmic barrier, we get the problem

$$(PP) \quad \max \quad \mathbf{c}'\mathbf{x} + \sigma \sum_j \log x_j + \sigma \sum_i \log w_i$$
$$\text{s.t.} \quad \mathbf{A}\mathbf{x} + \mathbf{w} = \mathbf{b}.$$

Since this problem has only equality constraints, we may dualize them by introducing the vector of Lagrange multipliers $\mathbf{y}$, yielding the Lagrangian function

$$\mathcal{L}(\mathbf{x}, \mathbf{w}, \mathbf{y}) = \mathbf{c}'\mathbf{x} + \sigma \sum_j \log x_j + \sigma \sum_i \log w_i + \mathbf{y}'(\mathbf{b} - \mathbf{A}\mathbf{x} - \mathbf{w}).$$

---
[8] The exposition here is based on [19].

The first-order stationarity conditions are then

$$\frac{\partial \mathcal{L}}{\partial x_j} = c_j + \sigma \frac{1}{x_j} - \sum_i y_i a_{ij} = 0 \quad \forall j$$

$$\frac{\partial \mathcal{L}}{\partial w_i} = \sigma \frac{1}{w_i} - y_i = 0 \quad \forall i$$

$$\frac{\partial \mathcal{L}}{\partial y_i} = b_i - \sum_j a_{ij} x_j - w_i = 0 \quad \forall i.$$

Using the notation

$$\mathbf{X} = \begin{bmatrix} x_1 & & & \\ & x_2 & & \\ & & \ddots & \\ & & & x_n \end{bmatrix}, \quad \mathbf{e} = \begin{bmatrix} 1 \\ 1 \\ \vdots \\ 1 \end{bmatrix},$$

the optimality equations may be rewritten in matrix form:

$$\mathbf{A}'\mathbf{y} - \sigma \mathbf{X}^{-1}\mathbf{e} = \mathbf{c}$$
$$\mathbf{y} = \sigma \mathbf{W}^{-1}\mathbf{e}$$
$$\mathbf{A}\mathbf{x} + \mathbf{w} = \mathbf{b}.$$

The addition of the auxiliary vector

$$\mathbf{z} = \sigma \mathbf{X}^{-1}\mathbf{e},$$

and a slight rearrangement yield the following set of optimality equations:

$$\begin{aligned} \mathbf{A}\mathbf{x} + \mathbf{w} &= \mathbf{b} \\ \mathbf{A}'\mathbf{y} - \mathbf{z} &= \mathbf{c} \\ \mathbf{X}\mathbf{Z}\mathbf{e} &= \sigma \mathbf{e} \\ \mathbf{Y}\mathbf{W}\mathbf{e} &= \sigma \mathbf{e}. \end{aligned}$$

These equations have a nice interpretation. We have just to recall that the starting problem has an LP dual:

$$\begin{aligned} \min \quad & \mathbf{b}'\mathbf{y} \\ \text{s.t.} \quad & \mathbf{A}'\mathbf{y} \geq \mathbf{c} \\ & \mathbf{y} \geq \mathbf{0}, \end{aligned}$$

or, adding slack variables $\mathbf{z}$,

$$\begin{aligned} \min \quad & \mathbf{b}'\mathbf{y} \\ \text{s.t.} \quad & \mathbf{A}'\mathbf{y} - \mathbf{z} = \mathbf{c} \\ & \mathbf{y}, \mathbf{z} \geq \mathbf{0}. \end{aligned}$$

Hence, the equations we arrived at are simply the conditions of primal feasibility, dual feasibility, and (if $\sigma = 0$) complementary slackness (see theorem 6.7). For $\sigma > 0$, they are a set of non-linear equations:

$$\mathbf{F}(\boldsymbol{\xi}) = \mathbf{0},$$

where

$$\boldsymbol{\xi} = \begin{bmatrix} \mathbf{x} \\ \mathbf{y} \\ \mathbf{w} \\ \mathbf{z} \end{bmatrix},$$

which may be tackled by Newton's method (section 3.4.2).

In principle, by solving this system of non-linear equations for different values of σ we get a path $(\mathbf{x}_\sigma, \mathbf{y}_\sigma, \mathbf{w}_\sigma, \mathbf{z}_\sigma)$. This path is called *central path* and for $\sigma \to 0$, it leads to the optimal solution of the original LP. From a computational point of view, it is not convenient to start with a too small σ, nor to solve the non-linear equations exactly for each σ. One idea is to reduce the value of the penalty parameter within the iterations of Newton's method, so that the central path is only a reference path leading to solution through the interior of the feasible set. It is worth noting the similarity between this path following approach and homotopy continuation methods described in section 3.4.5. In both cases we solve a difficult problem by a sequence of easier problems which converge to the original one.

Interior point methods have a polynomial computational complexity which is, theoretically, better than the complexity of the simplex method, which is exponential in pathological cases.[9] It should be stressed that many computational tricks are needed to implement both the simplex and interior point method in a very efficient way. These are beyond the scope of this book, but it should be clear that the two approaches may lead to qualitatively different, though cost-equivalent solutions, as illustrated in the next section.

6.5 CONSTRAINED OPTIMIZATION IN MATLAB

In this section, we briefly describe functions from the Optimization toolbox which can be used for constrained optimization. In particular we consider the functions `linprog` for linear programming, `quadprog` for quadratic programming, and `fmincon` for generic constrained optimization. The coverage is not complete really, but we will provide a couple of examples related to financial problems.

---

[9] Computational complexity has been introduced in section 3.1.3.

6.5.1 Linear programming in MATLAB

The Optimization toolbox includes a function, `linprog`, which solves LP problems of the form

$$\begin{aligned} \min \quad & \mathbf{c'x} \\ \text{s.t.} \quad & \mathbf{Ax} \leq \mathbf{b} \\ & \mathbf{A_{eq}x} = \mathbf{b_{eq}} \\ & \mathbf{l} \leq \mathbf{x} \leq \mathbf{u}. \end{aligned}$$

We have seen that alternative algorithms are available for linear programming. What happens in MATLAB, then? Consider the following rather trivial LP problem:

$$\begin{aligned} \max \quad & x_1 + x_2 \\ \text{s.t.} \quad & x_1 + x_2 \leq 1 \\ & x_1, x_2 \geq 0. \end{aligned}$$

It is easy to see that two basic optimal solutions are $(1,0)$ and $(0,1)$. All the solutions between these two extreme points are equivalent and optimal. We expect that the simplex algorithm should report one of the two extreme points. To use `linprog`, we have to change the sign of the coefficients in the objective function and to pass as null vectors the parameters we do not need:

```
>> x=linprog([-1 -1],[1 1],1,[],[],[0 0])
Optimization terminated successfully.
x =
    0.5000
    0.5000
```

We see that the reported solution is *on the center* of the face of equivalent solutions, and it is not basic. This happens since the default LP option in MATLAB is an interior point algorithm. Actually, `linprog` implements three alternative approaches:

- if the `LargeScale` option is `on`, then an interior point method is used;
- if the `LargeScale` option is `off`, then an active set method is used (see section 6.3.5), which by the way is the same used by `quadprog` ;
- when `LargeScale` option is `off`, we may select the simplex method by setting the `Simplex` option.

The three methods also differ in terms of using an initial solution or not. To get the point, we may play a little bit with the options.

```
>> options = optimset('LargeScale','off');
```

```
>> x=linprog([-1 -1],[1 1],1,[],[],[0 0],[],[],options)
Optimization terminated successfully.
x =
    0.5000
    0.5000
>> x=linprog([-1 -1],[1 1],1,[],[],[0 0],[],[0 0.5],options)
Optimization terminated successfully.
x =
    0.2500
    0.7500
```

Note that to set the options, we have to pass (possibly empty) vectors corresponding to upper bounds and initial points. We see that, starting from the initial solution, the search moves along the gradient until the constraint is reached, which is turned active and the process is stopped. With the active set method, the solution depends on the initial point. If we select the simplex method, we have a different behavior:

```
>> options = optimset('LargeScale','off', 'Simplex', 'on');
>> x=linprog([-1 -1],[1 1],1,[],[],[0 0],[],[],options)
Optimization terminated.
x =
    1
    0
>> x=linprog([-1 -1],[1 1],1,[],[],[0 0],[],[0 0.5],options)
Warning: Simplex method uses a built-in starting point;
    ignoring user-supplied X0.
> In linprog at 215
Optimization terminated.
x =
    1
    0
```

We see that a basic solution is obtained, and that the initial point is ignored. The initial point is ignored by the interior point method as well:

```
>> options = optimset('LargeScale','on');
>> x=linprog([-1 -1],[1 1],1,[],[],[0 0],[],[0 0.5],options)
Warning: Large scale (interior point) method uses
a built-in starting point; ignoring user-supplied X0.
> In linprog at 205
Optimization terminated.
x =
    0.5000
    0.5000
```

It is important to bear in mind that, unless the simplex method is selected, an optimal but non-basic solution may be obtained. This may have consequences if `linprog` is embedded within an algorithm that requires basic optimal solutions. For instance, in some problems with special structure, the simplex method yields an integer solution; this is the case when the feasible set is a polyhedron whose extreme points have integer coordinates. Indeed, the simplex method must be used when tackling a mixed-integer programming problem by a branch and bound strategy (see chapter 12).

6.5.2 A trivial LP model for bond portfolio management

In chapter 2 we have considered the immunization of a bond portfolio (see example 2.12 on page 63). We considered three bonds and we selected a portfolio with given value, duration, and convexity. The problem was set up in such a way that there was a unique solution (which may require selling a bond short). However, when many bonds are available, more than one solution can be found. In such a case, it might make sense to look for the "best" solution among the feasible ones. Defining "best" is not easy at all. We should probably include some explicit characterization of uncertainty in interest rates, and this leads to stochastic programming problems described in chapter 11. Furthermore, since this is likely to be an asset-liability management problem, rather than a simple portfolio management problem, we should also characterize uncertainty in liabilities. However, just to try a MATLAB programming exercise, let us consider a simple linear programming model.[10] One possible idea is maximizing the average yield of the portfolio, given that the portfolio must have duration D and convexity C; we also add the requirement that short sales are not allowed. This results in the following linear programming (LP) model:

$$\max \quad \sum_{i=1}^{N} Y_i w_i$$

$$\text{s.t.} \quad \sum_{i=1}^{N} d_i w_i = D$$

$$\sum_{i=1}^{N} c_i w_i = C$$

$$\sum_{i=1}^{N} w_i = 1$$

$$w_i \geq 0 \quad \forall i.$$

---

[10] We should probably say "simplistic" rather than simple. The model is somewhat inspired by a model discussed in [14] for a different purpose.

Note that, without the non-negativity constraints on the weights w_i, we may easily end up with an unbounded solution.[11] It is easy to write a MATLAB function solving this problem. The code is illustrated in figure 6.10.

Since all the functions dealing with bonds are able to cope with vector arguments, provided that they are of compatible size, we group the bond characteristics in vectors.[12] Here we assume that we know the clean price for each bond, and we use bndyield to compute the corresponding yield and bnddury and bndconvy to obtain the sensitivities. Note that, when calling linprog, we must change the sign of the coefficients in the objective function, because we want to maximize it; the next four arguments contain the coefficient matrix and the right-hand side of inequality and equality constraints (since we have only equality constraints in this model, the first two arguments are empty); finally, we have a vector of zeros representing the lower bound on the decision variables. First we consider a set of five bonds, and then we enlarge the set by adding five more bonds. Running this script, we get the following output:

```
>> LPbonds1
Optimization terminated.
weights1 =
    0.4955
    0.0000
    0.4361
    0.0684
    0.0000
Optimization terminated.
weights2 =
    0.0000
    0.0000
    0.3813
    0.0000
    0.0000
    0.0800
    0.0000
    0.5387
    0.0000
    0.0000
```

You may notice that in both cases only three bonds are included in the portfolio. This might appear a bit odd, since one would assume that considering

---

[11] See [14] for conditions ensuring the finiteness of the solution and for generalizations of the model.
[12] The reader is referred to section 2.3.4 for a description of MATLAB functions to deal with simple bonds.

382 CONVEX OPTIMIZATION

```
% SCRIPT LPBonds1.m
% BOND CHARACTERISTICS FOR SET 1
settle    = '19-Mar-2006';
maturity1 = ['15-Jun-2021' ; '02-Oct-2016' ; '01-Mar-2031' ; ...
    '01-Mar-2026' ; '01-Mar-2011'];
Face1     = [500  ;  1000  ;  250 ; 100 ; 100];
couponRate1 = [0.07 ;  0.066 ; 0.08 ; 0.06 ; 0.05];
cleanPrice1 = [ 549.42 ; 970.49 ; 264.00 ; 112.53 ; 87.93 ];
% COMPUTE YIELDS AND SENSITIVITIES
yields1 = bndyield(cleanPrice1, couponRate1, settle, maturity1, ...
          2, 0, [] , [] , [] , [], [] , Face1);
durations1 = bnddury(yields1, couponRate1, settle, maturity1, ...
          2, 0, [] , [] , [] , [], [] , Face1);
convexities1 = bndconvy(yields1, couponRate1, settle, maturity1, ...
          2, 0, [] , [] , [] , [], [] , Face1);
% SET UP AND SOLVE LP PROBLEM
A1 = [durations1'
      convexities1'
      ones(1,5)];
b = [ 10.3181 ; 157.6346 ; 1];
weights1 = linprog(-yields1,[],[],A1,b,zeros(1,5))
% BOND CHARACTERISTICS FOR SET 2
maturity2 = [maturity1 ; ...
    '15-Jan-2019' ; '10-Sep-2010' ; '01-Aug-2023' ; ...
    '01-Mar-2016' ; '01-May-2013'];
Face2      = [Face1 ; 100  ;   500  ;  200 ; 1000 ; 100];
couponRate2 = [couponRate1 ; 0.08 ;   0.07 ; 0.075 ; 0.07 ; 0.06];
cleanPrice2 = [ cleanPrice1 ; ...
      108.36 ; 519.36 ; 232.07 ; 1155.26 ; 89.29 ];
% COMPUTE YIELDS AND SENSITIVITIES
yields2 = bndyield(cleanPrice2, couponRate2, settle, maturity2, ...
          2, 0, [] , [] , [] , [], [] , Face2);
durations2 = bnddury(yields2, couponRate2, settle, maturity2, ...
          2, 0, [] , [] , [] , [], [] , Face2);
convexities2 = bndconvy(yields2, couponRate2, settle, maturity2, ...
          2, 0, [] , [] , [] , [], [] , Face2);
% SET UP AND SOLVE LP PROBLEM
A2 = [durations2'
      convexities2'
      ones(1,10)];
weights2 = linprog(-yields2,[],[],A2,b,zeros(1,10))
```

Fig. 6.10 Code to set up and solve a linear programming model for bond portfolio optimization.

more bonds leaves more space for diversification. Actually, this does not happen by chance, but it depends on the structure of the optimal solution of a linear programming problem. If we have only M equality constraints in a linear program, there is an optimal solution (provided that the problem is bounded and feasible) with at most M decision variables which take a non-zero value at optimality. Since here $M = 3$, the optimal portfolio will always include just three bonds, even if many more are available, unless there are alternative optima (in which case the solution would depend on the algorithm we select for linprog; for this problem instance, there are no alternative optima, and the interior point method returns the same solution we would obtain by selecting the simplex method). If we considered only duration constraints, we would include just two bonds, whose durations would bracket the target duration.

6.5.3 Using quadratic programming to trace efficient portfolio frontier

In section 2.4.3 we have considered some MATLAB functions to trace the set of mean-variance efficient portfolios. To that aim, we must solve a set of problems of the form (2.13), which we recall here for convenience:

$$\begin{aligned} \min \quad & \mathbf{w}'\Sigma\mathbf{w} \\ \text{s.t.} \quad & \mathbf{w}'\bar{\mathbf{r}} = \bar{r}_T \\ & \sum_{i=1}^{n} w_i = 1 \\ & w_i \geq 0, \end{aligned}$$

for different values of the target expected return $\bar{r}_T$. We see that this is a quadratic programming problem, which can be solved by quadprog. It is a useful exercise to write a function to do that, which will be a (very) simplified version of frontcon. The input arguments to this function, which we call NaiveMV and whose code is displayed in figure 6.11, are: ERet, the vector of expected return for the assets we are considering, ECov, the variance-covariance matrix, and NPts the number of efficient points (portfolios) we want to find on the frontier. Output arguments are: PRisk, the risk (standard deviation of return) for each portfolio we generate, PRoR, the rate of return, and PWts, the matrix portfolio weights (one vector for each portfolio).

To select target returns we have to spot both the maximum return achievable and the return associated with the minimum variance (minimum risk) portfolio. The first target return is obtained by solving

$$\begin{aligned} \max \quad & \mathbf{w}'\bar{\mathbf{r}} \\ & \sum_{i=1}^{n} w_i = 1 \\ & w_i \geq 0. \end{aligned}$$

```
function [PRisk, PRoR, PWts] = NaiveMV(ERet, ECov, NPts)
ERet = ERet(:);      % makes sure it is a column vector
NAssets = length(ERet); % get number of assets
% vector of lower bounds on weights
V0 = zeros(NAssets, 1);
% row vector of ones
V1 = ones(1, NAssets);
% set medium scale option
options = optimset('LargeScale', 'off');
% Find the maximum expected return
MaxReturnWeights = linprog(-ERet, [], [], V1, 1, V0);
MaxReturn = MaxReturnWeights' * ERet;
% Find the minimum variance return
MinVarWeights = quadprog(ECov,V0,[],[],V1,1,V0,[],[],options);
MinVarReturn = MinVarWeights' * ERet;
MinVarStd = sqrt(MinVarWeights' * ECov * MinVarWeights);
% check if there is only one efficient portfolio
if MaxReturn > MinVarReturn
    RTarget = linspace(MinVarReturn, MaxReturn, NPts);
    NumFrontPoints = NPts;
else
    RTarget = MaxReturn;
    NumFrontPoints = 1;
end
% Store first portfolio
PRoR = zeros(NumFrontPoints, 1);
PRisk = zeros(NumFrontPoints, 1);
PWts = zeros(NumFrontPoints, NAssets);
PRoR(1) = MinVarReturn;
PRisk(1) = MinVarStd;
PWts(1,:) = MinVarWeights(:)';
% trace frontier by changing target return
VConstr = ERet';
A = [V1 ; VConstr ];
B = [1 ; 0];
for point = 2:NumFrontPoints
    B(2) = RTarget(point);
    Weights = quadprog(ECov,V0,[],[],A,B,V0,[],[],options);
    PRoR(point) = dot(Weights, ERet);
    PRisk(point) = sqrt(Weights'*ECov*Weights);
    PWts(point, :) = Weights(:)';
end
```

Fig. 6.11 Simple MATLAB code to trace the mean-variance efficient frontier.

Actually, this is a trivial LP problem, whose optimal solution is clearly the maximum expected return. Nevertheless, if additional constraints are given on asset allocation, we may really have to solve an LP problem. This is why we use `linprog` in the code to get the `MaxReturn`. The second return is obtained by finding the minimum risk portfolio:

$$\min \quad \mathbf{w}'\Sigma\mathbf{w}$$
$$\sum_{i=1}^{n} w_i = 1$$
$$w_i \geq 0$$

and by computing its return (we take for granted that the solution of this problem is unique). These are the two extreme efficient portfolios. If they are equal, there is a unique portfolio maximizing return and minimizing risk: an unlikely event in practice, which is taken into account by the function (in this case the number `NumFrontPoints` of efficient points in the frontier is 1; otherwise it is the input number `NPts`). To find other efficient portfolios, we use the function `linspace` to specify the vector of `NPts` target returns between the two extremes. Then we solve a sequence of risk minimization problems, obtaining the risk/return characteristics and the composition of each portfolio. To that aim we must simply change one element, corresponding to target return, in the vector B containing the right-hand sides of linear equality constraints in the quadratic program.

We may check that `NaiveMV` yields the same results as `frontcon` for this simple problem:

```
>> ExpRet = [ 0.15 0.2 0.08];
>> CovMat = [ 0.2 0.05 -0.01 ; 0.05 0.3 0.015 ; ...
              -0.01 0.015 0.1];
>> [PRisk, PRoR, PWts] = naiveMV(ExpRet, CovMat, 10);
>> [PRoR , PRisk]
ans =
    0.1143    0.2411
    0.1238    0.2456
    0.1333    0.2588
    0.1428    0.2794
    0.1524    0.3060
    0.1619    0.3370
    0.1714    0.3714
    0.1809    0.4093
    0.1905    0.4682
    0.2000    0.5477
```

6.5.4 Non-linear programming in MATLAB

The most general function to deal with a non-linear programming problem is `fmincon`. How this function should be called depends on the problem at hand, as constraints are partitioned in linear and non linear constraints as follows:

$$\begin{aligned}\min\quad & f(\mathbf{x})\\ \text{s.t.}\quad & \mathbf{A}\mathbf{x}\leq\mathbf{b}\\ & \mathbf{A}_{eq}\mathbf{x}=\mathbf{b}_{eq}\\ & \mathbf{g}(\mathbf{x})\leq\mathbf{0}\\ & \mathbf{g}_{eq}(\mathbf{x})=\mathbf{0}\\ & \mathbf{l}\leq\mathbf{x}\leq\mathbf{u}.\end{aligned}$$

Matrices and both upper and lower bounds are passed as vector arguments, whereas the non-linear functions for inequality and equality constraints must be written as M-files or anonymous functions. For instance, to solve

$$\begin{aligned}\min\quad & e^{x_1}(4x_1^2+2x_2^2+4x_1x_2+2x_2+1)\\ \text{s.t.}\quad & x_1x_2-x_1-x_2\leq-1.5\\ & x_1x_2\geq 10\end{aligned}$$

we may write two M-files. The first one must return two vectors corresponding to non-linear constraints:

```
function [c, ceq] = confun(x)
% non-linear inequality constraints
c = [1.5 + x(1)*x(2) - x(1) - x(2);
     -x(1)*x(2) - 10];
% non-linear equality constraints
ceq = [];
```

Here the second vector is empty, since there are no equality constraints. Also note the change in sign for the second constraint. Another file is needed for the objective function:

```
function fval = objfun(x)
fval = exp(x(1)) * ( 4*x(1)^2 + 2*x(2)^2 + 4*x(1)*x(2) + 2*x(2) + 1);
```

These M-files may also return analytical values for the gradients of the involved functions. Then we may call fmincon:

```
>> x0 = [-1,1];
>> options = optimset('LargeScale','off');
>> [x, fval] = fmincon(@objfun,x0,[],[],[],[],[],[],@confun,options)
Optimization terminated: first-order optimality measure less
 than options.TolFun and maximum constraint violation is less
 than options.TolCon.
Active inequalities (to within options.TolCon = 1e-006):
  lower      upper      ineqlin    ineqnonlin
                                   1
                                   2
```

```
x =
   -9.5474    1.0474
fval =
   0.0236
```

We should note that fmincon is not always the best choice. For instance, model calibration may lead to optimization models of the form

$$\min \sum_{i=1}^{m} f_i^2(\mathbf{x}),$$

which are best solved as non-linear least squares problems by lsqnonlin.

6.6 INTEGRATING SIMULATION AND OPTIMIZATION

Simulation models are a convenient way to evaluate the performance of complex and stochastic systems for which analytical models may be very hard or even impossible to come up with. However, they are just able to evaluate a performance measure given a set of input parameters. In option pricing, this may be just what we need, but we could also be interested in finding the optimal set of parameters; in other words, in many settings, such as portfolio optimization, we would like to integrate simulation and optimization (see, e.g., [5]). Such an integration may certainly be worthwhile, as it provides us with a way to optimize complex and stochastic systems which cannot be dealt with by deterministic and even stochastic programming. However, we may have to face at least some of the following issues:

- The objective function may be non-convex.

- Some of the input parameters may be discrete rather than continuous.

- The evaluation of the objective function may be affected by noise.

- Using gradient-based methods may be difficult, as gradients must be estimated.

Let us start from the last point and assume for simplicity that we want to solve an unconstrained optimization whereby the objective function is the expected value of some random performance measure depending on a vector of parameters $\mathbf{x} \in \mathbb{R}^n$:

$$\min f(\mathbf{x}) = \mathrm{E}_\omega[h(\mathbf{x},\omega)].$$

For optimization purposes it would be useful to have a way to compute the gradient $\nabla f(\mathbf{x})$ at any point. As pointed out in section 4.5.2, a gradient could be estimated by finite differences, but this is made difficult by the noise in the estimates. Using common random numbers to reduce variance is the least we should do; an alternative is represented by using some form of regression. The idea is to use a simulation model to build a sort of empirical metamodel, the response surface, which yields an analytical approximation $g(\mathbf{x})$ of the unknown objective function $f(\mathbf{x})$ with respect to the input parameters. If we want to estimate the gradient at a certain point $\mathbf{x}$, we may consider a linear approximation, such as

$$g(\mathbf{x}) = \alpha + \sum_{i=1}^{n} \beta_i x_i = \alpha + \boldsymbol{\beta}'\mathbf{x}.$$

We may estimate α and $\boldsymbol{\beta}$ by evaluating f for a set of test values $\mathbf{x}^j$ and by minimizing a function of the regression errors. Let $\hat{f}_j$ be the estimate of f corresponding to the point $\mathbf{x}^j$ ($j = 1, \ldots, m$). We have

$$\hat{f}_j = \alpha + \boldsymbol{\beta}'\mathbf{x}^j + \epsilon_j,$$

where ϵ_j is an error term (or a residual, if you prefer; see section 3.3). Using least squares, we may find α and $\boldsymbol{\beta}$ in such a way that $\sum_j \epsilon_j^2$ is minimized. Let us define the matrix

$$\mathbf{X} = \begin{bmatrix} 1 & x_1^1 & x_2^1 & \cdots & x_n^1 \\ 1 & x_1^2 & x_2^2 & \cdots & x_n^2 \\ \vdots & \vdots & \vdots & \ddots & \vdots \\ 1 & x_1^m & x_2^m & \cdots & x_n^m \end{bmatrix},$$

where x_i^j is the jth setting of the parameter x_i. It can be shown that the sum of the squared errors is minimized by

$$\begin{bmatrix} \hat{\alpha} \\ \hat{\boldsymbol{\beta}} \end{bmatrix} = (\mathbf{X}'\mathbf{X})^{-1} \mathbf{X}'\hat{\mathbf{f}},$$

where $\hat{\mathbf{f}}$ is the vector of the m estimates of f. Then we may set $\hat{\nabla} f^{(k)} = \boldsymbol{\beta}$ and use it within a gradient optimization method. A first-order fit is suitable when we are not close to the optimum. When we are approaching the minimizer, a quadratic polynomial can be fitted:

$$f(\mathbf{x}) = \alpha + \boldsymbol{\beta}'\mathbf{x} + \frac{1}{2}\mathbf{x}'\boldsymbol{\Gamma}\mathbf{x},$$

where $\boldsymbol{\Gamma}$ is a square matrix, and quadratic programming may be used to find the optimal set of parameters for the metamodel, which is successively updated until some convergence criterion is met [7]. This results in a method resembling the quasi-Newton methods for non-linear programming.

An obvious disadvantage of an approach based on the response surface methodology is that it is likely to be quite expensive in computational terms.

Alternative methods, such as perturbation analysis, have been proposed to estimate sensitivities with a single simulation runs. An example of an application to estimate option sensitivities can be found in [4]; we will consider a simple case in section 8.5. A treatment of these methods require deep mathematical knowledge, so we refer, e.g., to [15] for a thorough treatment of these topics. We would only like to point out a subtle issue of using gradient-based methods for simulation optimization. In principle, we should evaluate

$$\nabla f(\mathbf{x}) = \nabla \mathrm{E}_\omega[h(\mathbf{x},\omega)],$$

but simulation actually yields something like

$$\mathrm{E}_\omega[\nabla h(\mathbf{x},\omega)].$$

That expectation can be commuted with differentiation is not granted at all. This issue is well explored in [8]; for a treatment oriented to financial engineering, see [9].

Given all of the considerations above, it's no surprise that non-linear programming methods that do not exploit derivatives in any way are of interest for simulation optimization. One such method is the simplex search procedure we have outlined in section 6.2.4; see [11] for a recent paper on this topic.[13]

Although using a simplex search procedure has its merit, it does not overcome the possible difficulties due to the non-convexity of the objective function or the discrete character of some decision parameters. For such cases the integration of simulation with metaheuristics such as tabu search or genetic algorithms, which we will describe in section 12.4, is probably the only practical solution approach. Indeed, this is the approach taken in some commercial stochastic simulation packages. The application of a population-based approach like genetic algorithms or their variants has the further advantage of making the noisy function evaluations less critical.

S6.1 ELEMENTS OF CONVEX ANALYSIS

Convexity is arguably the most important concept in optimization theory. In the next two sections we want first to recall the related concepts of convex set and convex function, and then to outline a few concepts in polyhedral theory which are important for linear and mixed-integer programming.

S6.1.1 Convexity in optimization

Convexity is a possible attribute of the feasible set S of an optimization problem.

---

[13] It may also be worth noting that MATLAB allows the integration of simplex search and other no-derivatives methods with the dynamic systems simulator SIMULINK.

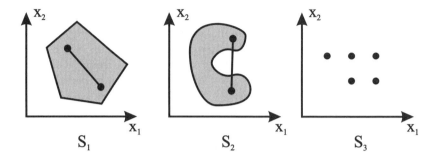

Fig. 6.12 Convex and non-convex sets.

Definition. A set $S \subseteq \mathbb{R}^n$ is a *convex set* if
$$\mathbf{x}, \mathbf{y} \in S \Rightarrow \lambda \mathbf{x} + (1-\lambda)\mathbf{y} \in S \quad \forall \lambda \in [0,1].$$

Example 6.13 The concept of convexity can be grasped intuitively by considering that the points of the form $\lambda \mathbf{x} + (1-\lambda)\mathbf{y}$, where $0 \leq \lambda \leq 1$, are simply the points on the straight line joining $\mathbf{x}$ and $\mathbf{y}$. A set S is convex if the line joining any pair of points $\mathbf{x}, \mathbf{y} \in S$ is contained in S. Consider the three subsets of $\mathbb{R}^2$ depicted in Figure 6.12. S_1 is convex, but S_2 is not. S_3 is a discrete set and it is not convex; this fact has important consequences for discrete optimization problems. □

The following property is easy to verify.

PROPERTY 6.10 *The intersection of convex sets is a convex set.*

Note that the union of convex sets need not be convex. The *convex combination* of p points $\mathbf{x}_1, \mathbf{x}_2, \ldots, \mathbf{x}_p \in \mathbb{R}^n$ is defined as

$$\mathbf{x} = \sum_{i=1}^{p} \mu_i \mathbf{x}_i, \quad \mu_1, \ldots, \mu_p \geq 0, \quad \sum_{i=1}^{p} \mu_i = 1.$$

Given a set $S \subset \mathbb{R}^n$, the set of points which are the convex combinations of points in S is the *convex hull* of S (denoted by $[S]$). If S is a convex set, then $S \equiv [S]$. The convex hull of a generic set S is the smallest convex set containing S; it can also be regarded as the intersection of all the convex sets containing S. Two non-convex sets and their convex hulls are shown in figure 6.13.

Definition. A scalar function $f: \mathbb{R}^n \to \mathbb{R}$, defined over a convex set $S \subseteq \mathbb{R}^n$, is a *convex function* on S if, for any $\mathbf{x}$ and $\mathbf{y}$ in S, for any $\lambda \in [0,1]$, we have

$$f(\lambda \mathbf{x} + (1-\lambda)\mathbf{y}) \leq \lambda f(\mathbf{x}) + (1-\lambda)f(\mathbf{y}).$$

If this condition is met with strict inequality for all $\mathbf{x} \neq \mathbf{y}$, the function is *strictly* convex.

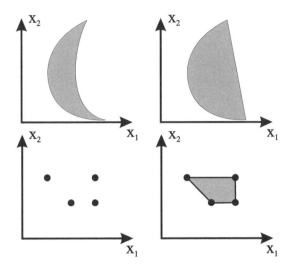

Fig. 6.13 Non-convex sets and their convex hulls.

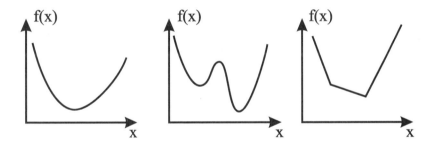

Fig. 6.14 Convex and non-convex functions.

Definition. A function f is *concave* if $(-f)$ is convex.

The concept of convex function is illustrated in figure 6.14. The first function is convex, whereas the second is not. Also, the third function is convex; a convex function need not be differentiable everywhere. The definition can be interpreted as follows. Given any two points $\mathbf{x}$ and $\mathbf{y}$, consider another point which is a convex combination of $\mathbf{x}$ and $\mathbf{y}$; then the function value in this point is overestimated by the convex combination of the function values $f(\mathbf{x})$ and $f(\mathbf{y})$, since the line segment joining $(\mathbf{x}, f(\mathbf{x}))$ and $(\mathbf{y}, f(\mathbf{y}))$ lies above the graph of the function between $\mathbf{x}$ and $\mathbf{y}$. In other words, a function is convex if its epigraph, i.e., the region above the function graph, is a convex set. A further link between convex sets and convex functions is that the set $S = \{\mathbf{x} \in \mathbb{R}^n \mid g(\mathbf{x}) \leq 0\}$ is convex if g is a convex function. Convexity of functions is preserved by some operations; in particular, a linear combination of convex functions f_i,

$$f(\mathbf{x}) = \sum_{i=1}^{n} \lambda_i f_i(\mathbf{x}),$$

is a convex function if $\lambda_i \geq 0$, for any i.

There are alternative characterizations of a convex function. For our purposes the most important is the following. If f is a differentiable function, it is convex (over S) if and only if

$$f(\mathbf{x}) \geq f(\mathbf{x}_0) + \nabla f'(\mathbf{x}_0)(\mathbf{x} - \mathbf{x}_0) \qquad \forall \mathbf{x}, \mathbf{x}_0 \in S. \tag{6.27}$$

Note that the hyperplane

$$z = f(\mathbf{x}_0) + \nabla f'(\mathbf{x}_0)(\mathbf{x} - \mathbf{x}_0)$$

is the usual tangent hyperplane, i.e., the first-order Taylor expansion of f at $\mathbf{x}_0$. For a differentiable function, convexity implies that the first-order approximation at a certain point $\mathbf{x}_0$ consistently underestimates the true value of the function at all the other points $\mathbf{x} \in S$. The concept of a tangent hyperplane applies only to differentiable convex functions, but it can be generalized by the concept of a support hyperplane.

Definition. Given a convex function f and a point $\mathbf{x}^0$, the hyperplane (in $\mathbb{R}^{n+1}$) given by $z = f(\mathbf{x}^0) + \gamma'(\mathbf{x} - \mathbf{x}^0)$, which meets the epigraph of f in $(\mathbf{x}^0, f(\mathbf{x}^0))$ and lies below it is called the *support hyperplane* of f at $\mathbf{x}^0$.

The concept of a support hyperplane is depicted in figure 6.15. A support hyperplane at $\mathbf{x}_0$ is essentially defined by a vector γ such that

$$f(\mathbf{x}) \geq f(\mathbf{x}_0) + \gamma'(\mathbf{x} - \mathbf{x}_0) \qquad \forall \mathbf{x} \in S. \tag{6.28}$$

The vector γ in inequality (6.28) plays the same role as the gradient does in inequality (6.27). If f is differentiable in $\mathbf{x}_0$, the support hyperplane is the usual tangent hyperplane and $\gamma = \nabla f(\mathbf{x}_0)$. This is why a vector γ such that inequality (6.28) holds is called a *subgradient* of f at $\mathbf{x}_0$. If f is non-differentiable, the support hyperplane need not be unique and there is a set of subgradients. The set of subgradients at a point $\mathbf{x}_0$ is called the *subdifferential* of f at $\mathbf{x}_0$, and it is denoted by $\partial f(\mathbf{x}_0)$. It can be shown that a convex function on a set S is subdifferentiable on the interior of S, i.e., we can always find a subgradient (on the boundary of the set S some difficulties may occur due, e.g., to discontinuities, but we need not be concerned with this technicality in the following).

A further characterization of convex functions can be given for twice-differentiable functions.

THEOREM 6.11 *If f is a twice-differentiable function, defined on a nonempty and open convex set S, then f is convex if and only if its Hessian matrix is positive semidefinite at any point in S.*

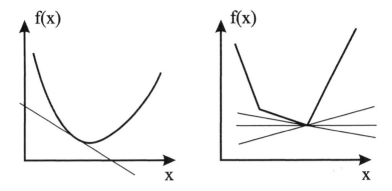

Fig. 6.15 Illustration of the support hyperplane for convex functions.

We recall that the Hessian matrix $\mathbf{H}(\mathbf{x})$ is the (symmetric) matrix of second-order derivatives of $f(\mathbf{x})$:

$$\mathbf{H}_{ij} = \frac{\partial^2 f}{\partial x_i \, \partial x_j}.$$

We also recall that a symmetric (hence square) matrix $\mathbf{A}(\mathbf{x})$ is positive semidefinite on S if

$$\mathbf{x}' \mathbf{A}(\mathbf{x}) \mathbf{x} \geq 0 \qquad \forall \mathbf{x} \in S.$$

The matrix is positive definite if the inequality above is strict for all $\mathbf{x} \neq \mathbf{0}$. If the Hessian matrix is positive definite, the function is strictly convex; however, the converse is not necessarily true. The definiteness of a matrix may be investigated by checking the sign of its eigenvalues; the matrix is positive semidefinite if all of its eigenvalues are non-negative, and it is positive definite if all of its eigenvalues are positive.

S6.1.2 Convex polyhedra and polytopes

Consider in $\mathbb{R}^n$ the *hyperplane* $\mathbf{a}_i' \mathbf{x} = b_i$, where $b_i \in \mathbb{R}$ and $\mathbf{a}_i, \mathbf{x} \in \mathbb{R}^n$ are column vectors.[14] A hyperplane divides $\mathbb{R}^n$ into two *half-spaces* expressed by the linear inequalities $\mathbf{a}_i' \mathbf{x} \leq b_i$ and $\mathbf{a}_i' \mathbf{x} \geq b_i$.

Definition. A *polyhedron* $P \subseteq \mathbb{R}^n$ is a set of points satisfying a finite collection of linear inequalities, i.e.,

$$P = \{ \mathbf{x} \in \mathbb{R}^n \mid \mathbf{A}\mathbf{x} \geq \mathbf{b} \}.$$

A polyhedron is therefore the intersection of a finite collection of half-spaces.

---

[14] Unless the contrary is stated, we assume that all vectors are columns.

PROPERTY 6.12 *A polyhedron is a convex set (it is the intersection of convex sets).*

Definition. A polyhedron is *bounded* if there exists a positive number M such that
$$P \subseteq \{\mathbf{x} \in \mathbb{R}^n \mid -M \leq x_j \leq M \; j = 1, \ldots, n\}.$$

A bounded polyhedron is called a *polytope*. A polytope and an unbounded polyhedron are shown in figure 6.16.

Definition. A point $\mathbf{x}$ is an *extreme point* of a polyhedron P if $\mathbf{x} \in P$ and it is not possible to express $\mathbf{x}$ as $\mathbf{x} = \frac{1}{2}\mathbf{x}' + \frac{1}{2}\mathbf{x}''$ with $\mathbf{x}', \mathbf{x}'' \in P$ and $\mathbf{x}' \neq \mathbf{x}''$.

A polytope P has a finite number of extreme points $\mathbf{x}^1, \ldots, \mathbf{x}^J$. Any point $\mathbf{x}$ in a polytope P can be expressed as a convex combination of its extreme points:
$$\mathbf{x} = \sum_{j=1}^{J} \lambda_j \mathbf{x}^j, \quad \sum_{j=1}^{J} \lambda_j = 1, \; \lambda_j \geq 0;$$

in other words, a polytope is the convex hull of its extreme points. In the case of an unbounded polyhedron, this is not true and we must introduce another concept.

Definition. A vector $\mathbf{r} \in \mathbb{R}^n$ is called a *ray* of the polyhedron
$$P = \{\mathbf{x} \in \mathbb{R}^n \mid \mathbf{Ax} \geq \mathbf{b}\}$$
if $\mathbf{Ar} \geq \mathbf{0}$.

If $\mathbf{x}_0$ is a point in a polyhedron P and $\mathbf{r}$ is a ray of P, then
$$\mathbf{y} = \mathbf{x}_0 + \lambda \mathbf{r} \in P \quad \forall \lambda \geq 0.$$

Clearly, only unbounded polyhedra have rays.

Definition. A ray $\mathbf{r}$ of a polyhedron P is called an *extreme ray* if it cannot be expressed as $\mathbf{r} = \frac{1}{2}\mathbf{r}_1 + \frac{1}{2}\mathbf{r}_2$ where $\mathbf{r}_1, \mathbf{r}_2$ are rays of P such that $\mathbf{r}_1 \neq \lambda \mathbf{r}_2$ for any number $\lambda > 0$.

A polyhedron P can be described in terms of its extreme rays and points, in the sense that any point $\mathbf{x} \in P$ can be expressed combining extreme rays and points:
$$\mathbf{x} = \sum_{j=1}^{J} \lambda_j \mathbf{x}^j + \sum_{k=1}^{K} \mu_k \mathbf{r}^k \quad \sum_{j=1}^{J} \lambda_j = 1, \; \lambda_j, \mu_k \geq 0.$$

ELEMENTS OF CONVEX ANALYSIS 395

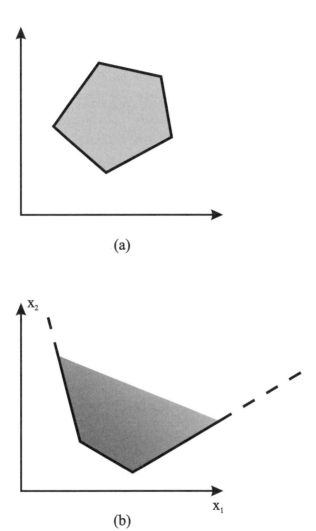

Fig. 6.16 Two-dimensional polytope (a) and unbounded polyhedron (b).

For further reading

In the literature

- A general and introductory book on optimization theory is [18].

- See, e.g., [3] for non-linear programming and [19] for linear programming.

- Interior point methods are dealt with in [20].

- If you are interested in the theory behind convex optimization, you should check [10] or [16]. If you are more interested in the numerical aspects of optimization, [6] is for you.

- For a text which presents non-linear programming methods is some detail, with applications to finance, see [1].

- Advanced issues in portfolio management are dealt with in [17].

- For a tutorial survey on the integration of simulation and optimization, see [7]. A deep mathematical treatment is given in [15].

- The use of simplex search to drive a simulator is explored in [2] and [11].

On the Web

- A good source for information on the practical application of optimization models and methods to a variety of problems is

  http://e-OPTIMIZATION.COM.

- Relevant academic societies in the field are:

  - http://www.informs.org (INFORMS: Institute for Operations Research and the Management Sciences)
  - http://www.siam.org (SIAM: Society for Industrial and Applied Mathematics)
  - http://www.caam.rice.edu/~mathprog (MPS: Mathematical Programming Society)

- A good pointer for interior point methods is

  http://www-unix.mcs.anl.gov/otc/InteriorPoint.

- Michael Trick's Web page lists several useful links to journals, societies, people, etc.; see http://mat.gsia.cmu.edu.

REFERENCES

1. M. Bartholomew-Biggs. *Nonlinear Optimization woith Financial Applications.* Kluwer Academic Publishers, New York, 2005.

2. R.R. Barton and J.S. Ivey, Jr. Nelder–Mead Simplex Modifications for Simulation Optimization. *Management Science*, 42:954–973, 1996.

3. M.S. Bazaraa, H.D. Sherali, and C.M. Shetty. *Nonlinear Programming. Theory and Algorithms (2nd ed.).* Wiley, Chichester, West Sussex, England, 1993.

4. M. Broadie and P. Glasserman. Estimating Security Price Derivatives Using Simulation. *Management Science*, 42:269–285, 1996.

5. A. Consiglio and S.A. Zenios. Designing Portfolios of Financial Products via Integrated Simulation and Optimization Models. *Operations Research*, 47:195–208, 1999.

6. R. Fletcher. *Practical Methods of Optimization (2nd ed.).* Wiley, Chichester, West Sussex, England, 1987.

7. M.C. Fu. Optimization by Simulation: A Review. *Annals of Operations Research*, 53:199–247, 1994.

8. P. Glasserman. *Gradient Estimation via Perturbation Analysis.* Kluwer Academic, Boston, MA, 1991.

9. P. Glasserman. *Monte Carlo Methods in Financial Engineering.* Springer-Verlag, New York, NY, 2004.

10. J.-B. Hiriart-Urruty and Claude Lemaréchal. *Convex Analysis and Minimization Algorithms (vols. 1 and 2).* Springer-Verlag, Berlin, 1993.

11. D.G. Humphrey and J.R. Wilson. A Revised Simplex Search Procedure for Stochastic Simulation Response Surface Optimization. *INFORMS Journal on Computing*, 12:272–283, 2000.

12. R. Korn. *Optimal Portfolios: Stochastic Models for Optimal Investment and Risk Management in Continuous Time.* World Scientific Publishing, Singapore, 1997.

13. R.C. Merton. *Continuous-Time Finance.* Blackwell Publishers, Malden, MA, 1990.

14. J. Paroush and E.Z. Prisman. On the Relative Importance of Duration Constraints. *Management Science*, 43:198–205, 1997.

15. G.C. Pflug. *Optimization of Stochastic Models: The Interface Between Simulation and Optimization.* Kluwer Academic, Dordrecht, The Netherlands, 1996.

16. R.T. Rockafellar. *Convex Analysis*. Princeton University Press, Princeton, NJ, 1970.

17. B. Scherer and D. Martin. *Introduction to Modern Portfolio Optimization with NuOPT, S-Plus, and S^+Bayes*. Springer, New York, 2005.

18. R.K. Sundaram. *A First Course in Optimization Theory*. Cambdridge University Press, Cambridge, 1996.

19. R.J. Vanderbei. *Linear Programming: Foundations and Extensions*. Kluwer Academic, Dordrecht, The Netherlands, 1996.

20. S.J. Wright. *Primal–Dual Interior-Point Methods*. Society for Industrial and Applied Mathematics, Philadelphia, 1997.

Part III
Pricing Equity Options

7

Option Pricing by Binomial and Trinomial Lattices

In this chapter we deal with binomial and trinomial lattices for option pricing. Binomial lattices were introduced in section 2.1 as a basic way to model uncertainty in prices. They rely on a discretization of the underlying stochastic process and exploit recombination to keep computational and memory requirements to a manageable level. We have also seen in section 2.6.1 that pricing options by a no-arbitrage argument is rather simple in a single step binomial lattice. In order to get a practical pricing procedure, we must extend the idea to a multistep lattice, but first we have to find a way to calibrate the lattice so that it reflects the underlying model which is a continuous-time, continuous-state stochastic differential equation. Then we can generalize to multidimensional binomial lattices and to trinomial lattices.

In section 7.1 we start by showing how a simple binomial lattice may be calibrated by matching moments of the discrete probability distribution of prices to drift and volatility of the stochastic process. From this point of view, it is important to understand the connection between lattice techniques and Monte Carlo simulation: Moment matching is a variance reduction strategy, and it can be regarded as a sort of clever sampling. Then we discuss how memory-efficient implementations may be devised. Pricing American options is the subject of section 7.2. Again, it is important to see connections with other techniques. What we do here is essentially a very simple application of the dynamic programming principle which is fully developed in chapter 10. In section 7.3 we consider the generalization to an option depending on two underlying assets; this is only the simplest case, but we see that efficient memory management is fundamental in this case. Another generalization is represented by trinomial lattices 7.4; trinomial lattices can be regarded as a

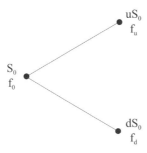

Fig. 7.1 Simple single-period binomial lattice.

particular case of the more general finite difference approach (this is discussed in section 9.2.1). Finally, we consider advantages and disadvantages of lattices in section 7.5.

7.1 PRICING BY BINOMIAL LATTICES

In section 2.6.1, we have considered arbitrage-free pricing of an option by a single step binomial lattice, which is recalled in figure 7.1 for convenience. The idea was to replicate the option with two assets, a risk-free asset and the underlying stock. With two assets, we may replicate any payoff defined over two states. If we model uncertainty with two possible multiplicative shocks u and d, we have seen that the fair option price f_0 is

$$f_0 = e^{-r \cdot \delta t} \left[p f_u + (1-p) f_d \right] \tag{7.1}$$

where f_u and f_d are the option payoffs in the up and down states, respectively, and p is the risk-neutral probability of the up step:

$$p = \frac{e^{r \cdot \delta t} - d}{u - d}.$$

To allow for a better model of uncertainty, we should increase the number of states; to replicate the option payoff, we can either use more assets or allow for trading at intermediate dates. The second possibility is more practical and it is essential, e.g., to price American options, which allow for early exercise at any time during option life. In the limit, this leads to a continuous time model and to the Black–Scholes framework. When the Black–Scholes framework does not lead to an analytical solution, we must resort to some discretization approach, which can be sampling by Monte Carlo simulation, to estimate the risk-neutral expectation, or setting up a grid an apply finite difference methods to solve the corresponding PDE. A multistage binomial lattice, like the one shown in figure 7.2, is an alternative discretization approach; we could also consider trees, but recombination keeps computational effort to a manageable level.

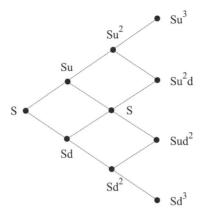

Fig. 7.2 Recombining binomial lattice.

Here we have adopted the convenient choice $u = 1/d$. This is not necessary, as we will see shortly, but in this way, an up step followed by a down step yields the same initial price:

$$S_0 ud = S_0 du = S_0.$$

As we may see form the figure, not only we have recombination, but the lattice uses a limited number of prices too. This may be an advantage when implementing the method. How can we select sensible values for u and d? We should calibrate the lattice in such a way that it approximates the underlying continuous-time process.

7.1.1 Calibrating a binomial lattice

The binomial lattice should be a good approximation of the risk-neutral process

$$dS = rS\,dt + \sigma S\,dW.$$

Hence, we should find parameters to set up the lattice, in such a way that some essential properties of the continuous-time model are preserved. This process is called *calibration*. Starting from S_t, after a small time interval δt, we know from section 2.5 that the new price is a random variable $S_{t+\delta t}$, such that:

$$\log(S_{t+\delta t}/S_t) \sim N\left((r - \sigma^2/2)\,\delta t,\ \sigma^2 \delta t\right).$$

Using properties of the lognormal distribution (see appendix B), we have

$$\mathrm{E}[S_{t+\delta t}/S_t] = e^{r\,\delta t} \tag{7.2}$$

and

$$\mathrm{Var}[S_{t+\delta t}/S_t] = e^{2r\,\delta t}\left(e^{\sigma^2 \delta t} - 1\right). \tag{7.3}$$

404   OPTION PRICING BY BINOMIAL AND TRINOMIAL LATTICES

A reasonable requirement on the discretized dynamics is that it should match these moments. Note that these are two conditions, but we have three parameters: p, u, and d. So we have one degree of freedom, and we may choose $u = 1/d$. This is a convenient choice from a computational point of view, but it is not the only possibility.

On the lattice, we have

$$E[S_{t+\delta t}] = pu \cdot S_t + (1-p)d \cdot S_t,$$

which, together with (7.2), yields

$$pu \cdot S_t + (1-p)d \cdot S_t = e^{r\,\delta t} S_t \quad \Rightarrow \quad p = \frac{e^{r\,\delta t} - d}{u - d}.$$

Note that p is a risk-neutral probability, which does not depend on the true drift. To match variance, we see that, on the lattice,

$$\text{Var}(S_{t+\delta t}) = E[S_{t+\delta t}^2] - E^2[S_{t+\delta t}] = S_t^2(pu^2 + (1-p)d^2) - S_t^2 e^{2r\,\delta t}.$$

From (7.3) we also see

$$\text{Var}[S_{t+\delta t}] = S_t^2 e^{2r\,\delta t}\left(e^{\sigma^2 \delta t} - 1\right),$$

and putting the last two equations together we get

$$S_t^2 e^{2r\,\delta t}\left(e^{\sigma^2 \delta t} - 1\right) = S_t^2(pu^2 + (1-p)d^2) - S_t^2 e^{2r\,\delta t},$$

which boils down to

$$e^{2r\,\delta t + \sigma^2 \delta t} = pu^2 + (1-p)d^2.$$

Substituting p in the right-hand side of the last equation and simplifying:

$$\begin{aligned}
&\frac{e^{r\,\delta t} - d}{u - d}u^2 + \frac{u - e^{r\,\delta t}}{u - d}d^2 \\
&= \frac{u^2 e^{r\,\delta t} - u^2 d + ud^2 - d^2 e^{r\,\delta t}}{u - d} \\
&= \frac{(u^2 - d^2)e^{r\,\delta t} - (u - d)}{u - d} = (u + d)e^{r\,\delta t} - 1,
\end{aligned}$$

we end up with the equation

$$e^{2r\,\delta t + \sigma^2\,\delta t} = (u + d)e^{r\,\delta t} - 1,$$

which, using $u = 1/d$, can be transformed into the quadratic equation:

$$u^2 e^{r\,\delta t} - u\left(1 + e^{2r\,\delta t + \sigma^2 \delta t}\right) + e^{r\,\delta t} = 0.$$

PRICING BY BINOMIAL LATTICES 405

A root of the equation is

$$u = \frac{\left(1 + e^{2r\,\delta t + \sigma^2 \delta t}\right) + \sqrt{\left(1 + e^{2r\,\delta t + \sigma^2 \delta t}\right)^2 - 4e^{2r\,\delta t}}}{2e^{r\,\delta t}}$$

Using first-order expansions, limited to powers of order δt, we may simplify the expression. Starting from the term under square root, we get

$$\left(1 + e^{2r\,\delta t + \sigma^2 \delta t}\right)^2 - 4e^{2r\,\delta t} \approx \left(2 + (2r + \sigma^2)\delta t\right)^2 - 4(1 + 2r\,\delta t) \approx 4\sigma^2 \delta t.$$

Hence,

$$\begin{aligned}
u &\approx \frac{2 + (2r + \sigma^2)\delta t + 2\sigma\sqrt{\delta t}}{2e^{r\,\delta t}} \\
&\approx \left(1 + r\,\delta t + \frac{\sigma^2}{2}\delta t + \sigma\sqrt{\delta t}\right)(1 - r\,\delta t) \\
&\approx 1 + r\,\delta t + \frac{\sigma^2}{2}\delta t + \sigma\sqrt{\delta t} - r\,\delta t = 1 + \sigma\sqrt{\delta t} + \frac{\sigma^2}{2}\delta t.
\end{aligned}$$

But this, to the second order, is the expansion of $e^{\sigma\sqrt{\delta t}}$. We end up with the parameterization

$$u = e^{\sigma\sqrt{\delta t}}, \qquad d = e^{-\sigma\sqrt{\delta t}}, \qquad p = \frac{e^{r\,\delta t} - d}{u - d}, \tag{7.4}$$

which is known as CRR (Cox, Ross, and Rubinstein).

It should be stressed that this is not the only plausible approach, and that alternative parameters are proposed in the literature. For instance, we could arbitrarily choose $p = 0.5$, which, after some calculations, leads to

$$p = \frac{1}{2}, \qquad u = e^{\left(r - \frac{\sigma^2}{2}\right)\delta t + \sigma\sqrt{\delta t}}, \qquad d = e^{\left(r - \frac{\sigma^2}{2}\right)\delta t - \sigma\sqrt{\delta t}},$$

which is known as Jarrow–Rudd parameterization. Furthermore, we have been grappling with rather involved calculations involving non-linear equations. By working on logarithms of price we may try to avoid these difficulties; we will pursue this approach later.

Assuming that the risk-free interest rate and volatility are constant in time, the parameters we have obtained apply to the entire lattice. To price an option, we should build (explicitly or implicitly) a lattice for the underlying asset prices, and then we should proceed backward in time. In fact, the option value is known at maturity, where it is given by the option payoff. Then we should apply equation (7.1) recursively, going backward one step at a time, until we reach the initial node. The binomial lattice approach is best illustrated by its application to a vanilla European call option.

Example 7.1 Suppose that we want to find the price of a vanilla European call with $S_0 = K = 50$, $r = 0.1$, $\sigma = 0.4$, and maturity in five months. From the Black–Scholes model, we know the solution:

```
>> call=blsprice(50,50,0.1,5/12,0.4)
call =
    6.1165
```

If we want to approximate the result by a binomial lattice, we must first set up the lattice parameters. Suppose that each time step is one month. Then

$$\delta t = 1/12 = 0.0833$$
$$u = e^{\sigma\sqrt{\delta t}} = 1.1224$$
$$d = 1/u = 0.8909$$
$$p = \frac{e^{r\,\delta t} - d}{u - d} = 0.5073.$$

The resulting lattices for the stock price and the option value are shown in figure 7.3. The rightmost layer in the call price lattice is obtained by computing the option payoff. To clarify the calculations, let us consider how the uppermost node in the second-to-last time layer is obtained:

$$e^{-r\cdot\delta t}\left[p \cdot 39.07 + (1-p) \cdot 20.77\right]$$
$$= e^{-0.1 \cdot 0.0833}\left[0.5073 \cdot 39.07 + 0.4927 \cdot 20.77\right] \approx 29.77.$$

Going on recursively, we see that the resulting option price is about 6.36, which is not too close to the exact price; a smaller time step is needed to get a good approximation.

To implement the approach in MATLAB we require an algebraic expression of the backward evaluation process. Let f_{ij} be the option value in node (i, j), where j refers to time instant $j\,\delta t$ ($j = 0, \ldots, N$) and i is the ith node in period j (node numbers increase going up in the lattice, $i = 0, \ldots, j$, so we should think of turning the lattice upside down). N is the number of time steps we consider; hence, there are $N + 1$ time layers in the lattice and $N\delta t = T$, the option maturity. With these conventions, the price of the underlying asset in node (i, j) is $Su^i d^{j-i}$. At maturity we have

$$f_{i,N} = \max\{0, Su^i d^{N-i} - K\}, \qquad i = 0, 1, \ldots, N.$$

Going backward in time (decreasing time subscript j), we get

$$f_{ij} = e^{-r\,\delta t}[pf_{i+1,j+1} + (1-p)f_{i,j+1}]. \tag{7.5}$$

The implementation in MATLAB is straightforward, and the resulting code is shown in figure 7.4. The only point worth noting is that matrix indexes start from 1 in MATLAB, which requires a little adjustment. The function

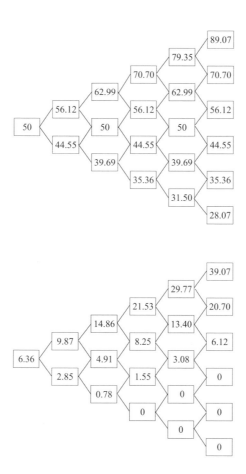

Fig. 7.3 Binomial lattices for the European call option of example 7.1.

```
function [price, lattice] = LatticeEurCall(S0,K,r,T,sigma,N)
deltaT = T/N;
u=exp(sigma * sqrt(deltaT));
d=1/u;
p=(exp(r*deltaT) - d)/(u-d);
lattice = zeros(N+1,N+1);
for i=0:N
   lattice(i+1,N+1)=max(0 , S0*(u^i)*(d^(N-i)) - K);
end
for j=N-1:-1:0
   for i=0:j
      lattice(i+1,j+1) = exp(-r*deltaT) * ...
          (p * lattice(i+2,j+2) + (1-p) * lattice(i+1,j+2));
   end
end
price = lattice(1,1);
```

Fig. 7.4 MATLAB code for pricing a European call by a binomial lattice.

`LatticeEurCall` receives the usual arguments, with the addition of the number of time steps N. By increasing the last parameter, we see that we get a more accurate price (with an increase in the computing time):

```
>> call=LatticeEurCall(50,50,0.1,5/12,0.4,5)
call =
    6.3595
>>call=LatticeEurCall(50,50,0.1,5/12,0.4,500)
call =
    6.1140
```

It is interesting to investigate how the price computed by the binomial lattice converges to the correct price. This may be accomplished by the script in figure 7.5, which produces the output shown in figure 7.6. In this case, the error exhibits an oscillatory behavior as the number of time steps increases.
□

The implementation we have just discussed has a number of weaknesses. To begin with, it uses a large matrix to store the lattice, almost half of which is left empty. We also return the whole lattice as an output argument, which may be useful to check the correspondence with figure 7.3, but may be useless in practice. Actually, we need only two consecutive time layers to store the required information, so some improvement can be obtained. Furthermore, we keep multiplying the discount factor times the risk-neutral probabilities inside the loop; time can be saved by moving this computation outside the

```
% CompLatticeBLS.m
S0 = 50;
K = 50;
r = 0.1;
sigma = 0.4;
T = 5/12;
N=50;
BlsC = blsprice(S0,K,r,T,sigma);
LatticeC = zeros(1,N);
for i=(1:N)
   LatticeC(i) = LatticeEurCall(S0,K,r,T,sigma,i);
end
plot(1:N, ones(1,N)*BlsC);
hold on;
plot(1:N, LatticeC);
```

Fig. 7.5 Script to check the accuracy of the binomial lattice for decreasing δt.

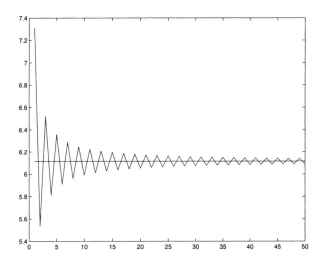

Fig. 7.6 Exact and approximate prices for increasing number of steps in a binomial lattice.

7.1.2 Putting two things together: pricing a pay-later option

We consider here a pay-later call option on a non-dividend paying stock.[1] The feature of the pay-later option is that no premium is paid up front, when the contract is entered; it will be paid later. If the option is in the money at expiration, the option must be exercised and a premium is paid to the writer. Otherwise, the option expires worthless and no premium is due. Note that the net payoff for the option holder can be negative, when the option is not deeply in the money, so that the payoff is smaller than the premium; it is easy to see by no-arbitrage arguments that if the net payoff were always non-negative, we could not have a contract with zero value at time $t = 0$. How can we find the fair premium value?

Given a premium P, the payoff will be

$$f(S_T, P) = \begin{cases} S_T - K - P & \text{if } S_T \geq K \\ 0 & \text{otherwise.} \end{cases}$$

For a given P we may find the value of the option using a binomial lattice. Now we must find a value P such that the risk-neutral expectation of the payoff, with respect to S_T, is zero:

$$\hat{\mathrm{E}}\left[f(S_T; P)\right] = 0.$$

Note that here the discount factor, provided interest rate is constant over time, does not play any role. To solve this equation for P, we may couple the binomial lattice with the bisection method to solve non-linear equations (see section 3.4.1). First we prepare a function to evaluate the expectation for given P; the MATLAB code is shown in figure 7.7. Let us consider an option on a stock whose current price is \$12, with volatility 20%; the risk-free rate is 10%; the strike price is \$14; maturity is 10 months. We use a binomial lattice with a time step corresponding to one month; hence the number of time steps is 10. We may build an anonymous function returning the discounted payoff when P is given, and then we apply bisection using `fzero` and a starting premium for the search:

```
>> f = @(P) L11(P,12,14,0.1, 0.2, 10/12, 10)
f =
    @(P) L11(P,12,14,0.1, 0.2, 10/12, 10)
>> fzero(f,2)
ans =
    2.0432
```

---
[1] This example is based on [5, chapter 13, exercise 11].

```
% exercise 11 chapter 13 from Luenberger, Investment Science
function ExpPayoff = L11(premium,S0,K,r,sigma,T,N)
deltaT = T/N;
u=exp(sigma * sqrt(deltaT));
d=1/u;
p=(exp(r*deltaT) - d)/(u-d);
lattice = zeros(N+1,N+1);
for i=0:N
   if (S0*(u^i)*(d^(N-i)) >= K)
      lattice(i+1,N+1)=S0*(u^i)*(d^(N-i)) - K - premium;
   end
end
for j=N-1:-1:0
   for i=0:j
      lattice(i+1,j+1) = p*lattice(i+2,j+2) + (1-p)*lattice(i+1,j+2);
   end
end
ExpPayoff = lattice(1,1);
```

Fig. 7.7 MATLAB code to price a pay-later option by a binomial lattice.

We see how `fzero` could be used in all those cases in which an analytical pricing formula is not known, even without relying on derivatives (which may be hard to compute for a binomial lattice, but it could be approximated numerically).

7.1.3 An improved implementation of binomial lattices

The implementation of binomial lattices we have used so far can be improved, both from the point of view of CPU time and memory requirements. To begin with, there is no need to repeat calculation of discounted probabilities in the `for` loop; we can multiply discount factor and probabilities once. Furthermore, we may also see that with the CRR lattice calibration, whereby $ud = 1$, we may save memory by using a vector to store the underlying asset prices, rather than a two-dimensional matrix. For instance, we see in figure 7.3 that only eleven different values are used for the underlying asset price. With this lattice calibration, if there are N time steps, we have $2N + 1$ different price values. Hence they can be stored in a single array, with considerable saving. If we require 1000 steps for an accurate evaluation, there is a big difference between requiring a matrix with 1000×1000 elements or a vector with 2001 entries. A possible scheme to store prices is shown in figure 7.8. The numbers shown in the picture are locations in the vector. In element 1 we store the lowest value, resulting from a sequence of down steps only. We see that odd-

412 OPTION PRICING BY BINOMIAL AND TRINOMIAL LATTICES

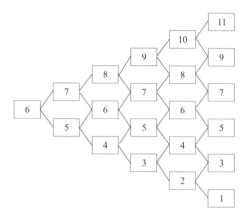

Fig. 7.8 Saving memory for binomial lattices.

numbered entries correspond to the last time layer, whereas even-numbered entries correspond to the second-to-last time layer. The root of the lattice may be even or odd-numbered depending on the number of time steps.

The same scheme may be adopted to store option values. In principle, we should use two vectors corresponding to two consecutive time layers; however, we may exploit the fact that even numbered elements belong to a layer, and odd-numbered elements belong to another one, in order to use one vector of $2N+1$ elements. The resulting code is shown in figure 7.9. A few comments are in order.

- We precompute invariant quantities, including discounted probabilities, in the first section of the code.

- When we write the vector SVals of underlying asset prices, we start with the smallest element, which is $S_0 d^N$; then we multiply by u; for numerical accuracy it would be somewhat better to store S_0 in element SVals(N+1), which is the mid-element, and then proceed both up and down.

- Note that when we work with call values (CVals) we step by two over the index, which amounts to alternating odd- and even-indexed values corresponding to consecutive time layers.

- When *time to maturity* is τ, we need to consider only the $2(N-\tau)+1$ innermost elements of the array CVals. The option price is stored in the root of the lattice, which corresponds to position N+1.

We may check that the computation here is a bit more efficient than with the previous version:

```
>> blsprice(50,50,0.1,5/12,0.4)
ans =
```

```
function price = SmartEurLattice(S0,K,r,T,sigma,N)
% Precompute invariant quantities
deltaT = T/N;
u=exp(sigma * sqrt(deltaT));
d=1/u;
p=(exp(r*deltaT) - d)/(u-d);
discount = exp(-r*deltaT);
p_u = discount*p;
p_d = discount*(1-p);
% set up S values
SVals = zeros(2*N+1,1);
SVals(1) = S0*d^N;
for i=2:2*N+1
    SVals(i) = u*SVals(i-1);
end
% set up terminal CALL values
CVals = zeros(2*N+1,1);
for i=1:2:2*N+1
    CVals(i) = max(SVals(i)-K,0);
end
% work backwards
for tau=1:N
    for i= (tau+1):2:(2*N+1-tau)
        CVals(i) = p_u*CVals(i+1) + p_d*CVals(i-1);
    end
end
price = CVals(N+1);
```

Fig. 7.9 Improved code for pricing a European call by a binomial lattice.

```
      6.1165
>> tic,LatticeEurCall(50,50,0.1,5/12,0.4,2000),toc
ans =
      6.1159
Elapsed time is 0.262408 seconds.
>> tic,SmartEurLattice(50,50,0.1,5/12,0.4,2000),toc
ans =
      6.1159
Elapsed time is 0.069647 seconds.
```

We could try looking for some further improvements by vectorizing code, or by taking a different approach. We will not pursue this in order to avoid obscure code, and to make further developments easier to grasp. The saving in CPU time may not look impressive, but this memory saving approach is essential when we deal with multidimensional options.

7.2 PRICING AMERICAN OPTIONS BY BINOMIAL LATTICES

Pricing an American option by the binomial lattice technique that we have illustrated in the last section is fairly easy. The only critical point is how we should account for early exercise. We deal here with a vanilla American-style put option on a non-dividend paying stock.[2] Consider a point (i, N) on the last time layer of the lattice. If the option is in the money at expiration, it is obviously optimal to exercise it. Hence, in the last time layer we have

$$f_{iN} = \max\{K - S_{iN}, 0\},$$

where $S_{iN} = Su^i d^{N-i}$ is the underlying asset price on that node. Now consider a point in the second-to-last time layer. If the option is not in the money, i.e., if $S_{i,N-1} > K$, we do not exercise. But if the option is in the money, we should wonder about the opportunity of taking an immediate profit $K - S_{i,N-1}$, rather than waiting for possibly better opportunities in the future. In other words, we have to solve an optimal stopping problem, whereby at each time step we must observe the state of a dynamical system and decide whether we should stop the game, and just grab the money we can immediately, or we should go on.

We do this in a simple way, by comparing the immediate payoff (the *intrinsic* value of the option) against the *continuation* value. If we continue and keep the option, we have an asset whose value is

$$f^c_{i,N-1} = e^{-r\delta t}(p_u f_{i+1,N} + p_d f_{i,N}),$$

---

[2]The corresponding call is not interesting, as it can be shown that it is never optimal to exercise it early, unless dividends are paid during the option life.

where p_u and p_d are risk-neutral probabilities. We should exercise if the intrinsic value exceeds the continuation value. Hence, the option value in each node in the second-to-last time layer is

$$f_{i,N-1} = \max\{K - S_{i,N-1},\ e^{-r\delta t}(p_u f_{i+1,N} + p_d f_{i,N})\}.$$

The same argument may be repeated in a recursive fashion for any time layer. This means that we should start from the last time ayer, where the option value is just the option payoff, and we should proceed backward in time using a slight modification of the usual discounted expectation scheme of equation (7.5):

$$f_{i,j} = \max\{K - S_{ij},\ e^{-r\delta t}(p f_{i+1,j+1} + (1-p) f_{i,j+1})\}. \tag{7.6}$$

This idea looks deceptively simple, but it is an application of a very general principle called dynamic programming. We will see in chapter 10 that the dynamic programming principle is extremely powerful in theory, but it is sometimes difficult to apply because of the "curse of dimensionality." In the binomial lattice case, we use a computationally cheap discretization of the underlying stochastic process, and dynamic programming looks almost trivial. However, the application of this principle should be carefully justified along the lines of section 2.6.6. In fact, the reasoning we have followed is somewhat misleading, as we have taken the point of view of the option holder who wants to exercise her option optimally. But we should wonder why we are just using expected values, ignoring risk aversion. A careful justification is not so trivial, and it should involve no-arbitrage arguments and the point of view of the option writer who should care about his worst case, which is when the option holder exercises optimally her rights.

Leaving theoretical issues aside, it is actually easy to adapt the code that we have developed for the European-style call to an American-style put. The resulting code is shown in figure 7.10. We initialize the lattice in a slightly different way, but the only significant change is in backward time-stepping, where we compare the hold value against intrinsic value.

The Financial toolbox provides us with a function, binprice, which prices vanilla American puts and calls, allowing for the possibility of continuous and lumpy dividends. We may compare binprice with AmPutLattice to check our implementation:

```
>> S0 = 50;
>> K = 50;
>> r = 0.05;
>> T = 5/12;
>> sigma = 0.4;
>> N = 1000;
>> price = AmPutLattice(S0,K,r,T,sigma,N)
price =
    4.6739
>> [p, o] = binprice(S0,K,r,T,T/N,sigma,0);
```

```
function price = AmPutLattice(S0,K,r,T,sigma,N)
% Precompute invariant quantities
deltaT = T/N;
u=exp(sigma * sqrt(deltaT));
d=1/u;
p=(exp(r*deltaT) - d)/(u-d);
discount = exp(-r*deltaT);
p_u = discount*p;
p_d = discount*(1-p);
% set up S values
SVals = zeros(2*N+1,1);
SVals(N+1) = S0;
for i=1:N
    SVals(N+1+i) = u*SVals(N+i);
    SVals(N+1-i) = d*SVals(N+2-i);
end
% set up terminal values
PVals = zeros(2*N+1,1);
for i=1:2:2*N+1
    PVals(i) = max(K-SVals(i),0);
end
% work backwards
for tau=1:N
    for i= (tau+1):2:(2*N+1-tau)
        hold = p_u*PVals(i+1) + p_d*PVals(i-1);
        PVals(i) = max(hold, K-SVals(i));
    end
end
price = PVals(N+1);
```

Fig. 7.10 MATLAB code for pricing an American put by a binomial lattice.

```
>> o(1,1)
ans =
    4.6739
```

The function `binprice` requires a flag indicating if the option is a put (flag set to 0) or a call (flag set to 1). This parameter is the last one in the snapshot above. Also note that `binprice` requires both option expiration date T and time step dt as inputs; we have set dt = T/N. We have omitted the optional parameters that may be used to account for dividends. The output from `binprice` is in the form of two lattices, one for the underlying asset price and one for the option value; it is important, when the time step is small, to use the semicolon to suppress output on the screen.

7.3 PRICING BIDIMENSIONAL OPTIONS BY BINOMIAL LATTICES

To illustrate the extension of lattice techniques to multidimensional options, we consider here an American spread option on two assets. The payoff of this option is

$$\max\{S_1 - S_2 - K, 0\}.$$

The basic approach can be extended to more general options, provided we do not include complex path dependencies. As a further generalization, we also consider continuous dividend yields q_1 and q_2. Actually this does not change the problem that much, as we have only to adjust the risk-neutral dynamics, which are given by the equations [see also equation (2.42)]:

$$dS_1 = (r - q_1)S_1 dt + \sigma_1 S_1 dW_1$$
$$dS_2 = (r - q_2)S_2 dt + \sigma_2 S_2 dW_2,$$

where the two Wiener processes are correlated, and the formal rule $dW_1 \, dW_2 = \rho \, dt$ applies (see section 2.5.5).

To avoid the difficulties we had with non-linearities in the calibration process, it is convenient to work with logarithms of asset prices. Setting $x_i = \log S_i$ and using Ito's lemma, we get the two stochastic differential equations:

$$dX_1 = \nu_1 S_1 \, dt + \sigma_1 \, dW_1$$
$$dX_2 = \nu_1 S_2 \, dt + \sigma_2 \, dW_2,$$

where $\nu_i = r - q_i - \sigma_i^2/2$, $i = 1, 2$.

Now, as typical in binomial lattices, we assume that both assets may go up or down by an amount δx_i, in terms of logarithm of prices. To calibrate the lattice, we match first- and second-order moments. We have two stocks which may jump up or down. Hence, each node in the lattice has four successors and we must also find four probabilities: p_{uu}, p_{ud}, p_{du}, and p_{dd}. We first require a matching condition on the expected values of the increments δX_i:

$$E[\delta X_1] = (p_{uu} + p_{ud})\delta x_1 - (p_{du} + p_{dd})\delta x_1 = \nu_1 \delta t$$
$$E[\delta X_2] = (p_{uu} + p_{du})\delta x_2 - (p_{ud} + p_{dd})\delta x_2 = \nu_2 \delta t,$$

where we distinguish between random variables δX_i and their realizations $\pm \delta x_i$. Then, we require a similar condition for second-order moments:

$$E[(\delta X_1)^2] = (p_{uu} + p_{ud} + p_{du} + p_{dd})(\delta x_1)^2 = \sigma_1^2 \delta t + \nu_1^2(\delta t)^2 \approx \sigma_1^2 \delta t$$
$$E[(\delta X_2)^2] = (p_{uu} + p_{ud} + p_{du} + p_{dd})(\delta x_2)^2 = \sigma_2^2 \delta t + \nu_2^2(\delta t)^2 \approx \sigma_2^2 \delta t,$$

where we have used the usual identity $\mathrm{Var}(X) = E[X^2] - E^2[X]$ and we have neglected higher-order terms in δt. These equations are immediately simplified, since probabilities must add up to 1:

$$\delta x_1 = \sigma_1 \sqrt{\delta t}, \qquad \delta x_2 = \sigma_2 \sqrt{\delta t}.$$

We should also account for covariance or, equivalently, for the cross product:

$$\begin{aligned} E[\delta X_1 \cdot \delta X_2] &= (p_{uu} - p_{ud} - p_{du} + p_{dd})\,\delta x_1\, \delta x_2 \\ &= \rho \sigma_1 \sigma_2\, \delta t + \nu_1 \nu_2 (\delta t)^2 \approx \rho \sigma_1 \sigma_2\, \delta t. \end{aligned}$$

Now we have a system of four equations with four unknown probabilities:

$$p_{uu} + p_{ud} - p_{du} - p_{dd} = \frac{\nu_1 \sqrt{\delta t}}{\sigma_1}$$

$$p_{uu} - p_{ud} + p_{du} - p_{dd} = \frac{\nu_2 \sqrt{\delta t}}{\sigma_2}$$

$$p_{uu} - p_{ud} - p_{du} + p_{dd} = \rho$$

$$p_{uu} + p_{ud} + p_{du} + p_{dd} = 1.$$

These equations may be solved by inverting the matrix numerically, or by taking suitable linear combinations of equations:

$$\begin{bmatrix} 1 & 1 & -1 & -1 \\ 1 & -1 & 1 & -1 \\ 1 & -1 & -1 & 1 \\ 1 & 1 & 1 & 1 \end{bmatrix}^{-1} = \frac{1}{4} \begin{bmatrix} 1 & 1 & 1 & 1 \\ 1 & -1 & -1 & 1 \\ -1 & 1 & -1 & 1 \\ -1 & -1 & 1 & 1 \end{bmatrix},$$

which yields:

$$p_{uu} = \frac{1}{4}\left\{1 + \sqrt{\delta t}\left(\frac{\mu_1}{\sigma_1} + \frac{\mu_2}{\sigma_2}\right) + \rho\right\}$$

$$p_{ud} = \frac{1}{4}\left\{1 + \sqrt{\delta t}\left(\frac{\mu_1}{\sigma_1} - \frac{\mu_2}{\sigma_2}\right) - \rho\right\}$$

$$p_{du} = \frac{1}{4}\left\{1 + \sqrt{\delta t}\left(-\frac{\mu_1}{\sigma_1} + \frac{\mu_2}{\sigma_2}\right) - \rho\right\}$$

$$p_{dd} = \frac{1}{4}\left\{1 + \sqrt{\delta t}\left(-\frac{\mu_1}{\sigma_1} - \frac{\mu_2}{\sigma_2}\right) + \rho\right\}.$$

These conditions have an intuitive interpretation. The probability of having two up jumps is large when the two drifts are large (with respect to the corresponding volatilities) and when correlation is positive. In the probability of an up jump in S_1 and a down jump in S_2, the drift μ_2 occurs with a minus sign (the larger the drift, the less likely a down jump), and negative correlation makes this joint movement more likely. A similar consideration applies to p_{du}, whereas p_{dd} is smaller when drifts are large and is larger when correlation is positive.

The implementation of this bidimensional lattice really requires careful memory management: We cannot simply store a large tridimensional matrix. Since up and down jumps in the two asset prices are the same in absolute value, we may exploit the same ideas we have used in section 7.1.3. The resulting code is displayed in figure 7.11 Input parameters are self-explanatory. First we compute invariant quantities. Note that in the lattice we work with prices, and not their logarithm. Hence, the up jumps are given by

$$u_i = e^{\delta x_i} = e^{\sigma_i \sqrt{\delta t}}$$

and $d_i = 1/u_i$, $i = 1, 2$. Probabilities are discounted outside the main loop. The values of the two underlying assets are stored in two vectors S1vals and S2vals, which work exactly like their counterpart in the vanilla option on one asset. The price of the option is stored in a bidimensional matrix Cvals, which is initialized with the option payoff; here subscript i refers to asset 1, and j refers to asset 2. We can use one matrix for two consecutive time layers because odd- and even-numbered positions are alternatively used for consecutive time layers. Since the option is American, we compute the continuation value hold as a risk-neutral expectation and we compare it against the intrinsic value.

To check the implementation we use the following example[3]:

```
>> S10 = 100;
>> S20 = 100;
>> K = 1;
>> r = 0.06;
>> T = 1;
>> sigma1 = 0.2;
>> sigma2 = 0.3;
>> rho = 0.5;
>> q1 = 0.03;
>> q2 = 0.04;
>> N = 3;
>> AmSpreadLattice(S10,S20,K,r,T,sigma1,sigma2,rho,q1,q2,N)
ans =
   10.0448
```

[3]This is the same example used in [1, pp. 47–51].

```
function price = AmSpreadLattice(S10,S20,K,r,T,sigma1,sigma2,rho,q1,q2,N)
% Precompute invariant quantities
deltaT = T/N;
nu1 = r - q1 - 0.5*sigma1^2;
nu2 = r - q2 - 0.5*sigma2^2;
u1 = exp(sigma1*sqrt(deltaT));
d1 = 1/u1;
u2 = exp(sigma2*sqrt(deltaT));
d2 = 1/u2;
discount = exp(-r*deltaT);
p_uu = discount*0.25*(1 + sqrt(deltaT)*(nu1/sigma1 + nu2/sigma2) + rho);
p_ud = discount*0.25*(1 + sqrt(deltaT)*(nu1/sigma1 - nu2/sigma2) - rho);
p_du = discount*0.25*(1 + sqrt(deltaT)*(-nu1/sigma1 + nu2/sigma2) - rho);
p_dd = discount*0.25*(1 + sqrt(deltaT)*(-nu1/sigma1 - nu2/sigma2) + rho);
% set up S values
S1vals = zeros(2*N+1,1);
S2vals = zeros(2*N+1,1);
S1vals(1) = S10*d1^N;
S2vals(1) = S20*d2^N;
for i=2:2*N+1
    S1vals(i) = u1*S1vals(i-1);
    S2vals(i) = u2*S2vals(i-1);
end
% set up terminal values
Cvals = zeros(2*N+1,2*N+1);
for i=1:2:2*N+1
    for j=1:2:2*N+1
        Cvals(i,j) = max(S1vals(i)-S2vals(j)-K,0);
    end
end
% roll back
for tau=1:N
    for i= (tau+1):2:(2*N+1-tau)
        for j= (tau+1):2:(2*N+1-tau)
            hold = p_uu * Cvals(i+1,j+1) + p_ud * Cvals(i+1,j-1) + ...
                p_du * Cvals(i-1,j+1) + p_dd * Cvals(i-1,j-1);
            Cvals(i,j) = max(hold, S1vals(i) - S2vals(j) - K);
        end
    end
end
price = Cvals(N+1,N+1);
```

Fig. 7.11 MATLAB code for pricing an American spread option by a bidimensional binomial lattice.

Clearly, three steps are not enough to get an acceptable approximation, but we may use this toy example to understand how the matrix Cvals is managed to store the lattice, by checking what happens layer by layer. In MATLAB, this can be done by stepping with the debugger, and we display here the essential information we get. The initial lattice is the following; for clarity, we have used an asterisk * to spot irrelevant data (when displaying Cvals with the debugger you would see some number there):

```
10.2473    *      0       *      0       *      0
   *       *      *       *      *       *      *
28.6198    *    3.9982    *      0       *      0
   *       *      *       *      *       *      *
51.7652    *   27.1436    *      0       *      0
   *       *      *       *      *       *      *
80.9233    *   56.3017    *   21.4873    *      0
```

After the first iteration, with one time step to maturity, the relevant data are

```
   *       *      *       *      *       *      *
   *     9.3123   *     0.5653   *       0      *
   *       *      *       *      *       *      *
   *    28.2778   *     5.3263   *       0      *
   *       *      *       *      *       *      *
   *    54.2561   *    25.8626   *    3.0381    *
   *       *      *       *      *       *      *
```

Note that the new values are obtained as averages of four neighboring values which store data for the next time layer. Then, going back one step, we have

```
   *       *      *       *      *       *      *
   *       *      *       *      *       *      *
   *       *    9.4563    *    0.9635    *      *
   *       *      *       *      *       *      *
   *       *   28.1353    *    6.7420    *      *
   *       *      *       *      *       *      *
   *       *      *       *      *       *      *
```

and the final result is, in the root of the lattice:

```
   *       *      *       *      *       *      *
   *       *      *       *      *       *      *
   *       *      *       *      *       *      *
   *       *      *   10.0448    *       *      *
   *       *      *       *      *       *      *
   *       *      *       *      *       *      *
   *       *      *       *      *       *      *
```

We may see that we are working with a sort of recursive pyramidal structure, which suffers from a small but acceptable memory waste.

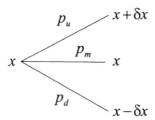

Fig. 7.12 Single-period trinomial lattice.

7.4 PRICING BY TRINOMIAL LATTICES

The idea of a trinomial lattice arises quite naturally as a generalization of binomial lattices. Each node has three successors, corresponding to the price going up, down, or staying the same (this is just one possible choice, actually). The lattice is calibrated in such a way to allow for recombination and to match the first two moments of the underlying continuous random variables. The additional degrees of freedom may be used to improve convergence or to impose additional conditions. A situation in which this may be useful is pricing a barrier option; in such a case we may require that the barrier price is *on* the lattice.

Here too it is convenient to work with the equation describing the stochastic process $X(t) = \log S(t)$. Over a small time step δt we may move in three directions, corresponding to increments $+\delta x$, 0, or $-\delta x$ in the logarithm of price, corresponding to multiplicative shocks on the price itself. The three alternatives occur with risk-neutral probabilities p_u, p_m, and p_d, respectively. The structure of the branching is shown in figure 7.12. Given the usual equation

$$dX = \nu\, dt + \sigma\, dW,$$

where $\nu = r - \sigma^2/2$, we write the moment matching equations:

$$E[\delta X] = p_u \delta x + p_m 0 - p_d \delta x = \nu\, \delta t$$
$$E[(\delta X)^2] = p_u (\delta x)^2 + p_m 0 + p_d (\delta x)^2 = \sigma^2 \delta t + \nu^2 (\delta t)^2$$
$$p_u + p_m + p_d = 1.$$

Solving this system yields

$$p_u = \frac{1}{2}\left(\frac{\sigma^2 \delta t + \nu^2 (\delta t)^2}{(\delta x)^2} + \frac{\nu \delta t}{\delta x}\right)$$
$$p_u = 1 - \frac{\sigma^2 \delta t + \nu^2 (\delta t)^2}{(\delta x)^2} \qquad (7.7)$$
$$p_u = \frac{1}{2}\left(\frac{\sigma^2 \delta t + \nu^2 (\delta t)^2}{(\delta x)^2} - \frac{\nu\, \delta t}{\delta x}\right),$$

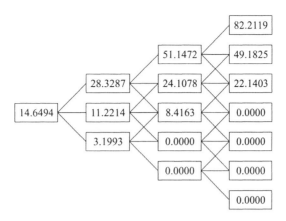

Fig. 7.13 Full example of trinomial lattice.

where we see that an additional degree of freedom is left to choose δx. In fact, it turns out that one cannot choose δx and δt independently. A common rule of thumb is $\delta x = 3\sqrt{\delta t}$. This relationship will be appreciated when we deal with stability of finite difference schemes. We should also note that a careless choice may result in negative probabilities. As an example, consider pricing a European call option on a non dividend paying stock with: $S_0 = 100$, $K = 100$, $r = 0.06$, $T = 1$, and $\sigma = 0.3$. If we build a three-step lattice, with $\delta x = 0.2$, we get the lattice in figure 7.13, where

$$p_u = 0.3878, \qquad p_m = 0.2494, \qquad p_d = 0.3628$$

MATLAB code to accomplish calculations on a trinomial lattice is shown in figure 7.14. As usual, discounted probabilities are computed outside the main for loops. There is just one observation needed here: unlike binomial lattices, we must store at least two consecutive time layers of the lattice, since there is no alternation between odd- and even-indexed entries in the arrays. Hence, we use a two-column array, with $2N + 1$ rows, where the roles of the columns may be "now" or "future." We use increments modulo 2 to swap the roles of the two layers, which are indexed by variables know and kthen, taking the values 1 and 2 alternatively. Here is the computation for the previous lattice:

```
>> S0=100;
>> K=100;
>> r=0.06;
>> T=1;
>> sigma=0.3;
>> N=3;
>> deltaX = 0.2;
>> EuCallTrinomial(S0,K,r,T,sigma,N,deltaX)
ans =
    14.6494
```

```
function price = EuCallTrinomial(S0,K,r,T,sigma,N,deltaX)
% Precompute invariant quantities
deltaT = T/N;
nu = r - 0.5*sigma^2;
discount = exp(-r*deltaT);
p_u = discount*0.5*((sigma^2*deltaT+nu^2*deltaT^2)/deltaX^2 + ...
    nu*deltaT/deltaX);
p_m = discount*(1 - (sigma^2*deltaT+nu^2*deltaT^2)/deltaX^2);
p_d = discount*0.5*((sigma^2*deltaT+nu^2*deltaT^2)/deltaX^2 - ...
    nu*deltaT/deltaX);
% set up S values (at maturity)
Svals = zeros(2*N+1,1);
Svals(1) = S0*exp(-N*deltaX);
exp_dX = exp(deltaX);
for j=2:2*N+1
    Svals(j) = exp_dX*Svals(j-1);
end
% set up lattice and terminal values
Cvals = zeros(2*N+1,2);
t = mod(N,2)+1;
for j=1:2*N+1
    Cvals(j,t) = max(Svals(j)-K,0);
end
for t=N-1:-1:0;
    know = mod(t,2)+1;
    knext = mod(t+1,2)+1;
    for j = N-t+1:N+t+1
        Cvals(j,know) = p_d*Cvals(j-1,knext)+p_m*Cvals(j,knext)+...
                        p_u*Cvals(j+1,knext);
    end
end
price = Cvals(N+1,1);
```

Fig. 7.14 MATLAB code for pricing a European call by a trinomial lattice.

We mentioned that proper choice of δx is an issue here. Playing with numbers as follows we see that the rule of thumb $\delta x = \sigma\sqrt{\delta t}$ does make sense:

```
>> blsprice(S0,K,r,T,sigma)
ans =
   14.7171
>> N=100;
>> deltaX = 0.2;
>> EuCallTrinomial(S0,K,r,T,sigma,N,deltaX)
ans =
   14.0715
>> deltaX = 0.5;
>> EuCallTrinomial(S0,K,r,T,sigma,N,deltaX)
ans =
   10.9345
>> deltaX = sigma*sqrt(T/N);
>> EuCallTrinomial(S0,K,r,T,sigma,N,deltaX)
ans =
   14.6869
>> N=1000;
>> deltaX = sigma*sqrt(T/N);
>> EuCallTrinomial(S0,K,r,T,sigma,N,deltaX)
ans =
   14.7141
```

7.5 SUMMARY

Binomial lattices are typically the first numerical method one meets when learning about option pricing, which is reasonable given the apparent simplicity of the approach. We have preferred to describe the approach at a later stage in order to place it within a more generic framework. Lattices are actually related to Monte Carlo and finite difference methods.

With respect to Monte Carlo methods, binomial and trinomial lattice represent a clever deterministic sampling based on moment matching; moment matching is one of the many variance reduction techniques which have been proposed over the years. An advantage of lattice techniques with respect to Monte Carlo simulation is computational speed, when the problem dimensionality is small. Lattice methods are not easily applied when complex path dependencies are built in the option. Clever techniques may be used and have been proposed, e.g., for lookback options, but they may suffer from poor convergence. Hence, for complex and/or high dimensional options, Monte Carlo simulation can well be the only practical approach. On the other side of the coin, lattice methods easily deal with early exercise features.

Some authors regard explicit finite difference schemes as a generalization of trinomial lattices. In fact, this will be apparent in section 9.2.1, where we see that numerical instability in an explicit scheme is linked to a bad discretization, essentially leading to a trinomial lattice with negative probabilities. Hence, it may be argued that the additional flexibility of grids and the possibility to use implicit and accurate schemes may supersede lattice techniques. But actually, as we have already pointed out, this is sometimes a matter of taste. With good calibrations (we have just scratched the surface here), accurate pricing may be obtained by lattice techniques in many practical cases.

We should also point out that we have worked under the idealized assumption of complete markets, deterministic volatility, etc. Furthermore, we have basically worked with the historical volatility, whereas we know that implied volatility is often considered as the relevant one. Lattice techniques have been proposed which are calibrated against market prices, resulting in the so-called *implied lattices*. We refer to the literature for more on this advanced topics, but we should keep in mind that the conceptual simplicity and the computational efficiency of lattice methods may be extremely useful to generalize option pricing beyond the Black–Scholes framework.

For further reading

- A very good source on lattice techniques is [1]. There you may also find a careful analysis of the relationship between finite differences and trinomial lattices.

- The classical reference [3] includes many variations on the basic techniques we have considered, including lattices for barrier and lookback options, adaptive node placement, etc. In the chapters on numerical methods you may also find the background on pricing options on stocks paying discrete dividends by binomial lattices. This is the basis of the implementation provided by the Financial toolbox function `binprice`.

- You may also consult [4], which also includes implied lattices and ideas for efficient implementations.

- If you want to dig deeply into the issue of implementing binomial lattices in MATLAB, you should have a look at [2].

REFERENCES

1. L. Clewlow and C. Strickland. *Implementing Derivatives Models*. Wiley, Chichester, West Sussex, England, 1998.

2. D.J. Higham. Nine Ways to Implement the Binomial Method for Option Valuation in MATLAB. *SIAM Review*, 44:661–677, 2002.

3. J.C. Hull. *Options, Futures, and Other Derivatives (5th ed.)*. Prentice Hall, Upper Saddle River, NJ, 2003.

4. G. Levy. *Computational Finance. Numerical Methods for Pricing Financial Instruments*. Elsevier Butterworth-Heinemann, Oxford, 2004.

5. D.G. Luenberger. *Investment Science*. Oxford University Press, New York, 1998.

8

Option Pricing by Monte Carlo Methods

Monte Carlo simulation is an important tool in computational finance: It may be used to evaluate portfolio management rules, to price options, to simulate hedging strategies, and to estimate Value at Risk. Its main advantages are generality, relative ease of use, and flexibility. It may take stochastic volatility and many complicating features of exotic options into account, and it lends itself to treating high-dimensional problems, where the lattice and PDE framework cannot be applied. The potential disadvantage of Monte Carlo simulation is its computational burden. An increasing number of replications is needed to refine the confidence interval of the estimates we are interested in. The problem may be partially solved by variance reduction techniques or by resorting to low-discrepancy sequences. The aim of this chapter is to illustrate the application of these techniques to a few examples, including some path-dependent options. This chapter is a direct extension of chapter 4, where we dealt with Monte Carlo integration. It must be emphasized that even if we use the more appealing terms "simulation" or "sampling,", Monte Carlo methods are conceptually a numerical integration tool. This must be kept in mind when applying low-discrepancy sequences rather than pseudo-random generators.

When possible, we will compare the results of simulation with analytical formulas. Clearly, our aim in doing so is a purely didactic one. If you have to compute the area of a rectangular room, you just multiply the room length times the room width; you would never count how many times a standard tile fits the surface. However, you should learn first to use simulation in easy cases, where we may check the consistency of results; moreover, we will also see that simulating options for which analytical formulas are available may yield powerful control variates for variance reduction purposes.

The starting point in the application of Monte Carlo simulation is sample path generation, given a stochastic differential equation describing the dynamics of a price (or an interest rate). In section 8.1 we illustrate path generation for geometric Brownian motion; two hedging strategies are simulated as a concrete example, and we also deal with Brownian bridge, which is an alternative to simulating sample paths by going forward in time. Section 8.2 deals with an exchange option, which is used as a simple illustration of how the approach can be extended to multidimensional processes. In section 8.3 we consider an example of weakly path-dependent option, a down-and-out put option; we apply both conditional Monte Carlo and importance sampling to reduce variance. A strongly path-dependent option is dealt with in section 8.4, where we show the application of control variates and low-discrepancy sequences to pricing an arithmetic average Asian option. We close the chapter by outlining the basic issues in estimating option sensitivities by Monte Carlo sampling; in section 8.5 we consider the simple case of the option Δ for a vanilla call.

Another application of stochastic simulation to option pricing is given in section 10.4, which is dedicated to American options; the early exercise feature makes a straightforward simulation approach infeasible, and the problem must be cast within the framework of stochastic dynamic optimization.

8.1 PATH GENERATION

The starting point for the application of Monte Carlo methods to option pricing is the generation of sample paths of the underlying factors. In vanilla options, there is really no need for path generation, as we have seen in chapter 4: only the price of the underlying asset at maturity is of concern. But if the option is path-dependent, we need the whole path or, at least, a sequence of values at given time instants. With geometric Brownian motion, we are facing a very lucky case. In fact, it must be understood that we have two potential sources of errors in path generation:

- sampling error,

- discretization error.

Sampling error is due to random nature of Monte Carlo methods, and it can be mitigated using variance reduction strategies. To understand what discretization error is, let us consider how we can discretize a typical continuous-time model, i.e., an Ito stochastic differential equation:

$$dS_t = a(S_t, t)\,dt + b(S_t, t)\,dW_t.$$

The simplest discretization approach, known as Euler scheme, yields the following discrete-time model:

$$\delta S_t = S_{t+\delta t} - S_t = a(S_t, t)\delta t + b(S_t, t)\sqrt{\delta t}\,\epsilon,$$

where δt is the discretization step and $\epsilon \sim \mathcal{N}(0, 1)$. This scheme is conceptually linked to finite differences and its application to a deterministic differential equation would yield a truncation error, which is arguably negligible when the discretization step is small.[1] Convergence is a critical concept in stochastic differential equations, as we are dealing with stochastic processes, but we may guess that, by sampling realizations of the random variable ϵ from the standard normal distribution, we should be able to simulate a discrete-time stochastic process which is well related to the solution of the continuous-time equation. Increasing the number of sample paths, or replications, we should also be able to reduce sampling error.

While the above reasoning can be justified more formally, we should realize that the discretization error may even change the probability distributions characterizing the solution. For instance, consider the geometric Brownian motion model:

$$dS_t = \mu S_t dt + \sigma S_t dW_t. \qquad (8.1)$$

The Euler scheme yields

$$S_{t+\delta t} = (1 + \mu\,\delta t)S_t + \sigma S_t \sqrt{\delta t}\,\epsilon.$$

This is very easy to grasp and to implement, but the marginal distribution of each value $S_i = S(i\,\delta t)$ is normal, rather than lognormal. Actually, taking a very small δt we may reduce the error, but this is time consuming. In this specific case, we may get rid of the discretization error altogether by a straightforward application of Ito's lemma, but this is not true in general. With complicated stochastic differential equations, we may have to generate the whole sample path, even if we are only interested in values at maturity, just to reduce the discretization error. In such a case, it may be advisable to use the more refined discretization schemes available in the literature.

8.1.1 Simulating geometric Brownian motion

Using Ito's lemma, we may transform (8.1) into the following form:

$$d\log S_t = \left(\mu - \frac{1}{2}\sigma^2\right)dt + \sigma\,dW_t. \qquad (8.2)$$

We also recall that, using properties of the lognormal distribution[2] and letting $\nu = \mu - \sigma^2/2$, we obtain:

---

[1] We have seen in chapter 5 that convergence of discretization schemes is not that trivial.
[2] See appendix B.

$$\begin{aligned}
E[\log(S(t)/S(0))] &= \nu t \\
\mathrm{Var}[\log(S(t)/S(0))] &= \sigma^2 t \\
E[S(t)/S(0)] &= e^{\mu t} \qquad (8.3)\\
\mathrm{Var}[S(t)/S(0)] &= e^{2\mu t}(e^{\sigma^2 t} - 1). \qquad (8.4)
\end{aligned}$$

Equation (8.2) is particularly useful as it can be integrated exactly, yielding:

$$S_t = S_0 \exp\left(\nu t + \sigma \int_0^t dW(\tau)\right).$$

To simulate the path of the asset price over an interval $(0, T)$, we must discretize time with a time step δt. From the last equation, and recalling the properties of the standard Wiener process (see section 2.5), we get

$$S_{t+\delta t} = S_t \exp\left(\nu\, \delta t + \sigma\sqrt{\delta t}\,\epsilon\right), \qquad (8.5)$$

where $\epsilon \sim N(0,1)$ is a standard normal random variable. Based on equation (8.5), it is easy to generate sample paths for the asset price.

A straightforward code to generate sample paths of asset prices following geometric Brownian motion is given in figure 8.1. The function `AssetPaths` yields a matrix of sample paths, where the replications are stored row by row and columns correspond to time instants. The first column contains the same value, the initial price, for all sample paths. We have to provide the function with the initial price `S0`, the drift `mu`, the volatility `sigma`, the time horizon `T`, the number of time steps `NSteps`, and the number of replications `NRepl`. Note that the function takes the drift parameter μ as input and then it computes the parameter ν.

For instance, let us generate and plot three one-year sample paths for an asset with an initial price $50, drift 0.1, and volatility 0.3 (on a yearly basis), assuming that the time step is one day[3]:

```
>> randn('state',0);
>> paths=AssetPaths(50,0.1,0.3,1,365,3);
>> plot(1:length(paths),paths(1,:))
>> hold on
>> plot(1:length(paths),paths(2,:))
>> hold on
>> plot(1:length(paths),paths(3,:))
```

The result is plotted in figure 8.2. If you start the random number generator

---

[3]We assume here that a year consists of 365 trading days. How to treat non-trading days is a bit controversial (see, e.g., [11, pp. 251–252]).

```
function SPaths=AssetPaths(S0,mu,sigma,T,NSteps,NRepl)
SPaths = zeros(NRepl, 1+NSteps);
SPaths(:,1) = S0;
dt = T/NSteps;
nudt = (mu-0.5*sigma^2)*dt;
sidt = sigma*sqrt(dt);
for i=1:NRepl
   for j=1:NSteps
      SPaths(i,j+1)=SPaths(i,j)*exp(nudt + sidt*randn);
   end
end
```

Fig. 8.1 MATLAB code to generate asset price paths by Monte Carlo simulation.

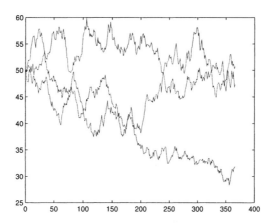

Fig. 8.2 Sample paths generated by Monte Carlo simulation.

```
function SPaths=AssetPathsV(S0,mu,sigma,T,NSteps,NRepl)
dt = T/NSteps;
nudt = (mu-0.5*sigma^2)*dt;
sidt = sigma*sqrt(dt);
Increments = nudt + sidt*randn(NRepl, NSteps);
LogPaths = cumsum([log(S0)*ones(NRepl,1) , Increments] , 2);
SPaths = exp(LogPaths);
Spaths(:,1) = S0;
```

Fig. 8.3 Vectorized code to generate asset price paths.

for standard normals `randn` with another seed state, you will get different results.

The code in figure 8.1 is based on two nested `for` loops. Sometimes, efficiency can be achieved in MATLAB by vectorizing code. In order to vectorize the code, it is convenient to rewrite equation (8.5) as

$$\log S_{t+\delta t} - \log S_t = \nu\,\delta t + \sigma\sqrt{\delta t}\,\epsilon.$$

We may generate the *differences* in the logarithm of the asset prices and then use the `cumsum` function with an optional parameter set to 2 in order to compute the cumulative sums over the rows (the default is summing over columns). The resulting function `AssetPathsV` is illustrated in figure 8.3. We should note that in the last line we write the initial asset price in the first column. To see why, check the following snapshot:

```
>> format long
>> exp(log(50))
ans =
   49.99999999999999
```

It is better to avoid this error (which is apparently negligible, but see later). We may compare the two implementations in terms of speed:

```
>> tic, paths=AssetPaths(50,0.1,0.3,1,100,1000);, toc
Elapsed time is 0.029226 seconds.
>> tic, paths=AssetPathsV(50,0.1,0.3,1,100,1000);, toc
Elapsed time is 0.034177 seconds.
```

In this case we do not see advantages in vectorizing code. We should keep in mind that the elapsed time returned by `tic` and `toc` is subject to some variability due to the background tasks carried out by the operating system, but when the first edition of this book was written, there was a striking advantage in vectorizing code. And, in fact, there are many situations in which this is true. The point is that improvements in hardware and in software (in this case, MATLAB interpreter has been arguably improved) may make certain programming practices obsolete. Sometimes a fully vectorized code requires huge matrices, which do not fit the main memory of the computer. In such a case, using disk space as virtual memory may slow the execution. So we should be aware of all possible tricks of the trade, but an empirical efficiency check has the ultimate say in the matter.

8.1.2 Simulating hedging strategies

Armed with a function to generate sample paths, we may try a first experiment in comparing hedging strategies for a vanilla European call option. We know from chapter 2 that the option price is essentially the cost of a delta-hedging strategy and that the continuous-time hedging strategy for the call option requires holding an amount Δ of the underlying asset. A simpler strategy is the stop-loss strategy.[4] The idea is that we should have a covered position (hold one share) when the option is in the money, and a naked position (hold no share) when it is out of the money. In practice, we could buy a share when the asset price goes above the strike price K, and we should sell it when it goes below. This strategy makes intuitive sense, but it is not that trivial to analyze in continuous time.[5] Nevertheless, we may evaluate its performance in discrete time by Monte Carlo simulation. The problem with an implementation in discrete time is that we cannot really buy or sell at the strike price: We buy at a price larger than K, when we detect that the price went above that critical value, and we sell at a price which is slightly lower. So, even without considering transaction costs, that would affect delta-hedging as well, we see a potential trouble with the stop-loss strategy.

A MATLAB function to estimate the average cost of a stop-loss strategy is given in figure 8.4. The function receives a matrix of sample Paths, possibly generated by function AssetPaths. Note that in this case, unlike option pricing, the real drift mu must be used in the simulation. In checking the code, we should note that the true number of steps (time intervals) is one less the number of columns in matrix Paths. If we need to buy shares of the underlying stock, we may need to borrow money, which should be taken into account. But, since we assume deterministic and constant interest rates, we will not account for borrowed money, since we can simply record cash flows from trading and discount them back to time $t = 0$, having precomputed discount factors in the vector DiscountFactors. We use a state variable Covered to detect when we cross the strike price going up or down. Since cash flow is negative when we buy, and positive when we sell, the option "price" is evaluated as the average total discounted cash flow, with a change in sign. We should also pay attention to what happens at maturity: If the

---

[4]See [11, pp. 300–302].
[5]See [5].

```
function P = StopLoss(S0,K,mu,sigma,r,T,Paths)
[NRepl,NSteps] = size(Paths);
NSteps = NSteps - 1; % true number of steps
Cost = zeros(NRepl,1);
dt = T/NSteps;
DiscountFactors = exp(-r*(0:1:NSteps)*dt);
for k=1:NRepl
    CashFlows = zeros(NSteps+1,1);
    if (Paths(k,1) >= K)
        Covered = 1;
        CashFlows(1) = -Paths(k,1);
    else
        Covered = 0;
    end
    for t=2:NSteps+1
        if (Covered == 1) & (Paths(k,t) < K)
            % Sell
            Covered = 0;
            CashFlows(t) = Paths(k,t);
        elseif (Covered == 0) & (Paths(k,t) > K)
            % Buy
            Covered = 1;
            CashFlows(t) = -Paths(k,t);
        end
    end
    if Paths(k,NSteps + 1) >= K
        % Option is exercised
        CashFlows(NSteps + 1) = ...
            CashFlows(NSteps + 1) + K;
    end
    Cost(k) = -dot(DiscountFactors, CashFlows);
end
P = mean(Cost);
```

Fig. 8.4 Evaluating the cost of a stop-loss hedging strategy.

option is in the money, the option holder will exercise her right and we will also get the strike price, which should be included in the cash flows.

Since we know that sometimes vectorizing code is beneficial, we also show a vectorized version of this code in figure 8.5. The main trick here is using a vector OldPrice, which is essentially a shifted copy of Paths to spot where the price crosses the critical level, going up or down. Times at which we go up are recorded in vector UpTimes, and there we have a negative cash flow; a similar consideration applies to DownTimes.

```
function P = StopLossV(S0,K,mu,sigma,r,T,Paths)
[NRepl,NSteps] = size(Paths);
NSteps = NSteps - 1;
Cost = zeros(NRepl,1);
CashFlows = zeros(NRepl,NSteps+1);
dt = T/NSteps;
DiscountFactors = exp(-r*(0:1:NSteps)*dt);
OldPrice = [zeros(NRepl,1), Paths(:,1:NSteps)];
UpTimes = find(OldPrice < K & Paths >= K);
DownTimes = find(OldPrice >= K & Paths < K);
CashFlows(UpTimes) = -Paths(UpTimes);
CashFlows(DownTimes) = Paths(DownTimes);
ExPaths = find(Paths(:,NSteps+1) >= K);
CashFlows(ExPaths,NSteps+1) = CashFlows(ExPaths,NSteps+1) + K;
Cost = -CashFlows*DiscountFactors';
P = mean(Cost);
```

Fig. 8.5 Vectorized code for the stop-loss hedging strategy.

Now we may check if the two function are actually consistent, and if there is any advantage in vectorization:

```
>> S0 = 50;
>> K = 50;
>> mu = 0.1;
>> sigma = 0.4;
>> r = 0.05;
>> T = 5/12;
>> NRepl =100000;
>> NSteps = 10;
>> randn('state',0);
>> Paths=AssetPaths(S0,mu,sigma,T,NSteps,NRepl);
>> tic, StopLoss(S0,K,mu,sigma,r,T,Paths), toc
ans =
    5.5780
Elapsed time is 3.100619 seconds.
>> tic, StopLossV(S0,K,mu,sigma,r,T,Paths), toc
ans =
    5.5780
Elapsed time is 0.735455 seconds.
```

Here, unlike asset path generation, we see an advantage from vectorization. We may also appreciate here why, with the vectorized function to generate asset paths, it may be important to assign the initial asset price correctly, as we did in the last line of the code in figure 8.3. In this case the option is at

```
function P = DeltaHedging(S0,K,mu,sigma,r,T,Paths)
[NRepl,NSteps] = size(Paths);
NSteps = NSteps - 1;
Cost = zeros(NRepl,1);
CashFlows = zeros(1,NSteps+1);
dt = T/NSteps;
DiscountFactors = exp(-r*(0:1:NSteps)*dt);
for i=1:NRepl
    Path = Paths(i,:);
    Position = 0;
    Deltas = blsdelta(Path(1:NSteps),K,r,T-(0:NSteps-1)*dt,sigma);
    for j=1:NSteps
        CashFlows(j) = (Position - Deltas(j))*Path(j);
        Position = Deltas(j);
    end
    if Path(NSteps+1) > K
        CashFlows(NSteps+1) = K - (1-Position)*Path(NSteps+1);
    else
        CashFlows(NSteps+1) = Position*Path(NSteps+1);
    end
    Cost(i) = -CashFlows*DiscountFactors';
end
P = mean(Cost);
```

Fig. 8.6 Evaluating the performance of delta-hedging.

the money, and we always buy the stock initially; but if the initial stock price is 49.9999 we don't do that, and an apparently negligible error has serious consequences in the analysis.

Now we should compare the cost of the stop-loss strategy with the cost of delta-hedging, and with the theoretical option price. A code to estimate the average cost of delta-hedging is displayed in figure 8.6. The code is similar to the stop-loss strategy, but it is not vectorized. The only vectorization we have done is in calling `blsdelta` once to get the option Δ for each point on the sample path. Note that Δ must be computed using the current asset price and the current time to maturity; we use the `blsdelta` function of the Financial toolbox. The current `Position` in the stock is updated given the new Δ, generating cash flows which are discounted.

Figure 8.7 displays a script to compare performances of the two hedging strategies. Running the script, we get the following:

```
>> HedgingScript
true price =   4.732837
cost of stop/loss (S) = 4.826756
cost of delta-hedging = 4.736975
```

```
% HedgingScript.m
S0 = 50;
K = 52;
mu = 0.1;
sigma = 0.4;
r = 0.05;
T = 5/12;
NRepl =10000;
NSteps = 10;
%
C = blsprice(S0,K,r,T, sigma);
fprintf(1, '%s %f\n', 'true price = ', C);
%
randn('state',0);
Paths=AssetPaths(S0,mu,sigma,T,NSteps,NRepl);
SL = StopLossV(S0,K,mu,sigma,r,T,Paths);
fprintf(1, 'cost of stop/loss (S) = %f\n', SL);
DC = DeltaHedging(S0,K,mu,sigma,r,T,Paths);
fprintf(1, 'cost of delta-hedging = %f\n', DC);
%
NSteps = 100;
randn('state',0);
Paths=AssetPaths(S0,mu,sigma,T,NSteps,NRepl);
SL = StopLossV(S0,K,mu,sigma,r,T,Paths);
fprintf(1, 'cost of stop/loss (S) = %f\n', SL);
DC = DeltaHedging(S0,K,mu,sigma,r,T,Paths);
fprintf(1, 'cost of delta-hedging = %f\n', DC);
```

Fig. 8.7 A script to compare hedging strategies.

```
cost of stop/loss (S) = 4.828571
cost of delta-hedging = 4.735174
```

where in the first pair of runs we use ten hedging steps and one hundred in the second pair. We see that the stop-loss strategy does not seem to converge to the true option price, unlike the cost of delta-hedging. Actually, the comparison should be made in different settings, and it should also involve the variability of the hedging cost.

8.1.3 Brownian bridge

In the previous sections we generated asset paths according to a natural process, which proceeds forward in time. Actually, the Wiener process enjoys some peculiar properties which allow us to generate the sample paths in a different way. Consider a time interval with left and right end points t_l and t_r, respectively, and an intermediate time instant s, such that $t_l < s < t_r$. In standard path generation, we would generate the Wiener process in the natural order: $W(t_l)$, $W(s)$, and finally $W(t_r)$. Using the so-called Brownian bridge, we may generate $W(s)$ conditional on the values $w_l = W(t_l)$ and $w_r = W(t_r)$. It can be shown that $W(s)$, conditional on those two values, is a normal variable with expected value

$$\frac{(t_r - s)w_l + (s - t_l)w_r}{t_r - t_l}$$

and variance

$$\frac{(t_r - s)(s - t_l)}{t_r - t_l}.$$

This is a consequence of some properties of the conditional distribution of a multivariate normal distribution. We will not prove the formulas above,[6] but they are fairly intuitive: The conditional expected value of $W(s)$ is obtained by linear interpolation through w_l and w_r; the variance is low near the two end points t_l and t_r, and is maximum in the middle of the interval.

Using Brownian bridge, we may generate sample paths by a sort of bisection strategy. Given $W(0) = 0$, we sample $W(T)$; then we sample $W(T/2)$. Given $W(0)$ and $W(T/2)$ we sample $W(T/4)$; given $W(T/2)$ and $W(T)$ we sample $W(3T/4)$, etc. Actually, we may generate sample paths in any order we wish, with non-homogeneous time steps. One could wonder why this complicated construction could be useful. There are at least two reasons.

1. It may help in using variance reduction by stratification. It is difficult to use stratification in multiple dimensions, but we may use stratification just on the terminal value of asset price, and maybe an intermediate point. Then we generate intermediate values using the bridge.

2. The Brownian bridge construction is also useful when used in conjunction with low-discrepancy sequences. We have seen in section 4.6 that application of simple low discrepancy sequences in high-dimensional domains may be difficult, because some dimensions are not well-covered. Using Brownian bridge, we may use high quality sequences to outline the paths of the Wiener process, by sampling points acting as milestones; then we can fill the trajectory using other sequences, or even Monte Carlo sampling.

In figure 8.8 we illustrate a MATLAB function to generate paths of the standard Wiener process using the Brownian bridge technique, but only in the specific case in which the time interval $[0, T]$ is bisected (i.e., the number of

---

[6]See, e.g., [8, pp. 82–84] for a readable proof.

intervals is a power of 2); the technique may be applied in more general settings. In this case, we may simplify the formulas above to sample $W(s)$; if we let $\delta t = t_l - t_r$, we have

$$W(S) = \frac{w_r + w_l}{2} + \frac{1}{2}\sqrt{\delta t}Z,$$

where Z is a standard normal variable. The function receives the length T of the time interval and the number NSteps of sub-intervals in which it must be partitioned, and it returns a vector containing one sample path. Assume that the number of intervals is 8 (it must be a power of two). Then we must carry out 3 bisections.

- Given the initial condition $W(t_0) = 0$, we must first sample $W(t_8)$, which means "jumping" over an interval of length T, which is TJump in the program. Since we store elements in a vector of nine elements (starting with index 1, and including $W(t_0)$), we must jump eight places in this vector to store the new value. The number of places to jump is stored in IJump.

- Then we start the first for loop. In the first pass we must only sample $W(t_4)$, given $W(t_0)$ and $W(t_8)$. Given positions left = 1 and right = IJump + 1 we must generate a new value and store that value in position i = IJump/2 + 1, which is 4+1 = 5 in this case. Here we generate only one value, and we divide both jumps by 2.

- In the second iteration we must sample $W(t_2)$, given $W(t_0)$ and $W(t_4)$, and $W(t_6)$, given $W(t_4)$ and $W(t_8)$. The nested loop will be executed twice, and indexes left, right, and i are incremented by 4.

- In the third and final iteration we generate the remaining four values.

We urge the reader to step through the function using the debugger to verify the pattern we have just described. In figure 8.8 we also give a script to check that the marginal distributions of the stochastic process we generate are the correct ones. Expected values should be zero, and standard deviation should be the square root of time:

```
>> CheckBridge
m =
    0.0025    0.0015    0.0028    0.0030
sdev =
    0.5004    0.7077    0.8646    0.9964
ans =
    0.5000    0.7071    0.8660    1.0000
```

We see that, apart from sampling errors, the result looks correct. Given a way to generate standard Wiener process, it is easy to simulate geometric Brownian motion. The function is given in figure 8.9, and it uses a similar approach as

```
function WSamples = WienerBridge(T, NSteps)
NBisections = log2(NSteps);
if round(NBisections) ~= NBisections
    fprintf('ERROR in WienerBridge: NSteps must be a power of 2\n');
    return
end
WSamples = zeros(NSteps+1,1);
WSamples(1) = 0;
WSamples(NSteps+1) = sqrt(T)*randn;
TJump = T;
IJump = NSteps;
for k=1:NBisections
    left = 1;
    i = IJump/2 + 1;
    right = IJump + 1;
    for j=1:2^(k-1)
        a = 0.5*(WSamples(left) + WSamples(right));
        b = 0.5*sqrt(TJump);
        WSamples(i) = a + b*randn;
        right = right + IJump;
        left = left + IJump;
        i = i + IJump;
    end
    IJump = IJump/2;
    TJump = TJump/2;
end
```

```
% CheckBridge.m
randn('state',0);
NRepl = 100000;
T = 1;
NSteps = 4;
WSamples = zeros(NRepl, 1+NSteps);
for i=1:NRepl
    WSamples(i,:) = WienerBridge(T, NSteps);
end
m = mean(WSamples(:,2:(1+NSteps)))
sdev = sqrt(var(WSamples(:,2:(1+NSteps))))
sqrt((1:NSteps).*T/NSteps)
```

Fig. 8.8 Implementing and checking path generation for the standard Wiener process by a Brownian bridge.

```
function SPaths = GBMBridge(S0, mu, sigma, T, NSteps, NRepl)
if round(log2(NSteps)) ~= log2(NSteps)
    fprintf('ERROR in GBMBridge: NSteps must be a power of 2\n');
    return
end
dt = T/NSteps;
nudt = (mu-0.5*sigma^2)*dt;
SPaths = zeros(NRepl, NSteps+1);
for k=1:NRepl
    W = WienerBridge(T,NSteps);
    Increments = nudt + sigma*diff(W');
    LogPath = cumsum([log(S0) , Increments]);
    SPaths(k,:) = exp(LogPath);
end
Spaths(:,1) = S0;
```

Fig. 8.9 Sampling geometric Brownian motion by a Brownian bridge.

the vectorized version **AssetPathsV**. One thing we should note is the use of the function diff to generate the vector of Increments in the logarithmic asset price. In fact, in standard Monte Carlo we generate the underlying Wiener process by successive increments; with the Brownian bridge construction we build directly the values of the process at different time instants, and we must use the function diff to obtain the relative differences. In some sense, diff works in the opposite way to cumsum, as shown by the following example:

```
>> diff([1 5 7 10 20])
ans =
     4     2     3    10
```

8.2 PRICING AN EXCHANGE OPTION

The purpose of this section is to show that Monte Carlo simulation may be easily adapted to multidimensional options. We will use a very simple example, so that we may compare our estimates against the exact value for illustration purposes. We want to price a European-style exchange option written on two assets whose price, under the risk neutral measure, is modeled as a bidimensional geometric Brownian motion:

$$dU(t) = rU(t)\,dt + \sigma_U U(t)\,dW_U(t)$$
$$dV(t) = rV(t)\,dt + \sigma_V V(t)\,dW_V(t),$$

```
function p = Exchange(V0,U0,sigmaV,sigmaU,rho,T,r)
sigmahat = sqrt(sigmaU^2 + sigmaV^2 - 2*rho*sigmaU*sigmaV);
d1 = (log(V0/U0) + 0.5*T*sigmahat^2)/(sigmahat*sqrt(T));
d2 = d1 - sigmahat*sqrt(T);
p = V0*normcdf(d1) - U0*normcdf(d2);
```

Fig. 8.10 Code to price an exchange option analytically.

where the two Wiener processes have instantaneous correlation ρ. The option payoff at maturity T is $\max(V_T - U_T, 0)$. We see that this option is a particular case of a spread option, whose payoff depends on the difference between two asset prices (we considered an American spread option in section 7.3). It is called "exchange" because it allows us to exchange one asset for the other at maturity. For instance, if we hold asset U and one exchange option, the payoff at maturity will be

$$U_T + \max(V_T - U_T, 0) = \max(V_T, U_T).$$

For this option, there is an analytical pricing formula which is a fairly straightforward generalization of the Black–Scholes formula:

$$\begin{aligned} P &= V_0 N(d_1) - U_0 N(d_2) \\ d_1 &= \frac{\ln(V_0/U_0) + \hat{\sigma}^2 T/2}{\hat{\sigma}\sqrt{T}} \\ d_2 &= d_1 - \hat{\sigma}\sqrt{T} \\ \hat{\sigma} &= \sqrt{\sigma_V^2 + \sigma_U^2 - 2\rho\sigma_V\sigma_U} \end{aligned}$$

The reason why we get this type of formula is that the payoff has a homogeneous form, which allows to simplify the corresponding partial differential equation by considering the ratio V/U of the two prices.[7] MATLAB code implementing this formula is shown in figure 8.10.

The only point we need to consider in order to apply Monte Carlo is how to generate sample paths for two correlated Wiener processes. We may apply the same idea we have seen in section 4.3.4 for the multivariate normal distribution. We should find the Cholesky factor for the covariance matrix corresponding to two standard normal variables with correlation ρ:

$$\Sigma = \begin{bmatrix} 1 & \rho \\ \rho & 1 \end{bmatrix}$$

---

[7] See, e.g., [2, pp. 184–188] for a proof.

PRICING AN EXCHANGE OPTION   445

```
function [p,ci] = ExchangeMC(V0,U0,sigmaV,sigmaU,rho,T,r,NRepl)
eps1 = randn(1,NRepl);
eps2 = rho*eps1 + sqrt(1-rho^2)*randn(1,NRepl);
VT = V0*exp((r - 0.5*sigmaV^2)*T + sigmaV*sqrt(T)*eps1);
UT = U0*exp((r - 0.5*sigmaU^2)*T + sigmaU*sqrt(T)*eps2);
DiscPayoff = exp(-r*T)*max(VT-UT, 0);
[p,s,ci] = normfit(DiscPayoff);
```

Fig. 8.11 Code to price an exchange option by Monte Carlo simulation.

It may be verified by straightforward multiplication that $\Sigma = \mathbf{LL}'$, where

$$\mathbf{L} = \begin{bmatrix} 1 & 0 \\ \rho & \sqrt{1-\rho^2} \end{bmatrix}.$$

Hence, to simulate bidimensional correlated Wiener processes, we must generate two independent standard normal variates Z_1 and Z_2 and use

$$\begin{aligned} \epsilon_1 &= Z_1 \\ \epsilon_2 &= \rho Z_1 + \sqrt{1-\rho^2} Z_2 \end{aligned}$$

to drive path generation.

In our case, we only need to generate joint samples of the two asset prices at maturity. The resulting MATLAB code is displayed in figure 8.11. We may check our results as usual:

```
>> V0 = 50;
>> U0 = 60;
>> sigmaV = 0.3;
>> sigmaU = 0.4;
>> rho = 0.7;
>> T = 5/12;
>> r = 0.05;
>> Exchange(V0,U0,sigmaV,sigmaU,rho,T,r)
ans =
    0.8633
>> NRepl = 200000;
>> randn('state', 0)
>> [p,ci] = ExchangeMC(V0,U0,sigmaV,sigmaU,rho,T,r,NRepl)
p =
    0.8552
ci =
    0.8444
    0.8660
```

```
% DOPutMC.m
function [P,CI,NCrossed] = DOPutMC(S0,K,r,T,sigma,Sb,NSteps,NRepl)
% Generate asset paths
[Call,Put] = blsprice(S0,K,r,T,sigma);
Payoff = zeros(NRepl,1);
NCrossed = 0;
for i=1:NRepl
   Path=AssetPaths(S0,r,sigma,T,NSteps,1);
   crossed = any(Path <= Sb);
   if crossed == 0
      Payoff(i) = max(0, K - Path(NSteps+1));
   else
      Payoff(i) = 0;
      NCrossed = NCrossed + 1;
   end
end
[P,aux,CI] = normfit( exp(-r*T) * Payoff);
```

Fig. 8.12 Crude Monte Carlo simulation for a discrete barrier option.

8.3 PRICING A DOWN-AND-OUT PUT OPTION

In this section, we consider an example of weakly path-dependent option, i.e., a down-and-out put option, under the assumption that the barrier is checked at the end of each trading day. We have seen in section 2.7.1 how the analytical formula for continuous monitoring can be adjusted to reflect discrete monitoring; we will use the function DOPut to check the result of Monte Carlo simulation. An important point is that barrier options in practice may be very sensitive to stochastic volatility; Monte Carlo simulation could be used together with a model of stochastic volatility to price a barrier option.

8.3.1 Crude Monte Carlo

A code implementing crude Monte Carlo simulation is given in figure 8.12. The parameter NSteps is used to determine how many times the stock price should be checked against the barrier level S_b. The payoff is set to 0 whenever the barrier is crossed. Note that we always simulate the complete path even if the barrier is crossed during the option life; some part of the path is actually useless, but doing so we can streamline code by using AssetPaths and the any vector operator. The DOPutMC function also returns the number NCrossed of paths in which the barrier has been crossed.

PRICING A DOWN-AND-OUT PUT OPTION

Let us price an option with two months to maturity, assuming that each month consists of 30 days and that the barrier is checked each day. The barrier S_b is $40:

```
>> DOPut(50,50,0.1,2/12,0.4,40*exp(-0.5826*0.4*sqrt(1/12/30)))
ans =
    1.3629
>> randn('seed',0)
>> [P,CI,NCrossed]=DOPutMC(50,50,0.1,2/12,0.4,40,60,50000)
P =
    1.3600
CI =
    1.3393
    1.3808
NCrossed =
        7392
```

8.3.2 Conditional Monte Carlo

From section 4.5.1 we know that antithetic sampling may be not very effective in this case, because the payoff is nonmonotonic with respect to the asset price at expiration. Things are more complicated here, as the complete asset price path matters. Control variates may also be used; a natural candidate as a control variate is the price of a vanilla put, which may be computed by the Black–Scholes formula. However, the strength of the correlation between the two options is questionable. Hence, we try a different approach, i.e., variance reduction by conditioning, which was explained in section 4.5.4. To this end, we will see that is convenient to consider the price P_{di} of the down-and-in put. Pricing this knock-in option is equivalent to pricing the corresponding knock-out option, since we know that

$$P_{\text{do}} = P - P_{\text{di}}.$$

Assume that we discretize the option life in time intervals of width δt (in our case, one day), so that $T = M\delta t$, and consider the asset price path for days i, $i = 1, \ldots, M$:

$$\mathbf{S} = \{S_1, S_2, \ldots, S_M\}.$$

Based on this path, we estimate the option price as

$$P_{\text{di}} = e^{-rT}\mathrm{E}[I(\mathbf{S})(K - S_M)^+],$$

where the indicator function I is

$$I(\mathbf{S}) = \begin{cases} 1 & \text{if } S_j < S_b \text{ for some } j \\ 0 & \text{otherwise.} \end{cases}$$

Now let j^* be the index of the time instant at which the barrier is first crossed; by convention, let $j^* = M + 1$ if the barrier is not crossed during the option life. At time $j^*\delta t$ the option is activated, and from now on it behaves just like a vanilla put. So, conditional on the crossing time $t^* = j^*\delta t$ and the price S_{j^*} at which we detect barrier crossing,[8] we may use the Black–Scholes formula to estimate the expected value of the payoff. Hence, if the barrier is crossed before maturity, we have

$$\mathrm{E}\left[I(\mathbf{S})(K-S_M)^+ \mid j^*, S_{j^*}\right] = e^{r(T-t^*)} B_p(S_{j^*}, K, T-t^*),$$

where $B_p(S_{j^*}, K, T-t^*)$ is the Black–Scholes price for a vanilla put with strike price K, initial underlying price S_{j^*}, and time to maturity $T-t^*$; the exponential term takes discounting into account, from maturity back to crossing time. Given a simulated path $\mathbf{S}$, this suggests using the following estimator:

$$I(\mathbf{S})e^{-rt^*} B_p(S_{j^*}, K, T-t^*).$$

Unlike antithetic sampling, conditional Monte Carlo exploits specific knowledge about the problem; the more we know, the less we leave to numerical integration. The function DOPutMCCond in figure 8.13 implements this variance reduction method. The only point worth noting is that, for efficiency reasons, it is advisable to call the blsprice function only once with a vector argument, rather than once per replication. So, when the barrier is crossed, we record the time Times at which the down-and-in put has been activated, and the stock price StockVals. When the barrier is not crossed, the estimator is simply 0. Also note that the vectors passed to blsprice have NCrossed elements, whereas the size of the vector Payoff containing the estimator values is NRepl.

```
>> DOPut(50,52,0.1,2/12,0.4,30*exp(-0.5826*0.4*sqrt(1/12/30)))
ans =
     3.8645
>> randn('seed',0)
>> [P,CI,NCrossed] = DOPutMC(50,52,0.1,2/12,0.4,30,60,200000)
P =
     3.8751
CI =
     3.8545
     3.8957
NCrossed =
     249
>> randn('seed',0)
>> [P,CI,NCrossed] = DOPutMCCond(50,52,0.1,2/12,0.4,30,60,200000)
P =
     3.8651
CI =
```

---

[8] With continuous monitoring, we would immediately detect crossing when $S(t^*) = S_b$, but this is not the case with discrete monitoring.

```
% DOPutMCCond.m
function [Pdo,CI,NCrossed] = ...
    DOPutMCCond(S0,K,r,T,sigma,Sb,NSteps,NRepl)
dt = T/NSteps;
[Call,Put] = blsprice(S0,K,r,T,sigma);
% Generate asset paths and payoffs for the down and in option
NCrossed = 0;
Payoff = zeros(NRepl,1);
Times = zeros(NRepl,1);
StockVals = zeros(NRepl,1);
for i=1:NRepl
   Path=AssetPaths(S0,r,sigma,T,NSteps,1);
   tcrossed = min(find( Path <= Sb ));
   if not(isempty(tcrossed))
      NCrossed = NCrossed + 1;
      Times(NCrossed) = (tcrossed-1) * dt;
      StockVals(NCrossed) = Path(tcrossed);
   end
end
if (NCrossed > 0)
   [Caux, Paux] = blsprice(StockVals(1:NCrossed),K,r,...
      T-Times(1:NCrossed),sigma);
   Payoff(1:NCrossed) = exp(-r*Times(1:NCrossed)) .* Paux;
end
[Pdo, aux, CI] = normfit(Put - Payoff);
```

Fig. 8.13 Conditional Monte Carlo simulation for a discrete barrier option.

450   OPTION PRICING BY MONTE CARLO METHODS

```
     3.8617
     3.8684
NCrossed =
    249
```

8.3.3 Importance sampling

The last run shows that variance reduction by conditioning may indeed be helpful, but we should not get too excited. To begin with, one lucky run does not prove anything. Even worse, we have run a huge number of replications (200,000), but the barrier has been crossed only in 249 replications. This means that most of the replications are a wasted effort.[9] In other words, with the data for this option, crossing the barrier is a rare event. This is a typical case in which importance sampling may help (see section 4.5.6).

One possible idea is changing the drift of the asset price in such a way that crossing the barrier is more likely.[10] We should go a step back and consider what we do in order to generate an asset price path $\mathbf{S}$. For each time step, we generate a normal variate Z_j with expected value

$$\nu = \left(r - \frac{\sigma^2}{2}\right)\delta t$$

and variance $\sigma^2\,\delta t$. All these variates are mutually independent, and the asset price is generated by setting

$$\log S_j - \log S_{j-1} = Z_j.$$

Let $\mathbf{Z}$ be the vector of the normal variates, and let $f(\mathbf{Z})$ be its joint density. If we use the modified expected value

$$\nu - b,$$

we may expect that the barrier will be crossed more often. Let $g(\mathbf{Z})$ be the joint density for the normal variates generated with this modified expected value. Then we must find out a correction term, the likelihood ratio, to come up with the correct importance sampling estimator. Combining importance sampling with the conditional expectation we have just described, we have (if the barrier is crossed before maturity):

$$E_g\left[\left.\frac{f(\mathbf{Z})I(\mathbf{S})(K-S_M)^+}{g(\mathbf{Z})}\right|j^*, S_{j^*}\right]$$

---

[9]It can also be argued that in this case we are doing a good job only because the option price is slightly less than the Black–Scholes price, which we use in conditioning, because crossing the barrier is unlikely.
[10]The treatment here follows the approach of [18].

$$= \frac{f(z_1,\ldots,z_{j^*})}{g(z_1,\ldots,z_{j^*})} E_g \left[\frac{f(Z_{j^*+1},\ldots,Z_M)}{g(Z_{j^*+1},\ldots,Z_M)} I(\mathbf{S})(K-S_M)^+ \Big| j^*, S_{j^*} \right]$$

$$= \frac{f(z_1,\ldots,z_{j^*})}{g(z_1,\ldots,z_{j^*})} E_f \left[I(\mathbf{S})(K-S_M)^+ \Big| j^*, S_{j^*} \right]$$

$$= \frac{f(z_1,\ldots,z_{j^*})}{g(z_1,\ldots,z_{j^*})} e^{r(T-t^*)} B_p(S_{j^*}, K, T-t^*).$$

In the expressions above, we should note the difference between z and Z; the first samples, given conditioning information, are actually numbers and are taken outside the expectation. In practice, we should generate the normal variates with expected value $(\nu - b)$, and multiply the conditional estimator by the likelihood ratio, which from the sampling point of view is a random variable.[11] The only open problem is how to compute the likelihood ratio. In appendix B we consider the joint distribution of a multivariate normal with expected value $\boldsymbol{\mu}$ and covariance matrix $\boldsymbol{\Sigma}$:

$$f(\mathbf{z}) = \frac{1}{(2\pi)^{n/2} |\boldsymbol{\Sigma}|^{1/2}} e^{-\frac{1}{2}(\mathbf{z}-\boldsymbol{\mu})^T \boldsymbol{\Sigma}^{-1}(\mathbf{z}-\boldsymbol{\mu})}.$$

In our case, due to the mutual independence of the random variates Z_j, the covariance matrix is a diagonal matrix with elements $\sigma^2 \delta t$, and the vector of the expected values has components

$$\mu = \left(r - \frac{\sigma^2}{2}\right) \delta t$$

for the density f and $\mu - b$ for the density g. So we have

$$\frac{f(z_1,\ldots,z_{j^*})}{g(z_1,\ldots,z_{j^*})}$$

$$= \exp\left\{-\frac{1}{2} \sum_{k=1}^{j^*} \left(\frac{z_k - \mu}{\sigma\sqrt{\delta t}}\right)^2\right\} \exp\left\{\frac{1}{2} \sum_{k=1}^{j^*} \left(\frac{z_k - \mu + b}{\sigma\sqrt{\delta t}}\right)^2\right\}$$

$$= \exp\left\{-\frac{1}{2\sigma^2 \delta t} \sum_{k=1}^{j^*} \left[(z_k - \mu)^2 - (z_k - \mu + b)^2\right]\right\}$$

$$= \exp\left\{-\frac{1}{2\sigma^2 \delta t} \sum_{k=1}^{j^*} \left[-2(z_k - \mu)b - b^2\right]\right\}$$

$$= \exp\left\{-\frac{1}{2\sigma^2 \delta t} \left[-2b \sum_{k=1}^{j^*} z_k + 2j^* \mu b - j^* b^2\right]\right\}$$

---

[11] Readers with a background in stochastic calculus will recall that the Radon–Nikodym derivative is a random variable.

$$= \exp\left\{\frac{b}{\sigma^2 \delta t}\sum_{k=1}^{j^*} z_k - \frac{j^*b}{\sigma^2}\left(r - \frac{\sigma^2}{2}\right) + \frac{j^*b^2}{2\sigma^2 \delta t}\right\}.$$

The resulting code is illustrated in figure 8.14. The function DOPutMCCondIS is similar to DOPutMCCond; the difference is that we must generate the asset price path and record the normal variates in vector vetZ, so that we may compute the likelihood ratio which is stored in the vector ISRatio. We compute the Black–Scholes price only at the end of the main loop. Finding the parameter b is a matter of trial and error. In the function DOPutMCCondIS we assume that the user provides a percentage bp, and the modified expected value is computed as

```
(1 - bp)(r-0.5*sigma^2)*dt
```

Thus the parameter b is given as a percentage of the correct expected value. Note that we may use a value for bp which is larger than 1, to lower the drift rate at will. Now we may experiment a bit with importance sampling.

```
>> randn('seed',0)
>> [P,CI,NCrossed] = DOPutMC(50,52,0.1,2/12,0.4,30,60,10000)
P =
    3.8698
CI =
    3.7778
    3.9618
NCrossed =
    12
>> randn('seed',0)
>> [P,CI,NCrossed] = DOPutMCCondIS(50,52,0.1,2/12,0.4,30,60,10000,0)
P =
    3.8661
CI =
    3.8513
    3.8810
NCrossed =
    12
>> randn('seed',0)
>> [P,CI,NCrossed] = DOPutMCCondIS(50,52,0.1,2/12,0.4,30,60,10000,20)
P =
    3.8651
CI =
    3.8570
    3.8733
NCrossed =
    43
>> randn('seed',0)
>> [P,CI,NCrossed] = DOPutMCCondIS(50,52,0.1,2/12,0.4,30,60,10000,50)
P =
```

```
% DOPutMCCondIS.m
function [Pdo,CI,NCrossed] = ...
      DOPutMCCondIS(S0,K,r,T,sigma,Sb,NSteps,NRepl,bp)
dt = T/NSteps;
nudt = (r-0.5*sigma^2)*dt;
b = bp*nudt;
sidt = sigma*sqrt(dt);
[Call,Put] = blsprice(S0,K,r,T,sigma);
% Generate asset paths and payoffs for the down and in option
NCrossed = 0;
Payoff = zeros(NRepl,1);
Times = zeros(NRepl,1);
StockVals = zeros(NRepl,1);
ISRatio = zeros(NRepl,1);
for i=1:NRepl
    % generate normals
    vetZ = nudt - b + sidt*randn(1,NSteps);
    LogPath = cumsum([log(S0), vetZ]);
    Path = exp(LogPath);
    jcrossed = min(find( Path <= Sb ));
    if not(isempty(jcrossed))
        NCrossed = NCrossed + 1;
        TBreach = jcrossed - 1;
        Times(NCrossed) = TBreach * dt;
        StockVals(NCrossed) = Path(jcrossed);
        ISRatio(NCrossed) = exp( TBreach*b^2/2/sigma^2/dt +...
        b/sigma^2/dt*sum(vetZ(1:TBreach)) - ...
        TBreach*b/sigma^2*(r - sigma^2/2));
    end
end
if (NCrossed > 0)
    [Caux, Paux] = blsprice(StockVals(1:NCrossed),K,r,...
                  T-Times(1:NCrossed),sigma);
    Payoff(1:NCrossed) = exp(-r*Times(1:NCrossed)) .* Paux ...
                 .* ISRatio(1:NCrossed);
end
[Pdo, aux, CI] = normfit(Put - Payoff);
```

Fig. 8.14 Using conditional Monte Carlo and importance sampling for a discrete barrier option.

```
        3.8634
CI =
        3.8596
        3.8671
NCrossed =
        225
>> randn('seed',0)
>> [P,CI,NCrossed] = DOPutMCCondIS(50,52,0.1,2/12,0.4,30,60,10000,200)
P =
        3.8637
CI =
        3.8629
        3.8645
NCrossed =
        8469
```

Calling DOPutMCCondIS with the parameter bp set to zero is just like calling DOPutMCCond; by increasing bp we see that the barrier is crossed in more and more replications, and the quality of the estimate is improved. Note that this does not necessarily imply that the larger b, the better; suggestions for setting this parameter are given in [18].

8.4 PRICING AN ARITHMETIC AVERAGE ASIAN OPTION

We consider here pricing an Asian average rate call option with discrete arithmetic averaging. The option payoff is

$$\max\left\{\frac{1}{N}\sum_{i=1}^{N}S(t_i) - K, 0\right\},$$

where the option maturity is T years, $t_i = i\,\delta t$, and $\delta t = T/N$. For the sake of simplicity we assume that the contract prescribes taking sample prices at equally spaced time instants, but this need not be the case. In a crude Monte Carlo approach, we must simply generate asset price paths and average the discounted payoff as usual. The code is illustrated in figure 8.15; the only thing worth noting is that NSamples is the number N of sampled points to compute the arithmetic average, which should not be confused with the number of replications NRepl. In this case, we have to generate whole sample paths; we need samples only at the time instants specified by the contract, but we may still have to generate a large amount of data. This is why the code is not vectorized: to avoid trouble with possibly large matrices. In the following sections we consider variance reduction by control variates and use of low discrepancy sequences.

```
function [P,CI] = AsianMC(S0,K,r,T,sigma,NSamples,NRepl)
Payoff = zeros(NRepl,1);
for i=1:NRepl
   Path=AssetPaths(S0,r,sigma,T,NSamples,1);
   Payoff(i) = max(0, mean(Path(2:(NSamples+1))) - K);
end
[P,aux,CI] = normfit( exp(-r*T) * Payoff);
```

Fig. 8.15 Monte Carlo simulation for an Asian option.

8.4.1 Control variates

This crude Monte Carlo sampling may be improved by using control variates. In this case, we have different possibilities.

- As a first control variate, we could use the sum of the asset prices[12]:

$$Y = \sum_{i=0}^{N} S(t_i). \quad (8.6)$$

This is a plausible control variate, because we are able to compute its expected value, and Y is clearly correlated to the option payoff. Note that the sum includes S_0, which is not random at all; we could eliminate that from the sum, but we prefer not doing that just to ease the following notation.

- A second possibility would be using the vanilla call option, whose analytical price is known. However, the option payoff of this control variate depends only on the price at maturity.

- A third, more sophisticated, control variate is the payoff of a geometric average option. This is also known analytically, and it looks much more promising than the vanilla call.

We will illustrate the application of the first and the third idea.

The expected value of the sum of the stock prices Y, as defined in (8.6), is (under the risk-neutral measure):

$$\begin{aligned} \mathrm{E}[Y] &= \mathrm{E}\left[\sum_{i=0}^{N} S(t_i)\right] = \sum_{i=0}^{N} \mathrm{E}[S(i\,\delta t)] \\ &= \sum_{i=0}^{N} S(0) e^{ri\,\delta t} = S(0) \sum_{i=0}^{N} [e^{r\,\delta t}]^i = S(0) \frac{1 - e^{r(N+1)\delta t}}{1 - e^{r\,\delta t}}, \end{aligned}$$

---

[12] This is the approach suggested in [17, chapter 9].

```
function [P,CI] = AsianMCCV(S0,K,r,T,sigma,NSamples,NRepl,NPilot)
% pilot replications to set control parameter
TryPath=AssetPaths(S0,r,sigma,T,NSamples,NPilot);
StockSum = sum(TryPath,2);
PP = mean(TryPath(:,2:(NSamples+1)) , 2);
TryPayoff = exp(-r*T) * max(0, PP - K);
MatCov = cov(StockSum, TryPayoff);
c = - MatCov(1,2) / var(StockSum);
dt = T / NSamples;
ExpSum = S0 * (1 - exp((NSamples + 1)*r*dt)) / (1 - exp(r*dt));
% MC run
ControlVars = zeros(NRepl,1);
for i=1:NRepl
   StockPath = AssetPaths(S0,r,sigma,T,NSamples,1);
   Payoff = exp(-r*T) * max(0, mean(StockPath(2:(NSamples+1))) - K);
   ControlVars(i) = Payoff + c * (sum(StockPath) - ExpSum);
end
[P,aux,CI] = normfit(ControlVars);
```

Fig. 8.16 Monte Carlo simulation with control variates for an Asian option.

where we have used the following formula:

$$\sum_{i=0}^{N} \alpha^i = \frac{1 - \alpha^{N+1}}{1 - \alpha}.$$

The MATLAB code in figure 8.16 implements this variance reduction strategy. The user must fix the number of pilot replications NPilot, needed to set the control parameter c in the control variates procedure. The following runs give an idea of the improvement we may obtain:

```
>> randn('state',0)
[P,CI] = AsianMC(50,50,0.1,5/12,0.4,5,50000)
P =
    3.9939
CI =
    3.9418
    4.0460
>> CI(2) - CI(1)
ans =
    0.1042
>> [P,CI] = AsianMCCV(50,50,0.1,5/12,0.4,5,45000,5000)
P =
    3.9562
CI =
```

```
                    3.9336
                    3.9789
>> CI(2) - CI(1)
ans =
                    0.0453
```

The alternative control variate is based on the exploitation of much deeper knowledge. The payoff of the discrete-time, geometric average Asian option is

$$\max\left\{\left(\prod_{i=1}^{N} S(t_i)\right)^{1/N} - K,\ 0\right\}.$$

Since the product of lognormal random variables is still lognormal, it is possible to find an analytical formula for the price of the geometric average option, which looks like a modified Black–Scholes formula. We report the formula as given in [6, pp. 118–119], where m is the last time at which we observed the price of the underlying asset, q is the continuous dividend yield, and G_t is the current geometric average:

$$P_{GA} = e^{-rT}\left[e^{a+\frac{1}{2}b}N(x) - KN\left(x - \sqrt{b}\right)\right],$$

where

$$a = \frac{m}{N}\log(G_t) + \frac{N-m}{N}\left[\log(S_0) + \nu(t_{m+1} - t) + \frac{1}{2}\nu(T - t_{m+1})\right]$$

$$b = \frac{(N-m)^2}{N^2}\sigma^2(t_{m+1} - t) + \frac{\sigma^2(T - t_{m+1})}{6N^2}(N-m)(2(N-m) - 1)$$

$$\nu = r - q - \frac{1}{2}\sigma^2$$

$$x = \frac{a - \log(K) + b}{\sqrt{b}}.$$

The formula gets considerably simplified if we just consider the option price at its inception, i.e., at time $t = 0$. In such a case $m = 0$, and the resulting MATLAB implementation is illustrated in figure 8.17.

Using the geometric average option as a control variate is fairly simple; we have to adapt the code in figure 8.16, obtaining the function displayed in figure 8.18. The figure also includes a script to compare crude Monte Carlo against the two control variates:

```
>> CompareAsian
P1 =
        3.6276
CI1 =
        3.4814
        3.7738
```

```
function P = GeometricAsian(S0,K,r,T,sigma,delta,NSamples)
dT = T/NSamples;
nu = r - sigma^2/2-delta;
a = log(S0)+nu*dT+0.5*nu*(T-dT);
b = sigma^2*dT + sigma^2*(T-dT)*(2*NSamples-1)/6/NSamples;
x = (a-log(K)+b)/sqrt(b);
P = exp(-r*T)*(exp(a+b/2)*normcdf(x) - K*normcdf(x-sqrt(b)));
```

Fig. 8.17 MATLAB code for the analytical pricing formula of a geometric average Asian option.

```
P2 =
    3.4694
CI2 =
    3.3907
    3.5480
P3 =
    3.4452
CI3 =
    3.4356
    3.4549
```

The advantage of a control variate embodying sophisticated knowledge is pretty evident.

8.4.2 Using Halton sequences

Another tool that we may use to improve pricing an Asian option is quasi-Monte Carlo simulation based on low-discrepancy sequences. We will use here Halton sequences to generate uniform "quasi-random" numbers and the inverse transform method to transform them to samples from the standard uniform distribution. This is just the simplest possibility, as we could use Sobol or other sequences, and maybe the Box–Muller transformation to generate normal variates.

The first issue to tackle is the generation of sample paths of geometric Brownian motion using Halton sequences. Say that we want to price an Asian option maturing in one year and we must sample price monthly. What is the dimension of the space over which we are integrating? We are integrating in a twelve-dimensional space, and we need a Halton sequence based on twelve Van der Corput sequences. It is very important to understand that each sequence must be assigned to a time instant. Sequences are *not* associated to sample paths. By the way, should we use Box–Muller approach to transform uniform numbers to standard normal variates, we would need twice as much sequences. Also note that we *cannot* use rejection-based approaches to gen-

```
function [P,CI] = AsianMCGeoCV(S0,K,r,T,sigma,NSamples,NRepl,NPilot)
% precompute quantities
DF = exp(-r*T);
GeoExact = GeometricAsian(S0,K,r,T,sigma,0,NSamples);
% pilot replications to set control parameter
GeoPrices = zeros(NPilot,1);
AriPrices = zeros(NPilot,1);
for i=1:NPilot
    Path=AssetPaths(S0,r,sigma,T,NSamples,1);
    GeoPrices(i)=DF*max(0,(prod(Path(2:(NSamples+1))))^(1/NSamples) - K);
    AriPrices(i)=DF*max(0,mean(Path(2:(NSamples+1))) - K);
end
MatCov = cov(GeoPrices, AriPrices);
c = - MatCov(1,2) / var(GeoPrices);
% MC run
ControlVars = zeros(NRepl,1);
for i=1:NRepl
   Path = AssetPaths(S0,r,sigma,T,NSamples,1);
   GeoPrice = DF*max(0, (prod(Path(2:(NSamples+1))))^(1/NSamples) - K);
   AriPrice = DF*max(0, mean(Path(2:(NSamples+1))) - K);
   ControlVars(i) = AriPrice + c * (GeoPrice - GeoExact);
end
[P,aux,CI] = normfit(ControlVars);
```

```
% CompareAsian.m
randn('state',0)
S0 = 50;
K = 55;
r = 0.05;
sigma = 0.4;
T = 1;
NSamples = 12;
NRepl = 9000;
NPilot = 1000;
[P1,CI1] = AsianMC(S0,K,r,T,sigma,NSamples,NRepl+NPilot)
[P2,CI2] = AsianMCCV(S0,K,r,T,sigma,NSamples,NRepl,NPilot)
[P3,CI3] = AsianMCGeoCV(S0,K,r,T,sigma,NSamples,NRepl,NPilot)
```

Fig. 8.18 Using the geometric average Asian option as a control variate.

```
function SPaths=HaltonPaths(S0,mu,sigma,T,NSteps,NRepl)
dt = T/NSteps;
nudt = (mu-0.5*sigma^2)*dt;
sidt = sigma*sqrt(dt);
% Use inverse transform to generate standard normals
NormMat = zeros(NRepl, NSteps);
Bases = myprimes(NSteps);
for i=1:NSteps
   H = GetHalton(NRepl,Bases(i));
   RandMat(:,i) = norminv(H);
end
Increments = nudt + sidt*RandMat;
LogPaths = cumsum([log(S0)*ones(NRepl,1) , Increments] , 2);
SPaths = exp(LogPaths);
SPaths(:,1) = S0;
```

Fig. 8.19 Generating asset price paths by Halton sequences.

```
function P = AsianHalton(S0,K,r,T,sigma,NSamples,NRepl)
Payoff = zeros(NRepl,1);
Path=HaltonPaths(S0,r,sigma,T,NSamples,NRepl);
Payoff = max(0, mean(Path(:,2:(NSamples+1)),2) - K);
P = mean( exp(-r*T) * Payoff);
```

Fig. 8.20 Pricing an Asian option by Halton sequences.

erate variates, as in that case the dimension of the space is not well-defined. For each dimension, we need a prime number to be used as the basis. To generate the first N prime numbers, we may use the myprimes function which is discussed in section A.3. The function HaltonPaths illustrated in figure 8.19 is an extension of the vectorized function AssetPathsV to generate random sample paths. The idea is generating each *column* of matrix NormMat using one dimension of the Halton sequence, corresponding to one prime number. We see replications along the rows of the matrix, and each column corresponds to a time instant. Given this, we compute increments in the natural logarithm of the asset price, which are then cumulated and transformed to asset prices.

Based on sample paths generated by HaltonPaths, it is very easy to write a function to price the arithmetic Asian option, as shown in figure 8.20. We may see the potential of low-discrepancy sequences from the following runs, in which we first compute a very accurate price using a large number of replications with crude Monte Carlo to have a reliable benchmark:

PRICING AN ARITHMETIC AVERAGE ASIAN OPTION

```
>> randn('state',0)
>> [P,CI] = AsianMC(50,50,0.1,5/12,0.4,5,500000)
P =
    3.9639
CI =
    3.9474
    3.9803
>> AsianHalton(50,50,0.1,5/12,0.4,5,1000)
ans =
    3.8450
>> AsianHalton(50,50,0.1,5/12,0.4,5,3000)
ans =
    3.9103
>> AsianHalton(50,50,0.1,5/12,0.4,5,10000)
ans =
    3.9461
>> AsianHalton(50,50,0.1,5/12,0.4,5,50000)
ans =
    3.9605
```

We cannot associate a confidence interval to the estimate obtained by the quasi-random approach,[13] but we see that with a limited number of replications we get an acceptable result. Here we have considered an option maturing in five months, with monthly sampling. Let us check what happens if we increase maturity to two years, with a corresponding increase in the number of monthly samples:

```
>> randn('state',0)
>> [P,CI] = AsianMC(50,50,0.1,2,0.4,24,500000)
P =
    8.3859
CI =
    8.3495
    8.4222
>> AsianHalton(50,50,0.1,2,0.4,24,1000)
ans =
    6.6219
>> AsianHalton(50,50,0.1,2,0.4,24,5000)
ans =
    7.9257
>> AsianHalton(50,50,0.1,2,0.4,24,50000)
ans =
    8.3424
```

We see that in this case the performance of Halton sequences is much worse. This is due to the fact that we need 24 bases, which are large prime numbers,

---

[13] The randomization of quasi-Monte Carlo scheme is one of the actively pursued research directions to get confidence bounds when using low discrepancy sequences.

and we have seen in section 4.6 that using large prime numbers yields poor results. We may expect that if the contract is characterized by more samples the situation will get even worse.

One possible solution is using more sophisticated approaches, such as Sobol sequences. Another idea, is using the Brownian bridge construction. Using the Brownian bridge, we associate the "good" small bases to time instants acting as milestones. Large bases are used to fill the sample paths, but we may hope that this will not have a too detrimental effect. In what follows we use Brownian bridge with Halton sequences, for the sake of simplicity, but of course the same idea can be used with any low-discrepancy sequence. We also consider the possibility of using low-discrepancy sequences for milestone time instants, and pseudo-random numbers to fill the sample paths.

The first step is simulating the standard Wiener process by Halton sequences and the Brownian bridge. We extend function `WienerBridge` of figure 8.8 to obtain the code in figure 8.21. The function `WienerHaltonBridge` differs from `WienerBridge` in a few basic features:

- It is partially vectorized, as it is convenient to generate all of the sample points on each time layer, in order to use Halton sequences in a more compact and readable way; the function returns a matrix, containing several replications, rather than only one.

- The matrix `NormMat` contains samples from standard normal distribution, which are used just like in function `HaltonPaths`; each column is associated to one prime number and one time instant.

- The input arguments also include the number of replications `NRepl` and a parameter `Limit`; this is used to limit the number of dimensions of the Halton sequence which are used; note how the variable `HUse` is incremented within the main `for` loop to pick successive dimensions, associated to increasingly large prime numbers; when `HUse` exceeds `Limit`, we switch to random sampling (just to fill sample paths which have been already outlined).

Please note that our function is very limited, in that we can only use Brownian bridge when the number of time instants is a power of two. This is a limitation of our implementation, but *not* of the technique in itself.

The second step is transforming the standard Wiener process to a geometric Brownian motion. The function `GBMHaltonBridge` of figure 8.22 works much like the function `GBMBridge` of figure 8.9. We should only note that it is vectorized and that this requires a different use of the `diff` function, which has to work horizontally on a matrix. In order to compute `Increments`, we call `diff(W,1,2)` on the matrix `W` containing the paths of the standard Wiener process: The argument 1 means that we want to compute first-order differences, and the argument 2 means that we want to work along the rows of the matrix, whereas the default is along columns (just like `mean` or `cumsum`).

```
function WSamples = WienerHaltonBridge(T, NSteps, NRepl, Limit)
NBisections = log2(NSteps);
if round(NBisections) ~= NBisections
    fprintf('ERROR in WienerHB: NSteps must be a power of 2\n');
    return
end
% Generate standard normal samples
NormMat = zeros(NRepl, NSteps);
Bases = myprimes(NSteps);
for i=1:NSteps
   H = GetHalton(NRepl,Bases(i));
   NormMat(:,i) = norminv(H);
end
% Initialize extreme points of paths
WSamples = zeros(NRepl,NSteps+1);
WSamples(:,1) = 0;
WSamples(:,NSteps+1) = sqrt(T)*NormMat(:,1);
% Fill paths
HUse = 2;
TJump = T;
IJump = NSteps;
for k=1:NBisections
    left = 1;
    i = IJump/2 + 1;
    right = IJump + 1;
    for j=1:2^(k-1)
        a = 0.5*(WSamples(:,left) + WSamples(:,right));
        b = 0.5*sqrt(TJump);
        if HUse <= Limit;
            WSamples(:,i) = a + b*NormMat(:,HUse);
        else
            WSamples(:,i) = a + b*randn(NRepl,1);
        end
        right = right + IJump;
        left = left + IJump;
        i = i + IJump;
    end
    IJump = IJump/2;
    TJump = TJump/2;
    HUse = HUse + 1;
end
```

Fig. 8.21 Simulating the standard Wiener process by Halton sequences and the Brownian bridge.

```
function Paths=GBMHaltonBridge(S0,mu,sigma,T,NSteps,NRepl,Limit)
if round(log2(NSteps)) ~= log2(NSteps)
    fprintf('ERROR in GBMBridge: NSteps must be a power of 2\n');
    return
end
dt = T/NSteps;
nudt = (mu-0.5*sigma^2)*dt;
W = WienerHaltonBridge(T,NSteps,NRepl,Limit);
Increments = nudt + sigma*diff(W,1,2);
LogPath = cumsum([log(S0)*ones(NRepl,1) , Increments], 2);
Paths = exp(LogPath);
Paths(:,1) = S0;
```

Fig. 8.22 Simulating geometric Brownian motion by Halton sequences and the Brownian bridge.

Now we should pause a little and use our knowledge of geometric Brownian motion, represented by equations (8.3) and (8.4) to check if we do generate a process with the correct expected values and variances at different time instants. In particular, we may wish to check the relative error of the sample mean and sample variance of the process generated by Monte Carlo and Halton sequences, with and without the Brownian bridge, against the theoretically correct values. In order to do so, it is convenient to use the function of figure 8.23. Given a matrix of sample paths, the function returns a two-column matrix; the first column contains, for each time instant (contained in vector Tvet), the relative percentage error in the mean, whereas the second column returns the error in variance. In the same figure we also provide the reader with a script to compare results. The script prints a table with six columns and sixteen rows.

```
>> CheckHaltonScript
ans =
    0.2510    0.0269    0.0927    0.9473    0.4045    1.0480
    0.4838    0.0701    0.0983    0.9765    0.8147    1.1005
    0.5893    0.1042    0.1685    0.4233    1.1434    1.9098
    0.3609    0.1651    0.1235    1.0490    1.9696    1.4138
    0.5580    0.2644    0.2351    0.9005    3.1095    2.7626
    0.4847    0.3787    0.2251    1.0232    4.2511    2.8336
    0.5960    0.4814    0.2826    3.7522    5.4619    3.3645
    0.8787    0.6607    0.2053    4.4059    7.5672    2.5914
    1.2209    0.8061    0.3353    5.3788    9.6047    4.1374
    1.1240    1.0044    0.3299    2.8125   11.1005    4.4781
    0.8548    1.2322    0.3945    0.0401   12.3976    5.2199
    1.0240    1.4891    0.2976    1.0730   14.0780    4.1875
    0.9923    1.6941    0.4268    0.7693   14.5632    5.9899
```

1.2271	1.9678	0.3922	3.2472	15.2210	5.8546
1.1193	2.2621	0.4274	0.8804	16.4125	6.2836
1.5650	2.6552	0.3018	0.1313	18.6872	4.9231

The first three columns give the relative errors in the estimate of expected value at each of the sixteen time instants, for Monte Carlo, Halton sequences without Brownian bridge, and Halton sequences with Brownian bridge, respectively. If we look at the second column, we see that the error tends to grow in time if we do not use the bridge; this makes sense, as we use large "bad" bases for later time intervals. If we compare the first and the third column, we see that the error with Halton sequences and the bridge compares favorably against the error with Monte Carlo. The last three columns display a similar pattern for variance, in the sense that there is a significant error that tends to increase over time if we use Halton sequences without the bridge. The error with Monte Carlo does not display a clear pattern. We may also see that Halton sequences with the bridge does not seem so superior to Monte Carlo in terms of matching variances. This is not that surprising given the simplicity of Halton sequences; nevertheless the reader is invited to verify that if we increase the number of sample paths we have a significant improvement.

After all of this work, it is easy to write a function to price the arithmetic Asian option based on Halton sequences and the Brownian bridge. The code is illustrated in figure 8.24, and we may check the advantage over a straightforward use of low-discrepancy sequences. To this purpose, we may use the script in figure 8.25. The idea here is

1. to evaluate first the option price by plain Monte Carlo with a large number of replications (500,000);

2. then to calculate the price with straightforward Halton sequences, on a limited number of replications (10,000);

3. to check what we obtain with plain Monte Carlo with the small number of replications, repeating the procedure twenty times and collecting the average price and its standard deviation;

4. to compare the two prices above with what we obtain using the Brownian bridge with different mixes of Halton and random sequences; when only Halton sequences are used, the experiment is not repeated and no standard deviation is reported, as there is no variability in that case.

The result is the following (the script takes some time to execute):

```
>> CompareAsianH
Extended MC 9.068486
Halton 8.800511
MC mean 9.135870 st.dev 0.135540
HB (limit: 1) mean 9.074675 st.dev 0.077153
```

```
function PercErrors = CheckGBMPaths(S0, mu, sigma, T, Paths);
[NRepl, NTimes] = size(Paths);
NSteps = NTimes-1;
Tvet = (1:NSteps).*T/NSteps;
SampleMean = mean(Paths(:,2:NTimes));
TrueMean = S0 * exp(mu*Tvet);
RelErrorM = abs((SampleMean - TrueMean)./TrueMean);
SampleVar = var(Paths(:,2:(1+NSteps)));
TrueVar =  S0^2 * exp(2*mu*Tvet) .* (exp((sigma^2) * Tvet) - 1) ;
RelErrorV = abs((SampleVar - TrueVar)./TrueVar);
PercErrors = 100*[RelErrorM', RelErrorV'];
```

```
% CheckHaltonScript.m
randn('state',0)
NRepl = 10000;
T = 5;
NSteps = 16;
Limit = NSteps;
S0 = 50;
mu = 0.1;
sigma = 0.4;
Paths = AssetPaths(S0, mu, sigma, T, NSteps, NRepl);
PercErrors1 = CheckGBMPaths(S0, mu, sigma, T, Paths);
Paths = HaltonPaths(S0, mu, sigma, T, NSteps, NRepl);
PercErrors2 = CheckGBMPaths(S0, mu, sigma, T, Paths);
Paths = GBMHaltonBridge(S0, mu, sigma, T, NSteps, NRepl, Limit);
PercErrors3 = CheckGBMPaths(S0, mu, sigma, T, Paths);
[PercErrors1(:,1), PercErrors2(:,1), PercErrors3(:,1), ...
    PercErrors1(:,2), PercErrors2(:,2), PercErrors3(:,2)]
```

Fig. 8.23 MATLAB function and script to evaluate sampling errors in the generation of geometric Brownian motion.

```
function P = AsianHaltonBridge(S0,K,r,T,sigma,NSamples,NRepl,Limit)
Payoff = zeros(NRepl,1);
Path=GBMHaltonBridge(S0,r,sigma,T,NSamples,NRepl,Limit);
Payoff = max(0, mean(Path(:,2:(NSamples+1)),2) - K);
P = mean( exp(-r*T) * Payoff);
```

Fig. 8.24 Pricing the arithmetic Asian option by Halton sequences and the Brownian bridge.

```
% CompareAsianH.m
randn('state',0)
S0 = 50;
K = 55;
r = 0.05;
sigma = 0.4;
T = 4;
NSamples = 16;
NRepl = 500000;
aux = AsianMC(S0,K,r,T,sigma,NSamples,NRepl);
fprintf(1,'Extended MC %f\n', aux);
NRepl = 10000;
aux = AsianHalton(S0,K,r,T,sigma,NSamples,NRepl);
fprintf(1,'Halton %f\n', aux);
for i=1:20
    aux(i) = AsianMC(S0,K,r,T,sigma,NSamples,NRepl);
end
fprintf(1,'MC mean %f st.dev %f\n', mean(aux), sqrt(var(aux)));
Limit = 1;
for i=1:20
    aux(i) = AsianHaltonBridge(S0,K,r,T,sigma,NSamples,NRepl,Limit);
end
fprintf(1,'HB (limit: %d) mean %f st.dev %f\n', ...
    Limit, mean(aux), sqrt(var(aux)));
Limit = 2;
for i=1:20
    aux(i) = AsianHaltonBridge(S0,K,r,T,sigma,NSamples,NRepl,Limit);
end
fprintf(1,'HB (limit: %d) mean %f st.dev %f\n', ...
    Limit, mean(aux), sqrt(var(aux)));
Limit = 4;
for i=1:20
    aux(i) = AsianHaltonBridge(S0,K,r,T,sigma,NSamples,NRepl,Limit);
end
fprintf(1,'HB (limit: %d) mean %f st.dev %f\n', ...
    Limit, mean(aux), sqrt(var(aux)));
Limit = 16;
aux = AsianHaltonBridge(S0,K,r,T,sigma,NSamples,NRepl,Limit);
fprintf(1,'HB (limit: %d) %f\n', Limit, aux);
```

Fig. 8.25 Comparing Monte Carlo and Halton sequences with Brownian bridge.

HB (limit: 2) mean 9.017819 st.dev 0.035962
HB (limit: 4) mean 9.307306 st.dev 0.010279
HB (limit: 16) 9.367783

We see that straightforward use of Halton sequences does not give a satisfactory result and that Monte Carlo with few replications is fairly acceptable. We should note that since the payoff is defined by an average, there is much less variability than with the corresponding vanilla option depending only on price at maturity. Using an Halton sequence only for the terminal price of the underlying asset, filling the trajectory by Brownian bridge and random variates yields good results with limited variability. Using more Halton sequences kills variability, of course, but it also tends to introduce a bias. In fact, using only Halton sequences with the Brownian bridge does not seem to work, and we overestimate the price. To understand why, we should carry out a more detailed analysis, which we do not report in detail, on the average price of the underlying asset generated by Halton sequences with the Brownian bridge. On the average, it is not too different from what we obtain by simple Monte Carlo sampling, but it is somewhat right-skewed which means that it tends to generate larger payoffs on the tail where the option is in-the-money.

To summarize this section, we see that Halton sequences are not very satisfactory, and we should look for alternatives. Nevertheless, the idea of the Brownian bridge looks like tool one should keep in mind. The best results we have obtained in the last experiment are actually due to a sort of stratification effect on the terminal price, and Brownian bridge allows to exploit such a mechanism.

8.5 ESTIMATING GREEKS BY MONTE CARLO SAMPLING

So far, we have only considered option pricing problems. However, estimating option sensitivities is another quite important task. We deal here with the estimation of Δ for a vanilla call, for the sake of simplicity. This section is linked to section 6.6, where we considered the interplay between simulation and optimization. We recall here the general framework. We have a function $f(S_0)$, which in our case is the price of an option depending on the initial underlying asset price S_0, and the sensitivity we want to estimate is

$$\Delta \equiv \frac{df(S_0)}{dS_0} = \lim_{\delta S_0 \to 0} \frac{f(S_0 + \delta S_0) - f(S_0)}{\delta S_0}.$$

Since we are estimating the option price by Monte Carlo simulation, the first approach coming to mind is to take sample paths and estimate Δ by the sample mean of finite differences between discounted payoffs. This approach can be implemented as illustrated in figure 8.26.

However, this idea is too naive. To begin with, some care is needed, since what we are doing is swapping an expectation and a limit. In fact, what we

```
function [Delta, CI] = BlsDeltaMCNaive(S0,K,r,T,sigma,dS,NRepl)
nuT = (r - 0.5*sigma^2)*T;
siT = sigma * sqrt(T);
Payoff1 = max(0, S0*exp(nuT+siT*randn(NRepl,1))-K);
Payoff2 = max(0, (S0+dS)*exp(nuT+siT*randn(NRepl,1))-K);
SampleDiff = exp(-r*T)*(Payoff2 - Payoff1)/dS;
[Delta, dummy, CI] = normfit(SampleDiff);
```

Fig. 8.26 Estimating the option Δ by crude Monte Carlo.

are interested in is

$$\lim_{\delta S_0 \to 0} \frac{\mathrm{E}_\omega[C(S_0 + \delta S_0, \omega)] - \mathrm{E}_\omega[C(S_0, \omega)]}{\delta S_0},$$

where $C(S_0, \omega)$ is the discounted payoff of the call with initial price S_0, for a sample path corresponding to event ω. But what we are really computing is:

$$\mathrm{E}_\omega \left[\frac{C(S_0 + \delta S_0, \omega) - C(S_0, \omega)}{\delta S_0} \right].$$

Even if we accept the approximation of the limit by a finite difference, we should not take for granted that swapping the two operators is legal. To see a potential trouble intuitively, we should think that we are interested in the derivative of a function defined by an integral (the expected value). But the integral is a "smoothing" operator; hence, even if the function we integrate is not quite regular, the derivative of the integral may be no trouble. However, if we integrate the derivative, we may run into difficulties. In statistical terms, commuting the two operators may result in a biased estimator.[14]

Even if we disregard these subtle issues, it is easy to see that the function above is far from satisfactory. If we compare the estimate we get against the exact value provided by taking the derivative of the Black–Scholes formula,[15] we see that the estimate is quite poor:

```
>> S0=50; K=52; r=0.05; T=5/12; sigma=0.4;
>> blsdelta(S0,K,r,T,sigma)
ans =
    0.5231
>> randn('state',0)
>> NRepl=50000;
>> dS = 0.5;
```

---

[14] See [8, chapter 7] for a full treatment.
[15] The function blsdelta is available in the Financial Toolbox.

```
function [Delta, CI] = BlsDeltaMCNaive(S0,K,r,T,sigma,dS,NRepl)
nuT = (r - 0.5*sigma^2)*T;
siT = sigma * sqrt(T);
Payoff1 = max(0, S0*exp(nuT+siT*randn(NRepl,1))-K);
Payoff2 = max(0, (S0+dS)*exp(nuT+siT*randn(NRepl,1))-K);
SampleDiff = exp(-r*T)*(Payoff2 - Payoff1)/dS;
[Delta, dummy, CI] = normfit(SampleDiff);
```

Fig. 8.27 Improving the estimate of the option Δ by Common Random Numbers.

```
>> [Delta, CI] = BlsDeltaMCNaive(S0,K,r,T,sigma,dS,NRepl)
Delta =
    0.3588
CI =
    0.1447
    0.5729
```

Actually, it is not too difficult to improve the estimator. From the theory of finite differences (section 5.2) we know that taking a central difference may be preferable:

$$\frac{C(S_0 + \delta S_0, \omega) - C(S_0 - \delta S_0, \omega)}{2\delta S_0}.$$

In our case, this may also reduce the effect of noise in our random sampling. Another point is that, to reduce variance, we may rely on common random numbers (section 4.5.2). In other words, we should use the same samples from the standard normal distribution when generating the two option payoffs. The related code is displayed in figure 8.27. We may verify that using these two tricks, we definitely improve the estimate of Δ:

```
>> randn('state',0)
>> [Delta, CI] = BlsDeltaMC(S0,K,r,T,sigma,dS,NRepl)
Delta =
    0.5296
CI =
    0.5241
    0.5350
```

We see that the least one should do to estimate option sensitivities is using central differences, when it makes sense, and using common random numbers. However, we see that if we are also interested in the option price, we basically have to repeat the same computations three times, for S_0 and $S_0 \pm \delta S_0$. The computational burden is actually larger, since we are typically interested in other sensitivities as well, and we have also to bother wondering about the right step δS_0. It would be much nicer if we could just use one run to estimate both the option price and Δ. In fact, this may be done in many cases, if we analyze more carefully what we are doing.[16]

Our discounted option payoff is a random variable

$$C = e^{-rT} \max\{S_T - K, 0\},$$

where

$$S_T = S_0 e^{(r-\sigma^2/2)T + \sigma\sqrt{T}Z}$$

and Z is a standard normal variable. Using the chain rule for differentiation, we have

$$\frac{dC}{dS_0} = \frac{dC}{dS_T}\frac{dS_T}{dS_0}.$$

The last derivative is easy:

$$\frac{dS_T}{dS_0} = \frac{S_T}{S_0}.$$

The first derivative is a bit more problematic, but we may see that

$$\frac{d}{dx}\max\{x - K, 0\} = \begin{cases} 0, & \text{if } x < K \\ 1, & \text{if } x > K \end{cases}.$$

There is some trouble when $x = K$, as the function as a kink there. It turns out that, since this event has probability zero, this difficulty can be disregarded. Hence, we may conclude

$$\frac{dC}{dS_T} = e^{-rT}\mathbf{I}\{S_T > K\}$$

where $\mathbf{I}$ is the usual indicator function. Putting everything together, we obtain the following estimator of Δ:

$$e^{-rT}\frac{S_T}{S_0}\mathbf{I}\{S_T > K\}.$$

This type of estimator, because of the way it is built, is called *pathwise* estimator. We should stress that this is not the only available approach, and that we have cut a few delicate corners in its explanation. However, the implementation is straightforward and is illustrated in figure 8.28.

```
>> randn('state',0)
>> [Delta, CI] = BlsDeltaMCPath(S0,K,r,T,sigma,NRepl)
Delta =
    0.5297
CI =
    0.5241
    0.5352
```

---

[16]The treatment here follows [8, pp. 388–389].

```
function [Delta, CI] = BlsDeltaMCPath(S0,K,r,T,sigma,NRepl)
nuT = (r - 0.5*sigma^2)*T;
siT = sigma * sqrt(T);
VLogn = exp(nuT+siT*randn(NRepl,1));
SampleDelta = exp(-r*T) .* VLogn .* (S0*VLogn > K);
[Delta, dummy, CI] = normfit(SampleDelta);
```

Fig. 8.28 Estimating the option Δ by a pathwise estimator.

This snapshot shows that the estimator actually works. The careful reader will notice that in this run the true value falls outside the confidence interval; this may actually happen, because of the way confidence interval are built. The reader is urged to run the experiment a few times in order to check that the true value usually falls within the bounds of the confidence interval.

For further reading

In the literature

- Path generation and numerical solution of stochastic differential equations are extensively treated in [14]. See also [10] for an introduction including MATLAB code.

- The main reference for Monte Carlo methods in finance is [8]. You may also see [6] and [12].

- An early paper on using Monte Carlo simulation in option pricing is [3]. An updated survey is given in [4].

- A nice collection of papers, gathered from the otherwise scattered literature, is [7].

- Interesting sources on the use of low-discrepancy sequences for derivatives pricing are [15] and [16]. See also [1] and [19] for specific issues such as path generation in high-dimensional problems and quantifying the estimation error.

- Another interesting paper on quasi-Monte Carlo simulation in finance is [13], where Faure low-discrepancy sequences, which we did not consider, are discussed.

- In this chapter we have only considered applications to option pricing. However, another important application field for Monte Carlo simulation is estimating Value at Risk. In [9], and related references, you may find some information on the use of variance reduction methods to speed up VaR computations.

On the Web

- A Web page related to Monte Carlo and quasi-Monte Carlo methods is http://www.mcqmc.org.

- Some information on using low-discrepancy sequences in finance can be also obtained by browsing the following pages:
  http://www.cs.columbia.edu/~traub
  http://www.cs.columbia.edu/~ap/html/information.html

REFERENCES

1. F. Åkesson and J.P. Lehoczky. Path Generation for Quasi-Monte Carlo Simulation of Mortgage-Backed Securities. *Management Science*, 46:1171–1187, 2000.

2. T. Björk. *Arbitrage Theory in Continuous Time (2nd ed.)*. Oxford University Press, Oxford, 2004.

3. P. Boyle. Options: A Monte Carlo Approach. *Journal of Financial Economics*, 4:323–338, 1977.

4. P. Boyle, M. Broadie, and P. Glasserman. Monte Carlo Methods for Security Pricing. *Journal of Economics Dynamics and Control*, 21:1267–1321, 1997.

5. P.P. Carr and R.A. Jarrow. The Stop-Loss Start-Gain Paradox and Option Valuation: a New Decomposition into Intrinsic and Time Value. *The Review of Financial Studies*, 3:469–492, 1990.

6. L. Clewlow and C. Strickland. *Implementing Derivatives Models*. Wiley, Chichester, West Sussex, England, 1998.

7. B. Dupire, editor. *Monte Carlo. Methodologies and Applications for Pricing and Risk Management*. Risk Books, London, 1998.

8. P. Glasserman. *Monte Carlo Methods in Financial Engineering*. Springer-Verlag, New York, NY, 2004.

9. P. Glasserman, P. Heidelberger, and P. Shahabuddin. Variance Reduction Techniques for Estimating Value-at-Risk. *Management Science*, 46:1349–1364, 2000.

10. D.J. Higham. An Algorithmic Introduction to Numerical Simulation of Stochastic Differential Equations. *SIAM Review*, 43:525–546, 2001.

11. J.C. Hull. *Options, Futures, and Other Derivatives (5th ed.)*. Prentice Hall, Upper Saddle River, NJ, 2003.

12. P. Jaeckel. *Monte Carlo Methods in Finance*. Wiley, Chichester, 2002.

13. C. Joy, P.P. Boyle, and K.S. Tan. Quasi-Monte Carlo Methods in Numerical Finance. *Management Science*, 42:926–938, 1996.

14. P.E. Kloeden and E. Platen. *Numerical Solution of Stochastic Differential Equations*. Springer-Verlag, Berlin, 1992.

15. S.H. Paskov. New Methodologies for Valuing Derivatives. In S.R. Pliska and M.A.H. Dempster, editors, *Mathematics of Derivative Securities*, pages 545–582. Cambridge University Press, Cambridge, 1997.

16. S.H. Paskov and J.F. Traub. Faster Valuation of Financial Derivatives. *Journal of Portfolio Management*, 22:113–120, Fall 1995.

17. S. Ross. *An Introduction to Mathematical Finance: Options and Other Topics*. Cambridge University Press, Cambridge, 1999.

18. S.M. Ross and J.G. Shanthikumar. Monotonicity in Volatility and Efficient Simulation. *Probability in the Engineering and Informational Sciences*, 14:317–326, 2000.

19. K.S. Tan and P.P. Boyle. Applications of Randomized Low Discrepancy Sequences to the Valuation of Complex Securities. *Journal of Economic Dynamics and Control*, 24:1747–1782, 2000.

9
Option Pricing by Finite Difference Methods

In this chapter we give a few simple examples of how the Partial Differential Equation (PDE) framework may be exploited in option pricing. The idea is applying the finite difference methods illustrated in chapter 5 to solve the Black–Scholes PDE. We start in section 9.1 by recalling derivatives approximation schemes and by pointing out how suitable boundary conditions may be set up in order to model a specific option. In section 9.2 we apply a straightforward explicit scheme to the pricing of a vanilla European option; as we already know, this scheme is prone to numerical instabilities, which we may also interpret from a financial point of view. In section 9.3 we see how a fully implicit method may overcome the instability issue. The Crank–Nicolson method, which may be regarded as a hybrid between the explicit and the fully implicit approach, is applied in section 9.4 to a barrier option. Finally, in section 9.5 we see how iterative overrelaxation methods may be exploited to tackle an American option with a fully implicit method, which is not trivial due to the presence of a free boundary due to the possibility of early exercise.

9.1 APPLYING FINITE DIFFERENCE METHODS TO THE BLACK–SCHOLES EQUATION

We have shown in section 2.6.2 that the value at time t of an option written on an underlying asset whose price is $S(t)$ is a function $f(S,t)$ satisfying the

476 OPTION PRICING BY FINITE DIFFERENCE METHODS

partial differential equation

$$\frac{\partial f}{\partial t} + rS\frac{\partial f}{\partial S} + \frac{1}{2}\sigma^2 S^2 \frac{\partial^2 f}{\partial S^2} = rf, \qquad (9.1)$$

with suitable boundary conditions that characterize the type of option. Different equations may be written if the hypotheses are changed and if path dependency is introduced, but this equation is the starting point to learn how to apply numerical methods based on finite differences for option pricing.

As we have seen in chapter 5, to solve a PDE by finite difference methods we must set up a discrete grid, in this case with respect to time and asset prices. Let T be option maturity and $S_{\max}$ a suitably large asset price, that cannot be reached by $S(t)$ within the time horizon we consider. We need $S_{\max}$, since the domain for the PDE is unbounded with respect to asset prices, but we must bound it in some way for computational purposes; $S_{\max}$ plays the role of $+\infty$. The grid consists of points (S, t) such that

$$S = 0, \delta S, 2\,\delta S, \ldots, M\,\delta S \equiv S_{\max},$$
$$t = 0, \delta t, 2\,\delta t, \ldots, N\,\delta t \equiv T.$$

We will use the grid notation $f_{i,j} = f(i\,\delta S, j\,\delta t)$.

Let us recall the different ways we have to approximate the partial derivatives in equation (9.1):

- Forward difference:

$$\frac{\partial f}{\partial S} = \frac{f_{i+1,j} - f_{i,j}}{\delta S}, \qquad \frac{\partial f}{\partial t} = \frac{f_{i,j+1} - f_{i,j}}{\delta t}$$

- Backward difference:

$$\frac{\partial f}{\partial S} = \frac{f_{i,j} - f_{i-1,j}}{\delta S}, \qquad \frac{\partial f}{\partial t} = \frac{f_{i,j} - f_{i,j-1}}{\delta t}$$

- Central (or symmetric) difference:

$$\frac{\partial f}{\partial S} = \frac{f_{i+1,j} - f_{i-1,j}}{2\,\delta S}, \qquad \frac{\partial f}{\partial t} = \frac{f_{i,j+1} - f_{i,j-1}}{2\,\delta t}$$

- As to the second derivative, we have

$$\frac{\partial^2 f}{\partial S^2} = \left(\frac{f_{i+1,j} - f_{i,j}}{\delta S} - \frac{f_{i,j} - f_{i-1,j}}{\delta S}\right) \bigg/ \delta S$$
$$= \frac{f_{i+1,j} - 2f_{i,j} + f_{i-1,j}}{\delta S^2}.$$

Depending on which combination of schemes we use in discretizing the equation, we end up with different approaches, explicit or implicit, which we experiment with in the following sections.

Another issue, which we must take care of, is setting the boundary conditions. The terminal condition at expiration is

$$f(S,T) = \max\{S - K, 0\} \quad \forall S$$

for a call with strike price K, and

$$f(S,T) = \max\{K - S, 0\} \quad \forall S$$

for a put. When we consider boundary conditions with respect to asset prices, the problem is not so trivial, since we have to solve the equation numerically on a bounded region, whereas the domain is unbounded with respect to asset prices. We may use a few examples to clarify this issue.

Example 9.1 Let us consider first a vanilla European put option. When the asset price $S(t)$ is very large, the option is worthless, since we may be (almost) sure that it will stay out-of-the-money:

$$f(S_{\max}, t) = 0.$$

The value of $S_{\max}$ must be relatively large for this boundary condition to work properly. When the asset price is $S(t) = 0$, we may say that, given our geometric Brownian motion model for asset dynamics, the asset price will remain zero. So the payoff at expiration will be K; discounting back to time t, we have

$$f(0,t) = Ke^{-r(T-t)}.$$

In grid notation:

$$f_{i,N} = \max[K - i\,\delta S, 0], \quad i = 0, 1, \ldots, M$$
$$f_{0,j} = Ke^{-r(N-j)\delta t}, \quad j = 0, 1, \ldots, N$$
$$f_{M,j} = 0, \quad j = 0, 1, \ldots, N.$$

$\square$

Example 9.2 We may deal with a vanilla European call by reasoning as in example 9.1. When the asset price is $S(t) = 0$, at any time t, the option will expire worthless:

$$f(0,t) = 0.$$

For a large asset price $S(t)$, we may be sure that it will be in-the-money at expiration and we will get a payoff $S(T) - K$. The value at time t requires discounting back the term K and considering that the arbitrage-free price at time t for the underlying asset is simply $S(t)$. Then a suitable boundary condition is

$$f(S_{\max}, t) = S_{\max} - Ke^{-r(T-t)}.$$

In grid notation:

$$f_{i,N} = \max[i\,\delta S - K, 0], \quad i = 0, 1, \ldots, M$$
$$f_{0,j} = 0, \quad j = 0, 1, \ldots, N$$
$$f_{M,j} = M\,\delta S - Ke^{-r(N-j)\delta t}, \quad j = 0, 1, \ldots, N.$$

An alternative boundary condition for large values of S would be requiring that the option Δ is 1; in such a case we have a boundary condition on the derivative of the unknown function, rather than the function itself. This is called a Neumann boundary condition and is common in mathematical physics. We will not pursue this approach, because it complicates the numerical solution a bit. □

When dealing with barrier options, things may be easier. In the case of a knock-out option, such as a down-and-out put, the option value is 0 on the barrier. The case of an up-and-out call is similar, with the additional advantage that the domain we must consider is naturally bounded. American options are more complex to deal with because of the early exercise boundary; we should take into account for which asset prices and at which times (if any) it is optimal to exercise the option. Thus we have a free boundary that must be discovered in the solution process. A variety of boundary conditions must be required for exotic options; figuring out the correct boundary conditions and approximating them within the numerical scheme is an option-dependent issue.

9.2 PRICING A VANILLA EUROPEAN OPTION BY AN EXPLICIT METHOD

As a first attempt to solve equation (9.1), let us consider a vanilla European put option. We approximate the derivative with respect to S by a central difference and the derivative with respect to time by a backward difference. This is not the only possibility, but any choice must be somehow compatible with the boundary conditions. The result is the following set of equations:

$$\frac{f_{i,j} - f_{i,j-1}}{\delta t} + ri\,\delta S\,\frac{f_{i+1,j} - f_{i-1,j}}{2\,\delta S}$$
$$+ \frac{1}{2}\sigma^2 i^2\,\delta S^2\,\frac{f_{i+1,j} - 2f_{i,j} + f_{i-1,j}}{\delta S^2} = rf_{i,j}, \qquad (9.2)$$

to be solved with the boundary conditions of example 9.1. It should be noted that, since we have a set of terminal conditions, the equations must be solved backward in time. Let $j = N$ in equation (9.2); given the terminal condition, we have one unknown quantity, $f_{i,N-1}$, expressed as a function of three known quantities. If we imagine going backward in time the same consideration holds for each time layer. Rewriting the equations, we get an explicit scheme:

$$f_{i,j-1} = a_i^* f_{i-1,j} + b_j^* f_{i,j} + c_j^* f_{i+1,j}$$
$$j = N-1, N-2, \ldots, 1, 0;\ i = 1, 2, \ldots, M-1, \qquad (9.3)$$

where

$$a_i^* = \frac{1}{2}\delta t(\sigma^2 i^2 - ri)$$

```
function price = EuPutExpl(S0,K,r,T,sigma,Smax,dS,dt)
% set up grid and adjust increments if necessary
M = round(Smax/dS);
dS = Smax/M;
N = round(T/dt);
dt = T/N;
matval = zeros(M+1,N+1);
vetS = linspace(0,Smax,M+1)';
veti = 0:M;
vetj = 0:N;
% set up boundary conditions
matval(:,N+1) = max(K-vetS,0);
matval(1,:) = K*exp(-r*dt*(N-vetj));
matval(M+1,:) = 0;
% set up coefficients
a = 0.5*dt*(sigma^2*veti - r).*veti;
b = 1- dt*(sigma^2*veti.^2 + r);
c = 0.5*dt*(sigma^2*veti + r).*veti;
% solve backward in time
for j=N:-1:1
   for i=2:M
      matval(i,j) = a(i)*matval(i-1,j+1) + b(i)*matval(i,j+1)+ ...
         c(i)*matval(i+1,j+1);
   end
end
% return price, possibly by linear interpolation outside the grid
price = interp1(vetS, matval(:,1), S0);
```

Fig. 9.1 MATLAB code to price a European vanilla put by a straightforward explicit scheme.

$$b_i^* = 1 - \delta t(\sigma^2 i^2 + r)$$
$$c_i^* = \frac{1}{2}\delta t(\sigma^2 i^2 + ri).$$

This scheme is rather straightforward to implement in MATLAB. The code is illustrated in figure 9.1, and it requires the value $S_{\max}$ as well as the two discretization steps. The only point requiring some care is that in the mathematical notation it is convenient to uses indexes starting from 0, whereas matrix indexes start from 1 in MATLAB. Moreover, if the initial asset price does not lie on the grid, we must interpolate between the two neighboring points. We have used here a crude linear interpolation; more sophisticated splines could be a better alternative, especially if we are interested in approximating option price sensitivities (as it is always the case in practice).

```
>> [c,p] = blsprice(50,50,0.1,5/12,0.4);
```

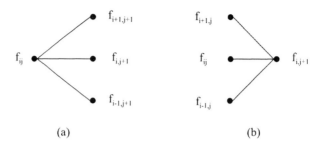

Fig. 9.2 View of explicit (a) and implicit (b) schemes to solve the Black–Scholes PDE.

```
>> p
p =
    4.0760
>> EuPutExpl(50,50,0.1,5/12,0.4,100,2,5/1200)
ans =
    4.0669
>> [c,p] = blsprice(50,50,0.1,5/12,0.3);
>> p
p =
    2.8446
>> EuPutExpl(50,50,0.1,5/12,0.3,100,2,5/1200)
ans =
    2.8288
```

We see that the numerical method gives fairly accurate results. We might try to improve them by using a finer grid.

```
>> EuPutExpl(50,50,0.1,5/12,0.3,100,1.5,5/1200)
ans =
    2.8597
>> EuPutExpl1(50,50,0.1,5/12,0.3,100,1,5/1200)
ans =
  -2.8271e+022
```

What we see here is another example of the numerical instability that we have analyzed in chapter 5. One possibility to avoid the trouble is to resort to implicit methods. Another one is to carry out a stability analysis and to derive bounds on the discretization steps. We will not pursue the second way here, which would be quite similar to what we have done in chapter 5 for the simpler transport and heat equations. Rather, in the next section we describe a financial interpretation of instability, which suggests still another possibility: rewriting the equation with a change of variables.

9.2.1 Financial interpretation of the instability of the explicit method

In the explicit scheme, we obtain an option value $f(S, t)$ as a combination of the values $f(S + \delta S, t + \delta t)$, $f(S, t + \delta t)$, and $f(S - \delta S, t + \delta t)$. This looks a bit like a trinomial lattice method, which we have described in section 7.4 (see figure 9.2a). We can make this interpretation clearer by deriving an alternative version of the explicit method. Following [1, chapter 18], we assume that the first- and second-order derivatives with respect to S at point (i, j) are equal to those at point $(i, j + 1)$:

$$\frac{\partial f}{\partial S} = \frac{f_{i+1,j+1} - f_{i-1,j+1}}{2\delta S}$$

$$\frac{\partial^2 f}{\partial S^2} = \frac{f_{i+1,j+1} - 2f_{i,j+1} + f_{i-1,j+1}}{\delta S^2}.$$

An alternative way to obtain the same scheme is substituting the right-hand term $f_{i,j}$ in equation (9.2) by $f_{i,j-1}$. This introduces an error which is bounded and tends to zero as the grid is refined.[1]

The finite difference equation is now

$$\frac{f_{i,j+1} - f_{i,j}}{\delta t} + r i \delta S \frac{f_{i+1,j+1} - f_{i-1,j+1}}{2\delta S}$$
$$+ \frac{1}{2}\sigma^2 i^2 \delta S^2 \frac{f_{i+1,j+1} - 2f_{i,j+1} + f_{i-1,j+1}}{\delta S^2} = r f_{i,j},$$

which may be rewritten (for $i = 1, 2, \ldots, M - 1$ and $j = 0, 1, \ldots, N - 1$) as

$$f_{i,j} = \hat{a}_i f_{i-1,j+1} + \hat{b}_i f_{i,j+1} + \hat{c}_i f_{i+1,j+1},$$

where

$$\hat{a}_i = \frac{1}{1 + r \delta t} \left(-\frac{1}{2} r i \, \delta t + \frac{1}{2}\sigma^2 i^2 \, \delta t \right) = \frac{1}{1 + r \delta t} \pi_d$$

$$\hat{b}_i = \frac{1}{1 + r \delta t} \left(1 - \sigma^2 i^2 \, \delta t \right) = \frac{1}{1 + r \delta t} \pi_0$$

$$\hat{c}_i = \frac{1}{1 + r \delta t} \left(\frac{1}{2} r i \, \delta t + \frac{1}{2}\sigma^2 i^2 \, \delta t \right) = \frac{1}{1 + r \delta t} \pi_u.$$

This scheme is again explicit and is subject to numerical instabilities as well. However, the coefficients $\hat{a}_i$, $\hat{b}_i$, and $\hat{c}_i$ lend themselves to a nice interpretation. Recall that, in a binomial or a trinomial lattice, we obtain an option value in a node as the discounted expected value of the values in the successor nodes, where expectation is taken with respect to a risk-neutral probability measure. In fact, the coefficients above include a $1/(1 + r \, \delta t)$ term, which

---

[1] A similar line of reasoning was used when deriving the ADI method in section 5.4.

may be interpreted as a discount factor over a time interval of length δt. Furthermore, we have

$$\pi_d + \pi_0 + \pi_u = 1.$$

This suggests interpreting the coefficients as probabilities, times a discount factor. Are they risk-neutral probabilities? We should first check the expected value of the increase in the asset price during the time interval δt:

$$\mathrm{E}[\Delta] = -\delta S \pi_d + 0 \pi_0 + \delta S \pi_u = ri\,\delta S\,\delta t = rS\,\delta t,$$

which is exactly what we would expect in a risk-neutral world. As to the variance of the increment, we have

$$\mathrm{E}[\Delta^2] = (-\delta S)^2 \pi_d + 0 \pi_0 + (\delta S)^2 \pi_u = \sigma^2 i^2 (\delta S)^2\,\delta t.$$

Hence, for small δt

$$\mathrm{Var}[\Delta] = \mathrm{E}[\Delta^2] - \mathrm{E}^2[\Delta] = \sigma^2 S^2\,\delta t - r^2 S^2 (\delta t)^2 \approx \sigma^2 S^2\,\delta t,$$

which is also coherent with geometric Brownian motion in a risk-neutral world. Thus we see that indeed the explicit method could be regarded as a trinomial lattice approach, except for a little problem. The "probabilities" π_d and π_0 may be negative. The careful reader will see a recurring pattern, since in chapter 5 we have already met stability conditions linked to the coefficients of a linear combination; in both transport and heat equations, we must make sure that this combination is convex, i.e., that the coefficients are positive and sum up to one, just like a discrete probability distribution.

One possibility to avoid the trouble, described in [1], is to change variables. By rewriting the Black–Scholes equation in terms of $Z = \ln S$, simple conditions for stability may be derived. However, a change of variables may not be a good idea for certain exotic options. In the next section we implement a fully implicit approach that avoids the stability issue altogether.

9.3 PRICING A VANILLA EUROPEAN OPTION BY A FULLY IMPLICIT METHOD

To overcome the stability issues of the explicit method, we may resort to an implicit method. This is obtained by using a forward difference to approximate the partial derivative with respect to time. We get the grid equations

$$\frac{f_{i,j+1} - f_{i,j}}{\delta t} + ri\,\delta S\,\frac{f_{i+1,j} - f_{i-1,j}}{2\,\delta S} + \frac{1}{2}\sigma^2 i^2\,\delta S^2\,\frac{f_{i+1,j} - 2f_{i,j} + f_{i-1,j}}{\delta S^2} = rf_{i,j},$$

which we may rewrite (for $i = 1, 2, \ldots, M-1$ and $j = 0, 1, \ldots, N-1$) as

$$a_i f_{i-1,j} + b_i f_{i,j} + c_i f_{i+1,j} = f_{i,j+1}, \tag{9.4}$$

where, for each i,

$$\begin{aligned} a_i &= \frac{1}{2} ri\,\delta t - \frac{1}{2}\sigma^2 i^2\,\delta t \\ b_i &= 1 + \sigma^2 i^2\,\delta t + r\,\delta t \\ c_i &= -\frac{1}{2} ri\,\delta t - \frac{1}{2}\sigma^2 i^2\,\delta t. \end{aligned}$$

Here we have three unknown values linked to one known value (see figure 9.2b). First note that, for each time layer, we have $M-1$ equations in $M-1$ unknowns; the boundary conditions yield the two missing values for each time layer and the terminal conditions give the values in the last time layer. As in the explicit case, we must go backward in time, solving a sequence of systems of linear equations for $j = N-1, \ldots, 0$. The system for time layer j is the following:

$$\begin{bmatrix} b_1 & c_1 & & & & \\ a_2 & b_2 & c_2 & & & \\ & a_3 & b_3 & c_3 & & \\ & & \ddots & \ddots & \ddots & \\ & & & a_{M-2} & b_{M-2} & c_{M-2} \\ & & & & a_{M-1} & b_{M-1} \end{bmatrix} \begin{bmatrix} f_{1,j} \\ f_{2,j} \\ f_{3,j} \\ \vdots \\ f_{M-2,j} \\ f_{M-1,j} \end{bmatrix}$$

$$= \begin{bmatrix} f_{1,j+1} \\ f_{2,j+1} \\ f_{3,j+1} \\ \vdots \\ f_{M-2,j+1} \\ f_{M-1,j+1} \end{bmatrix} - \begin{bmatrix} a_1 f_{0,j} \\ 0 \\ 0 \\ \vdots \\ 0 \\ c_{M-1} f_{M,j} \end{bmatrix}.$$

We may note that the matrix is tridiagonal and that it is constant for each time layer i. So we may speed up the computation by resorting to a LU-factorization.[2] All of this is accomplished by the MATLAB code in figure 9.3.

```
>> [c,p] = blsprice(50,50,0.1,5/12,0.4);
>> p
p =
```

---

[2] Due to the sparse structure of the matrix, it would be much better to write a specific code to solve the sequence of linear systems. Here we use just the ready-to-use MATLAB functionalities.

```
function price = EuPutImpl(S0,K,r,T,sigma,Smax,dS,dt)
% set up grid and adjust increments if necessary
M = round(Smax/dS);
dS = Smax/M;
N = round(T/dt);
dt = T/N;
matval = zeros(M+1,N+1);
vetS = linspace(0,Smax,M+1)';
veti = 0:M;
vetj = 0:N;
% set up boundary conditions
matval(:,N+1) = max(K-vetS,0);
matval(1,:) = K*exp(-r*dt*(N-vetj));
matval(M+1,:) = 0;
% set up the tridiagonal coefficients matrix
a = 0.5*(r*dt*veti-sigma^2*dt*(veti.^2));
b = 1+sigma^2*dt*(veti.^2)+r*dt;
c = -0.5*(r*dt*veti+sigma^2*dt*(veti.^2));
coeff = diag(a(3:M),-1) + diag(b(2:M)) + diag(c(2:M-1),1);
[L,U] = lu(coeff);
% solve the sequence of linear systems
aux = zeros(M-1,1);
for j=N:-1:1
   aux(1) = - a(2) * matval(1,j); % other term from BC is zero
   matval(2:M,j) = U \ (L \ (matval(2:M,j+1) + aux));
end
% return price, possibly by linear interpolation outside the grid
price = interp1(vetS, matval(:,1), S0);
```

Fig. 9.3 MATLAB code to price a vanilla European option by a fully implicit method.

```
    4.0760
>> EuPutImpl(50,50,0.1,5/12,0.4,100,0.5,5/2400)
ans =
    4.0718
```

The results are fairly accurate and may be improved by a refined grid without the risk of running into numerical instabilities. Another way to improve accuracy is to exploit the Crank–Nicolson method; we will do this in the next section for a barrier option.

9.4 PRICING A BARRIER OPTION BY THE CRANK–NICOLSON METHOD

The Crank–Nicolson method has been introduced in section 5.3.3 as a way to improve accuracy by combining the explicit and implicit methods. Applying this idea to the Black–Scholes equation leads to the following grid equation:

$$\frac{f_{ij} - f_{i,j-1}}{\delta t} + \frac{ri\,\delta S}{2}\left(\frac{f_{i+1,j-1} - f_{i-1,j-1}}{2\,\delta S}\right) + \frac{ri\,\delta S}{2}\left(\frac{f_{i+1,j} - f_{i-1,j}}{2\,\delta S}\right)$$
$$+ \frac{\sigma^2 i^2 (\delta S)^2}{4}\left(\frac{f_{i+1,j-1} - 2f_{i,j-1} + f_{i-1,j-1}}{(\delta S)^2}\right)$$
$$+ \frac{\sigma^2 i^2 (\delta S)^2}{4}\left(\frac{f_{i+1,j} - 2f_{i,j} + f_{i-1,j}}{(\delta S)^2}\right)$$
$$= \frac{r}{2}f_{i,j-1} + \frac{r}{2}f_{ij}.$$

These equations may be rewritten as

$$-\alpha_i f_{i-1,j-1} + (1 - \beta_i) f_{i,j-1} - \gamma_i f_{i+1,j-1} = \alpha_i f_{i-1,j} + (1 + \beta_i) f_{ij} + \gamma_i f_{i+1,j}, \tag{9.5}$$

where

$$\alpha_i = \frac{\delta t}{4}(\sigma^2 i^2 - ri)$$
$$\beta_i = -\frac{\delta t}{2}(\sigma^2 i^2 + r)$$
$$\gamma_i = \frac{\delta t}{4}(\sigma^2 i^2 + ri).$$

We consider here the down-and-out put option, that we have introduced in section 2.7.1, assuming continuous barrier monitoring. In this case we need only to consider the domain $S_b \leq S \leq S_{\max}$; the boundary conditions are

$$f(S_{\max}, t) = 0, \qquad f(S_b, t) = 0.$$

Taking these boundary conditions into account, we may rewrite equation (9.5) in matrix form:

$$\mathbf{M}_1 \mathbf{f}_{j-1} = \mathbf{M}_2 \mathbf{f}_j, \tag{9.6}$$

where

$$M_1 = \begin{bmatrix} 1-\beta_1 & -\gamma_1 \\ -\alpha_2 & 1-\beta_2 & -\gamma_2 \\ & -\alpha_3 & 1-\beta_3 & -\gamma_3 \\ & & \ddots & \ddots & \ddots \\ & & & -\alpha_{M-2} & 1-\beta_{M-2} & -\gamma_{M-2} \\ & & & & -\alpha_{M-1} & 1-\beta_{M-1} \end{bmatrix}$$

$$M_2 = \begin{bmatrix} 1+\beta_1 & \gamma_1 \\ \alpha_2 & 1+\beta_2 & \gamma_2 \\ & \alpha_3 & 1+\beta_3 & \gamma_3 \\ & & \ddots & \ddots & \ddots \\ & & & \alpha_{M-2} & 1+\beta_{M-2} & \gamma_{M-2} \\ & & & & \alpha_{M-1} & 1+\beta_{M-1} \end{bmatrix}$$

$$\mathbf{f}_j = [f_{1j}, f_{2j}, \ldots, f_{M-1,j}]^T.$$

The MATLAB code is displayed in figure 9.4. The result may be compared with those obtained by the analytical pricing formula of section 2.7.1:

```
>> DOPut(50,50,0.1,5/12,0.4,40)
ans =
    0.5424
>> DOPutCK(50,50,0.1,5/12,0.4,40,100,0.5,1/1200)
ans =
    0.5414
```

Barrier options come in a variety of forms; more on the application of PDEs to barrier options may be found in [9].

9.5 DEALING WITH AMERICAN OPTIONS

While pricing a vanilla European option by finite differences is certainly instructive, it is not very practical. We may apply the idea to American options, for which exact formulas are not available. The main difficulty in pricing an American option is the existence of a free boundary due to the possibility of early exercise. To avoid arbitrage, the option value at each point in the (S, t) space cannot be less than the intrinsic value (i.e., the immediate payoff if the option is exercised). For a vanilla American put, this means

$$f(S, t) \geq \max\{K - S(t), 0\}.$$

From a strictly practical point of view, taking this condition into account is not very difficult, at least in an explicit scheme. We could simply apply the

```
function price = DOPutCK(S0,K,r,T,sigma,Sb,Smax,dS,dt)
% set up grid and adjust increments if necessary
M = round((Smax-Sb)/dS);
dS = (Smax-Sb)/M;
N = round(T/dt);
dt = T/N;
matval = zeros(M+1,N+1);
vetS = linspace(Sb,Smax,M+1)';
veti = vetS / dS;
vetj = 0:N;
% set up boundary conditions
matval(:,N+1) = max(K-vetS,0);
matval(1,:) = 0;
matval(M+1,:) = 0;
% set up the coefficients matrix
alpha = 0.25*dt*( sigma^2*(veti.^2) - r*veti );
beta = -dt*0.5*( sigma^2*(veti.^2) + r );
gamma = 0.25*dt*( sigma^2*(veti.^2) + r*veti );
M1 = -diag(alpha(3:M),-1) + diag(1-beta(2:M)) - diag(gamma(2:M-1),1);
[L,U] = lu(M1);
M2 = diag(alpha(3:M),-1) + diag(1+beta(2:M)) + diag(gamma(2:M-1),1);
% solve the sequence of linear systems
for j=N:-1:1
   matval(2:M,j) = U \ (L \ (M2*matval(2:M,j+1)));
end
% return price, possibly by linear interpolation outside the grid
price = interp1(vetS, matval(:,1), S0);
```

Fig. 9.4 MATLAB code to price a down-and-out put option by the Crank–Nicolson method.

procedure of section 9.2 with a small modification. After computing f_{ij}, we should check for the possibility of early exercise, and set

$$f_{ij} = \max[f_{ij}, K - i\delta S],$$

just like we do with binomial lattices. Due to instability issues, we might prefer adopting an implicit scheme. In this case, there is an additional complication, as the relationship above requires knowing f_{ij} already, which is not the case in an implicit scheme. To get past this difficulty, we may resort to an iterative method to solve the linear system rather than to a direct method based on LU-factorization. In section 3.2.5 we considered the Gauss–Seidel method with overrelaxation. We recall the idea here for convenience. Given a system of linear equations such as

$$\mathbf{Ax} = \mathbf{b},$$

we should apply the following iterative scheme, starting from an initial point $\mathbf{x}^{(0)}$:

$$x_i^{(k+1)} = x_i^{(k)} + \frac{\omega}{a_{ii}}\left(b_i - \sum_{j=1}^{i-1} a_{ij} x_j^{(k+1)} - \sum_{j=i}^{N} a_{ij} x_j^{(k)}\right), \quad i = 1, \ldots, N,$$

where k is the iteration counter and ω is the overrelaxation parameter, until a convergence criterion is met, such as

$$\|\mathbf{x}^{(k+1)} - \mathbf{x}^{(k)}\| < \epsilon,$$

where ϵ is a tolerance parameter.

Now, suppose that we want to apply the Crank–Nicolson method to price an American put option. We have to solve more or less the same system as (9.6), but here the boundary conditions are a bit different, since there is no barrier on which the option value is zero. The systems we should solve backward in time look like

$$\mathbf{M}_1 \mathbf{f}_{j-1} = \mathbf{r}_j,$$

where the right-hand side is

$$\mathbf{r}_j = \mathbf{M}_2 \mathbf{f}_j + \alpha_1 \begin{bmatrix} f_{0,j-1} + f_{0,j} \\ 0 \\ \vdots \\ 0 \end{bmatrix}.$$

The additional term takes the customary boundary conditions for a put into account. The overrelaxation scheme should take into account the tridiagonal nature of the matrix $\mathbf{M}_1$, and it should also be adjusted for early exercise. Let g_i, $i = 1, \ldots, M - 1$, be the intrinsic value when $S = i\delta S$. For each time layer j, we have the iterative scheme

$$f_{1j}^{(k+1)} = \max\left\{g_1,\; f_{1j}^{(k)} + \frac{\omega}{1-\beta_1}\left[r_1 - (1-\beta_1)f_{1j}^{(k)} + \gamma_1 f_{2j}^{(k)}\right]\right\}$$

$$f_{2j}^{(k+1)} = \max\left\{g_2,\; f_{2j}^{(k)} + \frac{\omega}{1-\beta_2}\left[r_2 + \alpha_2 f_{1j}^{(k+1)} - (1-\beta_2)f_{2j}^{(k)} + \gamma_2 f_{3j}^{(k)}\right]\right\}$$

$$\vdots$$

$$f_{M-1,j}^{(k+1)} = \max\left\{g_{M-1},\; f_{M-1,j}^{(k)} + \frac{\omega}{1-\beta_{M-1}}\left[r_{M-1} + \alpha_{M-1} f_{M-2,j}^{(k+1)} - (1-\beta_{M-1})f_{M-1,j}^{(k)}\right]\right\}.$$

When passing from a time layer to the next one, it may be reasonable to initialize the iteration with a starting vector equal to the outcome of the previous time layer. The resulting code is displayed in figure 9.5. The code is a bit tricky because MATLAB starts indexing vectors from 1, but it should be clear enough. In this case we have not set up a matrix to contain all of the f_{ij} values, and the sparse matrix M_1 has not been stored; the iterations above are best carried out by using the vectors α, β, and γ directly.

The code may be compared with the binprice function, available in the Financial toolbox, which prices American options by a binomial lattice method (see section 7.1).

```
>> tic,[pr,opt] = binprice(50,50,0.1,5/12,1/1200,0.4,0);,toc
Elapsed time is 0.408484 seconds.
>> opt(1,1)
ans =
    4.2830
>> tic,AmPutCK(50,50,0.1,5/12,0.4,100,1,1/600,1.5,0.001),toc
ans =
    4.2815
Elapsed time is 0.031174 seconds.
>> tic,AmPutCK(50,50,0.1,5/12,0.4,100,1,1/600,1.8,0.001),toc
ans =
    4.2794
Elapsed time is 0.061365 seconds.
>> tic,AmPutCK(50,50,0.1,5/12,0.4,100,1,1/600,1.2,0.001),toc
ans =
    4.2800
Elapsed time is 0.023053 seconds.
>> tic,AmPutCK(50,50,0.1,5/12,0.4,100,1,1/1200,1.2,0.001),toc
ans =
    4.2828
```

```
function price = AmPutCK(S0,K,r,T,sigma,Smax,dS,dt,omega,tol)
M = round(Smax/dS); dS = Smax/M; % set up grid
N = round(T/dt); dt = T/N;
oldval = zeros(M-1,1); % vectors for Gauss-Seidel update
newval = zeros(M-1,1);
vetS = linspace(0,Smax,M+1)';
veti = 0:M; vetj = 0:N;
% set up boundary conditions
payoff = max(K-vetS(2:M),0);
pastval = payoff; % values for the last layer
boundval = K*exp(-r*dt*(N-vetj)); % boundary values
% set up the coefficients and the right hand side matrix
alpha = 0.25*dt*( sigma^2*(veti.^2) - r*veti );
beta = -dt*0.5*( sigma^2*(veti.^2) + r );
gamma = 0.25*dt*( sigma^2*(veti.^2) + r*veti );
M2 = diag(alpha(3:M),-1) + diag(1+beta(2:M)) + diag(gamma(2:M-1),1);
% solve the sequence of linear systems by SOR method
aux = zeros(M-1,1);
for j=N:-1:1
   aux(1) = alpha(2) * (boundval(1,j) + boundval(1,j+1));
   % set up right hand side and initialize
   rhs = M2*pastval(:) + aux;
   oldval = pastval;
   error = realmax;
   while tol < error
      newval(1) = max ( payoff(1), ...
         oldval(1) + omega/(1-beta(2)) * (...
         rhs(1) - (1-beta(2))*oldval(1) + gamma(2)*oldval(2)));
      for k=2:M-2
         newval(k) = max ( payoff(k), ...
            oldval(k) + omega/(1-beta(k+1)) * (...
            rhs(k) + alpha(k+1)*newval(k-1) - ...
            (1-beta(k+1))*oldval(k) + gamma(k+1)*oldval(k+1)));
      end
      newval(M-1) = max( payoff(M-1),...
         oldval(M-1) + omega/(1-beta(M)) * (...
         rhs(M-1) + alpha(M)*newval(M-2) - ...
         (1-beta(M))*oldval(M-1)));
      error = norm(newval - oldval);
      oldval = newval;
   end
   pastval = newval;
end
newval = [boundval(1) ; newval ; 0]; % add missing values
% return price, possibly by linear interpolation outside the grid
price = interp1(vetS, newval, S0);
```

Fig. 9.5 MATLAB code to price an American put option by Crank–Nicolson method.

```
Elapsed time is 0.036693 seconds.
>> tic,AmPutCK(50,50,0.1,5/12,0.4,100,1,1/100,1.2,0.001),toc
ans =
    4.2778
Elapsed time is 0.009989 seconds.
```

From these examples we see that the overrelaxation parameter ω has a significant effect on the convergence of the iterative methods. In terms of computational speed, the finite difference approach seems even faster than the binomial lattice approach, but we must be very careful here. We are comparing *implementations* of approaches, and both could be improved. Furthermore, the CPU requirements are possibly affected by the way the MATLAB interpreter works.[3] Anyway, having a whole grid of values, rather than nodes on a binomial lattice, allows us to obtain better estimates of some of the sensitivities (those involved in the Black–Scholes equation). Furthermore, the finite difference approach may be preferable when dealing with complex exotic options.

For further reading

- Many examples of how the PDE approach may be exploited in financial engineering are given in [6] or [7], which include interesting chapters on finite difference methods. You may also find [2] useful.

- We have used the finite difference approach on the Black–Scholes equation directly; however, a change of variables may be helpful in analyzing stability. See, e.g., the related chapters in [3]. In that book you also find a treatment on finite element methods, which are considerably more refined than simple-minded finite difference schemes.

- Books aimed specifically at finite differences in financial engineering are [4] and [8].

- See also [5] if you are interested in the finite element method.

REFERENCES

1. J.C. Hull. *Options, Futures, and Other Derivatives (5th ed.)*. Prentice Hall, Upper Saddle River, NJ, 2003.

---

[3]Reader owning a copy of the first edition of this book will find that quite different computational results were reported there: `binprice` took 14.17 seconds and the first run of `AmPutCK` took 59.48 seconds. We see that there is a speed-up which cannot be attributed only to faster hardware; the improvement in the MATLAB interpreter has been impressive too.

2. Y.K. Kwok. *Mathematical Models of Financial Derivatives.* Springer-Verlag, Berlin, 1998.

3. R. Seydel. *Tools for Computational Finance.* Springer-Verlag, Berlin, 2002.

4. D. Tavella and C. Randall. *Pricing Financial Instruments: The Finite Difference Method.* Wiley, New York, 2000.

5. J. Topper. *Financial Engineering with Finite Elements.* Wiley, New York, 2005.

6. P. Wilmott. *Derivatives: The Theory and Practice of Financial Engineering.* Wiley, Chichester, West Sussex, England, 1999.

7. P. Wilmott. *Quantitative Finance (vols. I and II).* Wiley, Chichester, West Sussex, England, 2000.

8. Y.-I. Zhu, X. Wu, and I.-L. Chern. *Derivative Securities and Difference Methods.* Springer, New York, 2004.

9. R. Zvan, K.R. Vetzal, and P.A. Forsyth. PDE Methods for Pricing Barrier Options. *Journal of Economic Dynamics and Control*, 24:1563–1590, 2000.

Part IV
Advanced Optimization Models and Methods

10
Dynamic Programming

Dynamic programming is arguably the most powerful principle in optimization and it can be applied to a wide range of problems with radically different features. As its name suggests, dynamic programming was originally conceived as a method to solve dynamic optimization models over time. As such, it can be applied to discrete- and continuous-time models, deterministic and stochastic models, and finite- and infinite-horizon models. Actually, with a little creativity, it can also be applied to non-dynamic problems. For instance, it can be used to tackle a combinatorial optimization problem like the knapsack model.[1] All of this potential comes with a price. To begin with, dynamic programming is a *principle*, rather than a well-defined and ready-to-use algorithm. It must be customized to the problem at hand. Furthermore, it may be computationally quite expensive. This is not always true: in some cases, application of the principle yields quite efficient numerical algorithms, or even analytical solutions. Even when dynamic programming does not yield the solution itself, it can be most valuable in characterizing its qualitative properties, which can provide us with very valuable insights. But in many practical cases, straightforward application of the principle is not possible because of the so-called "curse-of-dimensionality."

Some tricks of the trade can be used to reduce problem dimensionality, but dynamic programming is often considered a typically academic concept. Nevertheless, there are very good reasons why we have decided to include a chapter on this topic.

---

[1] See, e.g., [16].

- Having a basic grasp of dynamic programming is needed to understand recently developed approaches to price high-dimensional American options by Monte Carlo methods. Finite difference and lattice based methods are very well suited to price American options, but they do not cope well with high-dimensionality. On the other hand, Monte Carlo methods deal easily with high-dimensionality, but not with early exercise. Exploiting early exercise opportunities optimally requires going backward in time, since at each point in the state space we must compare the value of immediate exercise with the value of keeping the option, which is simply the price of the option at that point. Hence, it would seem that one has to chase her tail a bit, since while running a simulation forward in time, we should already know the option value. Indeed, until a few years ago, it was a common belief that simulation could not be applied to American-style options, but the situation has changed.

- While a literal application of dynamic programming may be overly difficult, approximate strategies have been developed which are very promising in terms of their ability to tackle real problems. Clearly, the increase in computational power of hardware plays a role here, but it is not the only factor.

- Understanding dynamic programming also sheds some light on stochastic programming, which is the topic of next chapter.

A comprehensive treatment of dynamic programming in one chapter is out of the question. Our main aim is to illustrate Monte Carlo methods to price American options. A secondary aim is to outline how the idea can be applied to portfolio optimization over a finite time horizon. Given our limited scope, we will only cover discrete-time and finite-horizon models. In section 10.1 we first illustrate the principle behind dynamic programming with the simplest example, the shortest path problem in a network. Then, in section 10.2, we show the connection between this simple example and more general deterministic sequential decision processes. In this section we get acquainted with the dynamic programming principle in a deterministic setting. In section 10.3 we illustrate how the principle can be extended to stochastic problems. Finally, a regression-based Monte Carlo method to price American options is illustrated in section 10.4.

10.1 THE SHORTEST PATH PROBLEM

The easiest way to introduce dynamic programming is by considering one of its most natural applications, i.e., finding the shortest path in a network. Graph and network optimization are not customary topics in Finance or Economics, but a quick look at figure 10.1 is enough to understand what we are talking about. A network consists of a set of nodes (numbered from 0 to 7 in our

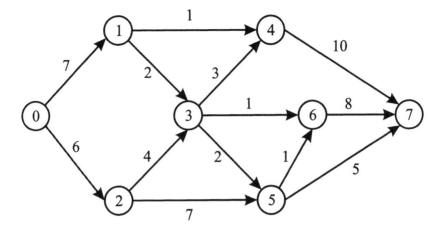

Fig. 10.1 A shortest path problem.

toy example) and a set of arcs joining pairs of nodes. Arcs are labeled by a number which can be interpreted as the arc length (or cost). Our purpose is finding a path in the network, starting from node 0 and leading to node 7, such that the path has total minimal length. For instance, summing the arc lengths we visit on the path $(0, 1, 4, 7)$, we see that its total length is 18, whereas path $(0, 1, 3, 5, 7)$ has length 16. At each node, we must choose the next node to visit. We may immediately appreciate that this problem bears some resemblance to dynamic decision making; given some state we are in, we should decide what to do in order to optimize an outcome that depends on the whole path. A greedy decision need not be the optimal one; for instance, the closest node to the starting point 0 in our network is node 2, but there is no guarantee that this arc is on an optimal path.

Of course, we could simply enumerate all the possible paths to spot the optimal one; here we have just a finite set of alternatives and there is no uncertainty involved, so the approach is conceptually feasible. However, this approach becomes quickly infeasible in practice, as the network size increases. So we must come up with some clever way to avoid exhaustive enumeration. Dynamic programming is one possible approach to accomplish this aim. It is worth noting that more efficient algorithms are available for the shortest path problem, but the idea we illustrate here can be extended to problems featuring infinite state spaces (countable or not) and uncertain data.

Let $\mathcal{N} = \{0, 1, 2, \ldots, N\}$ be the node set and $\mathcal{A}$ be the arc set; let the start and final nodes be 0 and N, respectively. For simplicity, we assume that the network is acyclic and that the arc lengths c_{ij}, $i, j \in \mathcal{N}$, are non-negative: If we had the possibility of getting trapped in a loop of negative length arcs, the optimal cost would be $-\infty$ and we do not want to consider such pathological cases.

The starting point is to find a *characterization* of the optimal solution, that can be translated into a constructive algorithm. Let V_i be the length of the shortest path from node $i \in \mathcal{N}$ to node N (denoted by $i \xrightarrow{*} N$). Assume that, for a specific $i \in \mathcal{N}$, node j lies on the optimal path $i \xrightarrow{*} N$. Then the following property holds: $j \xrightarrow{*} N$ is a subpath of $i \xrightarrow{*} N$. In other words, the optimal solution for a problem is obtained by assembling optimal solutions for subproblems. To understand why, consider the decomposition of $i \xrightarrow{*} N$ into the subpaths $i \to j$ and $j \to N$. The length of $i \xrightarrow{*} N$ is the sum of the lengths of the two subpaths:

$$V_i = L(i \to j) + L(j \to N). \tag{10.1}$$

Note that the second subpath is not affected by *how* we go from i to j. This is strongly related to the concept of state in Markovian dynamic systems: how we get to state j has no influence on the future. Now, assume that the subpath $j \to N$ is not the optimal path from j to N. Then we could improve the second term of (10.1) by considering the path consisting of $i \to j$ followed by $j \xrightarrow{*} N$. The length of this new path would be

$$L(i \to j) + L(j \xrightarrow{*} N) < L(i \to j) + L(j \to N) = V_i,$$

which is a contradiction, as we assumed that V_i was the optimal path length.

This observation leads to the following recursive equation for the shortest path from a generic node i to the terminal node N:

$$V_i = \min_{(i,j) \in \mathcal{A}} \{c_{ij} + V_j\} \qquad \forall j \in \mathcal{N}. \tag{10.2}$$

In other words, to find the optimal path from node i to node N, we should consider the immediate cost c_{ij} of going from i to all of its immediate successors j, plus the optimal cost of going from j to the terminal node. Note that we do not only consider the *immediate* cost, as in a greedy decision rule. We also add the future cost of the optimal sequence of decisions starting from each state we can visit next; this is what makes the approach non myopic. The function V_i is called **cost-to-go** or **value function** and is defined recursively by equation (10.2). The value function, for each point in the state space, tells us what the future optimal cost would be, if we reach that state and go on with an optimal policy. This kind of recursive equation, whose exact form depends on the problem at hand, is the heart of dynamic programming and is an example of a **functional equation**. In the shortest path problem, we have a finite set of states, and the value function is a vector; in an continuous-state model, the value function is an infinite-dimensional object.

Solving the problem requires finding the value function V_0 for the initial node, and to do that we should go backward in time.[2] We can associate a

---

[2] We are considering here only the *backward* version of dynamic programming. For the shortest path, and other deterministic combinatorial optimization problems, we could also

terminal condition $V_N = 0$ to our functional equation. Then we unfold the recursion by considering the immediate predecessors i of the terminal node N; for each of them, finding the optimal path length is trivial, as this is just c_{iN}. Then we proceed backward, labeling each node with the corresponding value function. In this unstructured network, we may label a node only when all of its successors have been labeled; we can always find the correct ordering in acyclic networks.

Example 10.1 Let us find the shortest path for the network depicted in Figure 10.1. We have the terminal condition $V_7 = 0$ for the terminal node, and we look for its immediate predecessors 4 and 6 (we cannot label node 5 yet, because node 6 is one of its successors). We have

$$V_4 = c_{47} + V_7 = 10 + 0 = 10$$
$$V_6 = c_{67} + V_7 = 8 + 0 = 8.$$

Now we may label node 5:

$$V_5 = \min \left\{ \begin{array}{c} c_{56} + V_6 \\ c_{57} + V_7 \end{array} \right\} = \min \left\{ \begin{array}{c} 1+8 \\ 5+0 \end{array} \right\} = 5.$$

Then we consider node 3 and its immediate successors 4, 5, and 6:

$$V_3 = \min \left\{ \begin{array}{c} c_{34} + V_4 \\ c_{35} + V_5 \\ c_{36} + V_6 \end{array} \right\} = \min \left\{ \begin{array}{c} 3+10 \\ 2+5 \\ 1+8 \end{array} \right\} = 7.$$

By the same token we have:

$$V_1 = \min \left\{ \begin{array}{c} c_{13} + V_3 \\ c_{14} + V_4 \end{array} \right\} = \min \left\{ \begin{array}{c} 2+7 \\ 1+10 \end{array} \right\} = 9$$

$$V_2 = \min \left\{ \begin{array}{c} c_{23} + V_3 \\ c_{25} + V_5 \end{array} \right\} = \min \left\{ \begin{array}{c} 4+7 \\ 7+5 \end{array} \right\} = 11$$

$$V_0 = \min \left\{ \begin{array}{c} c_{01} + V_1 \\ c_{02} + V_2 \end{array} \right\} = \min \left\{ \begin{array}{c} 7+9 \\ 6+11 \end{array} \right\} = 16.$$

Apart from getting the optimal length, which is 16, we may find the optimal path by looking for the nodes optimizing each single decision, starting from node 0:

$$0 \to 1 \to 3 \to 5 \to 7.$$

□

This might seem like a clumsy approach, but even in our simple shortest path problem this is better than an exhaustive enumeration of the alternatives. Furthermore, the same idea may be applied when uncertainty is involved and the value function is defined as an expected value.

---

apply a *forward* equation (see, e.g., [3, appendix D]). We consider only backward DP because of its relevance in stochastic decision making.

10.2 SEQUENTIAL DECISION PROCESSES

In this section we generalize the functional equation approach that we have just introduced for the shortest path problem. Consider a discrete-time dynamic system modeled by the state equation:

$$\mathbf{x}_{t+1} = \mathbf{h}_t(\mathbf{x}_t, \mathbf{u}_t), \qquad t = 0, 1, 2, \ldots, \tag{10.3}$$

where $\mathbf{x}_t$ is the vector of the state variables *at the beginning* of time interval t and $\mathbf{u}_t$ is the vector of the control variables applied *during* time interval t. No uncertainty is considered here: Given the current value of the state variable $\mathbf{x}_t$, after selecting the control variable $\mathbf{u}_t$ we know exactly what the future state will be, according to the time-varying dynamics described by the $\mathbf{h}_t$ functions. If system dynamics does not change in time, we can drop the subscript t from $\mathbf{h}_t$. The initial state $\mathbf{x}_0$ is given and we consider a finite time horizon from $t = 0$ to $t = T$. We want to find an optimal sequence of controls $(\mathbf{u}_0^*, \mathbf{u}_1^*, \ldots, \mathbf{u}_{T-1}^*)$, to which an optimal trajectory $(\mathbf{x}_0^*, \mathbf{x}_1^*, \ldots, \mathbf{x}_T^*)$ corresponds, in such a way as to minimize the objective function:

$$\sum_{t=1}^{T-1} f_t(\mathbf{x}_t, \mathbf{u}_t) + F_T(\mathbf{x}_T). \tag{10.4}$$

We have assumed an additive form, which makes the application of the dynamic programming principle easier, but other forms lend themselves to a decomposition approach. The objective function consists of a trajectory cost and a cost linked to the terminal state. The optimization must be carried out subject to the dynamic constraints (10.3) and, possibly, to constraints on the control variables and/or the state variables.

Example 10.2 As an example of deterministic sequential decision process we consider a stylized consumption–saving problem. We have an initial wealth W_0, and we must decide how much to save and how much to consume, at time instants $t = 0, 1, 2 \ldots, T-1$. What we save can be invested at a risk-free interest rate r. Furthermore, we have an income stream over the planning horizon. The state variable is W_t, the current wealth level. The control variable is immediate consumption C_t; if we rule out borrowing money, we must have $C_t \leq W_t$. The (exogenous) income stream is I_t. The state dynamics is

$$W_{t+1} = (W_t - C_t)(1 + r) + I_t, \qquad t = 0, 1, \ldots, T - 1.$$

We may want to maximize an additive utility function including a time discount factor $\beta < 1$:

$$\max \sum_{t=0}^{T-1} \beta^t u(C_t) + \beta^T B(W_T),$$

where u is some concave utility function and $B(\cdot)$ is the utility from bequest, valuing terminal wealth. If we do not consider the utility from bequest, the last decision is clearly to consume all available wealth. There is no uncertainty in this model, but the concavity of the utility function tends to enforce some regularity in the consumption stream.[3]

In some cases, the terminal valuation function B must be selected in such a way to overcome myopic behavior due to end-of-horizon effects. This happens when our planning horizon is truncated to make the problem manageable or to avoid planning for time periods so far that we cannot even characterize uncertainty in probabilistic terms. However, infinite time horizon models are often used in Economics:

$$\max \sum_{t=0}^{\infty} \beta^t u(C_t).$$

Here, discounting is essential to get a bounded objective function. The average cost/profit criterion may be also be used:

$$\lim_{T \to \infty} \frac{1}{T} \sum_{t=0}^{T-1} u(C_t),$$

but this more common in Engineering applications. □

This sequential decision problem can be solved by ordinary mathematical programming techniques, such as those discussed in chapter 6. However, understanding how we can tackle it by dynamic programming is helpful to develop approaches which can also be applied in more general settings.

10.2.1 The optimality principle and solving the functional equation

The objective function (10.4) is *separable*, in the sense that, for a given number r, the contribution of the last r decision stages depends only on the current state $\mathbf{x}_{T-r}$ and the r controls $\mathbf{u}_{T-r}, \ldots, \mathbf{u}_{T-1}$. Furthermore, a similar separation property (known as *Markovian state property*) holds for the trajectory, in the sense that the state $\mathbf{x}_{t+1}$ reached from $\mathbf{x}_t$ by applying the control $\mathbf{u}_t$ depends only on $\mathbf{x}_t$ and $\mathbf{u}_t$, and not on the past history $\mathbf{x}_0, \ldots, \mathbf{x}_{t-1}$. As a consequence of such separation properties, we obtain the **optimality principle**.

> An optimal policy $(\mathbf{u}_0^*, \mathbf{u}_1^*, \ldots, \mathbf{u}_{T-1}^*)$ is such that, whatever the initial state $\mathbf{x}_0$ and the first control $\mathbf{u}_0^*$, the next controls $(\mathbf{u}_1^*, \ldots, \mathbf{u}_{T-1}^*)$ are an optimal policy for the $(T-1)$-stage problem with initial state $\mathbf{x}_1$, obtained by applying the first control $\mathbf{u}_0^*$.

---

[3]One may actually argue that such a simple additive function does not capture habit formation effects.

Therefore, we may write a recursive functional equation to obtain the optimal policy:
$$V_t(\mathbf{x}_t) = \min_{\mathbf{u}_t} \left\{ f_t(\mathbf{x}_t, \mathbf{u}_t) + V_{t+1}(\mathbf{h}_t(\mathbf{x}_t, \mathbf{u}_t)) \right\}, \quad (10.5)$$
where the minimization is possibly carried out taking into account constraints on the control variable. This equation is known as **Bellman equation**, after the pioneer in dynamic programming. The value function $V_t(\mathbf{x}_t)$ is the total cost we incur by applying the optimal policy starting from state $\mathbf{x}_t$ at time t. This is a again a backward functional equation which must be solved to obtain the initial value function $V_0(\mathbf{x}_0)$.

The functional equation has a boundary condition that helps to start unfolding the recursion:
$$V_T(\mathbf{x}_T) = F_T(\mathbf{x}_T).$$
Then we step back to $t = T - 1$ and, for each possible state $\mathbf{x}_{T-1}$, we solve the following optimization problem:
$$V_{T-1}(\mathbf{x}_{T-1}) = \min_{\mathbf{u}_{T-1}} \left\{ f_{T-1}(\mathbf{x}_{T-1}, \mathbf{u}_{T-1}) + F_T(\mathbf{h}_{T-1}(\mathbf{x}_{T-1}, \mathbf{u}_{T-1})) \right\}.$$
This is, for each value of the state variable $\mathbf{x}_{T-1}$, a possibly constrained optimization problem: we have eliminated the dynamic constraint (10.3), but we could have constraints on state and/or control variables. Assuming we know the value function $V_{T-1}(\cdot)$, we may step back to build the value function $V_{T-2}(\cdot)$, by solving:
$$V_{T-2}(\mathbf{x}_{T-2}) = \min_{\mathbf{u}_{T-2}} \left\{ f_{T-2}(\mathbf{x}_{T-2}, \mathbf{u}_{T-2}) + V_{T-1}(\mathbf{h}_{T-2}(\mathbf{x}_{T-2}, \mathbf{u}_{T-2})) \right\}.$$
Going backward to the initial state $\mathbf{x}_0$, we solve the overall problem, one stage at a time. Note that if we knew the whole set of value functions, we could find the optimal control at each decision stage, given the current state we observe before making our decision.

We should wonder where this sequential decision problem differs from the previous shortest path problem:

- the state space is continuous
- there is an explicit time dynamics
- the set of available controls *can* be continuous, whereas in the shortest path problem the set of control actions was finite, as there was a finite set of successor nodes

Having a continuous state space means that, in principle, we should solve an infinite set of optimization problems for each time period. This can be avoided in a few lucky cases where we can find an analytical solution, but this is the exception rather than the rule. A possible approach is to discretize the state space. If we imagine doing that for each time period, we may see some

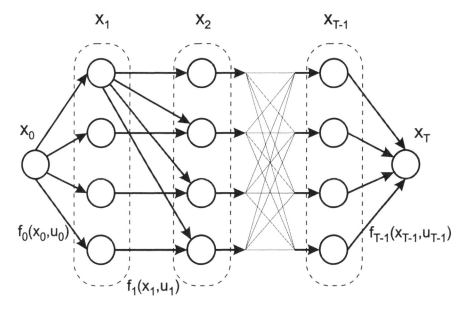

Fig. 10.2 A shortest path representation of a finite sequential decision process (for clarity, not all transitions are shown); the final state is assumed fixed.

similarity with the shortest path problem by looking at figure 10.2, where a network of discrete states is drawn. In order to emphasize the similarity with the shortest path problem, the network has been drawn under the assumption that the terminal state $\mathbf{x}_T$ is fixed. In this case we see that we have no difficulty with labeling nodes as the network is layered. The arc lengths are given by the cost of the corresponding state transitions.

Clearly, if we may find a suitable discretization of the state space, we have a computationally feasible approach. But we already know, from chapter 4, that discretizing high-dimensional state spaces with regular grids may be difficult. This is known as the curse of dimensionality in dynamic programming. Nevertheless, we may also see that the real issue is approximating the value function. If we know the set of value functions, or a suitable approximation, we can find the optimal control at any point of the state space. Using concepts introduced in section 3.3, we can approximate each value function as a linear combination of a set of basis functions:

$$V_t(\mathbf{x}_t) \approx \sum_{k=1}^{M} \alpha_{k,t}\psi_k(\mathbf{x}_t), \qquad (10.6)$$

where we have assumed that the set of basis functions does not change over time, but the set of weights $\alpha_{k,t}$ does. Hence, an infinite-dimensional problem boils down to the finite-dimensional problem of finding a suitable set of weights, possibly determined by interpolation or least squares. The quality of the solution we find depends on our choice of basis functions and on the choice of nodes in the state space, which we use to solve the function approximation problem. This is not easy and it is rather problem-dependent, but we see that the numerical techniques we have considered in the previous chapters, such as function approximation and numerical optimization, are building blocks in numerical dynamic programming. If we also introduce uncertainty, numerical integration comes into play as well.

10.3 SOLVING STOCHASTIC DECISION PROBLEMS BY DYNAMIC PROGRAMMING

In the deterministic setting, we may find the optimal control sequence and the corresponding state trajectory. But in a stochastic problem, the current state $\mathbf{x}_t$ and the control $\mathbf{u}_t$ we apply do not determine the next state, but only its conditional probability distribution. In a discrete-state setting, we may introduce a set of controlled transition probabilities. To ease the notation, let us assume that the transition probabilities are time-independent:

$$q_{ij}(\mathbf{u}) = P\{\mathbf{X}_{t+1} = j \mid \mathbf{x}_t = j, \mathbf{u}_t = \mathbf{u}\}.$$

where $\mathbf{X}_{t+1}$ is the next state (a random variable) and we have indexed states by integer numbers for conveniency. In the continuous-state case, we may think of dynamic equations such as

$$\mathbf{X}_{t+1} = \mathbf{h}(\mathbf{x}_t, \mathbf{u}_t, \epsilon_{t+1}), \qquad (10.7)$$

where ϵ_{t+1} is a random shock; this random variable has a subscript $t+1$ to emphasize that it is realized *after* we decide the control action $\mathbf{u}_t$. We cannot anticipate the control sequence, which is implicitly determined by the solution of recursive equations such as

$$V_t(\mathbf{x}_t) = \min_{\mathbf{u}_t} \{f(\mathbf{x}_t, \mathbf{u}_t) + E_t[V_{t+1}(\mathbf{h}(\mathbf{x}_t, \mathbf{u}_t, \epsilon_{t+1}))]\}, \qquad (10.8)$$

This is a straightforward generalization of equation (10.5). In the stochastic case, the future cost term is a *conditional* expectation; the notation E_t points out that expectation is carried out with respect to what we know now (the current state).

Example 10.3 To illustrate (10.7) and (10.8), we may generalize example 10.2 by including a risky asset in the set of investment opportunities. Assume we have a risky asset, whose price S_t follows, in continuous-time, the familiar geometric Brownian motion with drift μ and volatility σ. In discrete-time, we have:

$$\tilde{S}_{t+1} = S_t e^{\tilde{Y}}$$

where $\tilde{Y} \sim \mathcal{N}\left((\mu - \sigma^2/2)\delta t, \sigma\sqrt{\delta t}\right)$ and δt is the length of the time step. We use here the notation $\tilde{Y}$ to point out what is random at time t, and what is not. If we denote by $\alpha_t \in [0,1]$ the fraction of saved wealth that is invested in the risky asset, the wealth dynamics is

$$\tilde{W}_{t+1} = (W_t - C_t)\left[\alpha\frac{\tilde{S}_{t+1}}{S_t} + (1-\alpha)(1+r)\right] + I_t.$$

The recursive Bellman equation is, at time t,

$$V_t(W_t) = \max_{C_t, \alpha_t}\left\{u(C_t) + \beta E_t\left[V_{t+1}(\tilde{W}_{t+1})\right]\right\},$$

with terminal condition

$$V_T(W_T) = B(W_T).$$

□

In deterministic sequential processes, we want the optimal control path. In stochastic dynamic programming, what we really need is the set of value functions, one for each decision stage. Given the value function, at each decision stage we observe the current state and, given the value function, find the optimal control by solving a one-step optimization problem. The value function is what we need to avoid myopic decisions. Hence, we implicitly obtain the optimal control in feedback form: $\mathbf{u}_t = \boldsymbol{\phi}_t(\mathbf{x}_t)$.

As we already mentioned, in a continuous-state model the value function is an infinite-dimensional object, and we must somehow reduce it to a finite-dimensional object. Interpolation or approximation by a set of basis function are typically used to this aim. As we have seen in section 3.3, placing nodes is important in function approximation. This means that we should devise a grid in the state space, and using an evenly spaced one need not be the best idea.[4] In the stochastic case, an additional difficulty is given by the conditional expectation in (10.8). If the random shocks are continuously distributed random variables, our recursive equation involves a numerical integration problem. As we have seen in chapter 4, we may use deterministic or stochastic approaches, such as Gaussian quadrature or Monte Carlo sampling. It is important to note that here we want to approximate a *function* defined by an expectation, and not just an expected value as typical in option pricing. The following example shows how Gaussian quadrature can be extremely valuable in the discretization of conditional expectation.[5]

---

[4]See, e.g., [5] for numerical tricks useful in solving discrete-time DP models.
[5]This is strongly linked to scenario generation issues in stochastic programming with recourse; see section 11.3.

Example 10.4 Let us consider an extremely stylized asset allocation problem. An investor has a current wealth W_0 that she can invest at a continuously compounded risk free rate r, locking a total return $R = e^{rT}$ over a time horizon of length T. As an alternative, she can consider a risky stock whose current price is S_0. The risky asset price at T will be a random variable $\tilde{S}_T$; assuming geometric Brownian motion, we can express this future price as

$$\tilde{S}_T = S_0 e^{\tilde{Y}},$$

where $\tilde{Y}$ is normally distributed with expected value $(\mu - \sigma^2/2)T$ and variance $\sigma^2 T$. In section 4.1.2 we have described how Gauss–Hermite quadrature can be used to discretize such a random variable, and we have also implemented a MATLAB function, GaussHermite, to this aim. As an alternative, we may adopt plain Monte Carlo sampling.

We consider in this example a buy-and-hold strategy, with no intermediate consumption. Hence, the only decision variable is the fraction δ of wealth that our investor should allocate to the risky stock; we do not consider either borrowing or short-selling, hence δ must lie in the interval $[0, 1]$. Assuming a concave utility function $u(\cdot)$ the problem is

$$\max_{0 \le \delta \le 1} E\left[u(\tilde{W}_T)\right],$$

where future wealth $\tilde{W}_T$ is

$$\tilde{W}_T = W_0\left[\delta \frac{\tilde{S}_T}{S_0} + (1-\delta)R\right] = W_0\left[\delta\left(e^{\tilde{Y}} - R\right) + R\right]$$

and the term $e^{\tilde{Y}} - R$ can be interpreted as an *excess return* over the risk-free (total) return R. To discretize the problem, we should generate K scenarios, characterized by a realization Y_k and a probability π_k. If we use Monte Carlo sampling, we have $\pi_k = 1/K$; if we use Gauss–Hermite quadrature, the probability is the weight in the quadrature formula. The resulting problem is

$$\max_{0 \le \delta \le 1} \sum_{k=1}^{K} \pi_k u\left[W_0\left(\delta(e^{Y_k} - R) + R\right)\right].$$

Such a simple optimization problem can be tackled by the MATLAB function fminbnd. MATLAB code implementing Monte Carlo sampling is displayed in figure 10.3. The function receives self-explanatory arguments including a function argument utilf which is the utility function. If we assume logarithmic utility, here is the solution we may get:

```
>> randn('state',0)
>> share = OptFolioMC(1000,50,0.1,0.4,0.05,1,10000,@log)
share =
```

```
function share = OptFolioMC(W0,S0,mu,sigma,r,T,NScen,utilf)
muT = (mu - 0.5*sigma^2)*T;
sigmaT = sigma*sqrt(T);
R = exp(r*T);
NormSamples = muT + sigmaT*randn(NScen,1);
ExcessRets = exp(NormSamples) - R;
MExpectedUtility = @(x) -mean(utilf(W0*((x*ExcessRets) + R)));
share = fminbnd(MExpectedUtility, 0, 1);
```

Fig. 10.3 Simple asset allocation problem under uncertainty: Monte Carlo sampling.

```
        0.3092
>> share = OptFolioMC(1000,50,0.1,0.4,0.05,1,10000,@log)
share =
        0.3246
>> share = OptFolioMC(1000,50,0.1,0.4,0.05,1,10000,@log)
share =
        0.3112
>> share = OptFolioMC(1000,50,0.1,0.4,0.05,1,10000,@log)
share =
        0.3763
>> share = OptFolioMC(1000,50,0.1,0.4,0.05,1,10000,@log)
share =
        0.3341
>> share = OptFolioMC(1000,50,0.1,0.4,0.05,1,10000,@log)
share =
        0.3436
>> share = OptFolioMC(1000,50,0.1,0.4,0.05,1,10000,@log)
share =
        0.2694
```

There is a striking variability in the solution, which is due to sampling variability in scenario generation. Even 10000 samples do not seem reliable. If we increase the number of scenarios, the solution does stabilize:

```
>> randn('state',0)
>> share = OptFolioMC(1000,50,0.1,0.4,0.05,1,5000000,@log)
share =
        0.3049
>> share = OptFolioMC(1000,50,0.1,0.4,0.05,1,5000000,@log)
share =
        0.3067
>> share = OptFolioMC(1000,50,0.1,0.4,0.05,1,5000000,@log)
share =
        0.3074
```

```
function share = OptFolioGauss(W0,S0,mu,sigma,r,T,NScen,utilf)
muT = (mu - 0.5*sigma^2)*T;
sigmaT = sigma*sqrt(T);
R = exp(r*T);
[x,w] = GaussHermite(muT,sigmaT^2,NScen);
ExcessRets = exp(x) - R;
MExpectedUtility = @(x) -dot(w, utilf(W0*((x*ExcessRets) + R)));
share = fminbnd(MExpectedUtility, 0, 1);
```

Fig. 10.4 Simple asset allocation problem under uncertainty: Gauss–Hermite quadrature.

However, we cannot afford such a huge number of scenarios in a complex problem, even less when we have to solve such a problem repeatedly within a numerical dynamic programming scheme. Hence, we may try to improve things using Gauss–Hermite quadrature. Using the function GaussHermite from chapter 4, we get the code in figure 10.4. Using clever scenario generation, we need much less scenarios to get a reliable solution:

```
>> share = OptFolioGauss(1000,50,0.1,0.4,0.05,1,2,@log)
share =
    0.3139
>> share = OptFolioGauss(1000,50,0.1,0.4,0.05,1,3,@log)
share =
    0.3061
>> share = OptFolioGauss(1000,50,0.1,0.4,0.05,1,4,@log)
share =
    0.3064
>> share = OptFolioGauss(1000,50,0.1,0.4,0.05,1,5,@log)
share =
    0.3064
>> share = OptFolioGauss(1000,50,0.1,0.4,0.05,1,100,@log)
share =
    0.3064
```

This little experiment just shows that Gaussian quadrature is a most valuable tool for numerical dynamic programming. Of course, apart from playing with numbers, one should try to understand some qualitative properties of the optimal solution by more analytical approaches. For instance, we have seen in example 2.14 (page 71) that logarithmic utility is a CRRA (constant relative risk aversion) function. Hence, we should expect that the solution does not depend on the current wealth W_0. Numerical experimentation confirms (but *does not* prove) this:

```
>> share = OptFolioGauss(100,50,0.1,0.4,0.05,1,5,@log)
```

```
share =
    0.3064
>> share = OptFolioGauss(10,50,0.1,0.4,0.05,1,5,@log)
share =
    0.3064
```

We may also play with different utility functions, such as power utility $u(W) = W^{1-\gamma}/(1-\gamma)$, to see the effect of risk aversion:

```
>> gamma = 0.3;, powU = @(W) W.^(1-gamma)/(1-gamma);
>> share = OptFolioGauss(1000,50,0.1,0.4,0.05,1,5,powU)
share =
    0.9999
>> gamma = 0.4;, powU = @(W) W.^(1-gamma)/(1-gamma);
>> share = OptFolioGauss(1000,50,0.1,0.4,0.05,1,5,powU)
share =
    0.7887
>> gamma = 0.5;, powU = @(W) W.^(1-gamma)/(1-gamma);
>> share = OptFolioGauss(1000,50,0.1,0.4,0.05,1,5,powU)
share =
    0.6295
```

Note the use of the dot (.) operator in the definition of powU and the fact that, if we change gamma, we have to redefine the function, because the function is bound to the current value of gamma when the function is defined. □

In the example above, we have just played with numbers on a possible subproblem of dynamic programming. We may also take this opportunity to stress the fact that, by analyzing the Bellman equations, we may obtain important insights into the *structure* of the optimal solution. For instance, applying dynamic programming to a consumption-saving problem like the one we have described in example 10.3, it can be shown that logarithmic utility implies that a fixed fraction of wealth is consumed at each decision stage.[6]

Generalizing the example, if we apply Gaussian quadrature to discretize conditional expectation, equation (10.8) becomes

$$V_t(\mathbf{x}_t) = \min_{\mathbf{u}_t} \left\{ f(\mathbf{x}_t, \mathbf{u}_t) + \sum_{k=1}^{K} \pi_k V_{t+1}(\mathbf{h}(\mathbf{x}_t, \mathbf{u}_t, \epsilon_k)) \right\}.$$

Even though Gaussian quadrature is very helpful, it does not solve all of our difficulties. In high-dimensional problems, we may still be forced to use Monte Carlo sampling. Furthermore, we have to discretize the state space and to solve possibly difficult optimization problems. But all of this is very easy if we

---

[6]See, e.g., [8, chapter 11] for a careful analysis of intertemporal consumption and portfolio choices with logarithmic and power utilities.

are able to find a suitable discretization of the state space and if the control decision is very simple, as the following example shows.

Example 10.5 Now that we are acquainted with dynamic programming, it is very useful to reinterpret the binomial lattice approach to price American options (see section 7.2). Indeed, equation (7.6), which we recall here for convenience, is a very simple case of a dynamic programming recursion:

$$f_{i,j} = \max\{K - S_{ij},\; e^{-r\delta t}(pf_{i+1,j+1} + (1-p)f_{i,j+1})\}.$$

It is so easy because we have a finite state space, arising from a moment matching discretization of geometric Brownian motion, and because the set of control decisions is finite: either we continue, or we exercise the option. The value function is $f_{i,j}$, i.e., the option value for asset price i at time j. Maximization over control decisions just requires to choose if we want to exercise, and grab the intrinsic value, or we want to continue. In the second case, the continuation value is the discounted expected value function, computed over the two successor states of the current one, under the risk neutral measure. □

We close this section by giving a few clues about how one can handle infinite-horizon dynamic programs, assuming a discount factor is used.[7] A fairly natural guess is that the recursive equation (10.8) can be applied by dropping the time subscripts:

$$V(\mathbf{x}) = \min_{\mathbf{u}} \left\{ f(\mathbf{x}, \mathbf{u}) + \mathrm{E}\left[V(\mathbf{h}(\mathbf{x}, \mathbf{u}, \tilde{\boldsymbol{\epsilon}}])\right] \right\}. \qquad (10.9)$$

The intuition here is that in an infinite-horizon problem we may look for a stationary policy, i.e., a policy such that a control decision is associated to each state; on the contrary, in a finite-horizon problem the policy can change when we are approaching the end of the time horizon. Existence of a stationary optimal policy should not be taken for granted,[8] but the approach can be rigorously justified under some hypotheses. It is also interesting to note that, in this case, solving the Bellman equation calls for finding a fixed point of an operator; iterative methods are available to this purpose.

In the finite-dimensional case the Bellman equation boils down to a set of non-linear equations:

$$V_i = \min_{\mathbf{u}} \left[f(\mathbf{x}, \mathbf{u}) + \beta \sum_{j=1}^{N} q_{ij}(\mathbf{u}) V_j \right], \qquad (10.10)$$

---

[7] The average cost/profit case is more difficult; see, e.g., [1].
[8] A rather odd case may occur when chance constraints are enforced on states, i.e., when we require that the probability of visiting a subset of "bad" states is low. It may happen that the optimal policy is randomized, i.e., when we are in certain states we should select the control action according to a probability distribution. See, e.g., [14, pp. 255–257].

where $q_{ij}(\mathbf{u})$ is an element of the (control dependent) transition probability matrix. This system can be tackled by iterative methods, including variants of Newton's method. In the infinite-dimensional case, we may resort to the collocation method that we have introduced in section 3.4.4. This requires choosing a set of basis functions and collocation nodes to approximate the value function:

$$V(\mathbf{x}) \approx \sum_{j=1}^{M} \alpha_j \psi_j(\mathbf{x}).$$

If we consider M basis functions, we should select M collocation nodes $\mathbf{x}_1, \ldots, \mathbf{x}_M$. We should also discretize the random shocks. Assume we adopt Gaussian quadrature with weights π_k and nodes ϵ_k, $k = 1, \ldots, K$. Then, Bellman equation for each state $\mathbf{x}_i$ reads

$$\sum_{j=1}^{M} \alpha_j \psi_j(\mathbf{x}_i) = \min_{\mathbf{u}} \left\{ f(\mathbf{x}_i, x) + \beta \sum_{k=1}^{K} \sum_{j=1}^{M} \pi_k \alpha_j \psi_j \left(\mathbf{h}(\mathbf{x}_i, \mathbf{u}, \epsilon_k) \right) \right\}.$$

This is a set of non-linear equation in the unknown weights α_j. It can be tackled, e.g., by Newton's method.[9]

10.4 AMERICAN OPTION PRICING BY MONTE CARLO SIMULATION

Example 10.5 shows that if we discretize geometric Brownian motion using a lattice, dynamic programming boils down to a simple pricing approach. However, discretization with respect to time means that we are actually pricing a Bermudan option; since the exercise opportunities are restricted to a set of discrete times, what we get is actually a lower bound on the option price. We may actually apply a dynamic programming framework in continuous-time, but this essentially leads to the Black–Scholes partial differential equation with a free boundary. This may be tackled, e.g., by finite differences. Both lattices and finite differences are limited in their ability to cope with multiple stochastic factors, which is what Monte Carlo simulation is good at. Hence, it is natural to wonder if Monte Carlo simulation can be applied to option pricing with early exercise features. The answer is that indeed we can apply Monte Carlo, within a stochastic dynamic optimization framework. In this section we describe an approach due to Longstaff and Schwartz [10], which should be interpreted as a way to approximate the value function of dynamic programming by linear regression against a set of basis functions. Since we

---

[9] We refer the reader to [12] for more details, a set of examples, and a MATLAB-based toolbox accomplishing this task.

512 DYNAMIC PROGRAMMING

approximate the value function, what we expect is a suboptimal solution; furthermore, time is discretized; hence, we should expect some low bias in our estimate of price. Approaches to get high-biased estimators are described in the literature, and are useful to bound the price.

For the sake of simplicity, we will just consider a vanilla American put option on a single, non-dividend paying stock. Clearly, the approach makes sense in more complex settings. As usual with Monte Carlo simulation, we generate sample paths $(S_0, S_1, \ldots, S_j, \ldots, S_N)$, where we use j as a discrete time index, $S_j = S(j\,\delta t)$, and $T = M\,\delta t$ is the expiration time of the option. If we denote by $I_j(S_j)$ the intrinsic value of the option at time j, the dynamic programming recursion for the value function $V_j(S_j)$ is

$$V_j(S_j) = \max\left\{I_j(S_j),\, \mathrm{E}_j^Q\left[e^{-r\,\delta t}V_{j+1}(\tilde{S}_{j+1})\middle|\, S_j\right]\right\}. \qquad (10.11)$$

In the case of a vanilla American put, we have $I_j(S_j) = \max\{K - S_j, 0\}$. This is the generalization to a continuous-state model of the recursive equation in example 10.5. Having to cope with continuous prices is the only difficulty we have here, as time is discretized and the set control actions is finite: either exercise, or continue. It is important to realize that we *cannot* take this decision along individual sample paths; if we are at a given point of a sample path generated by Monte Carlo sampling, we cannot exploit knowledge of future prices along that path, as this would imply clairvoyance.[10] What we can do is using our set of scenarios to build an approximation of the conditional expectation in equation (10.11), for some choice of basis functions $\psi_k(S_j)$, $k = 1,\ldots,K$. The simplest choice we can think of is regressing the conditional expectation against a basis of monomials: $\psi_1(S) = 1$, $\psi_2(S) = S$, $\psi_3(S) = S^2$, etc. In practice, orthogonal polynomials can also be used. Note that we are using the same set of basis function for each time instant, but the weights in the linear combination will depend on time:

$$\mathrm{E}_j^Q\left[e^{-r\,\delta t}V_{j+1}(\tilde{S}_{j+1})\middle|\, S_j\right] \approx \sum_{k=1}^{K} \alpha_{kj} S_j^{k-1}.$$

The weights α_{kj} can be found by linear regression, going backward in time; the approximation is non-linear in S_j, but it is linear in terms of the weights.

In order to illustrate the method, we should start from the last time period. Assume we have generated N sample paths, and let us denote by S_{ji} the price at time j on sample path $i = 1,\ldots,N$. When $j = M$, i.e., at expiration, the value function is trivially:

$$V_M(S_{Mi}) = \max\{K - S_{Mi}, 0\}$$

---

[10] This point will also be appreciated in section 11.2, where we discuss the role of non-anticipativity in multistage stochastic programming. See also section 4.5.4 where we use Monte Carlo simulation to price a chooser option.

for each sample path i. These values can be used, in a sense, as the Y-values in a linear regression, where the X values are the prices at time $j = M - 1$. More precisely, we may consider the regression model:

$$e^{-r\,\delta t}\max\{K - S_{Mi}, 0\} = \sum_{k=1}^{K} \alpha_{k,M-1} S_{M-1,i}^{k-1} + e_i, \qquad i = 1, \ldots, N,$$

where e_i is the residual for each sample path. We may find the weights $\alpha_{k,M-1}$ by the usual least squares approach, minimizing the sum of squared residuals. Note that we are considering the discounted payoff, so that we may then compare it directly against the intrinsic value.

In the regression above, we have considered all of the generated sample paths. Actually, it is much better to consider only the subset of sample paths for which we have a decision to take at time $j = M - 1$. This subset is simply the set of sample paths in which the option is in the money at time $j = M - 1$. In fact, if the option is not in the money, we have no reason to exercise; using only the sample paths for which the option is in the money is called the "moneyness" criterion and it improves the performance of the overall approach. Denoting this subset by $\mathcal{I}_{M-1}$ and assuming $K = 3$, we would have to solve the following least squares problem:

$$\begin{aligned}
\min \quad & \sum_{i \in \mathcal{I}_{M-1}} e_i^2 \\
\text{s.t.} \quad & \alpha_{1,M-1} + \alpha_{2,M-1} S_{M-1,i} + \alpha_{3,M-1} S_{M-1,i}^2 + e_i \\
& = e^{-r\,\delta t}\max\{K - S_{Mi}, 0\}, \qquad i \in \mathcal{I}_{M-1}. \quad (10.12)
\end{aligned}$$

The output of this problem is a set of weights, which allow us to approximate the continuation value. Note that the weights are linked to the time period, and not to sample paths. Using the same approximation for each sample path in $\mathcal{I}_{M-1}$, we may decide if we exercise or not.

We should pause and illustrate what we have seen so far by a little numerical example. We will use the same example as the original reference [10], where the eight sample paths given in table 10.1 are considered for a vanilla American put with strike price $K = 1.1$. For each sample path, we also have a set of cash flows at expiration; cash flows are positive where the option is in the money. Cash flows are discounted back to time $j = 2$ and used for the first linear regression. Assuming a risk free rate of 6% per period, the discount factor is $e^{-0.06} = 0.94176$. The data for the regression are given in table 10.2; X corresponds to current underlying asset price and Y corresponds to discounted cash flows in the future. We see that only the sample paths in which the option is in the money at time $j = 2$ are used. The following approximation is obtained:

$$E[Y \mid X] \approx -1.070 + 2.983X - 1.813X^2.$$

Table 10.1 Sample path and cash flows at option expiration for a vanilla American put.

Path	$j=0$	$j=1$	$j=2$	$j=3$	Path	$j=1$	$j=2$	$j=3$
1	1.00	1.09	1.08	1.34	1	-	-	.00
2	1.00	1.16	1.26	1.54	2	-	-	.00
3	1.00	1.22	1.07	1.03	3	-	-	.07
4	1.00	0.93	0.97	0.92	4	-	-	.18
5	1.00	1.11	1.56	1.52	5	-	-	.00
6	1.00	0.76	0.77	0.90	6	-	-	.20
7	1.00	0.92	0.84	1.01	7	-	-	.09
8	1.00	0.88	1.22	1.34	8	-	-	.00

Table 10.2 Regression data for time $j=2$.

Path	Y	X
1	.00 × .94176	1.08
2	-	-
3	.07 × .94176	1.07
4	.18 × .94176	0.97
5	-	-
6	.20 × .94176	0.77
7	.09 × .94176	0.84
8	-	-

Table 10.3 Comparing intrinsic and continuation value at time $j = 2$, and resulting cash flow matrix.

Path	Exercise	Continue	Path	$j=1$	$j=2$	$j=3$
1	.02	.0369	1	-	.00	.00
2	-	-	2	-	.00	.00
3	.03	.0461	3	-	.00	.07
4	.13	.1176	4	-	.13	.00
5	-	-	5	-	.00	.00
6	.33	.1520	6	-	.33	.00
7	.26	.1565	7	-	.26	.00
8	-	-	8	-	.00	.00

Now, based on this approximation, we may compare at time $j = 2$ the intrinsic value and the continuation value. This is carried out in table 10.3. Given the exercise decisions, we update the cash flow matrix. Note that the exercise decision does not exploit knowledge of the future. Consider sample path 4: we exercise, making $0.13; on that sample path, we would regret our decision, because we could make $0.18 at time $j = 3$. We should also note that on some paths we exercise at time $j = 2$, and this is reflected by the updated cash flow matrix in the table.

The process is repeated going backward in time. To carry out the regression, we must consider the cash flows on each path, resulting from the early exercise decisions. Say we are at time step j, and consider path i. For each path i, there will be an exercise time j_e^*, which we set conventionally to $M+1$ if the option will never be exercised in the future. Then the regression problem (10.12) should be rewritten, for the generic time period j, as:

$$\min \quad \sum_{i \in \mathcal{I}_j} e_i^2$$
$$\text{s.t.} \quad \alpha_{1j} + \alpha_{2j} S_{ji} + \alpha_{3j} S_{ji}^2 + e_i \qquad (10.13)$$
$$= \begin{cases} e^{-r(j_e^* - j)\delta t} \max\{K - S_{j_e^*, i}, 0\} & \text{if } j_e^* \leq M \\ 0 & \text{if } j_e^* = M+1 \end{cases} \quad i \in \mathcal{I}_j.$$

Since there can be at most one exercise time for each path, it may be the case that after comparing the intrinsic value with the continuation value on a path, the exercise time j_e^* is reset to a previous period. Stepping back to time $j = 1$, we have the regression data of table 10.4. The discount factor $e^{-2 \cdot 0.06} = 0.88692$ is applied on paths 1 and 8. Since the cash flow there is zero, the discount factor is irrelevant, but we prefer using this to point out that we are discounting cash flows from time period $j = 3$; if we had a positive cash flow at $j = 3$ and zero cash flow at $j = 2$, this is the discount factor we

DYNAMIC PROGRAMMING

Table 10.4 Regression data for time $j = 1$.

Path	Y	X
1	.00 × .88692	1.09
2	-	-
3	-	-
4	.13 × .94176	0.93
5	-	-
6	.33 × .94176	0.76
7	.26 × .94176	0.92
8	.00 × .88692	0.88

Table 10.5 Comparing intrinsic and continuation value at time $j = 1$, and resulting cash flow matrix.

Path	Exercise	Continue	Path	$j = 1$	$j = 2$	$j = 3$
1	.01	.0139	1	.00	.00	.00
2	-	-	2	.00	.00	.00
3	-	-	3	.00	.00	.07
4	.17	.1092	4	.17	.00	.00
5	-	-	5	.00	.00	.00
6	.34	.2866	6	.34	.00	.00
7	.18	.1175	7	.18	.00	.00
8	.22	.1533	8	.22	.00	.00

should use. Least squares yield the approximation:

$$E[Y \mid X] \approx 2.038 - 3.335X + 1.356X^2.$$

This approximation may seem unreasonable, as we expect smaller payoffs for larger asset prices, yet the highest power of the polynomial has a positive coefficient here. It can be verified that, for the range of X values we are considering, the function is decreasing. Based on this approximation of the continuation value, we obtain the exercise decisions illustrated in table 10.5. Discounting all cash flows back to time $j = 0$ and averaging over the eight sample paths, we get an estimate of the continuation value of $0.1144, which is larger than the intrinsic value $0.1; hence, the option should not be exercised immediately. In the next section we illustrate how MATLAB can be used to implement this procedure.

10.4.1 A MATLAB implementation of the least squares approach

To carry out linear regression, there are at least two possibilities. One is to use the regress function from the Statistics toolbox. This function also returns a lot of statistically relevant information; however, since we are using regression only as a function approximation tool, and not all readers have access to that toolbox, we will use the familiar backslash \ operator. When used with a square matrix **A** and a correspondingly sized vector **b**, this operator solves the system $\mathbf{Ax} = \mathbf{b}$. Otherwise, it returns a least squares solution, which is what we are looking for.

A first step is writing a function which replicates the toy example we have just considered. The MATLAB code is displayed in figure 10.5; it is written as a function, but in fact it is a script. The sample paths from the example are assigned to matrix SPaths, where we do not include the initial price S_0. The cash flow matrix is stored in the vector CashFlows. We use a vector, since there can be at most one positive entry on each row in this matrix; we use another vector, ExerciseTime, to store the times at which the option is exercised on each path; this corresponds to time subscript j_e^* above, and is used to select the appropriate discount factor in the vector discountVet. If the option is never exercised along a sample path, we can set j_e^* to the number of steps, since we are discounting zero for that path. The main for loop proceeds backward in time. The vector InMoney contains the indexes of sample paths which are in the money at the time step we are considering; we carry out regression by least squares using the relevant data, obtaining the coefficient vector alpha which is used to compute the continuation value for each point. The vector Index contains the indexes of the in-the-money sample paths on which we exercise; these indexes are relative to the subset of these sample paths (which are in one-to-one correspondence to the rows of matrix RegrMat) and do not correspond to the original sample path indexes; these are recovered in vector ExercisePaths. After carrying out all regressions, we average the discounted cash flows to get the continuation value at time $j = 0$; this should be checked against the immediate intrinsic value to yield the option price. The reader is urged to step through this function using the debugger to check the calculations in the toy example.

Now it is fairly easy to extend this function to price an American put option using an arbitrary set of basis functions. The code for GenericLS and a script to check it against binomial lattices are given in figure 10.6. Sample paths are generated by function AssetPaths from section 8.1.1, and we get rid of the initial price. The function is much like ExampleLS, and the only difference is that we use a cell array, fhandles, of function handles to contain the set of basis functions. Each element in the set of basis function is used to evaluate a column in the regression matrix. To this aim, we use the feval MATLAB function; this is, in some sense, a higher-order function taking as arguments another function and a set of arguments on which this should be evaluated. Function handles are built in the script using the @ operator and can be stored

```
function price = ExampleLS;
% this function replicates example 1 on pages 115-120 of the
% original paper by Longstaff and Schwartz
S0 = 1; K = 1.1; r = 0.06; T = 3;
NSteps = 3; dt = T/NSteps;
discountVet = exp(-r*dt*(1:NSteps)');
% generate sample paths
NRepl = 8;
SPaths = [
    1.09 1.08 1.34
    1.16 1.26 1.54
    1.22 1.07 1.03
    0.93 0.97 0.92
    1.11 1.56 1.52
    0.76 0.77 0.90
    0.92 0.84 1.01
    0.88 1.22 1.34
];
%
alpha = zeros(3,1); % regression parameters
CashFlows = max(0, K - SPaths(:,NSteps));
ExerciseTime = NSteps*ones(NRepl,1);
for step = NSteps-1:-1:1
    InMoney = find(SPaths(:,step) < K);
    XData = SPaths(InMoney,step);
    RegrMat = [ones(length(XData),1), XData, XData.^2];
    YData = CashFlows(InMoney).*discountVet(ExerciseTime(InMoney)-step);
    alpha = RegrMat \ YData;
    IntrinsicValue = K - XData;
    ContinuationValue = RegrMat * alpha;
    Index = find(IntrinsicValue>ContinuationValue);
    ExercisePaths = InMoney(Index);
    CashFlows(ExercisePaths) = IntrinsicValue(Index);
    ExerciseTime(ExercisePaths) = step;
end % for
price = max( K-S0, mean(CashFlows.*discountVet(ExerciseTime)) );
```

Fig. 10.5 MATLAB function to replicate example 1 from [10].

either in cell arrays or structs, not in ordinary arrays; we have chosen the first possibility.

Now we may check the results we obtain by least squares Monte Carlo against those provided by lattice based `binprice` function:

```
>> CheckLS
priceLS =
    6.8074
priceBIN =
    6.8129
```

10.4.2 Some remarks and alternative approaches

In the previous example, we have used a simple quadratic polynomial. In more complex cases, we should be careful in the selection of basis functions. In the case of multiple assets, say S_1 and S_2, one could consider regressing against polynomials involving cross-products such as $S_1 S_2$, $S_1^2 S_2$, $S_1 S_2^2$, etc. There is a non-trivial trade-off between accuracy and complexity.

The reader may also have noticed that we did not evaluate a confidence interval for the price. Actually, this can and should be done, but we must be careful in considering the bias in our estimator. Least-squares Monte Carlo, when properly used, yields a low-biased estimator. As we have already noted, one source of bias, for truly American options, comes from the fact that we are considering a subset of the available exercise opportunities. This is not a problem for Bermudan options, and Richardson extrapolation has been proposed to improve accuracy for American options. Another source of bias comes from suboptimality. We have seen a similar issue when pricing a chooser option in section 4.5.4. But to have a clear bias, we should actually use least squares Monte Carlo first to generate an early exercise strategy; then we should simulate the application of that (suboptimal) strategy to estimate the average discounted payoff. Alternative, more sophisticated, approaches have been proposed to compute high-biased estimators. One way to do so could be simulating early exercise with clairvoyance: along each path, we take exercise decisions knowing what comes next along each sample path. This is not feasible in practice, and corresponds to relaxing obvious non-anticipativity constraints on our decisions, and it results in a upper bound on the option price. However, this bound may be rather weak, i.e., too large. Having confidence intervals from low- and high-biased estimators, we may build an overall confidence interval for the price.

In least squares Monte Carlo, we have built an exercise strategy based on the value of continuation. A possible alternative is trying to find the exercise boundary directly, e.g., using splines or a suitably parameterized family of functions. This is clearly feasible for simple options; for the vanilla put, we should get something like figure 2.22 on page 118. But this is not easy in general, since the early exercise region need not be connected: We may have to find multiple surfaces describing a complicated region.

```
function price = GenericLS(S0,K,r,T,sigma,NSteps,NRepl,fhandles)
dt = T/NSteps;
discountVet = exp(-r*dt*(1:NSteps)');
NBasis = length(fhandles); % number of basis functions
alpha = zeros(NBasis,1); % regression parameters
RegrMat = zeros(NRepl,NBasis);
% generate sample paths
SPaths=AssetPaths(S0,r,sigma,T,NSteps,NRepl);
SPaths(:,1) = []; % get rid of starting prices
%
CashFlows = max(0, K - SPaths(:,NSteps));
ExerciseTime = NSteps*ones(NRepl,1);
for step = NSteps-1:-1:1
    InMoney = find(SPaths(:,step) < K);
    XData = SPaths(InMoney,step);
    RegrMat = zeros(length(XData), NBasis);
    for k=1:NBasis
        RegrMat(:, k) = feval(fhandles{k}, XData);
    end
    YData = CashFlows(InMoney).*discountVet(ExerciseTime(InMoney)-step);
    alpha = RegrMat \ YData;
    IntrinsicValue = K - XData;
    ContinuationValue = RegrMat * alpha;
    Index = find(IntrinsicValue > ContinuationValue);
    ExercisePaths = InMoney(Index);
    CashFlows(ExercisePaths) = IntrinsicValue(Index);
    ExerciseTime(ExercisePaths) = step;
end % for
price = max(K-S0, mean(CashFlows.*discountVet(ExerciseTime)));
```

```
% CheckLS.m
S0 = 50; K = 50; r = 0.05;
sigma = 0.4; T = 1; NSteps = 50;
NRepl = 10000;
randn('state',0)
fhandles = {@(x)ones(length(x),1), @(x)x, @(x)x.^2};
priceLS = GenericLS(S0,K,r,T,sigma,NSteps,NRepl,fhandles)
[LatS, LatPrice]=binprice(S0,K,r,T,T/NSteps,sigma,0);
priceBIN = LatPrice(1,1)
```

Fig. 10.6 MATLAB function to price a vanilla American put by least squares Monte Carlo and a script to check it.

In recent years, many alternative approaches have been proposed for pricing American or Bermudan options by random sampling. In section 4.5.4 we have seen a simple case in which we build a bushy tree; given the need to generate a large number of samples, the approach may be feasible when a limited number of exercise opportunities are given. Alternative discretization strategies based on a recombining mesh have been proposed; for all of this we refer the reader to the specific literature.

For further reading

In the literature

- Dynamic programming is arguably the most powerful concept in optimization, and its many potential applications are well illustrated in [1].

- To overcome the curse of dimensionality, a great deal of effort has been devoted to the development of approximate solution methods [1, 2], which also include simulation-based methods. This has paved the way to simulation-based pricing for high-dimensional American-style options. An example of this line of research is [15].

- In the original paper [10], the reader may also find some treatment of convergence issues, which we have neglected.

- The best treatment of Monte Carlo for American options is [6, chapter 8]. See also [13, chapter 6] or [7].

- Numerical dynamic programming for applications in Economics is dealt with in [9] and [12].

- Continuous-time models are quite useful when the model is reasonably simple and an analytical solution can be found, usually yielding valuable insights into the nature of the problem. An excellent reference on continuous-time dynamic programming in finance is [11].

- Another valuable reference is [4], where stylized models are used to gain insights into household long-term saving behavior. There, the value of approximate analytical solutions is also emphasized.

On the Web

- The MATLAB toolbox for computational economics, which is associated to [12], can be downloaded from

    http://www4.ncsu.edu/~pfackler/compecon/

- Useful lecture notes on numerical dynamic programming, and some (Mathematica) code, can be downloaded from

  http://www.econ.jhu.edu/people/ccarroll/index.html

REFERENCES

1. D.P. Bertsekas. *Dynamic Programming and Optimal Control (2nd ed., Vols. 1 and 2)*. Athena Scientific, Belmont, MA, 2001.

2. D.P. Bertsekas and J.N. Tsitsiklis. *Neuro-Dynamic Programming*. Athena Scientific, Belmont, MA, 1996.

3. P. Brandimarte and A. Villa. *Advanced Models for Manufacturing Systems Management*. CRC Press, Boca Raton, FL, 1995.

4. J.Y. Campbell and L.M. Viceira. *Strategic Asset Allocation*. Oxford University Press, Oxford, 2002.

5. C. Carroll. Solving Microeconomic Dynamic Stochastic Optimization Problems. Lecture Notes downloadable from
http://www.econ.jhu.edu/people/ccarroll/index.html.

6. P. Glasserman. *Monte Carlo Methods in Financial Engineering*. Springer-Verlag, New York, NY, 2004.

7. P. Jaeckel. *Monte Carlo Methods in Finance*. Wiley, Chichester, 2002.

8. J.E. Ingersoll, Jr. *Theory of Financial Decision Making*. Rowman & Littlefield, Totowa, NJ, 1987.

9. K.L. Judd. *Numerical Methods in Economics*. MIT Press, Cambridge, MA, 1998.

10. F.A. Longstaff and E.S. Schwartz. Valuing American Options by Simulation: a Simple Least-Squares Approach. *The Review of Financial Studies*, 14:113–147, 2001.

11. R.C. Merton. *Continuous-Time Finance*. Blackwell Publishers, Malden, MA, 1990.

12. M.J. Miranda and P.L. Fackler. *Applied Computational Economics and Finance*. MIT Press, Cambridge, MA, 2002.

13. D. Tavella. *Quantitative Methods in Derivatives Pricing: Introduction to Computational Finance*. Wiley, New York, 2002.

14. H.C. Tijms. *A First Course in Stochastic Models.* Wiley, Chichester, 2003.

15. J.N. Tsitsiklis and B. Van Roy. Optimal Stopping of Markov Processes: Hilbert Space Theory, Approximation Algorithms, and an Application to Pricing High-Dimensional Financial Derivatives. *IEEE Transactions on Automatic Control*, 44:1840–1851, 1999.

16. L.A. Wolsey. *Integer Programming.* Wiley, New York, 1998.

11
Linear Stochastic Programming Models with Recourse

In the last chapter we have considered dynamic programming as a way to tackle dynamic stochastic optimization problems. Dynamic programming is, in principle, a very powerful framework, which is able to cope with a wide variety of problems, but it is plagued by the curse of dimensionality. An alternative framework is represented by stochastic programming models with recourse. Among economists, stochastic programming models are arguably much less widespread than dynamic programming approaches. Nevertheless, there is a rich literature concerning financial applications, and we do believe that having at least some familiarity with this modeling framework is useful, even if we cannot dwell too deeply in the severe computational challenges stochastic programming must face. We will only consider linear models; this is a limitation, but non-linear models can often be approximated using linear programming modeling tricks.

Stochastic programming models are introduced in section 11.1 as an extension of the linear programming models we have described in chapter 6. We will see that stochastic programming with recourse is just one possible modeling framework; however, since it is arguably the most common one, we will identify this subclass of models with "stochastic programming models" for the sake of brevity. We will see a few toy portfolio management models in section 11.2, just to show the potential for applications. A fundamental issue in stochastic programming is scenario generation, which is outlined in section 11.3. A potentially large scenario tree is needed to represent uncertainty, resulting in a large-scale optimization model. Sometimes, special-purpose methods can be applied, which rely on the special structure of stochastic programming models to devise decomposition approaches. We will outline the basic method in this vein, L-shaped decomposition, in section 11.4. This will also shed some light on the differences and similarities between stochastic programming with recourse and dynamic programming, an issue which is briefly discussed in section 11.5.

11.1 LINEAR STOCHASTIC PROGRAMMING MODELS

We have introduced linear programming (LP) models in chapter 6. An LP model in canonical form is

$$\begin{aligned} \min \quad & \mathbf{c}'\mathbf{x} \\ \text{s.t.} \quad & \mathbf{A}\mathbf{x} \geq \mathbf{b} \\ & \mathbf{x} \geq \mathbf{0}. \end{aligned}$$

When we formulate a model like this, we assume that we have exact knowledge of all the model parameters embedded in matrix $\mathbf{A}$ and in vectors $\mathbf{c}$ and $\mathbf{b}$. However, in finance there are several sources of uncertainty, and this modeling framework may be insufficient to tackle general optimization problems, such as portfolio optimization. One naive attempt to extend LP models to cope with uncertainty would be to replace the given parameters with random variables, yielding the model below:

$$\begin{aligned} \text{``min''} \quad & \mathbf{c}(\omega)'\mathbf{x} \\ \text{s.t.} \quad & \mathbf{A}(\omega)\mathbf{x} \geq \mathbf{b}(\omega) \\ & \mathbf{x} \geq \mathbf{0}. \end{aligned} \quad (11.1)$$

Here the data $\mathbf{c}(\omega)$, $\mathbf{A}(\omega)$, and $\mathbf{b}(\omega)$ depend on random events ω. The "min" notation is used to point out that this problem actually does not make sense, since minimizing a random variable has no meaning. We could define a sensible objective function by taking its expected value:

$$\mathrm{E}[\mathbf{c}(\omega)'\mathbf{x}] = \mathrm{E}[\mathbf{c}(\omega)]'\mathbf{x}.$$

An objection here may concern risk neutrality, but the real trouble is feasibility of the solution. Finding a solution $\mathbf{x}$ such that the constraints (11.1) are always satisfied may be impossible, or it could lead to a poor solution. By the way, this is why we did not consider LP problems in standard form, i.e., involving equality constraints. A possible approach is to relax the constraints a bit and to accept the fact that, in some cases, the constraints could not be met; we might just ask that this undesirable event is unlikely enough. This leads to *chance-constrained models* such as

$$\begin{aligned} \min \quad & \mathbf{c}'\mathbf{x} \\ \text{s.t.} \quad & \mathbf{A}\mathbf{x} \geq \mathbf{b} \\ & \mathrm{P}\{\mathbf{G}(\omega)\mathbf{x} \geq \mathbf{h}(\omega)\} \geq \alpha \\ & \mathbf{x} \geq \mathbf{0}, \end{aligned}$$

where we have separated the deterministic constraints from those involving uncertainty. Such models trade off the cost of the solution with its reliability, or robustness. We will not consider the computational challenge of solving such a model. This task may be relatively easy, if the problem above turns out to be a convex model. This may happen, depending on the probability distribution of the uncertain parameter. In general, the problem may be non-convex, which makes it much more difficult to cope with.

But even if we leave computational issues aside, there is another potential difficulty. Chance-constrained models may fully capture decision making under uncertainty in many cases of practical relevance, but they lack the ability of modeling a dynamic decision process in which decisions are revised when more and more information is acquired. In a truly dynamic decision process, we take a set of decisions here-and-now, based on limited information, but then we may adjust the decisions when the uncertainty is resolved. Of course, adjusting the decisions will imply some additional costs, and we would like to take good decisions minimizing the immediate costs as well as the expected value of the adjustment costs we will pay in the future. This idea leads to stochastic programming models with recourse. As an example, we may consider a two-stage stochastic linear programming model, which is usually stated as follows. The first-stage problem, involving the decisions $\mathbf{x}$ that we must take here and now, is

$$\begin{aligned} \min \quad & \mathbf{c}'\mathbf{x} + \mathrm{E}[h(\mathbf{x},\omega)] \\ \text{s.t.} \quad & \mathbf{A}\mathbf{x} = \mathbf{b} \\ & \mathbf{x} \geq \mathbf{0}. \end{aligned}$$

The first-stage problem involves a set of deterministic constraints and the expected cost of adjusting the solution at the second stage. The second-stage problem, involving the adjustments, or recourse variables $\mathbf{y}$, defines the function $h(\mathbf{x},\omega)$:

$$\begin{aligned} h(\mathbf{x},\omega) \equiv \min \quad & \mathbf{q}(\omega)'\mathbf{y} \\ \text{s.t.} \quad & \mathbf{W}(\omega)\mathbf{y} = \mathbf{r}(\omega) - \mathbf{T}(\omega)\mathbf{x} \\ & \mathbf{y} \geq \mathbf{0}. \end{aligned}$$

There are a few things to point out as far as the second-stage problem is concerned.

- We have written the problem in its most general form, allowing randomness in all the parameters, but this need not be the case. For instance, if the recourse matrix $\mathbf{W}$ is deterministic, we have a *fixed recourse problem*. Some algorithms may only be applied if the recourse is fixed and if the recourse cost vector $\mathbf{q}$ is deterministic as well; other solution algorithms have no such limitations.

- The overall problem can be thought of as a non-linear programming problem involving a recourse function $H(\mathbf{x}) \equiv \mathrm{E}[h(\mathbf{x},\omega)]$. Such a func-

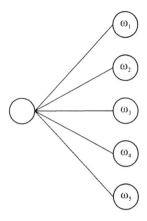

Fig. 11.1 Scenario tree for a two-stage stochastic optimization problem.

tion may seem intractable, as it involves the multidimensional integration of a function implicitly defined through an optimization problem. However, it may be shown that in the relevant cases the recourse function is convex. So, even if we do not know how to express $H(\mathbf{x})$ in a simple analytical form, we may still be able both to evaluate (or estimate) its value and to find a subgradient at a given point $\mathbf{x}$. On the contrary, chance-constrained problems are not convex problems in general.

- Depending on its structure, the second-stage problem may have a feasible solution for any first-stage vector $\mathbf{x}$ and for any random event ω, or not. In the second case, the second-stage problem implicitly defines some further constraints on $\mathbf{x}$.

- The approach may be generalized to multiple stages. We will see how in the next section.

In principle, we can define a stochastic programming model based on a continuous distribution of the uncertain parameters. However, although there are methods which are devised to solve approximately such problems, they are beyond the scope of this introduction. A natural alternative, given our knowledge of Monte Carlo sampling, is to approximate the continuous distribution by a discrete scenario tree like the one depicted in figure 11.1. We repeat that picture here, but we have already met this type of representation in figure 2.2 and we know that the idea can be generalized to multiple stages as shown in figure 2.3. The root node of the tree represents the present state of the world, from which different future states branch, corresponding to possible realizations of the uncertain data. We have to take first-stage decisions here and now, i.e., in the root of the tree; then, when the uncertainty is revealed, we will have the chance to take second-stage decisions to adapt to the circumstances; each possible contingency is represented by a leaf node in the tree. The overall problem entails taking a good first-stage decision, which should be robust, in that it should leave room for not too costly adaptations at the second-stage. Assume that we have a set of scenarios, indexed by $s \in S$, each with associated probability p_s. Then the two-stage stochastic LP problem boils down to a large-scale LP problem:

$$\begin{aligned} \min \quad & \mathbf{c}'\mathbf{x} + \sum_{s \in S} p_s \mathbf{q}'_s \mathbf{y}_s \\ \text{s.t.} \quad & \mathbf{A}\mathbf{x} = \mathbf{b} \\ & \mathbf{T}_s\mathbf{x} + \mathbf{W}_s\mathbf{y}_s = \mathbf{r}_s \quad \forall s \in S \\ & \mathbf{x}, \mathbf{y}_s \geq \mathbf{0}. \end{aligned}$$

In principle, This problem could be simply tackled by standard LP techniques; however, its size and its peculiar structure suggest the adoption of more specific approaches, one of which is described in section 11.4. Now a natural question is: Since solving a stochastic LP looks like a non-trivial task, why bother? Shouldn't we simply take the expected values of the data and solve a much simpler deterministic problem? Indeed, in some cases, solving a stochastic LP is a wasted effort. To characterize the cases in which the added effort is worthwhile, we may consider the VSS (*value of the stochastic solution*) concept.

Let us define the individual scenario problem

$$\begin{aligned} \min \quad & z(\mathbf{x}, \omega) = \mathbf{c}'\mathbf{x} + \min\{\mathbf{q}_\omega \mathbf{y} \mid \mathbf{W}_\omega = \mathbf{r}_\omega - \mathbf{T}_\omega \mathbf{x}, \mathbf{y} \geq \mathbf{0}\} \\ \text{s.t.} \quad & \mathbf{A}\mathbf{x} = \mathbf{b} \\ & \mathbf{x} \geq \mathbf{0}. \end{aligned}$$

Note that this scenario problem assumes knowledge of the future event ω. The recourse problem we have just considered amounts to solving

$$\text{RP} = \min_{\mathbf{x}} \mathrm{E}_\omega[z(\mathbf{x}, \omega)].$$

Solving a deterministic problem, based on the expected values $\bar{\omega} = \mathrm{E}[\omega]$ of the data, corresponds to the expected value problem:

$$\text{EV} = \min_{\mathbf{x}} z(\mathbf{x}, \bar{\omega}),$$

which yields a solution $\bar{\mathbf{x}}(\bar{\omega})$. However, this solution should be checked in the real context; this means that we should evaluate the expected cost of using the EV solution, which calls for some adjustments anyway:

$$\text{EEV} = \mathrm{E}_\omega[z(\bar{\mathbf{x}}(\bar{\omega}), \omega)].$$

The VSS is defined as[1]
$$\text{VSS} = \text{EEV} - \text{RP}.$$
It can be shown that VSS ≥ 0. A large VSS value suggests that solving the stochastic problem is well worth the effort; a small value suggests the opportunity to take the much simpler deterministic approach. As expected, it turns out that finance is a typical field in which the stochastic character of the problem cannot be neglected. Furthermore, by a proper choice of the recourse function, different risk attitudes of the decision makers may be represented.

11.2 MULTISTAGE STOCHASTIC PROGRAMMING MODELS FOR PORTFOLIO MANAGEMENT

The best way to introduce multistage stochastic models is by using a simple asset–liability management model. We use the same basic problem and data as [2, pp. 20–28]. We have an initial wealth W_0 now, and in the future we will have to pay an amount L, which is our only liability. We should devise an investment strategy to meet the liability; if possible, we would like to end up with a final wealth larger than L; however, we should account properly for risk aversion, since there could be some chance to end up with a terminal wealth which is not sufficient to pay for the liability, in which case we will have to borrow some money. A non-linear, strictly concave utility function of the difference between the terminal wealth and the liability would do the job, but this would lead to a non-linear programming model. As an alternative, we may build a piecewise linear utility function like that illustrated in figure 11.2. The utility is zero when the terminal wealth W matches the liability exactly. If the slope r penalizing the shortfall is larger than q, this function is concave, but not strictly.

The portfolio consists of a set of I assets. For simplicity, we assume that we may rebalance it only at a discrete set of time instants $t = 1, \ldots, T$, with no transaction cost; the initial portfolio is chosen at time $t = 0$, and the liability must be paid at time $T+1$. Time period t is the period between time instants $t-1$ and t. In order to represent uncertainty, we may build a tree like that in figure 11.3, which is a generalization of the two-stage tree of figure 11.1. Each node n_k corresponds to an event, where we should take some decision. We have an initial node n_0 corresponding to time $t = 0$. Then, for each event node, we have two branches; each branch is labeled by a conditional probability of occurrence, $P\{n_k \mid n_i\}$, where $n_i = a(n_k)$ is the immediate predecessor of node n_k. Here, we have two nodes at time $t = 1$ and four at time $t = 2$, where we may rebalance our portfolio on the basis of the previous

---

[1] A related but different concept is the expected value of perfect information (EVPI); see, e.g., [2, chapter 4].

MULTISTAGE STOCHASTIC PROGRAMMING MODELS   531

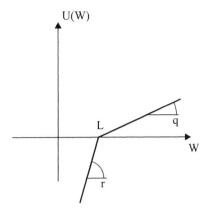

Fig. 11.2 Piecewise linear concave utility function.

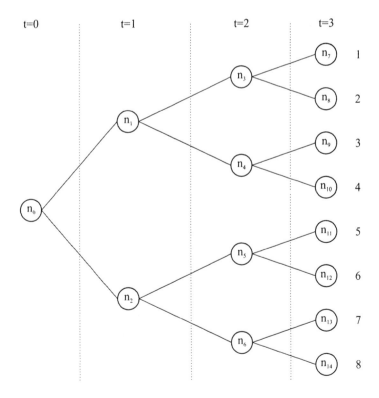

Fig. 11.3 Scenario tree for a simple asset and liability management problem.

asset returns. Finally, in the eight nodes corresponding to $t = 3$, we just compare our final wealth to the liability and we evaluate our utility function. Each node of the tree is associated with the set of asset returns during the corresponding time period. A scenario consists of an event sequence, i.e., a sequence of asset returns. We have eight scenarios in figure 11.3. For instance, scenario 2 consists of the node sequence (n_0, n_1, n_3, n_8). The probability of each scenario depends on the conditional probability of each node on its path. If each branch at each node is equiprobable, i.e., the conditional probability is $1/2$, each scenario in the figure has probability $1/8$. The branching factor may be arbitrary in principle; the more branches we use, the better our ability to model uncertainty; unfortunately, the number of nodes grows exponentially with the number of stages, as well as the computational effort.

At each node in the tree, we must take a set of decisions. In practice, we are interested in the decisions that must be implemented here and now, i.e., those corresponding to the first node of the tree; the other (recourse) decision variables are instrumental to the aim of devising a robust plan, but they are not implemented in practice, as the multistage model is solved on a rolling horizon basis. This suggests that, in order to model the uncertainty as accurately as possible with a limited computational effort, a possible idea is to branch many paths from the initial node, and less from the subsequent nodes. Each decision at each stage may depend on the information gathered so far, but not on the future; this requirement is called *non-anticipativity condition*. There are two basic ways to build a multistage stochastic programming model: the split-variable and the compact formulations, which are described in the next sections. They depend on how the non-anticipativity requirement is modeled. The suitability of each modeling approach also depends on the solution algorithm.

The numerical parameters, which are common to both model formulations, are as follows:

- The initial wealth is 55.
- The target liability is 80.
- There are two assets, stocks and bonds.
- In the scenario tree of figure 11.3 we have up and down branches; in the up (lucky) branches, the (total) return is 1.25 for stocks and 1.14 for bonds; in the down (bad) branches, the (total) return is 1.06 for stocks and 1.12 for bonds.
- The reward for excess wealth above the target liability is 1.
- The penalty for the shortfall below the target liability is 4.

11.2.1 Split-variable model formulation

In the split-variable approach, the decision variables are defined as follows:

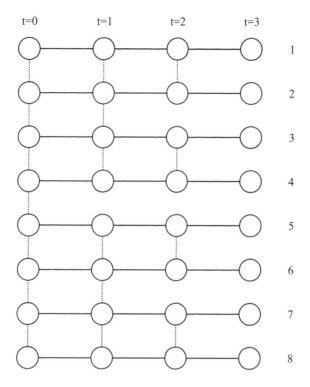

Fig. 11.4 Split-variable view of a scenario tree.

- x_{it}^s is the amount invested in asset i at the beginning of time period t in scenario s.

By the same token, R_{it}^s is the (total) return of asset i in scenario $s = 1, \ldots, S$ during time period t. It is important to understand that, if we define the decision variables in this way, we must enforce the non-anticipativity constraint explicitly. The issue may be understood by looking at figure 11.4. We have a set of decision variables for each node; however, the decision variables corresponding to different scenarios at the same time t must be equal if the two scenarios are indistinguishable at time t. This is represented by the dotted lines in figure 11.4. To begin with, the initial portfolio must be the same for all scenarios. Hence:

$$x_{i0}^s = x_{i0}^{s'}, \qquad i = 1, \ldots, I; \; s, s' = 1, \ldots, S.$$

Now consider time $t = 1$ and node n_1 of the original event tree as depicted in figure 11.3; the scenarios $s = 1, 2, 3, 4$ pass through this node and are indistinguishable at time $t = 1$. Hence, we must have

$$x_{i1}^1 = x_{i1}^2 = x_{i1}^3 = x_{i1}^4, \qquad i = 1, \ldots, I.$$

In fact, node n_1 corresponds to four nodes in the split view of the tree. By the same token, at time $t = 2$ we have constraints like

$$x_{i2}^5 = x_{i2}^6, \qquad i = 1, \ldots, I.$$

More generally, it is customary to denote by $\{s\}_t$ the set of scenarios which are not distinguishable from s up to time t. For instance:

$$\begin{aligned} \{1\}_0 &= \{1,2,3,4,5,6,7,8\} \\ \{2\}_1 &= \{1,2,3,4\} \\ \{5\}_2 &= \{5,6\}. \end{aligned}$$

Then the non-anticipativity constraints may be written as

$$x_{it}^s = x_{it}^{s'} \qquad \forall i, t, s, s' \in \{s\}_t.$$

This is not the only way of expressing the non-anticipativity requirement, and the best approach depends on the chosen solution algorithm. Now we may write the following model for the basic asset–liability management problem:

$$\max \sum_s p^s (q w_+^s - r w_-^s) \qquad (11.2)$$

$$\text{s.t.} \quad \sum_{i=1}^{I} x_{i0}^s = W_0 \qquad \forall s \in S \qquad (11.3)$$

$$\sum_{i=1}^{I} R_{it}^s x_{i,t-1}^s = \sum_{i=1}^{I} x_{it}^s \qquad \forall s \in S; \; t = 1, \ldots, T \qquad (11.4)$$

$$\sum_{i=1}^{I} R_{i,T+1}^s x_{iT}^s = L + w_+^s - w_-^s \qquad \forall s \in S \qquad (11.5)$$

$$x_{it}^s = x_{it}^{s'} \qquad \forall i, t, s, s' \in \{s\}_t$$

$$x_{it}^s, w_+^s, w_-^s \geq 0.$$

Here w_+^s is the surplus at the end of the planning horizon, with reward q, and w_-^s is the shortfall, with penalty r. The objective function (11.2) is the expected value of the utility function; p^s is the probability of each scenario; the utility function is concave if $r > q$. Equation (11.3) states that our initial wealth W_0 is allocated among the different assets. The portfolio rebalancing constraints (11.4) say that the wealth at time t is reallocated. In equation (11.5) we evaluate how we did, by comparing the final wealth with the liability L, and setting the proper surplus and shortfall values. Then we add non-anticipativity and non-negativity constraints. Note that, since the variables w_+^s and w_-^s are restricted by non-negativity constraints, we will have $w_+^s \cdot w_-^s = 0$ in the optimal solution (i.e., only one variable may be different than 0 in each scenario). The non-negativity requirements on x_{it}^s may be relaxed if we allow short selling.

In this modeling approach, we introduce a large set of variables, which are then linked by non-anticipativity constraints. Hence, one could wonder if this really makes sense. The answer depends on the solution algorithm. If one wants to adopt an algorithm like the L-shaped decomposition, the compact formulation explained in the following section must be used. The split-variable approach may be exploited with interior point methods aimed at stochastic programming. Furthermore, relaxing the non-anticipativity constraints by a set of Lagrange multipliers, we obtain a set of independent subproblems, one per scenario (much in the same vein as example 6.10). Pursuing this idea leads to scenario aggregation algorithms.

Representing the split-variable formulation in AMPL The split-variable formulation is easily expressed in an algebraic language like AMPL, which is introduced in appendix C, to which we refer the reader interested in a quick tour. It is customary to set up two files: The first one contains the model structure, which is illustrated in figure 11.5, and the other one contains the data for a particular model instance, as illustrated in figure 11.6.

The way we express a model in AMPL is almost self-explanatory. All the characters after the # character are treated as a comment; note also that in an algebraic language, one prefers longer names than in the usual mathematical notation. As is customary in AMPL models, we have to define sets, parameters, decision variables, the objective function, and the constraints. Most of the following reflects what we illustrate in the simple introductory models described in the appendix, but there are a few new things. Let us check the model file first (figure 11.5).

- The sets involved in our formulation are introduced by the keyword set. Here we have a simple set, assets.

- The numerical parameters are introduced by the keyword param. Most of them are scalar values, with the exception of scenario probabilities, which are contained in the vector parameter prob, and returns, which are contained in the tridimensional array return.

- A new element is the indexed collection of sets links. For each time period, we have a set of pairs; each pair consists of two scenarios, which are not distinguishable up to that time period. As we have said, this is where we enforce non-anticipativity.

- The decision variables are introduced by the var keyword and correspond clearly to the variables in the mathematical statement of the model.

- Then the objective function is expressed and the solver is instructed to maximize its value. Note how the sum notation is used to express sums over an index in a very natural way.

```
set assets;            # set of available assets
param initwealth;      # initial wealth
param scenarios;       # number of scenarios
param T;               # number of time periods
param target;   # target value (liability) at time T
param reward;   # reward for wealth beyond target value
param penalty;  # penalty for not meeting the target
# return of each asset during each period in each scenario
param return{assets, 1..scenarios, 1..T};
param prob{1..scenarios}; # probability of each scenario
# the indexed set points out which scenarios
# are linked at each period t in 0..T-1
set links{0..T-1} within {1..scenarios, 1..scenarios};

# DECISION VARIABLES
# amount invested in each asset at each period of time
# in each scenario
var invest{assets,1..scenarios,0..T-1} >= 0;
var above_target{1..scenarios}>=0; # amount above final target
var below_target{1..scenarios}>=0; # amount below final target

# OBJECTIVE FUNCTION
maximize exp_value:
   sum{i in 1..scenarios} prob[i]*(reward*above_target[i]
       - penalty*below_target[i]);

# CONSTRAINTS
# initial wealth is allocated at time 0
subject to budget{i in 1..scenarios}:
   sum{k in assets} (invest[k,i,0]) = initwealth;
# portfolio rebalancing at intermediate times
subject to balance{j in 1..scenarios, t in 1..T-1} :
   (sum{k in assets} return[k,j,t]*invest[k,j,t-1]) =
   sum{k in assets} invest[k,j,t];
# check final wealth against liability
subject to scenario_value{j in 1..scenarios} :
   (sum{k in assets} return[k,j,T]*invest[k,j,T-1])
       - above_target[j] + below_target[j] = target;
# this makes all investments non-anticipative
subject to linkscenarios
   {k in assets, t in 0..T-1, (s1,s2) in links[t]} :
   invest[k,s1,t] = invest[k,s2,t];
```

Fig. 11.5 AMPL model for the split variable formulation (SplitALM.mod).

```
set assets := stocks bonds;
param initwealth := 55;
param scenarios := 8;
param T := 3;

set links[0] := (1,2) (2,3) (3,4) (4,5) (5,6) (6,7) (7,8);
set links[1] := (1,2) (2,3) (3,4) (5,6) (6,7) (7,8);
set links[2] := (1,2) (3,4) (5,6) (7,8);

param target := 80;
param reward := 1;
param penalty := 4;

param return :=
[stocks, 1, *] 1 1.25 2 1.25 3 1.25
[stocks, 2, *] 1 1.25 2 1.25 3 1.06
[stocks, 3, *] 1 1.25 2 1.06 3 1.25
[stocks, 4, *] 1 1.25 2 1.06 3 1.06
[stocks, 5, *] 1 1.06 2 1.25 3 1.25
[stocks, 6, *] 1 1.06 2 1.25 3 1.06
[stocks, 7, *] 1 1.06 2 1.06 3 1.25
[stocks, 8, *] 1 1.06 2 1.06 3 1.06
[bonds, 1, *] 1 1.14 2 1.14 3 1.14
[bonds, 2, *] 1 1.14 2 1.14 3 1.12
[bonds, 3, *] 1 1.14 2 1.12 3 1.14
[bonds, 4, *] 1 1.14 2 1.12 3 1.12
[bonds, 5, *] 1 1.12 2 1.14 3 1.14
[bonds, 6, *] 1 1.12 2 1.14 3 1.12
[bonds, 7, *] 1 1.12 2 1.12 3 1.14
[bonds, 8, *] 1 1.12 2 1.12 3 1.12;

param prob default 0.125;
```

Fig. 11.6 AMPL data file for the split variable model formulation (SplitALM.dat).

- The constraints are introduced by the subject to keywords. For each constraint we list a name first (which may be used to get the dual variables for each constraint after solving the model); then, we specify the index values for which the constraint should be replicated (which corresponds to universal quantification, such as $\forall s$, in mathematical notation); finally, we express the constraints themselves.

- An interesting piece of syntax is the last constraint, where we model non-anticipativity by enforcing the constraint for each time period and for each scenario pair in the indexed collection for that time period. We have a small glimpse of how powerful the AMPL syntax is to work with sets.

Now let us check the data file (figure 11.6).

- The set of assets and the scalar parameters are specified with a simple syntax.

- With respect to what we illustrate in the appendix on AMPL, one new element is how we specify the time-indexed collection links of sets of pairs. Again the syntax is rather natural and self-explanatory.

- Another new element is how we list asset returns, indexed by asset, scenario, and time period. In this case, what we have illustrated in the appendix for vector and matrix data is not enough, as we have a tridimensional array. We basically "slice" the tridimensional array in two matrices. A notation like [stocks, 1, *] means that values of the third index, to which the wildcard corresponds, will be listed together with the corresponding entries: Given an asset and a scenario, we list the return for each time period.

- The last parameter, prob, is assigned using a shorthand notation; since the probability for all the scenarios is 0.125, we use the default keyword to streamline notation.

Now we are ready to load the two files, solve the model, and display the solution:

```
ampl: model SplitALM.mod;
ampl: data SplitALM.dat;
ampl: solve;
CPLEX 9.1.0: optimal solution; objective -1.514084643
20 dual simplex iterations (13 in phase I)
ampl: display invest;
invest [bonds,*,*]
:         0         1         2      :=
1    13.5207    2.16814    0
2    13.5207    2.16814    0
```

```
3    13.5207    2.16814    71.4286
4    13.5207    2.16814    71.4286
5    13.5207    22.368     71.4286
6    13.5207    22.368     71.4286
7    13.5207    22.368      0
8    13.5207    22.368      0

[stocks,*,*]
:      0         1          2       :=
1    41.4793   65.0946    83.8399
2    41.4793   65.0946    83.8399
3    41.4793   65.0946     0
4    41.4793   65.0946     0
5    41.4793   36.7432     0
6    41.4793   36.7432     0
7    41.4793   36.7432    64
8    41.4793   36.7432    64
;
```

We see quite clearly the non-anticipative nature of the solution: The first column of each table shows one number, since the initial decision, in the root of the tree, is common to all scenarios; the second column shows two values, corresponding to the decisions in nodes n_1 and n_2; at time period 2, we have four nodes, and four different values. We may notice that in the last period the portfolio is not diversified, since the whole wealth is allocated to one asset, and we should wonder if this makes sense. Actually, it is a consequence of two features of this toy model:

- We are approximating a non-linear utility function by a piecewise linear function, and this may imply "local" risk neutrality; we should either use a non-linear programming model or a more accurate representation with more pieces.

- The scenario tree has a very low branching factor, and this does not represent uncertainty accurately.

However, the portfolio allocation in the last time period is not necessarily a critical output of the model: the real stuff is the *initial* portfolio allocation. The decision variables for future stages have the purpose of avoiding a myopic policy, but they are not meant to be implemented. Nevertheless, the possible impact of poor modeling in the last stages should be assessed; in fact, for problems involving a short time horizon, end-effects may be detrimental. Unlike dynamic programming, we do not get the solution in feedback form: We do not have a good recipe to take optimal decisions in the future, as a multistage stochastic program should be re-run in a rolling horizon fashion whenever we need taking more decisions. More on this in section 11.5.

Finally, we should note that the solution has been obtained using CPLEX as a solver, but this need not be the case. If you have the AMPL student demo version, you could also use MINOS. By the way, MINOS should be used if you want to use a truly non-linear utility function. Other linear and non-linear solvers are available for use with AMPL.[2]

11.2.2 Compact model formulation

The split-variable formulation is based on a large number of variables, which are then linked together by the non-anticipativity constraints. This may be useful for algorithms based on decomposition into independent scenarios, which could be accomplished by dualizing non-anticipativity constraints. But if we want to apply a generalization of the L-shaped method (section 11.4) to multistage stochastic programs, the model must be written in a different way. A more compact formulation may be obtained directly by associating decision variables to the nodes in the tree. Let us introduce the following notation:

- N is the set of event nodes, in our case

$$N = \{n_0, n_1, n_2, \ldots, n_{14}\}.$$

- Each node $n \in N$, apart from the root node n_0, has a unique direct predecessor node, denoted by $a(n)$: for instance, $a(n_3) = n_1$.

- There is a set $S \subset N$ of leaf (terminal) nodes, in our case

$$S = \{n_7, \ldots, n_{14}\};$$

for each node $s \in S$ we have surplus and shortfall variables w_+^s and w_-^s.

- There is a set $T \subset N$ of intermediate nodes, where portfolio rebalancing may occur after the initial allocation in node n_0; in our case

$$T = \{n_1, \ldots, n_6\};$$

for each node $n \in \{n_0\} \cup T$ there is an investment variable x_{in}, corresponding to the amount invested in asset i at node n.

With this notation, the model may be written as follows:

$$\max \quad \sum_{s \in S} p^s (q w_+^s - r w_-^s)$$

$$\text{s.t.} \quad \sum_{i=1}^{I} x_{i,n_0} = W_0$$

$$\sum_{i=1}^{I} R_{i,n} x_{i,a(n)} = \sum_{i=1}^{I} x_{in} \quad \forall n \in T$$

---

[2] See http:www.ampl.com and the other web sites listed at the end of appendix C.

$$\sum_{i=1}^{I} R_{is} x_{i,a(s)} = L + w_+^s - w_-^s \qquad \forall s \in S$$
$$x_{in}, w_+^s, w_-^s \geq 0,$$

where $R_{i,n}$ is the total return for asset i during the period that *leads to* node n, and p^s is the probability of reaching the terminal node $s \in S$; this probability is the product of all the conditional probabilities on the path that leads from node n_0 to s.

Representing the compact formulation in AMPL The compact formulation can also be easily expressed in AMPL. The structure of the model file is similar to the split-variable formulation. The main differences are:

- We introduce the three sets of nodes: the set of initial nodes, init_node, which is actually a singleton; the set of intermediate nodes interm_nodes; and, finally, the set of terminal nodes term_nodes, which correspond to the eight scenarios.

- For each node, apart from n_0, we have a predecessor; we use an array pred of singleton sets to store the predecessor; this is needed if we want to treat nodes as sets of symbols, but we could also use an array of numerical values to index nodes.

- Return and decision variables are now indexed by nodes, rather than by a (scenario, time) pair as we did in the split-variable model formulation.

- The objective function and the constraints are a straightforward translation of the mathematical model.

The data file is also fairly self-explanatory. We may see how the three node subsets are listed. The only noteworthy point is the use of transposition (keyword tr) to assign the return table; in fact, return is defined in the model file as a table indexed by assets and nodes, and we must transpose the table if we want to swap the two indexes in order to improve readability in the data file.

Now we are ready to solve the model and to check that we get the same solution we obtained by the alternative model formulation:

```
ampl: model CompactALM.mod;
ampl: data CompactALM.dat;
ampl: solve;
CPLEX 9.1.0: optimal solution; objective -1.514084643
20 dual simplex iterations (13 in phase I)
ampl: display invest;
invest :=
bonds   n0     13.5207
bonds   n1      2.16814
```

```
set assets;              # available investment options
param initwealth;        # initial wealth
param target;    # target liability at time T
param reward;    # reward for excess wealth beyond target value
param penalty; # shortfall penalty

# NODE SETS
set init_node;    # initial node
set interm_nodes;       # intermediate nodes
set term_nodes;         # terminal nodes
# immediate predecessor node
set pred{interm_nodes union term_nodes}
          within {init_node union interm_nodes};
param prob{term_nodes}; # probability of each scenario
# return of each investment option at the end of time periods
param return{assets, interm_nodes union term_nodes};

# DECISION VARIABLES
# amount invested in trading nodes
var invest{assets,init_node union interm_nodes} >= 0;
var above_target{term_nodes}>=0;      # amount above final target
var below_target{term_nodes}>=0;      # amount below final target

# OBJECTIVE FUNCTION
maximize exp_value:
    sum{s in term_nodes} prob[s]*(reward*above_target[s]
       - penalty*below_target[s]);

# CONSTRAINTS
# initial wealth is allocated in the root node
subject to budget{n0 in init_node} :
    sum{k in assets} (invest[k,n0]) = initwealth;
# portfolio rebalancing at intermediate nodes
subject to balance{n in interm_nodes, a in pred[n]} :
    (sum{k in assets} return[k,n]*invest[k,a]) =
    sum{k in assets} invest[k,n];
# check final wealth against target
subject to scenario_value{s in term_nodes, a in pred[s]} :
    (sum{k in assets} return[k,s]*invest[k,a])
    - above_target[s] + below_target[s] = target;
```

Fig. 11.7 AMPL model for the compact formulation (CompactALM.mod).

```
set assets := stocks bonds;
param initwealth := 55;
param target := 80;
param reward := 1;
param penalty := 4;

set init_node := n0;
set interm_nodes := n1 n2 n3 n4 n5 n6;
set term_nodes := n7 n8 n9 n10 n11 n12 n13 n14;

param return (tr):
      stocks   bonds :=
n1    1.25     1.14
n2    1.06     1.12
n3    1.25     1.14
n4    1.06     1.12
n5    1.25     1.14
n6    1.06     1.12
n7    1.25     1.14
n8    1.06     1.12
n9    1.25     1.14
n10   1.06     1.12
n11   1.25     1.14
n12   1.06     1.12
n13   1.25     1.14
n14   1.06     1.12  ;

param prob default 0.125;

# immediate predecessors
set pred[n1]  := n0;
set pred[n2]  := n0;
set pred[n3]  := n1;
set pred[n4]  := n1;
set pred[n5]  := n2;
set pred[n6]  := n2;
set pred[n7]  := n3;
set pred[n8]  := n3;
set pred[n9]  := n4;
set pred[n10] := n4;
set pred[n11] := n5;
set pred[n12] := n5;
set pred[n13] := n6;
set pred[n14] := n6;
```

Fig. 11.8 AMPL data file for the compact model formulation (CompactALM.dat).

544   *LINEAR STOCHASTIC PROGRAMMING MODELS WITH RECOURSE*

```
bonds   n2   22.368
bonds   n3   0
bonds   n4   71.4286
bonds   n5   71.4286
bonds   n6   0
stocks  n0   41.4793
stocks  n1   65.0946
stocks  n2   36.7432
stocks  n3   83.8399
stocks  n4   0
stocks  n5   0
stocks  n6   64
;
```

It is worth noting that writing the data file manually, in particular the information representing the scenario tree structure, is out of the question for realistically sized problem instances. One possibility is writing a MATLAB function to do that. A few modeling tools for stochastic programming have also been developed; although they are mostly research products at present, the situation is likely to change in the future.

11.2.3 Asset and liability management with transaction costs

To give the reader an idea of how to build non-trivial financial planning models, we generalize a bit the compact formulation of the preceding section. The assumptions and the limitations behind the model are the following:

- We are given a set of initial holdings for each asset; this is a more realistic assumption, since we should use the model to rebalance the portfolio periodically according to a rolling horizon strategy.

- We take linear transaction costs into account; the transaction cost is a percentage c of the traded value, both for buying and selling.

- We want to maximize the expected utility of the terminal wealth.

- There is a stream of uncertain liabilities that we have to meet.

- We do not consider the possibility of borrowing money; we assume all of the available wealth at each rebalancing period is invested in the available assets; actually, the possibility of investing in a risk-free asset is implicit in the model.

- We do not consider the possibility of investing new cash at each rebalancing date (as would be the case, e.g., for a pension fund).

Some of the limitations of the model may easily be relaxed. The important point we make is that when transaction costs are involved, we have to introduce new decision variables to express the amount of assets held, sold, and bought at each rebalancing date. We use a notation which is similar to that used in the compact formulation:

- N is the set of nodes in the tree; n_0 is the initial node.
- The (unique) predecessor of node $n \in N \backslash \{n_0\}$ is denoted by $a(n)$; the set of terminal nodes is denoted by S; as in the previous formulation, each of these nodes corresponds to a scenario, which is the unique path leading from n_0 to $s \in S$, with probability p^s.
- $T = N \backslash (\{n_0\} \cup S)$ is the set of intermediate trading nodes.
- L^n is the liability we have to meet in node $n \in N$.
- c is the percentage transaction cost.
- $\overline{h}_i^{n_0}$ is the initial holding for asset $i = 1, \ldots, I$ at the initial node.
- P_i^n is the price for asset i at node n.
- z_i^n is the amount of asset i purchased at node n.
- y_i^n is the amount of asset i sold at node n.
- x_i^n is the amount of asset i we hold at node n, after rebalancing.
- W^s is the wealth at node $s \in S$.
- $U(W)$ is the utility for wealth W.

Based on this notation, we may write the following model:

$$\max \sum_{s \in S} p_s U(W^s) \qquad (11.6)$$

$$\text{s.t.} \quad x_i^{n_0} = \overline{h}_i^{n_0} + z_i^{n_0} - y_i^{n_0} \qquad \forall i \qquad (11.7)$$

$$x_i^n = x_i^{a(n)} + z_i^n - y_i^n \qquad \forall i, \forall n \in T \qquad (11.8)$$

$$(1-c) \sum_{i=1}^{I} P_i^n y_i^n - (1+c) \sum_{i=1}^{I} P_i^n z_i^n = L^n \qquad \forall n \in T \cup \{n_0\}$$

$$(11.9)$$

$$W^s = \sum_{i=1}^{I} P_i^s x_i^{a(s)} - L^s \qquad \forall s \in S \qquad (11.10)$$

$$x_i^n, z_i^n, y_i^n, W^s \geq 0. \qquad (11.11)$$

The objective (11.6) is the expected utility of the terminal wealth; if we approximate this non-linear concave function by a piecewise linear concave function, we get an LP problem (as we did in section 12.1.1). Equation (11.7) expresses the initial asset balance, taking the current holdings into account;

the asset balance at intermediate trading dates is taken into account by equation (11.8). Equation (11.9) makes sure that enough cash is generated by selling assets in order to meet the liabilities; we may also reinvest the proceeds of what we sell in new asset holdings; note how the transaction costs are expressed for selling and purchasing. Equation (11.10) is used to estimate the final wealth; note that here we have not taken into account the need to sell assets to generate cash to meet the last liability. If we assume that the entire portfolio is liquidated at the end of the planning horizon, we could rewrite equation (11.10) as

$$W^s = (1-c) \sum_{i=1}^{I} P_i^s x_i^{a(s)} - L^s.$$

In practice, we would repeatedly solve the model on a rolling horizon basis, so the exact expression of the objective function is a bit debatable.

This model can be generalized in a number of ways, which are left as an exercise to the reader. The most important point is that we have assumed that the liabilities must be met. This may be a very hard constraint; if extreme scenarios are included in the formulation, as they should be, it may well be the case that the model above is infeasible. So the formulation should be relaxed in a sensible way; we could consider the possibility of borrowing cash; we could also introduce suitable penalties for not meeting the liabilities. In principle, we could also require that the probability of not meeting the liabilities is small enough; this leads to chance-constrained formulations, for which we refer the reader to the literature.

11.3 SCENARIO GENERATION FOR MULTISTAGE STOCHASTIC PROGRAMMING

The quality of the solution obtained by solving a multistage stochastic program depends on how well the scenario tree represents the inherent uncertainty influencing the decision problem. To generate scenarios in the financial domain, the necessary starting point is a sensible model describing the evolution of relevant quantities, such as interest rates, stock prices, inflation, etc. Stochastic differential equations are a possible modeling framework, in which case we should discretize time according to the structure of our scenario tree. Alternatively, discrete-time models may be built directly, such as time series models. A class of simple discrete-time models are vector autoregressive models (VAR, which should not be confused with Value at Risk). Let $\mathbf{h}_t$ be a vector of economic and financial variables at time t. An example of a VAR model is

$$\mathbf{h}_t = \mathbf{c} + \mathbf{\Omega} \mathbf{h}_{t-1} + \boldsymbol{\epsilon}_t, \qquad t = 1, \ldots, T,$$

where $\mathbf{c}$ and $\mathbf{\Omega}$ are model parameters, and $\boldsymbol{\epsilon} \sim N(\mathbf{0}, \boldsymbol{\Sigma})$ is a vector of jointly normal random variables with zero mean and covariance matrix $\boldsymbol{\Sigma}$.

Given a dynamic model in some form, generating a scenario tree requires some form of sampling. However, especially in multistage problems, there is the danger of an exponential growth in size of the tree. Note that we cannot exploit recombination, as we did with binomial lattices, because we have to take path-dependent decisions at each stage. Hence, due attention must be paid to scenario generation. In this section we first review clever mechanisms that have been proposed to keep the size of the tree limited. We should bear in mind that the purpose of scenario trees is not really to yield a 100% faithful representation of the underlying uncertainty over the whole planning horizon, as there is little hope to achieve this goal while keeping the optimization model to a computationally tractable size. The real aim is to get robust first-stage decisions. Then we illustrate issues related to arbitrage, which is obviously relevant in a financial domain.

11.3.1 Sampling for scenario tree generation

The first decision to take is the shape of the scenario tree, i.e., the branching factor which is applied at each node. A typical approach is to have a larger branching factor at early stages, as representing uncertainty there accurately may be more important in getting robust first stage decisions. A further observation is that the time step need not be the same for each stage; it may be reasonable to use larger time steps in later time periods, where aggregate decisions may be considered.

Given a scenario tree structure, we have to decide which outcomes we should associate to nodes in the tree, and possibly the (conditional) probabilities associated to each branch in the tree. The techniques we have already met in chapter 4 can be used here.

- The first possibility that we may think of is naive Monte Carlo sampling. In this case, the probability distribution for future nodes branching from current node is uniform. This approach may be sensible for two-stage models, but it is not quite feasible for multistage models due to the number of nodes we need to capture uncertainty. Variance reduction techniques may be useful. Antithetic sampling is the simplest option; importance sampling has been proposed in [5] and [13]. In the last case, probabilities should be adjusted to reflect the change in measure. Stratified sampling may also be used.

- Numerical integration methods are an alternative. In particular, Gaussian quadrature is a suitable way to discretize a continuous probability distribution; we have seen in example 10.4 how Gaussian quadrature may capture uncertainty much more efficiently than crude Monte Carlo sampling. Low-discrepancy sequences may also be used, but this again looks feasible for two-stage models. See, e.g., [16].

- Antithetic sampling, in the case of symmetric distributions, leads to a sample that matches odd moments of the underlying density; for instance, expected value is matched, and the symmetric sampling leads to zero skewness. It is natural to consider sampling in such a way that other moments are matched as well, such as variances, covariances, and kurtosis. In general, matching all moments exactly will be impossible with a limited number of samples, but we can try to match them as well as possible, in a least squares sense, This leads to an approach to generate a set of "optimized" scenarios. To illustrate the idea, consider a random variable $\mathbf{X}$ which has a multivariate normal distribution. Assume that we know the expected values μ_i of each component X_i, as well as the variance σ_i^2 and the set of covariances σ_{ij} for each pair (i,j) of variables ($\sigma_{ii} = \sigma_i^2$). Furthermore, since we are dealing with a normal distribution, we know that skewness $\xi = E[(\tilde{d} - \mu)^3/\sigma^3]$ should be zero and that kurtosis $\chi = E[(\tilde{d} - \mu)^4/\sigma^4]$ should be 3 (here we are considering the marginal distribution of each random variable).

Let us denote by x_i^s the sample of X_i in node s belonging to a certain branching of size S. For the sake of simplicity, we assume that all conditional probabilities of branches are equal, but we know from Gaussian quadrature that there are potential advantages in setting such probabilities with care. Natural requirements are

$$\frac{1}{S}\sum_s x_i^s \approx \mu_i \quad \forall i$$

$$\frac{1}{S}\sum_s (x_i^s - \mu_i)(x_j^s - \mu_j) \approx \sigma_{ij} \quad \forall i,j$$

$$\frac{1}{S}\sum_s \frac{(x_i^s - \mu_i)^3}{\sigma_i^3} \approx 0 \quad \forall i$$

$$\frac{1}{S}\sum_s \frac{(x_i^s - \mu_i)^4}{\sigma_i^4} \approx 3 \quad \forall i.$$

We should point out, e.g., in the second requirement related to covariance, that we divide by number of sample S, and not by $S-1$ as typical with sample variance, since the parameters are known a priori and not estimated from the data. Approximate moment matching is obtained by minimizing the following squared error:

$$w_1 \sum_i \left[\frac{1}{S}\sum_s x_i^s - \mu_i\right]^2$$

$$+ w_2 \sum_{i,j} \left[\frac{1}{S}\sum_s (x_i^s - \mu_i)(x_j^s - \mu_j) - \sigma_{ij}\right]^2$$

$$+w_3 \sum_i \left[\frac{1}{S}\sum_s \left(\frac{x_i^s - \mu_i}{\sigma_i}\right)^3\right]^2$$

$$+w_4 \sum_i \left[\frac{1}{S}\sum_s \left(\frac{x_i^s - \mu_i}{\sigma_i}\right)^4 - 3\right]^2. \tag{11.12}$$

The objective function includes four weights w_k, which may be used to fine tune performance. It should be mentioned that the resulting scenario optimization problem need not be convex. However, if we manage to find any solution with a low value of the "error" objective function, this is arguably a satisfactory solution, even though it is not necessarily the globally optimal one [12].

- The moment matching approach is a flexible and intuitively appealing way of generating scenarios. Nevertheless, it has been argued that it lacks a sound theoretical background. Indeed, counterexamples can be built, showing that quite different probability distributions may share the first moments [11]. In order to find a scenario generation approach resting on a sound basis, some researchers have proposed formal approaches relying on stability concepts and the definition of probability metrics. These methods require a high level of mathematical sophistication; hence, in this introductory chapter, we limit ourselves to provide the reader with a basic feeling for the overall idea (see, e.g., [20], for a thorough treatment). To begin with, we should try to formalize the concept of stability. To this aim, let us consider an abstract view of a stochastic optimization problem:

$$v(P) \equiv \inf_{\mathbf{x}\in \mathbf{X}} \int_\Xi f_0(\mathbf{x}, \xi) P(d\xi)$$

Here $\mathbf{x}$ is the set of decision variables, constrained on a set $\mathbf{X}$. The random data are represented by ξ, which belongs to set Ξ on which a probability measure P is defined. The optimal value of this stochastic program depends on the probability measure P, as pointed out by the notation $v(P)$. What happens if we perturb the measure P? A possible reason for the perturbation is that we have unreliable data, which means that we actually ignore the "true" measure P and we consider another measure Q instead. Alternatively, we may be forced to resort to an approximate measure Q, in the sense that we use a scenario tree which approximates the true measure P. Whatever the reason, we must first define a probability metric in order to quantify the distance between two probability measures.

There are many ways to do so. One possibility has its roots in the Monge transportation problem, which asks for the optimal way of transporting mass (e.g., soil, when we are building a road). The problem has a probabilistic interpretation, which was pointed out by Kantorovich, when

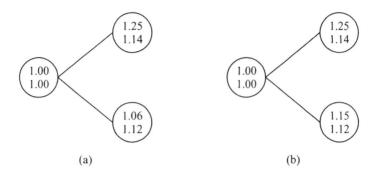

Fig. 11.9 Two simple scenario trees for asset price paths.

we interpret mass in a probabilistic sense (see [19], for more details). In order to define a concept of distance between two probability measures, we may define a transportation functional:

$$\mu_c(P, Q) \equiv \inf \left\{ \int_{\Xi \times \Xi} c(\xi, \tilde{\xi}) \eta(d\xi, d\tilde{\xi}) : \pi_1 \eta = P, \; \pi_2 \eta = Q. \right\}$$

Here $c(\cdot, \cdot)$ is a suitably chosen cost function; the problem calls for finding the minimum of the integral over all joint measures η, defined on the Cartesian product $\Xi \times \Xi$, whose marginals coincide with P and Q, respectively (π_1 and π_2 represent projection operators). In the case of two discrete measures P and Q, this boils down to the classical transportation problem with a linear programming formulation. Under some technical conditions, a form of Lipschitz continuity can be proved:

$$\mid v(P) - v(Q) \mid \leq L \mu_c(P, Q).$$

In practical terms, what one can do is selecting a cost function $c : \Xi \times \Xi \to \mathbb{R}$ in order to define a probability metric. Then we look for an approximate distribution P_{tree}, i.e., the scenario tree, such that $\mu_c(P, P_{tree}) < \epsilon$. This leads to algorithms to reduce the scenario tree. In [9] a scenario reduction procedure is described, based on the theoretical concepts above. The idea is sampling a large tree, and then reducing its size to a manageable level.

11.3.2 Arbitrage free scenario generation

The considerations we have done so far apply to a generic stochastic program. When we deal with an application in finance, there is still another issue: arbitrage. Consider the data of the toy problem we have solved in section 11.2. Are they sensible data? To understand the issue, consider the two simple trees depicted in figure 11.9. The first one corresponds to the scenarios we have used in the example. If we assume that the initial prices are 1 for both assets, the total returns we used in the toy example can be regarded as prices in the two scenarios. Sensible scenarios should not only reflect the information we have, but they should also rule out arbitrage opportunities. One way to define an arbitrage opportunity is the following. We have an arbitrage opportunity if there exists a portfolio which is guaranteed to have a non-negative value at the end of the holding period in any scenario, but which has a negative value at the beginning. Formally, let $\mathbf{p} \in \mathbb{R}^n_+$ be the vector of the initial prices for n assets, $\mathbf{x} \in \mathbb{R}^n$ the portfolio holdings for each asset, and $\mathbf{R} \in \mathbb{R}^{m,n}$ the return of each asset in each of the m scenarios (i.e., R_{ij} is the return of asset j in scenario i). Then an arbitrage opportunity is a portfolio $\mathbf{x}$ such that

$$\mathbf{Rx} \geq 0 \quad \text{and} \quad \mathbf{p'x} < 0. \tag{11.13}$$

Another form of arbitrage opportunity is the following[3]:

$$\mathbf{Rx} \geq 0 \quad \text{and} \quad \mathbf{p'x} = 0, \tag{11.14}$$

where at least one inequality is strict. In other words, we are sure that we will not lose any money in any scenario and there is at least one scenario in which we gain something.

In order to exploit an arbitrage opportunity to gain an infinite profit, we should be able to do some short selling; if the optimization model forbids short selling, we will not see such a blatant error as an unbounded solution, but what we get could be not very sensible anyway.

It is easy to see that the scenario tree in figure 11.9b leads to an arbitrage opportunity like (11.14). With those asset prices, an initial portfolio has zero value if

$$x_1 + x_2 = 0.$$

We may use this condition to express the final portfolio value in the two scenarios:

$$1.25 x_1 + 1.14 x_2 = (1.25 - 1.14) x_1$$
$$1.15 x_1 + 1.12 x_2 = (1.15 - 1.12) x_1.$$

It is easy to see that we should sell the second asset short, so that $x_1 > 0$, to get an arbitrage opportunity. The same does not hold in the case of figure 11.9a.

But how can we be sure that a set of scenarios is arbitrage-free? An answer is given by the following theorem.

THEOREM 11.1 *There is no arbitrage opportunity of the form (11.13) if and only if there exists a vector $\mathbf{y}$ such that*

$$\mathbf{R'y} = \mathbf{p} \quad \text{and} \quad \mathbf{y} \geq 0.$$

---

[3]See [14, chapter 2] for a discussion about the relationships between the two forms of arbitrage.

Proof. Consider the following linear programming problem:

$$\max \quad \mathbf{0}'\mathbf{y}$$
$$\text{s.t.} \quad \mathbf{R}'\mathbf{y} = \mathbf{p}$$
$$\mathbf{y} \geq \mathbf{0}.$$

If this problem is solvable, so is its dual:

$$\min \quad \mathbf{p}'\mathbf{x}$$
$$\text{s.t.} \quad \mathbf{R}\mathbf{x} \geq \mathbf{0}.$$

But in this case, the optimal objective values are both equal to zero. Then we see that if there exists a feasible vector $\mathbf{y}$ for the primal problem, we cannot have $\mathbf{p}'\mathbf{x} < 0$. □

On the one hand, the theorem suggests a way to make scenarios arbitrage free. We could simply add a node in such a way that the conditions of the theorem are met. The full details of this idea are given in [6]. It should be noted that finding the best way to generate scenarios is still an open issue, as we may well generate arbitrage-free scenarios which do not fit the assumed distributions at all. On the other hand, by reasoning along the lines of the theorem, we may get a grasp on the relationships between the absence of arbitrage opportunities and the existence of risk-neutral probability measures.[4]

To begin with, we should note that if a vector $\mathbf{p}$ of initial prices satisfies theorem 11.1, then any vector $\lambda \mathbf{p}$, $\lambda > 0$, does, too. So there is a degree of freedom in pricing; in fact, we have only considered risky assets. What if we consider a risk-free asset with a risk-free rate r? To characterize arbitrage when a risk-free asset is available, let us consider a two-stage scenario tree: the initial node is 0 and there are N nodes at the second stage. Let P_{i0} be the current price of asset i, $i = 1, \ldots, I$, and P_{in} the price if scenario n, $n = 1, \ldots, N$, occurs. For each asset, we may define the discounted gain for asset i in scenario n, with respect to the risk-free asset:

$$R_{in}^* = \frac{P_{in}}{1+r} - P_{i0} \quad \forall i, n.$$

Note that if a discounted gain is positive, it means that the risky asset has performed better than the risk-free asset. Given a set of portfolio holdings x_i, we may define the overall discounted gain in node n:

$$g_n^* = \sum_{i=1}^{I} R_{in}^* x_i,$$

---

[4] The rest of this section can be safely skipped; we include this topic to point out another use of linear programming duality, but it is not essential for the following. The treatment follows [17], to which we refer the interested reader for more details.

which is the realization of a random variable G^* in scenario n. Now it is intuitive that an arbitrage opportunity may be characterized by the conditions

$$g_n^* \geq 0 \quad \forall n$$
$$E[G^*] > 0.$$

This means that the portfolio is expected to gain more than the risk-free asset on the average, but it cannot gain less in any possible scenario. To find a condition ruling out arbitrage, we may try to reason as in theorem 11.1. We may rewrite the arbitrage conditions as

$$\sum_{n=1}^{N}\sum_{i=1}^{I} R_{in}^* x_i = 1$$

$$\sum_{i=1}^{I} R_{in}^* x_i \geq 0 \quad \forall n.$$

The first condition may look a bit arbitrary, but its purpose is to make sure that at least one of the g_n^* is strictly positive; since an arbitrage opportunity may be scaled arbitrarily, setting the double sum value to 1 serves the purpose. Now, to apply linear programming duality, we should rewrite these conditions in the standard form:

$$\mathbf{Ax} = \mathbf{b}$$
$$\mathbf{x} \geq \mathbf{0}.$$
(11.15)

We may simply express each portfolio holding, which may be negative if short-selling is allowed, as

$$x_i = x_i^+ - x_i^-, \quad x_i^+, x_i^+ \geq 0,$$

and introduce a set of non-negative auxiliary variables x_{I+n}, $n = 1, \ldots, N$:

$$x_{I+n} = \sum_{i=1}^{I} R_{in}^* x_i = \sum_{i=1}^{I} \left(R_{in}^* x_i^+ - R_{in}^* x_i^- \right) \quad \forall n.$$

So we have a vector of non-negative decision variables:

$$\mathbf{x} = \begin{bmatrix} x_1^+ & x_1^- & x_2^+ & \cdots & x_I^- & x_{I+1} & \cdots & x_{I+N} \end{bmatrix}'.$$

Now the existence of an arbitrage is linked to the existence of a solution to the system (11.15), where

$$\mathbf{A} = \begin{bmatrix} 0 & 0 & 0 & \cdots & 0 & 1 & 1 & \cdots & 1 \\ R_{11}^* & -R_{11}^* & R_{21}^* & \cdots & -R_{I1}^* & -1 & 0 & \cdots & 0 \\ R_{12}^* & -R_{12}^* & R_{22}^* & \cdots & -R_{I2}^* & 0 & -1 & \cdots & 0 \\ \vdots & \vdots & \vdots & \ddots & \vdots & \vdots & \vdots & \ddots & \vdots \\ R_{1N}^* & -R_{1N}^* & R_{2N}^* & \cdots & -R_{IN}^* & 0 & 0 & \cdots & -1 \end{bmatrix}$$

and
$$\mathbf{b} = [1, 0, \ldots, 0]'.$$

If there is a feasible solution of (11.15), there cannot be a solution of the following system:

$$\mathbf{A}'\mathbf{y} \leq \mathbf{0} \qquad (11.16)$$
$$\mathbf{b}'\mathbf{y} > 0.$$

This is a direct consequence of linear programming duality. In fact, the existence of a solution of system (11.16) would imply that there is direction $\hat{\mathbf{y}}$ along which we may arbitrarily increase the objective function $\mathbf{b}'\mathbf{y}$ without violating the constraints $\mathbf{A}'\mathbf{y} \leq \mathbf{c}$, for an arbitrary vector $\mathbf{c}$. Hence, the dual linear program would be unbounded, and the primal could not be feasible.

Seeing it the other way around, if there is a solution to system (11.16), there is no arbitrage opportunity. It is possible to find an important interpretation of system (11.16), taking the forms of $\mathbf{A}$ and $\mathbf{b}$ into account. Let us denote the dual variable corresponding to the first primal constraint by y_0; we also have a dual variable y_n for each primal constraint corresponding to scenario n. Now let us write the dual constraints $\mathbf{A}'\mathbf{y} \leq \mathbf{0}$ explicitly. For each asset i, we have a pair of inequalities:

$$\sum_{n=1}^{N} R_{in}^* y_n \leq 0$$
$$-\sum_{n=1}^{N} R_{in}^* y_n \leq 0.$$

Together, they imply that for all assets i we have

$$\sum_{n=1}^{N} R_{in}^* y_n = 0. \qquad (11.17)$$

Furthermore, considering the last n columns of matrix $\mathbf{A}$, we also have

$$y_0 - y_n \leq 0 \qquad \forall n.$$

This, together with second condition in system (11.16), has the following implications:

$$\mathbf{b}'\mathbf{y} > 0 \;\Rightarrow\; y_0 > 0 \;\Rightarrow\; y_n > 0 \qquad \forall n.$$

Let us rescale the dual solution as follows:

$$\pi_n = \frac{y_n}{\sum_{k=1}^{N} y_k} \qquad \forall n. \qquad (11.18)$$

We see that the vector π may be interpreted as a probability measure, since the components are non-negative and their sum is 1. Moreover, it is a *risk-neutral* probability measure, according to which any scenario is possible (it has strictly positive probability) and any asset gains the risk-free return on the average. To see this, we may plug equation (11.18) into equation (11.17) to obtain

$$\sum_{n=1}^{N} R_{in}^* \pi_n = 0.$$

This means that, under this probability measure, the expected discounted gain for any asset is zero, which in turn implies

$$E_\pi[P_i] = (1+r)P_{i0}.$$

Now we may see a little better why risk-neutral probability measures play a role in option pricing under the no-arbitrage assumption, at least in a two-period economy with discrete states of the world. Rigorous treatment with continuous time and continuous asset prices requires the tools of stochastic calculus.

11.4 L-SHAPED METHOD FOR TWO-STAGE LINEAR STOCHASTIC PROGRAMMING

In the first sections of this chapter, we have formulated a few simple stochastic LP models, and we have seen that they can be tackled by the simplex method; interior point methods are a possible alternative. In other words, by using a discretized representation of uncertainty we obtain a deterministic equivalent program. However, given the number of scenarios we need to generate, the sheer size of the resulting model may be overwhelming and it can exceed the capabilities of the best available solvers. This is why clever scenario generation is so important. Another difficulty which is not so evident, is that even moderately sized stochastic programs may be difficult to solve because of their structure: it may happen that the progress made by the simplex method is very slow. Interior point methods may be a suitable alternative in some cases, and another possibility is the development of specific solution methods which take advantage of the structure of stochastic programs. This is a very active and technically challenging area of research. What we would like to do is to give an idea of how structure can be exploited to devise solution algorithms based on decomposition. We will describe a simplified version of L-shaped decomposition, which was the first specific algorithm developed to cope with large-scale two-stage stochastic programs.

Consider a two-stage problem with a fixed recourse matrix $\mathbf{W}$:

$$\min \quad \mathbf{c}'\mathbf{x} + \sum_{s \in S} p_s \mathbf{q}_s' \mathbf{y}^s$$

s.t. $\quad \mathbf{Ax} = \mathbf{b}$
$\quad\quad\quad \mathbf{Wy}_s + \mathbf{T}_s\mathbf{x} = \mathbf{r}_s \quad \forall s \in S$
$\quad\quad\quad \mathbf{x}, \mathbf{y}_s \geq \mathbf{0},$

where p_s is the probability of scenario s. It may be seen that the problem lends itself to a decomposition approach: in fact, once the first-stage decisions $\mathbf{x}$ are fixed, the problem is decomposed into a set of small subproblems, one for each scenario s. This point may be appreciated by looking at the sparse structure of the overall technological matrix for this problem:

$$\begin{bmatrix} \mathbf{A} & 0 & 0 & \cdots & 0 \\ \mathbf{T}_1 & \mathbf{W} & 0 & \cdots & 0 \\ \mathbf{T}_2 & 0 & \mathbf{W} & \cdots & 0 \\ \vdots & \vdots & \vdots & \ddots & \vdots \\ \mathbf{T}_S & 0 & 0 & \cdots & \mathbf{W} \end{bmatrix}.$$

This matrix is almost block-diagonal. The recourse function is

$$H(\mathbf{x}) = \sum_{s \in S} p_s h_s(\mathbf{x}),$$

where

$$h_s(\mathbf{x}) \equiv \min \quad \mathbf{q}_s'\mathbf{y}^s$$
$$\text{s.t.} \quad \mathbf{Wy}_s = \mathbf{r}_s - \mathbf{T}_s\mathbf{x} \quad (11.19)$$
$$\mathbf{y}_s \geq \mathbf{0}.$$

Evaluating the recourse function for a given first-stage decision $\hat{\mathbf{x}}$ entails solving a set of independent LP problems. For simplicity, we assume here that all these problems are solvable, i.e., $h_s(\hat{\mathbf{x}}) < +\infty$ for any scenario s, for any $\hat{\mathbf{x}}$ that is feasible with respect to the first-stage constraints. We say in this case that the problem has relatively complete recourse. This may be a reasonable assumption in financial problems. Consider, for instance, an asset–liability management problem; if we include extreme and pessimistic financial scenarios in our model, it might be the case that some liabilities are not always met; in such a case, we may relax the constraints by suitable shortfall penalties (like we did section 11.2). These penalties make the recourse complete. If the recourse is not complete, the approach we describe here may easily be extended.

It can be shown that the recourse function $H(\mathbf{x})$ is convex; hence we may consider the application of Kelley's cutting plane algorithm, which was illustrated in section 6.3.4. To this end, let us rewrite the two-stage problem as

$$\min \quad \mathbf{c}'\mathbf{x} + \theta$$

$$\text{s.t.} \quad \mathbf{Ax} = \mathbf{b}$$
$$\theta \geq H(\mathbf{x}) \tag{11.20}$$
$$\mathbf{x} \geq \mathbf{0}.$$

We may relax the constraint (11.20), obtaining a relaxed master problem, and then add cutting planes of the form

$$\theta \geq \alpha' \mathbf{x} + \beta.$$

The coefficients of each cut are obtained by solving the scenario subproblems for given first-stage decisions. To see how, let $\hat{\mathbf{x}}$ be the optimal solution of the initial master problem. Consider the dual of problem (11.19):

$$h_s(\hat{\mathbf{x}}) \equiv \max \quad (\mathbf{r}_s - \mathbf{T}_s \hat{\mathbf{x}})' \boldsymbol{\pi}_s$$
$$\text{s.t.} \quad \mathbf{W}' \boldsymbol{\pi}_s \leq \mathbf{q}_s. \tag{11.21}$$

Given an optimal dual solution $\hat{\boldsymbol{\pi}}_s$, it is easy to see that the following relationships hold:

$$h_s(\hat{\mathbf{x}}) = (\mathbf{r}_s - \mathbf{T}_s \hat{\mathbf{x}})' \hat{\boldsymbol{\pi}}_s \tag{11.22}$$
$$h_s(\mathbf{x}) \geq (\mathbf{r}_s - \mathbf{T}_s \mathbf{x})' \hat{\boldsymbol{\pi}}_s \quad \forall \mathbf{x}. \tag{11.23}$$

The inequality (11.23) derives from the fact that $\hat{\boldsymbol{\pi}}_s$ is the optimal dual solution for $\hat{\mathbf{x}}$, but not for a generic $\mathbf{x}$. Summing (11.23) over the scenarios, we get

$$H(\mathbf{x}) = \sum_{s \in S} p_s h_s(\mathbf{x}) \geq \sum_{s \in S} p_s (\mathbf{r}_s - \mathbf{T}_s \mathbf{x})' \hat{\boldsymbol{\pi}}_s.$$

Hence, we may add the cutting plane

$$\theta \geq \sum_{s \in S} p_s (\mathbf{r}_s - \mathbf{T}_s \mathbf{x})' \hat{\boldsymbol{\pi}}_s.$$

The L-shaped decomposition algorithm is obtained by iterating the solution of the relaxed master problem, which yields $\hat{\theta}$ and $\hat{\mathbf{x}}$, and of the corresponding scenario subproblem. At each iteration, cuts are added to the master problem. The algorithm stops when the optimal solution of the master problem satisfies

$$\hat{\theta} \leq H(\hat{\mathbf{x}}).$$

This condition may be relaxed if a near-optimal solution is good enough for our purposes.

If the recourse is not complete, some of scenario subproblems may be infeasible for certain first-stage decisions. In this case we may again exploit the dual of the scenario subproblem. Note that the feasibility region of this dual does not depend on the first-stage decisions, since $\hat{\mathbf{x}}$ does not enter constraints (11.21). Thus, if a dual problem is infeasible, it means that the second-stage problem for the corresponding scenario will be infeasible for any first-stage decision. Ruling out this case, which is likely to be due to a modeling error, when the primal problem is infeasible, the dual will be unbounded. Hence, there is an extreme ray of the dual feasible set along which the optimal solution goes to infinity. In this case we may easily add an infeasibility cut to the master, cutting the first-stage decisions which lead to an infeasible second-stage problem. Thus, at any iteration, we discover either an extreme point or an extreme ray of the dual feasible sets of each second-stage subproblem. The finite convergence of the method derives from the fact that any polyhedron has a finite number of extreme points and extreme rays (see supplement S6.1.2).

We have just outlined the basic principles of one possible approach to cope with stochastic programs. Other approaches have been pursued, but we would like to point out that this idea can also be generalized to multistage stochastic programs. Furthermore, the idea of cutting planes is the foundation of some methods which are able to cope with continuous distributions. In the modeling approach we have pursued we first sample a set of scenarios, and then we solve an optimization model. It is also possible to integrate sampling within the optimization algorithm, to generate cutting planes, in such a way that a problem with continuously distributed parameters can be tackled (see [10]).

11.5 A COMPARISON WITH DYNAMIC PROGRAMMING

In the last chapter, we have considered dynamic programming as a framework to tackle dynamic decision making under uncertainty, and it is natural to wonder about connections or differences between that approach and stochastic programming with recourse. Indeed, the concept of recourse function looks quite similar to the concept of value function or cost-to-go in dynamic programming. While the two approaches are clearly related, they are actually complementary.

- Dynamic programming approaches require finding the value function, as a function of state variables, for each decision stage. Stochastic programming methods based on L-shaped decomposition aim at finding only a *local* approximation of the recourse function.

- Dynamic programming methods, after computing the value functions, allow for a simulation of the whole decision process over the planning horizon. Stochastic programming methods aim at finding the solution for the first stage only, even though in principle further stage decision variables represent a feedback policy. In this sense stochastic programming is a more operational approach. Indeed, the use of dynamic programming models is the rule whenever one wants to use an optimization model to gain insights in a problem, possibly by a stylized model, rather than actually solving it in operational terms. This is quite common in

Economics. For instance, dynamic programming has been used to investigate strategic allocation between risky and risk-free assets for a long-term investor, for varying income profiles over time [3]. This is certainly important in Pension Economics, but it is probably not what the manager of a pension fund would use for operational decisions.

- Dynamic programming methods are able to cope with infinite-horizon problems, whereas stochastic programming methods are not. Again, this is typical of dynamic models in Economics.

- Dynamic programming models, in some cases, may be solved analytically, maybe approximately. The usefulness of insights from approximate analytical solutions is illustrated, e.g., in [3]. On the contrary, stochastic programming approaches are numerical in nature.

- Dynamic programming models assume some condition on the underlying uncertainty, since the disturbance process should be Markovian (actually, often one can get around this difficulty by augmenting the set of state variables). In principle, any type of uncertainty and any type of intertemporal dependence can be tackled by stochastic programming, provided we are able to generate a scenario tree.

Given these differences, it is no surprise that dynamic programming is more common in the Economics community, whereas stochastic programming is more familiar to the Operations Research community. However, a broader knowledge of pro and cons of both approaches is most valuable. For instance, the regression-based approach to pricing American options by Monte Carlo simulation can be better understood if we interpret the procedure as a way to enforce non-anticipativity of decisions under uncertainty.

For further reading

In the literature

- An early reference on stochastic programming with recourse is [4].
- Introductions to modeling with stochastic programming can be found in [21] and [23].
- Textbook treatments, covering also solution methods, are available in [2] and [15].
- A survey about solution methods can also be found in [1].
- The L-shaped method is described in the original reference [24].
- We have only covered stochastic programming models with recourse. For an introduction to chance-constrained models, see [18].

- Since scenario generation is only an approximate way to represent uncertainty, we should wonder how errors may affect the solution. Theoretical results are surveyed in [20]; a sensitivity analysis approach based on "contamination" between different scenario trees is described in [7].
- Scenario generation is one of the topics covered in [6] and [12]. The first reference also addresses arbitrage issues in financial scenario generation.
- For a thorough discussion on arbitrage and risk-neutral probability measures, see [14] and [17].
- The AMPL language is described in [8].
- A reference describing many portfolio optimization models, including stochastic programming models, is the two-volume set [25] and [26].
- Stochastic programming for portfolio management is also covered by [22].

On the Web

- The AMPL site is http://www.ampl.com.
- The main web reference for stochastic programming is http://stoprog.org.
- Other pointers to stochastic programming, including financial applications, can be found by browsing http://mat.gsia.cmu.edu.

REFERENCES

1. J.R. Birge. Stochastic Programming Computation and Applications. *INFORMS Journal of Computing*, 9:111–133, 1997.
2. J.R. Birge and F. Louveaux. *Introduction to Stochastic Programming*. Springer-Verlag, New York, 1997.
3. J.Y. Campbell and L.M. Viceira. *Strategic Asset Allocation*. Oxford University Press, Oxford, 2002.
4. G.B. Dantzig. Linear Programming under Uncertainty. *Management Science*, 1:197–206, 1955.
5. M.A.H. Dempster and R.T. Thompson. EVPI-Based Importance Sampling Solution Procedures for Multistage Stochastic Linear Programmes on Parallel MIMD Architectures. *Annals of Operations Research*, 90:161–184, 1999.

REFERENCES 561

6. C. Dert. *Asset Liability Management for Pension Funds: A Multistage Chance Constrained Programming Approach.* Ph.D. thesis, Erasmus University, Rotterdam, The Netherlands, 1995.

7. J. Dupačová. Stability and Sensitivity Analysis for Stochastic Programming. *Annals of Operations Research*, 27, 1990.

8. R. Fourer, D.M. Gay, and B.W. Kernighan. *AMPL: A Modeling Language for Mathematical Programming.* Boyd and Fraser, Danvers, MA, 1993.

9. H. Heitsch and W. Roemisch. Scenario Reduction Algorithms in Stochastic Programming. *Computational Optimization and Applications*, 24:187–206, 2003.

10. J.L. Higle and S. Sen. *Stochastic Decomposition.* Kluwer Academic Publishers, Dordrecht, 1996.

11. R. Hochreiter and G.Ch. Pflug. Scenario Tree Generation as a Multidimensional Facility Location Problem. Aurora Technical Report, University of Wien, 2002 (paper downloadable from http://www.vcpc.univie.ac.at/aurora/publications/).

12. K. Hoyland and S.W. Wallace. Generating Scenario Trees for Multistage Decision Problems. *Management Science*, 47:296–307, 2001.

13. G. Infanger. *Planning under Uncertainty: Solving Large-Scale Stochastic Linear Programs.* Boyd and Fraser, Danvers, MA, 1994.

14. J.E. Ingersoll, Jr. *Theory of Financial Decision Making.* Rowman & Littlefield, Totowa, NJ, 1987.

15. P. Kall and S.W. Wallace. *Stochastic Programming.* Wiley, Chichester, 1994.

16. M. Koivu. Variance reduction in sample approximations of stochastic programs. *Mathematical Programming*, 103:463–485, 2005.

17. S.R. Pliska. *Introduction to Mathematical Finance: Discrete Time Models.* Blackwell Publishers, Malden, MA, 1997.

18. A. Prékopa. Probabilistic Programming. In A. Ruszczyński and A. Shapiro, editors, *Stochastic Programming.* Elsevier, Amsterdam, 2003.

19. S.T. Rachev. *Probability Metrics and the Stability of Stochastic Models.* Wiley, Chichester, 1991.

20. W. Roemisch. Stability of Stochastic Programming Problems. In A. Ruszczyński and A. Shapiro, editors, *Stochastic Programming.* Elsevier, Amsterdam, 2003.

21. A. Ruszczyński and A. Shapiro. Stochastic Programming Models. In A. Ruszczyński and A. Shapiro, editors, *Stochastic Programming*. Elsevier, Amsterdam, 2003.

22. B. Scherer and D. Martin. *Introduction to Modern Portfolio Optimization with NuOPT, S-Plus, and S^+ Bayes*. Springer, New York, 2005.

23. S. Sen and J.L. Higle. An Introductory Tutorial on Stochastic Programming Models. *Interfaces*, 29:33–61, 1999.

24. R. Van Slyke and R.J-B. Wets. L-Shaped Linear Programs with Application to Optimal Control and Stochastic Programming. *SIAM Journal on Applied Mathematics*, 17:638–663, 1969.

25. S. Zenios, editor. *A Library of Financial Optimization Models*. Blackwell Publishers, Oxford, 2006.

26. S. Zenios. *Practical Financial Optimization*. Blackwell Publishers, Oxford, 2006.

12
Non-Convex Optimization

All of the optimization models we have considered so far have a common characteristic: They are convex, which means that we are minimizing a convex objective function (or maximizing a concave one) over a convex feasible set. In principle, convex optimization problems are easy. In practice, they can prove numerically difficult to deal with because of hard non-linearities or because of their sheer size (as is the case with large-scale stochastic programming models). Nevertheless, the optimal solution of convex problems is characterized by some relatively simple properties. Hence, if we are handed a solution by someone claiming it is the optimal one, it is usually easy to check the claim. In non-convex problems, even checking optimality is a hard task. Hence, solution methods for non-convex problems are far less efficient and much less standardized. Many of them are actually heuristics aimed at finding a good solution with a reasonable computational effort, without any claim about optimality.

In fact, non-convex optimization methods are typically outside the bag of customary tools of people in Economics and Finance. Despite all of these difficulties, there are good reasons why we should have at least a grasp of them. There are a variety of issues in portfolio management, which are neglected in classical mean-variance models, that could be tackled fruitfully within an integer programming framework:

- Limited diversification portfolio
- Minimum portfolio weights for assets
- Minimum transaction lots

- Fixed or piecewise linear transaction costs

While the resulting models were very hard to solve some years ago, astonishing progress both in computing hardware and commercially available solvers has made their practical use feasible.

Non-convexity can arise because the feasible region is non-convex. The most common case arises because some decision variables are restricted to integer values, possibly the set $\{0, 1\}$. This happens when decision variables model logical decisions, which are by their very nature discrete: Either I do something or I do not. In section 12.1 we introduce mixed-integer programming models. First we show the most common "modeling tricks" based on logical decision variables; then we outline portfolio optimization models including logical variables.

Another way non-convexity can arise is in the objective function. For instance, an objective function represented by a polynomial is likely to have a lot of local minima. This is why such problems are known as *global optimization* problems. In section 12.2 we show a portfolio optimization model based on a fixed-mix, which gives rise to a non-convex problem over continuous decision variables.

Then we consider solution methods for non-convex models. We will actually only consider branch-and-bound methods in section 12.3. Branch and bound is the standard approach for mixed-integer models, and it is available in most commercial solvers. MATLAB, at present, has a limited ability to cope with such models, and this is why we will mainly use AMPL and CPLEX to illustrate how models can be solved. Branch and bound can also be applied to some continuous global optimization models. However, global optimization methods are far less standardized. There is a wide variety of methods which are specific to subclasses of global optimization models. Apart from their conceptual difficulty, most of them are not available in commercial packages, and this is another reason why we will not deal with them in detail.

Finally, in section 12.4 we will cover some general-purpose principles that can be used to devise heuristics. In fact, non-convex problems may be a very hard nut to crack. In extreme cases, a practical alternative is to give up optimality and to look for a reasonably good solution. We will consider local search heuristics, such as simulated annealing, tabu search, and genetic algorithms. They are fairly general and flexible approaches, and indeed they have been implemented in commercial solvers which have been successfully integrated with simulation packages to tackle some optimization problems, where complexity precludes the mathematical formulation of an objective function.

12.1 MIXED-INTEGER PROGRAMMING MODELS

We have already met an integer programming model in example 1.2 on page 15, where we introduced the knapsack problem as a very rough representation of capital budgeting:

$$\max \quad \sum_{i=1}^{n} R_i x_i$$
$$\text{s.t.} \quad \sum_{i=1}^{N} C_i x_i \leq W$$
$$x_i \in \{0, 1\}.$$

This is actually a linear programming model with an additional restriction on the decision variables, which may take values only within a discrete set; this is what makes the problem non-convex.

A more general form of a **mixed-integer linear programming** model is

$$\min \quad \mathbf{c'x + d'y}$$
$$\text{s.t.} \quad \mathbf{Ax + Dy \geq b}$$
$$\mathbf{x} \geq 0, \quad \mathbf{y} \in \mathbb{Z}_+ = \{0, 1, 2, 3, \ldots\}$$

The name stems from the fact that we mix continuous variables **x** and integer variables **y**. When all of the decision variables must take integer values, we speak of **pure** integer programming models. A very common case arises when a decision variable is *binary*, i.e., it must take values within the set $\{0, 1\}$. This is typical of logical decision variables, as we will illustrate in the next section. Moreover, using binary variables is a very powerful modeling trick to represent non-trivial constraints. When all the decision variables are binary, we have a **pure binary** programming model. The knapsack problem is such a case.

General integer variables may arise, e.g., when an asset must be purchased in multiples of a base lot. If the number of such multiples is large, then a continuous approximation is reasonable; otherwise the discrete nature of the investment must be properly reflected in the optimization model. However, the most common model is a linear mixed-integer model in which all integer variables are actually logical. Non-linear mixed-integer programming models can be formulated, but efficient solvers are not that widespread commercially, although they are actually available. An exception is quadratic mixed-integer programming. Recent releases of ILOG CPLEX are able to tackle this class of models, which allow us to generalize mean-variance portfolio optimization models, as we shall see.

12.1.1 Modeling with logical variables

It is useful to point out a few situations that require the introduction of binary decision variables.

Logical constraints Consider a set of N activities, perhaps investment opportunities. Starting an activity or not is modeled by a corresponding binary decision variable x_i, $i = 1, \ldots, N$. You might wish to enforce some logical constraints involving subsets of activities. Here are a few examples:

- Exactly one activity within a subset S must start (exclusive "or"):

$$\sum_{j \in S} x_j = 1.$$

- At least one activity within a subset S must start (inclusive "or"):

$$\sum_{j \in S} x_j \geq 1.$$

- At most one activity within a subset S may start:

$$\sum_{j \in S} x_j \leq 1.$$

- If activity j is started, then activity k must start, too:

$$x_j \leq x_k.$$

All the constraints above may be generalized to more complex situations, which are relevant, for instance, if you want to enforce qualitative constraints on a portfolio of investments.

Fixed-charge problem and semicontinuous decision variables We obtain LP models when we assume, among other things, that the cost of carrying out a set of activities depend linearly on the activity levels. In some cases, the cost structure is more complex; the *fixed-charge problem* is one such case. We are given a set of activities, indexed by $i = 1, \ldots, N$. The level of activity i is measured by a non-negative continuous variable x_i; the activity levels are subject to a set of constraints, formally expressed as $\mathbf{x} \in S$. Each activity has a cost proportional to the level x_i and a fixed cost f_i, which is paid whenever $x_i > 0$. The fixed cost does not depend on the activity level. It is interesting to note that the cost function is in this case discontinuous at the origin, but a simple modeling trick allows us to build a mixed-integer model.

Assume that we know an upper bound M_i on the level of activity i, and introduce a set of binary variables y_i such that

$$y_i = \begin{cases} 1 & \text{if } x_i > 0 \\ 0 & \text{otherwise.} \end{cases}$$

We can build the following model:

$$\begin{align} \min \quad & \sum_{i=1}^{N} (c_i x_i + f_i y_i) \\ \text{s.t.} \quad & x_i \leq M_i y_i \quad \forall i \tag{12.1} \\ & \mathbf{x} \in S \\ & y_i \in \{0, 1\} \quad \forall i. \end{align}$$

The inequality (12.1) is a common way to model fixed-charge costs. If $y_i = 0$, necessarily $x_i = 0$; if $y_i = 1$, then we obtain $x_i \leq M_i$, which is a non-binding constraint if M_i is large enough. Apparently, the constraint (12.1) allows a non-logical choice: pay the fixed charge, but let $x_i = 0$. However, this is ruled out by the minimization of the objective function.

Another common requirement on the level of an activity is that, if it is undertaken, its level should be in the interval $[m_i, M_i]$. Note that this is *not* equivalent to requiring that $m_i \leq x_i \leq M_i$. Rather, we want something like

$$x_i \in \{0\} \cup [m_i, M_i],$$

which is a non-convex set (recall that the union of convex sets need not be convex). Using the same trick as above, we may just write

$$x_i \geq m_i y_i, \qquad x_i \leq M_i y_i.$$

These constraints define a *semicontinuous decision variable*. Semicontinuous variables may be used when the amount of an asset in a portfolio must be above a minimum threshold if the asset is included in the portfolio.

Piecewise linear functions Sometimes we have to model a non-linear dependency between two variables; to name one case, transaction costs may depend in a non-obvious way on the trading volume. Although it is possible to adopt non-linear programming methods to cope with this case, it may be advisable to avoid the issue by approximating the non-linear function by a piecewise linear function; in other words, we may try a linear interpolation (see section 3.3). Piecewise linear functions may arise quite naturally in applications. A few examples are shown in figure 12.1, where the points $x^{(i)}$ are the breakpoints separating the linearity intervals. There are different reasons for doing so. If the non-linear function occurs in an equality constraints, the problem is non-convex; the practical implication is that a non-linear optimizer may get stuck in a local optimum. The same happens if the objective function is non-linear and non-convex. Here we show that these cases may be transformed into mixed-integer programming problems which are non-convex but can be solved by branch and bound methods yielding a global optimum. Furthermore, it may be the case that the model involves integer decision variables, in which case it may be preferable to keep the model linear, as non-linear mixed-integer programming problems may be overly difficult to solve.

Consider a function like

$$f(x) = \begin{cases} c_1 x, & 0 \leq x \leq x^{(1)} \\ c_2(x - x^{(1)}) + c_1 x^{(1)}, & x^{(1)} \leq x \leq x^{(2)} \\ c_3(x - x^{(2)}) + c_1 x^{(1)} + c_2(x^{(2)} - x^{(1)}), & x^{(2)} \leq x \leq x^{(3)}. \end{cases}$$

If $c_1 < c_2 < c_3$ (increasing marginal costs), then $f(x)$ is convex (figure 12.1a); if $c_1 > c_2 > c_3$ (decreasing marginal costs), the function is concave (figure

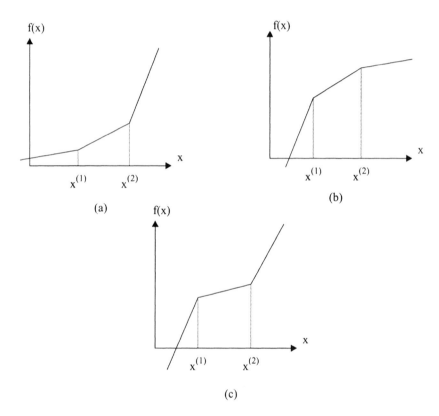

Fig. 12.1 Piecewise linear functions: (a) convex, (b) concave, (c) neither convex nor concave.

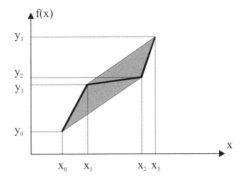

Fig. 12.2 Modeling a piecewise linear function.

12.1b); for arbitrary slopes c_i the function is neither convex nor concave (figure 12.1c).

The convex case is easy and it can be coped with by continuous LP models. The function $f(x)$ can be converted to a linear form by introducing three auxiliary variables y_1, y_2, y_3 and substituting:

$$x = y_1 + y_2 + y_3$$
$$0 \leq y_1 \leq x^{(1)}$$
$$0 \leq y_2 \leq (x^{(2)} - x^{(1)})$$
$$0 \leq y_3 \leq (x^{(3)} - x^{(2)}).$$

Then we can express

$$f(x) = c_1 y_1 + c_2 y_2 + c_3 y_3,$$

since $c_1 < c_2$, y_2 is positive in the optimal solution only if y_1 is set to its upper bound. Similarly, y_3 is activated only if both y_1 and y_2 are saturated to their upper bounds. If the function is not convex, this is not guaranteed, and we must come up with a modeling trick based on binary decision variables.

To get a clue on how a general piecewise linear function may be modeled, assume that the function is described by the knots (x_i, y_i), $y_i = f(x_i)$, $i = 0, 1, 2, 3$, as in figure 12.2. Any point on the line from (x_i, y_i) to (x_{i+1}, y_{i+1}) can be expressed as a convex combination:

$$x = \lambda x_i + (1 - \lambda) x_{i+1}$$
$$y = \lambda y_i + (1 - \lambda) y_{i+1},$$

where $0 \leq \lambda \leq 1$. Now what about forming a convex combination of the four knots?

$$x = \sum_{i=0}^{3} \lambda_i x_i$$

$$y = \sum_{i=0}^{3} \lambda_i y_i$$

$$\sum_{i=0}^{3} \lambda_i = 1, \qquad \lambda_i \geq 0.$$

This is not really what we want, since this is the convex hull of the four knots (the shaded area in figure 12.2; see supplement S6.1). However, we are close; we have just to allow only pairs of adjacent coefficients λ_i to be strictly positive. This is accomplished by introducing a binary decision variable s_i, $i = 1, 2, 3$, for each line segment $(i-1, i)$:

$$x = \sum_{i=0}^{3} \lambda_i x_i$$

$$y = \sum_{i=0}^{3} \lambda_i y_i$$

$$0 \leq \lambda_0 \leq s_1$$

$$0 \leq \lambda_1 \leq s_1 + s_2$$

$$0 \leq \lambda_2 \leq s_2 + s_3$$

$$0 \leq \lambda_3 \leq s_3$$

$$\sum_{i=1}^{3} s_i = 1 \qquad s_i \in \{0, 1\}.$$

In practice, optimization software packages and languages, such as AMPL, provide the user with an easier but equivalent way to express piecewise linear functions.

Example 12.1 Assume we want to model a piecewise linear objective function like those depicted in figure 12.1, where we have two breakpoints and three slopes. To express this in AMPL, we should use the keyword **param** to declare parameters corresponding to breakpoints and slopes, say x1, x2, c1, c2, and c3, and the keyword **var** to introduce a decision variable, say x. In AMPL, the objective function would include a term like:

`<<x1, x2; c1, c2, c3>> x`

We see that slopes are always one more than breakpoints: the first slope c1 applies to values of x smaller than x1, and c3 applies to values larger than x2. AMPL detects automatically if, given the characteristics of the function and the sense of optimization (min or max), a continuous or a discrete optimization method is required. ☐

12.1.2 Mixed-integer portfolio optimization models

An efficient mean-variance portfolio may include a large set of assets, and some of them may account for a tiny part of the overall asset allocation. While this is, at least in principle, beneficial for diversification, there are a few downsides in a too diversified portfolio. One issue is the amount of transaction costs we have to pay, making small transactions unattractive. Another issue is the effort that is required in analyzing the historical data for too many assets, in order to control the portfolio risk. These requirements are particularly important for passively-managed funds, which cannot be expensive, since they just aim at tracking some target. We could extend the mean-variance model by constraining the portfolio cardinality, i.e., the number of assets included. Writing a constraint stating that at most k assets out of the I available may be included in the portfolio is easily accomplished by introducing, for each asset $i = 1, \ldots, I$, the following binary variable:

$$\delta_i = \begin{cases} 1 & \text{if asset } i \text{ is included in the portfolio,} \\ 0 & \text{otherwise.} \end{cases}$$

Then all we have to do is to add the following constraints to the model:

$$x_i \leq M_i \delta_i \qquad \forall i \tag{12.2}$$

$$\sum_{i=1}^{I} \delta_i \leq k, \tag{12.3}$$

where M_i is an upper bound on the weight of asset i. This is actually the same trick we have just described to model fixed costs in the fixed-charge problem. Another requirement could be enforcing a minimal limit to an asset weight if positive. This requirement cannot be enforced within a continuous linear or quadratic programming model. However, it is easy to extend constraint (12.2):

$$m_i \delta_i \leq x_i \leq M_i \delta_i \qquad \forall i.$$

This is an example of a semicontinuous variable. By the way, x_i need not be the weight in a portfolio; it could be the amount of stock traded, in which case m_i would be the minimal tradeable lot. We could even go further and require, in such a case, that x_i is a general integer variable, in order to avoid the additional costs involved in trading odd lots. Putting all of this together, we can trace the efficient frontier by solving a set of mixed-integer quadratic programs like the following:

$$\min \quad \sum_{i=1}^{I} \sum_{j=1}^{I} \sigma_{ij} x_i x_j$$

$$\text{s.t.} \quad \sum_{i=1}^{I} \bar{r}_i x_i \geq \bar{r}_T$$

572 NON-CONVEX OPTIMIZATION

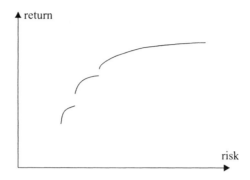

Fig. 12.3 Qualitative sketch of a cardinality-constrained efficient frontier.

$$\sum_{i=1}^{I} x_i = 1$$
$$m_i \delta_i \leq x_i \leq M_i \delta_i \qquad \forall i$$
$$\sum_{i=1}^{I} \delta_i \leq k$$
$$w_i \geq 0, \quad \delta_i \in \{0,1\} \qquad \forall i,$$

where $\bar{r}_i$ is the expected return of asset i, σ_{ij} is the covariance between the returns of assets i and j, and $\bar{r}_T$ is a target return. By varying the target return we would trace the efficient frontier. It is also important to realize that the efficient frontier will be qualitatively different from the usual one, which was illustrated in figure 2.12. A qualitative sketch of the cardinality-constrained efficient frontier is illustrated in figure 12.3. This plot may be understood by imagining of tracing the efficient sets for each portfolio consisting of a subset of cardinality k, and then patching all of them together. One difficulty with the formulation above is that it is a mixed-integer *quadratic*, rather than linear, problem. In principle, and in practice as well, it can be solved by the same branch and bound algorithm illustrated in section 12.3.1; the only difference is that the lower bounds are computed by solving a quadratic programming problem. Nowadays, commercial codes are available to tackle such problems efficiently; however, the computational requirements could turn out to be prohibitive for a large-scale application. Still, different alternatives may be tried.

- We may trace only the relevant part of the efficient set, given our risk aversion.

- In [3] ad hoc methods are discussed for mixed-integer quadratic programming; taking a route like this may be advantageous, but it requires writing our own code.

- Another possibility is to simplify the model by reducing the data requirements, e.g., by assuming that all the correlations are equal. See [17] for an approach like this, and for additional references as well.

- Metaheuristics such as genetic algorithms and simulated annealing (section 12.4) may also be used [4].

- If one wants to use MILP codes, it is also possible to devise a different representation of risk. In [9] the use of the mean absolute deviation has been advocated:

$$\mathrm{E}\left[\left|\sum_{i=1}^{I} R_i x_i - \mathrm{E}\left[\sum_{j=1}^{I} R_j x_j\right]\right|\right],$$

where R_i is the random return of asset i. This definition is quite similar to variance; an absolute deviation is used rather than a squared deviation. This objective may be translated in linear terms, and MILP methods, exact or heuristic, may be applied. Suppose in fact that we have a set of historical returns r_{it} for each asset in time periods $t = 1, \ldots, T$. Then we may estimate $\mathrm{E}[R_i] = \bar{r}_i = (1/T) \sum_{t=1}^{T} r_{it}$ and set

$$\mathrm{E}\left[\sum_{j=1}^{N} R_j x_j\right] = \sum_{j=1}^{N} \bar{r}_j x_j.$$

By the same token, we may approximate the objective function as

$$\mathrm{E}\left\{\left|\sum_{i=1}^{N} R_i x_i - \sum_{j=1}^{N} \bar{r}_j x_j\right|\right\} = \frac{1}{T} \sum_{t=1}^{T} \sum_{i=1}^{N} |(r_{it} - \bar{r}_i) x_i|.$$

This objective function may be expressed in linear form by introducing a set of auxiliary variables y_t. The model will include, among other things, the following objective function and constraints:

$$\min \quad \frac{1}{T} \sum_{t=1}^{T} y_t$$

$$\text{s.t.} \quad y_t + \sum_{i=1}^{N} (r_{it} - \bar{r}_i) x_i \geq 0 \quad \forall t$$

$$y_t - \sum_{i=1}^{N} (r_{it} - \bar{r}_i) x_i \geq 0 \quad \forall t.$$

For instance, this approach is taken in [11], where minimum transaction lots are dealt with. This approach does not require any statistical modeling, but we should mention that there is a risk of overfitting with respect to historical data.

- Finally, the MILP model may not really be aimed at building a portfolio from scratch. Rather, one could devise a target portfolio by whatever technique, subject to variety of constraints related to critical market exposure and liquidity. Then the target is approximated by enforcing some practical requirements, such as minimizing the number of assets included in the real portfolio. This is the approach taken in [2] to cope with a real-life case.

A final important remark is that the difficulty of solving a mixed-integer problem depends on the strength of its relaxation (see section 12.3.1). The least one should do is to reduce the M_i bounds in constraints like (12.2). Thanks to careful modeling, computational times on the order of a few minutes are reported in [2] for problems involving something like 1500 assets (using what is now an old version of CPLEX).

A last point is that classical mean-variance models neglect transaction costs. This is debatable in a single-period model, and is even more questionable in a multiple-period model, since excessive trading may disrupt any advantage gained by optimizing the portfolio. The simplest idea is to use a linear model of the transaction cost; i.e., if we trade an amount x_i of an asset, we pay a proportional cost $\alpha_i x_i$, where the proportionality constant may depend on the asset liquidity. This results in a linear programming model, and one such formulation was given in section 11.2.3. However, a linear model fails to account for the dependence of transaction costs on the volume traded. Different assumptions can be made, depending on the nature of the traded asset, leading to different model formulations. In the case of fixed transaction costs, we may simply adopt the binary variable trick used earlier and treat it as a fixed cost. If transaction costs are non-linear, they may be approximated by piecewise linear functions, along the lines we illustrated at the beginning of this chapter. If we assume that transaction costs increase marginally with the traded volume (maybe because the asset is highly illiquid and it is difficult to deal with the sale/purchase order), the function is convex and can be dealt with by ordinary LP methods. However, in the case of concave costs, this is no longer the case, and mixed-integer models must be used. See also [10] for an example of how a model involving fixed transaction costs may be tackled.

Example 12.2 We illustrate here how AMPL can be used to express risk minimization subject to constraints on maximum cardinality of the portfolio and on target expected return. This is a fairly simple extension of the mean-variance model we illustrated in section C.2, in appendix C. The model file is illustrated in figure 12.4. We see that binary decision variables delta are introduced and linked to portfolio weights by the constraint LogicalLink. The maximum cardinality MaxAssets is enforced in constraint MaxCardinality. The corresponding data file is given, for a toy problem instance, in figure 12.5.

```
param NAssets > 0, integer;
param MaxAssets > 0, integer;
param ExpRet{1..NAssets};
param CovMat{1..NAssets, 1..NAssets};
param TargetRet;

var W{1..NAssets} >= 0;
var delta{1..NAssets} binary;

minimize Risk:
    sum {i in 1..NAssets, j in 1..NAssets} W[i]*CovMat[i,j]*W[j];

subject to SumToOne:
    sum {i in 1..NAssets} W[i] = 1;
subject to MinReturn:
    sum {i in 1..NAssets} ExpRet[i]*W[i] = TargetRet;
subject to LogicalLink {i in 1..NAssets}:
    W[i] <= delta[i];
subject to MaxCardinality:
    sum {i in 1..NAssets} delta[i] <= MaxAssets;
```

Fig. 12.4 AMPL model file for limited cardinality portfolio (MeanVarCard.mod).

```
param NAssets = 3;
param MaxAssets = 2;
param ExpRet :=
    1 0.15
    2 0.2
    3 0.08;
param CovMat:
       1          2         3        :=
1    0.2000    0.0500    -0.0100
2    0.0500    0.3000     0.0150
3   -0.0100    0.0150     0.1000;

param TargetRet := 0.1;
```

Fig. 12.5 AMPL data file for limited cardinality portfolio (MeanVarCard.dat).

Using AMPL, we may compare what happens here against what we have obtained in section C.2:

```
AMPL Version 20021038 (x86_win32)
ampl: model MeanVarCard.mod;
ampl: data MeanVarCard.dat;
ampl: option cplex_options 'mipdisplay 2';
ampl: solve;
CPLEX 9.1.0: mipdisplay 2
MIP emphasis: balance optimality and feasibility
Root relaxation solution time =    0.05 sec.

         Nodes                                  Cuts/
   Node  Left   Objective  IInf  Best Integer  Best Node   ItCnt     Gap

      0     0      0.0631     1                   0.0631       7
*    0+     0                 0        0.0633     0.0631       7   0.27%
CPLEX 9.1.0: optimal integer solution; objective 0.06326530612
9 MIP simplex iterations
1 branch-and-bound nodes
ampl: display W;
W [*] :=
1  0.285714
2  0
3  0.714286
;
```

The first thing we should notice is that branch and bound is invoked, rather than just a barrier solver. We also see that asset 2 does not enter the portfolio, and that the cardinality constraint also implies an increase in risk. This increase is moderate here, but it should traded off against simplified portfolio management and the reduction in transaction costs in a real setting. ☐

12.2 FIXED-MIX MODEL BASED ON GLOBAL OPTIMIZATION

In the multistage stochastic programming models we have illustrated in section 11.2, we have assumed that the portfolio could be freely rebalanced at specified time instants. A different type of model is obtained if we assume that the asset mix is held constant over the whole period. This means that the proportion of wealth that we allocate to each asset is kept constant; thus, we trade according a sell-high/buy-low strategy. Using the same notation as in section 11.2.1, we have a discrete set of scenarios, each with a probability p_s, $s = 1 \ldots, S$, where the returns are represented by R_{it}^s. Now, the decision variables are simply the proportion of wealth allocated to each asset, denoted by x_i; note that since there is no recourse action, the scenarios need not be structured according to a tree, as the non-anticipativity condition is immediately satisfied, given the definition of the decision variables. The model we describe here is due to [12], to which we refer the reader for further information and for computational experiments, and is basically an extension of the mean-variance framework; no liability is considered, and we base our objective function on the terminal wealth.

Let W_0 be the initial wealth. Then the wealth at the end of time period 1 in scenario s will be

$$W_1^s = W_0 \sum_{i=1}^{I} R_{i1}^s x_i.$$

Note that the wealth is scenario-dependent, but the asset allocation is not. In general, when we consider two consecutive time periods, we have

$$W_t^s = W_{t-1}^s \sum_{i=1}^{I} R_{it}^s x_i \qquad \forall t, s.$$

The wealth at the end of the planning horizon is

$$W_T^s = W_0 \prod_{t=1}^{T} \left(\sum_{i=1}^{I} R_{it}^s x_i \right) \qquad \forall s.$$

Within a mean-variance framework, we may build a quadratic utility function depending on the terminal wealth. Given a parameter λ linked to our risk aversion, the objective function will be something like

$$\max \ \lambda \operatorname{E}[W_T] - (1-\lambda) \operatorname{Var}(W_T).$$

To express the objective function, we must recall that $\operatorname{Var}(X) = \operatorname{E}[X^2] - \operatorname{E}^2[X]$, and we may write the model as

$$\max \quad \lambda W_0 \sum_{s=1}^{S} p^s \left[\prod_{t=1}^{T} \left(\sum_{i=1}^{I} R_{it}^s x_i \right) \right]$$

$$+ (1-\lambda) W_0^2 \left\{ \left[\sum_{s=1}^{S} p^s \left[\prod_{t=1}^{T} \left(\sum_{i=1}^{I} R_{it}^s x_i \right) \right] \right]^2 \right.$$

$$\left. - \sum_{s=1}^{S} p^s \left[\prod_{t=1}^{T} \left(\sum_{i=1}^{I} R_{it}^s x_i \right) \right]^2 \right\}$$

$$\text{s.t.} \quad \sum_{i=1}^{I} x_i = 1$$

$$0 \leq x_i \leq 1.$$

This looks like a very complex problem; however, while the objective function is a bit messy, the constraints are quite simple. The real difficulty is that this is a non-convex problem. To see why, just note that the objective turns out to be a polynomial in the decision variables; since polynomials may have many minima and maxima, we have a non-linear non-convex problem.

The problem may be tackled by the branch and bound methods described in section 12.3. In particular, the idea of bounding a non-convex function by a convex underestimator is used in [12]. If complicating features are added to the model, this may turn out a quite difficult mixed-integer non-linear problem; in this case, the use of metaheuristics such as tabu search may be the best option [6].

It is useful to interpret this approach within an integration framework of simulation and optimization. Actually, simulation is separated from optimization, since scenarios are generated beforehand; we evaluate the solutions on the same set of scenarios, which is consistent with variance reduction by common random numbers. After the optimization, simulation could be used to evaluate the solution we obtain on a larger set of scenarios, possibly including stress test scenarios; in other words, we may carry out an out-of-sample analysis to check the robustness of the solution. This is easily accomplished for a fixed-mix policy, but not for a dynamic policy, as this would require the repeated solution of difficult multistage stochastic programs. In fact, even if a fixed-mix policy is in principle an inferior policy with respect to a dynamic one, it may be more robust in practice; what's more important, it is easier to prove its robustness with respect to an arbitrary set of scenarios, and to persuade a manager to adopt it.

Selection of the best portfolio management policy is actually an open issue, but it is worth noting that the fixed-mix policy is only the simplest policy structure that we may consider for the integration of simulation and optimization. More complex policies could be devised, depending on a set of numerical parameters, whose value may be set by the integration of simulation and optimization methods.

12.3 BRANCH AND BOUND METHODS FOR NON-CONVEX OPTIMIZATION

Consider a generic optimization problem

$$P(S): \quad \min_{\mathbf{x} \in S} f(\mathbf{x}),$$

and assume that it is a difficult one, as either the objective function or the feasible set is non-convex. Consider figure 12.6; in the first case, the objective function has local minima; in the second case, the feasible set is discrete, and hence non-convex. While solving non-convex problems is very difficult in general, in some cases it could be made a straightforward task if a suitable convexification were available. For instance, if S is convex but f is not, we could take the convex hull of the epigraph of f, as illustrated in figure 12.7. Taking the convex hull of the epigraph of f yields a function h such that:

BRANCH AND BOUND METHODS FOR NON-CONVEX OPTIMIZATION   579

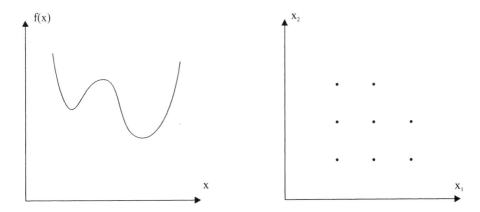

Fig. 12.6 Non-convex objective function and discrete non-convex feasible set.

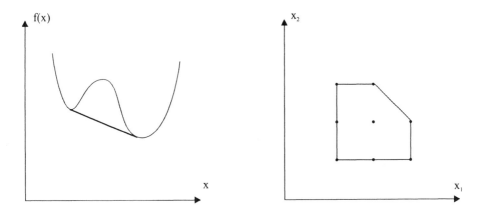

Fig. 12.7 Convexification of a non-convex objective function and a discrete non-convex feasible set.

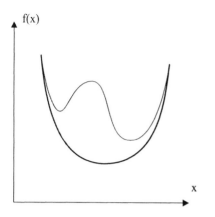 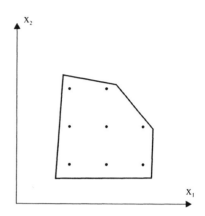

Fig. 12.8 Convex lower bounding function and a relaxation of a discrete feasible set.

- h is convex on S.
- $h(\mathbf{x}) \leq f(\mathbf{x})$ for any $\mathbf{x} \in S$.
- If g is a convex function such that $g(\mathbf{x}) \leq f(\mathbf{x})$ for any $\mathbf{x} \in S$, then $g(\mathbf{x}) \leq h(\mathbf{x})$ for any $\mathbf{x} \in S$.

In this case, we could think of replacing f by h and solve the problem by convex optimization techniques. By the same token, consider a linear integer programming problem:

$$\begin{aligned}
(\text{PI}) \quad & \min \quad \mathbf{c}'\mathbf{x} \\
& \text{s.t.} \quad \mathbf{A}\mathbf{x} \leq \mathbf{b} \\
& \qquad \mathbf{x} \in \mathbb{Z}_+^n.
\end{aligned}$$

The feasible set is a discrete set much like that in figure 12.6. If we knew its convex hull, illustrated in figure 12.7, we could simply tackle the problem as an ordinary LP problem by the simplex method. In fact, the convex hull of a discrete set of points is a polyhedron; if the points have integer coordinates, then the extreme points of the convex hull will be integer too, and one of them will turn out to be the optimal solution returned by the simplex method.[1]

Unfortunately, we are rarely in the lucky position of being able to find such a convexification easily. However, we might be able to find weaker convex objects, as illustrated in figure 12.8. They are exploited to define a relaxation of the original problem.

DEFINITION 12.1 *An optimization problem,*

$$\text{RP}(T): \qquad \min_{\mathbf{x} \in T} h(\mathbf{x}),$$

---

[1] We recall from section 6.5.1 that interior point methods, when alternative optima exist, tend to yield a solution on the center of a face of the polyhedron defining the feasible set.

is a relaxation of problem P(S) if:

- $S \subseteq T$.
- $h(\mathbf{x}) \leq f(\mathbf{x})$, *for any* $\mathbf{x} \in S$.

Solving a relaxation does not yield the optimal solution of the original problem in general, but it gives a lower bound for its optimal value.

Example 12.3 Consider a non-convex function $f(\mathbf{x})$ on a hyperrectangle S defined by the bounds

$$l_j \leq x_j \leq u_j, \qquad j = 1, \ldots, n.$$

Assume that f is twice continuously differentiable. In supplement S6.1.1 we stated that a twice continuously differentiable function is convex if its Hessian matrix is positive semidefinite, which is equivalent to requiring that its eigenvalues are non-negative. We may build a convex underestimating function for f by adding an additional term and considering

$$h(\mathbf{x}) = f(\mathbf{x}) + \alpha \sum_{i=1}^{n}(l_i - x_i)(u_i - x_i)$$

for some $\alpha > 0$. It is easy to see that the additional term is nonpositive on the region S and that it is zero on its boundary. Thus h is an underestimator for f. It will be convex if α is large enough. To see this, consider how the Hessian $\mathbf{H}$ of h is related to the Hessian $\mathbf{H}_f$ of the original objective f:

$$\frac{\partial^2 h}{\partial x_i^2} = \frac{\partial^2 f}{\partial x_i^2} + 2\alpha, \qquad i = 1, \ldots, n$$

$$\frac{\partial^2 h}{\partial x_i \partial x_j} = \frac{\partial^2 f}{\partial x_i \partial x_j}, \qquad i, j = 1, \ldots, n; i \neq j.$$

The eigenvalues of h are the solution of the following equation:

$$\det(\mathbf{H}_f + 2\alpha\mathbf{I} - \mu\mathbf{I}) = \det(\mathbf{H}_f - (\mu - 2\alpha)\mathbf{I}) = 0.$$

It is easy to see that, if the eigenvalues of $\mathbf{H}_f$ are λ_i, $i = 1, \ldots, n$, then the eigenvalues of the Hessian of h are simply

$$\mu_i = \lambda_i + 2\alpha,$$

which may be made positive by choosing a suitably large value of α. We will see shortly that a relaxation should be as tight as possible. This means that the underestimating function should be as large as possible and that α should be as small as possible. Guidelines for the selection of α are given in the original reference [13]. □

Example 12.4 Consider the integer programming problem (IP). A convex relaxation of the feasible set

$$S = \{\mathbf{x} : \mathbf{Ax} \leq \mathbf{b}; \mathbf{x} \in \mathbb{Z}_+^n\}$$

can be obtained by dropping the integrality requirement:

$$T = \{\mathbf{x} : \mathbf{Ax} \leq \mathbf{b}; \mathbf{x} \in \mathbb{R}_+^n\}.$$

This yields an LP problem which is readily solved by the simplex method. In general, some components of the solution of the relaxed problem will be fractional; this implies that the solution we obtain is not feasible, but we get a lower bound on the optimal value of the objective function. □

We have seen, in the two examples above, that when the relaxed problem is convex, it is easily solved, but it will only yield a lower bound on the optimal value of the objective function.

A possible solution strategy is to decompose the original problem $P(S)$ by splitting the feasible set S into a collection of subsets $S_1, \ldots, S_q$ such that

$$S = S_1 \cup S_2 \cup \cdots \cup S_q;$$

then we have

$$\min_{\mathbf{x} \in S} f(\mathbf{x}) = \min_{i=1,\ldots,q} \left\{ \min_{\mathbf{x} \in S_i} f(\mathbf{x}) \right\}.$$

The rationale behind this decomposition of the feasible set is that we may expect that solving the problems over smaller sets is easier; or, at least, the lower bounds obtained by solving the relaxed problems will be tighter. For efficiency reasons it is advisable, but not strictly necessary, to partition the set S in such a way that

$$S_i \cap S_j = \emptyset, \qquad i \neq j.$$

This type of decomposition is called *branching*.

Example 12.5 Consider the binary programming problem:

$$\begin{aligned} \min \quad & \mathbf{c}'\mathbf{x} \\ \text{s.t.} \quad & \mathbf{x} \in S = \{\mathbf{x} \mid \mathbf{Ax} \geq \mathbf{b}; x_j \in \{0, 1\}\}. \end{aligned}$$

The problem may be decomposed in two subproblems by picking a variable x_p and fixing it to 1 and 0:

$$\begin{aligned} S_1 &= \{\mathbf{x} \in S;\ x_p = 0\} \\ S_2 &= \{\mathbf{x} \in S;\ x_p = 1\}. \end{aligned}$$

The resulting problems $P(S_1)$ and $P(S_2)$ can be decomposed in turn, until eventually all the variables have been fixed. The branching process can be pictorially represented as a *search tree*, as shown in figure 12.9. □

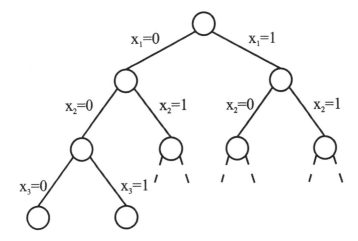

Fig. 12.9 Search tree for a binary programming problem.

The branching process leads to easier problems. In the example, the leaves of the search tree are trivial problems, since all variables are fixed to one of the two feasible values; actually, the search tree is, in this case, just a way to enumerate the possible solutions. Unfortunately, there are a large number of leaves; if $\mathbf{x} \in \{0,1\}^N$, there are 2^N possible solutions. Actually, the constraints $\mathbf{Ax} \geq \mathbf{b}$ rule out many of them, but a brute-force enumeration is not feasible except for the smallest problems.

To reduce the computational burden, one can try to eliminate a subproblem $P(S_k)$ or, equivalently, a node of the tree, by showing that it cannot lead to the optimal solution of $P(S)$. This can be accomplished if it is possible to compute a lower bound for each subproblem by a convex relaxation or by whatever method. Let $\nu[P(S_k)]$ denote the optimal value of problem $P(S_k)$. The lower bound $\beta[P(S_k)]$ is such that

$$\beta[P(S_k)] \leq \nu[P(S_k)].$$

Now assume that we know a feasible, but not necessarily optimal solution $\hat{\mathbf{x}}$ of $P(S)$. Such a solution, if any exists, is eventually found while searching the tree (with the exception of pathological cases). The value $f(\hat{\mathbf{x}})$ is an upper bound on the optimal value $\nu^* = \nu[P(S)]$. Clearly, there is no point in solving a subproblem $P(S_k)$ if

$$\beta[P(S_k)] \geq f(\hat{\mathbf{x}}). \tag{12.4}$$

In fact, solving this subproblem cannot yield an improvement with respect to feasible solution $\hat{\mathbf{x}}$ that we already know. In this case, we can eliminate $P(S_k)$ from further consideration; this elimination, called *fathoming*, corresponds to pruning a branch of the search tree. Note that $P(S_k)$ can be fathomed only by comparing the lower bound $\beta[P(S_k)]$ with an upper bound on $\nu[P(S)]$. It is

not correct to fathom $P(S_k)$ on the basis of a comparison with a subproblem $P(S_i)$ such that
$$\beta[P(S_i)] < \beta[P(S_k)].$$
The branching and fathoming mechanism is the foundation of a wide class of algorithms known as *branch and bound methods*. In the next subsection we outline the basic structure of branch and bound methods for mixed-integer linear programming (MILP) problems. These methods are widely available in commercial optimization software libraries. On the contrary, branch and bound methods for non-convex continuous problems require ad hoc coding in practice.

12.3.1 LP-based branch and bound for MILP models

The fundamental branch and bound algorithm can be outlined as follows. At each step we work on a list of open subproblems, corresponding to nodes of a search tree, and we try to generate a sequence of improving incumbent solutions until we can prove that an incumbent solution is the optimal one. At intermediate steps, the incumbent solution is the best feasible (integer) solution found so far; the incumbent solution, for a minimization problem, provides us with an upper bound on the value of the optimal solution. We give the algorithm for a minimization problem; it is easy to adapt the algorithm to a maximization problem.

Fundamental branch and bound algorithm

1. *Initialization.* The list of open subproblems is initialized to $P(S)$; the value of the incumbent solution ν^* is set to $+\infty$.

2. *Selecting a candidate subproblem.* If the list of open subproblems is empty, stop: the incumbent solution $\mathbf{x}^*$, if any has been found, is optimal; if $\nu^* = +\infty$, the original problem was infeasible. Otherwise, select a subproblem $P(S_k)$ from the list.

3. *Bounding.* Compute a lower bound $\beta(S_k)$ on $\nu[P(S_k)]$ by solving a relaxed problem $P(\overline{S}_k)$. Let $\overline{\mathbf{x}}_k$ be the optimal solution of the relaxed subproblem.

4. *Prune by optimality.* If $\overline{\mathbf{x}}_k$ is feasible, prune subproblem $P(S_k)$. Furthermore, if $f(\overline{\mathbf{x}}_k) < \nu^*$, update the incumbent solution $\mathbf{x}^*$ and its value ν^*. Go to step 2.

5. *Prune by infeasibility.* If the relaxed subproblem $P(\overline{S}_k)$ is infeasible, eliminate $P(S_k)$ from further consideration. Go to step 2.

6. *Prune by bound.* If $\beta(S_k) \geq \nu^*$, eliminate subproblem $P(S_k)$ and go to step 2.

7. *Branching.* Replace $P(S_k)$ in the list of open subproblems with a list of child subproblems $P(S_{k1})$, $P(S_{k2}),\ldots, P(S_{kq})$, obtained by partitioning S_k; go to step 2.

To apply this algorithm successfully, we must cope with the following issues:

- How to compute a strong lower bound efficiently
- How to branch to generate subproblems
- How to select the right candidate from the list of open subproblems

The last issue is very important and calls for selecting a strategy to explore the tree. One possibility is to explore the most promising node first, in terms of lower bound; this yields the best-bound strategy. Another possibility is the depth-first strategy, whereby the last generated node is explored first; this strategy may have the merit of limiting the memory space required to store the search tree. In practice, we should also pay attention to how far the solution of a relaxed problem is from integrality. In example 12.8 below we will also check the effect of these choices.

Commercial branch and bound procedures compute bounds by the following LP-based (continuous) relaxation. Given a MILP problem

$$P(S) \quad \min \quad \mathbf{c'x + d'y}$$
$$\text{s.t.} \quad \mathbf{Ax + Ey \le b}$$
$$\mathbf{x} \in \mathbb{R}_+^{n_1}, \quad \mathbf{y} \in \mathbb{Z}_+^{n_2},$$

the continuous relaxation is obtained by relaxing the integrality constraints:

$$P(\overline{S}) \quad \min \quad \mathbf{c'x + d'y}$$
$$\text{s.t.} \quad \mathbf{Ax + Ey \le b}$$
$$\begin{bmatrix}\mathbf{x}\\\mathbf{y}\end{bmatrix} \in \mathbb{R}_+^{n_1+n_2}.$$

Ideally, the relaxed region $\overline{S}$ should be as close as possible to the convex hull of S; the smaller $\overline{S}$, the larger the lower bound. Tighter lower bounds make pruning by bound easier. To this end, careful model formulation may help.

Example 12.6 Consider a fixed-charge model in which the level of activity i is measured by the continuous decision variable x_i and the decision of starting that activity is modeled by the binary decision variable $\delta_i \in \{0,1\}$. To relate the two decision variables, we may write the constraint

$$x_i \le M_i \delta_i,$$

where M_i is an upper bound on the level x_i. When we solve the continuous relaxation, we drop the integrality constraint on δ_i, and we replace it by

$\delta_i \in [0,1]$. In principle, M_i may be a very large number, but to get a tight relaxation, we should select M_i as small as possible. □

Example 12.7 In example 1.2 we have considered how the basic knapsack model can be extended to deal with interactions among activities: in the example, activity 0 may be started only if all of the activities within a certain subset may be started. A possible constraint to model this requirement is

$$Nx_0 \leq \sum_{i=1}^{N} x_i,$$

where $x_0 \in \{0,1\}$ models the decision of starting activity 0, and $x_i \in \{0,1\}$ is related to the N activities in the subset conditioning activity 0. An alternative and equivalent formulation is

$$x_0 \leq x_i, \quad i = 1, \ldots, N.$$

On the one hand, this disaggregated form entails more constraints and probably require more work in solving the continuous relaxation. However, when we consider the continuous relaxation, all the points that are feasible for the disaggregate formulation are feasible for the aggregate constraint, but not vice versa. Hence, the feasible set for the relaxation of the disaggregate formulation is smaller, and the lower bound is tighter. Such a reformulation, as well as others, is carried out automatically by some packages (e.g., CPLEX) and may cut the computational effort of a branch and bound algorithm considerably.
□

As to branching, the following strategy is commonly applied to general integer variables. Assume that an integer variable y_j takes a non-integer value $\bar{y}_j$ in the optimal solution of the relaxed subproblem (one must exist; otherwise, we would prune by feasibility). Then two subproblems are generated; in the *down-child* we add the constraint

$$y_j \leq \lfloor \bar{y}_j \rfloor$$

to the formulation; in the *up-child* we add

$$y_j \geq \lfloor \bar{y}_j \rfloor + 1.$$

For instance, if $\bar{y}_j = 4.2$, we generate two subproblems with the addition of constraints $y_j \leq 4$ (for the down-child) and $y_j \geq 5$ (for the up-child).

A thorny issue is which variable we should branch on. Similarly, we should decide which subproblem we select from the list at step 2 of the branch and bound algorithm. As is often the case, there is no general answer; software packages offer different options to the user, and some experimentation may be required to come up with the best strategy.

BRANCH AND BOUND METHODS FOR NON-CONVEX OPTIMIZATION

Quite impressive improvements have been made in commercial branch and bound packages. Despite this, some large-scale problems cannot be solved to optimality within a reasonable amount of time. If this is the case, one possibility is to run branch and bound with a suboptimality tolerance. Instead of pruning a subproblem $P(S_k)$ only if the lower bound is larger than or equal to the incumbent, $\beta(S_k) \geq \nu^*$, we may introduce a tolerance parameter ϵ and eliminate a node in the tree whenever

$$\beta(S_k) \geq (1-\epsilon)\nu^*.$$

Doing so, we have only the guarantee of finding a near-optimal solution, but we have a bound on the level of suboptimality. In exchange, we may considerably reduce the computational effort. Whet we get is a mathematically motivated heuristic. Of course, heuristics need not be based on mathematical principles, but before considering heuristics, we would like to illustrate branch and bound in some detail.

Example 12.8 In section C.3 (page 652) we show how the following knapsack problem can be solved by AMPL:

$$\begin{aligned} \max \quad & 10x_1 + 7x_2 + 25x_3 + 24x_4 \\ \text{s.t.} \quad & 2x_1 + 1x_2 + 6x_3 + 5x_4 \leq 7 \\ & x_i \in \{0,1\}. \end{aligned}$$

The same problem can be solved by MATLAB using `bintprog`. A script for doing so is illustrated in figure 12.10. The script is very simple; the only noteworthy point is strategy selection. In the first run we use the depth-first exploration strategy, whereas the second run uses best-node. Strategies are selected as usual in the Optimization toolbox, building a `option` structure by `optimset`. We may also see that there is some difference between the two strategies:

```
>> knapsack
Optimization terminated.
Optimization terminated.
Optimal solution: 1 0 0 1
Value: 34
Nodes with depth-first: 9
Nodes with best-node: 7
```

A fair number of nodes is explored to find the optimal solution and prove its optimality. It is very instructive to try doing branch and bound manually using `linprog`. We must use the simplex algorithm in this case, because of its tendency to yield extreme solutions, when multiple ones exist, which means that they tend to be integer.

We first solve the root problem (P_0) in the tree, which is the continuous relaxation of the binary problem:

```
% Knapsack.m
A = [2 1 6 5];
b = 7;
c = - [10 7 25 24];
options = optimset('NodeSearchStrategy','df');
[x, value, exitflag, outputdf] = bintprog(c,A,b,[],[],[],options);
options = optimset('NodeSearchStrategy','bn');
[x, value, exitflag, outputbn] = bintprog(c,A,b,[],[],[],options);
fprintf(1,'Optimal solution: ', x');
fprintf(1,'%d ', x');
fprintf(1,'\nValue: %d\n', -value);
fprintf(1,'Nodes with depth-first: %d\n', outputdf.nodes);
fprintf(1,'Nodes with best-node: %d\n', outputbn.nodes);
```

Fig. 12.10 MATLAB script to solve a simple knapsack problem.

```
>> options = optimset('LargeScale', 'off', 'Simplex', 'on');
>> A = [2 1 6 5];
>> b = 7;
>> c = - [10 7 25 24];
>> lb = zeros(4,1);
>> ub = ones(4,1);
>> [x, val] = linprog(c,A,b,[],[],lb,ub,[],options)
Optimization terminated.
x =
    1.0000
    1.0000
         0
    0.8000
val =
  -36.2000
```

We see that the value of the objective is 36.2, which is an upper bound on the optimal value 34 (recall that we are maximizing and that there is a change in the sign of the objective), and that x_4 is fractional. We may branch on this variable by generating subproblems P_1, where $x_4 = 0$, and P_2, where $x_4 = 1$. Let us solve $P1$ first:

```
>> Aeq = [0 0 0 1];
>> beq = 0;
>> [x, val] = linprog(c,A,b,Aeq,beq,lb,ub,[],options)
Optimization terminated.
x =
    1.0000
    1.0000
    0.6667
```

```
                  0
val =
   -33.666
```

We see that the solution is getting worse because of the additional constraints. Solving P_2, we get

```
>> Aeq = [0 0 0 1];
>> beq = 1;
>> [x, val] = linprog(c,A,b,Aeq,beq,lb,ub,[],options)
Optimization terminated.
x =
    0.5000
    1.0000
         0
    1.0000
val =
   -36
```

This relaxation looks more promising, so we branch from here, generating subproblems P_3, where $x_1 = 0$, and P_4, where $x_1 = 1$. It is easy to see that P_4 yields the integer solution $x_1 = x_4 = 1$, $x_2 = x_3 = 0$, with value 34. Now we may eliminate P_1, since its bound shows that this subproblem cannot yield the optimal solution. However, we have not finished yet, because subproblem P_3 yields a promising fractional solution:

```
>> Aeq = [0 0 0 1; 1 0 0 0];
beq = [1;0];
[x, val] = linprog(c,A,b,Aeq,beq,lb,ub,[],options)
Optimization terminated.
x =
         0
    1.0000
    0.1667
    1.0000
val =
   -35.1667
```

We leave to the reader the task of verifying that branching on $x_3 = 0$, we get a solution with value 32, whereas $x_3 = 1$ yields an unfeasible problem (we have three items in the knapsack, exceeding its capacity). Hence, we have proven that the optimal solution has value 34, after exploring a few nodes. It is important to notice that a brute force enumeration strategy would require the exploration of $2^4 = 16$ possible solutions. Now what about AMPL? Well, you can see from the appendix that AMPL/CPLEX uses *zero* branch and bound nodes:

```
ampl: model Knapsack.mod;
```

590   NON-CONVEX OPTIMIZATION

```
ampl: data Knapsack.dat;
ampl: options cplex_options 'mipdisplay 2';
ampl: solve;
CPLEX 9.1.0: mipdisplay 2
Clique table members: 2
MIP emphasis: balance optimality and feasibility
Root relaxation solution time =    0.02 sec.

          Nodes                              Cuts/
    Node  Left   Objective  IInf  Best Integer  Best Node   ItCnt    Gap

      0    0     36.2000     1                   36.2000      1
*     0+   0                 0      32.0000      36.2000      1   13.12%
*          34.0000           0      34.0000      Cuts: 3      3    0.00%
Cover cuts applied:  1
Implied bound cuts applied:  1
CPLEX 9.1.0: optimal integer solution; objective 34
3 MIP simplex iterations
0 branch-and-bound nodes
```

How is this possible? If we check the budget constraint, it is easy to see that item 1 and 3 cannot be both selected, as their total capacity is 8 and it exceeds the available budget. Hence we might add the constraint:

$$x_1 + x_3 \leq 1,$$

which is obviously redundant in the discrete domain, but is *not* redundant in the continuous relaxation. By the same token, we could add the following constraints

$$x_3 + x_4 \leq 1$$
$$x_1 + x_2 + x_4 \leq 2$$

Such additional constraints are called *cover inequalities* and may contribute to strengthen the bound from the LP relaxation, cutting the CPU time considerably. If we try solving the LP relaxation in MATLAB, adding the three cover inequalities, we get

```
>> A1 = [2 1 6 5; 1 0 1 0; 0 0 1 1; 1 1 0 1];
b1 = [7;1;1;2];
c = - [10 7 25 24];
lb = zeros(4,1);
ub = ones(4,1);
[x, val] = linprog(c,A1,b1,[],[],lb,ub,[],options)
Optimization terminated.
x =
    0.3333
    1.0000
```

```
     0.3333
     0.6667
val =
   -34.6667
```

We see how cover inequalities my strengthen the relaxation. Now we may conclude that the optimal solution cannot be worth more than 34, since all the coefficients in the model are integer. AMPL/CPLEX is able to exploit this and other type of inequalities to reduce the computational requirements of branch and bound. The automatic generation of inequalities is also called cut generation, as we aim at cutting the relaxed feasible region in order to get as close as possible to the convex hull of integer solutions. Efficient cut generation is not trivial, as it is important to generate only the effective cuts; the reader may play with MATLAB to check that in the toy example above, not all the cover inequalities are really helpful as some are actually redundant.

◻

In the example above, we may appreciate the sophistication of state-of-the-art packages for mixed-integer programming. We should also stress that heuristics may actually be integrated within a branch-and-bound procedure. The role of heuristics is to generate, given a nearly-integer solution, a feasible solution; if this is of good quality, it will improve the incumbent solution and the upper bound against which we compare lower bounds. In the ILOG CPLEX trace above whenever you see an asterisk (*) in a row, it means that the search process has found a new incumbent. When you also see a plus (+), it means that it was found by a heuristic. One possibility to devise such heuristics is clever rounding; rounding does not work in general, if we use it to find an optimal solution, but when the continuous relaxation is tight enough, it may yield very good solutions. Another principle that can be exploited is local search, which is introduced in the next section.

12.4 HEURISTIC METHODS FOR NON-CONVEX OPTIMIZATION

When a branch and bound method is not able to yield an optimal or near-optimal solution with a reasonable effort, we may settle for a quick heuristic method able to provide us with a good solution. For any specific problem it is possible to devise an ad hoc method. However, it is interesting to consider relatively general principles which, with some adaptation, may yield good heuristics for a wide class of problems. Local search metaheuristics[2] are quite popular and have also been proposed for financial problems. They were originally developed for discrete optimization problems; however, they may

---

[2]This name reflects the relatively general nature of the principle. In practice, a good deal of customization is needed to come up with a truly effective method for a specific problem.

also be applied to continuous non-linear programming when the objective is non-convex.

Local search algorithms are similar to the gradient method for non-linear programming. The basic idea is to improve a known solution by applying a set of local perturbations. Consider a generic optimization problem

$$\min_{x \in S} f(x),$$

defined over a discrete set S. Given a feasible solution x, a neighborhood $\mathcal{N}(x)$ is defined as the set of solutions obtained by applying a set of simple perturbations to x. Different perturbations yield different *neighborhood structures*.

The simplest local search algorithm is *local improvement*. Given a current, or incumbent, solution $\bar{x}$, an alternative (candidate) solution x° is searched for in the neighborhood of the incumbent, such that

$$f(x^\circ) = \min_{x \in \mathcal{N}(\bar{x})} f(x).$$

If the neighborhood structure $\mathcal{N}(\cdot)$ is simple enough, the minimization above can be performed by an exhaustive search; we speak of a *best-improving method* since we try to find the best solution in the neighborhood. Clearly, there is a trade-off between the effectiveness of the neighborhood structure (the larger the better) and the efficiency of the algorithm. If $f(x^\circ) < f(\bar{x})$, then x° is set as the new current solution and the process is iterated. If $f(x^\circ) \geq f(\bar{x})$, the algorithm is stopped. A possible variation is to *partially* explore the neighborhood of the current solution until an improving solution is found; this approach is known as *first-improving*, since we do not explore the entire neighborhood before committing to a new current solution.

The neighborhood structure is problem dependent. In the case of discrete optimization problems, devising a neighborhood structure may be relatively straightforward. For instance, in a capital budgeting problem, the solution is represented by the subset of selected projects. The neighborhood might be generated by exchanging a project within the current subset with a project not included in it. In a general programming problem with binary variables, one might consider complementing each variable in turn. Actually, devising a clever and effective neighborhood is not as trivial as it might seem, since due attention must be paid to constraints. In the case of continuous variables, a further complication arises; we may generate neighboring points by moving along a set of directions, but we must find a way to select the step size. To this aim, dynamic strategies have been devised (see, e.g., [6] for a financial application).

This basic idea is generally easy to apply, but it has one major drawback: The algorithm usually stops in a *locally* (with respect to the neighborhood structure) optimal solution. This is the same difficulty we face when applying the gradient method to a non-convex objective function; the reason behind the trouble is that only improving perturbations [i.e., such that $\Delta f = f(x^\circ) - f(\bar{x}) < 0$] are accepted. To avoid getting stuck in a local optimum, we must relax this assumption.

In the following we describe three local search approaches that have been proposed to overcome the limitations of local improvement: simulated annealing, tabu search, and genetic algorithms.

Simulated annealing It has been pointed out that to overcome the problem of local minima, we have to accept, in some disciplined way, non-improving perturbations, i.e., perturbations for which $\Delta f > 0$. Simulated annealing is based on an analogy between cost minimization in discrete optimization and energy minimization in physical systems. The local improvement strategy behaves much like physical systems do, according to classical mechanics. It is impossible for a system to have a certain energy at a certain time and to increase it without external input: If you place a ball in a hole, it will stay there. This is not true in thermodynamics and statistical mechanics; according to these physical models, at a temperature above absolute zero, thermal noise makes an increase in the energy of a system possible. An increase in energy is more likely to occur at high temperatures. The probability P of this upward jump depends on the amount of energy ΔE acquired and the temperature T, according to the Boltzmann distribution

$$P(\Delta E, T) = \exp\left(-\frac{\Delta E}{KT}\right),$$

where K is the Boltzmann constant.

Annealing is a metallurgical process by which a melted material is slowly cooled in order to obtain good (low-energy) solid-state configurations. If the temperature is decreased too fast, the system gets trapped in a local energy minimum, and a glass is produced. But if the process is slow enough, random kinetic fluctuations due to thermal noise allow the system to escape from local minima, reaching a point very close to the global optimum.

In strict analogy with statistical mechanics, in the simulated annealing method a perturbation of the current solution yielding $\Delta f < 0$ is always accepted; a perturbation with $\Delta f > 0$ is accepted with a probability given by a Boltzmann-like probability distribution

$$P(\Delta f, T) = \exp\left(-\frac{\Delta f}{T}\right).$$

This probability distribution is a decreasing exponential in Δf, whose shape depends on the parameter T, acting as a temperature (see figure 12.11). The probability of accepting a non-improving perturbation decreases as the deterioration of the solution increases. For a given Δf, the acceptance probability is higher at high temperatures. For $T \to 0$ the probability collapses into a step function, and the method behaves like local improvement. For $T \to +\infty$

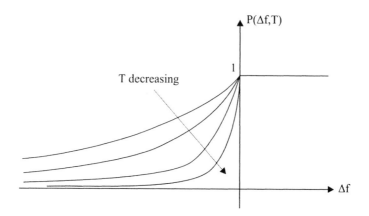

Fig. 12.11 Acceptance probabilities as a function of cost increase for different temperatures.

the probability is 1 everywhere, and we have a random exploration of the solutions space. The parameter T allows balancing the need to *exploit* the solution at hand by improving it and the need to *explore* the solution space.

The simulated annealing method simply substitutes the deterministic acceptance rule of local improvement with a probabilistic rule. The temperature is set to a relatively high initial value T_1, and the algorithm is iterated using at step k a temperature T_k until some termination criterion is satisfied. The strategy by which the temperature is decreased is called the *cooling schedule*. The simplest cooling schedule is

$$T_k = \alpha T_{k-1}, \quad 0 < \alpha < 1.$$

In practice, it is advisable to keep the temperature constant for a certain number of steps, in order to reach a thermodynamic equilibrium before changing the control parameter. More sophisticated adaptive cooling strategies have been proposed, but the increase in complexity does not always seem justified. A very simple implementation of the annealing algorithm could be the following:

Step 1. Choose an initial solution x_{old}, an initial temperature T_1, and a decrease parameter α; let $k = 1$, $f_{\text{old}} = f(x_{\text{old}})$; let $\hat{f} = f_{\text{old}}$ and $\hat{x} = x_{\text{old}}$ be the current optimal value and optimal solution, respectively.

Step 2. Randomly choose a candidate solution x_{new} from the neighborhood of x_{old}, and compute its value f_{new}.

Step 3. Set the acceptance probability

$$P = \min\left\{1, \exp\left(-\frac{f_{\text{new}} - f_{\text{old}}}{T_k}\right)\right\}.$$

Step 4. Accept the new solution with probability P; if accepted, set $x_{\text{old}} = x_{\text{new}}$ and $f_{\text{old}} = f_{\text{new}}$; if necessary, update $\hat{f}$ and $\hat{x}$.

Step 5. If some termination condition is met, stop; otherwise, set $k = k + 1$, set the new temperature according to the cooling schedule, and go to step 2.

The probabilistic acceptance is easy to implement. P is evaluated according to the Boltzmann distribution; then a pseudorandom number U, uniformly distributed between 0 and 1, is generated and the move is accepted if $U \leq P$ (pseudorandom number generation is dealt with in section 4.3).

The termination condition could be related to a maximum iteration number, to a minimum temperature, or to a maximum number of steps in which the current solution remains unchanged. Note that we do not explore the entire neighborhood of the current solution; the method is of the first-improving type. If a candidate solution is rejected, we select another candidate in the neighborhood of the current solution. In principle, it is possible to visit the same solution twice; if the neighborhood structure is rich enough, this is unlikely. It is necessary to save the best solution found, since the freezing point (the last current solution) need not be the best solution visited.

An implementation of the annealing algorithm is therefore characterized by the solution space, the neighborhood structure, the rule by which the neighborhood is explored, and the cooling schedule. It can be shown that under some conditions, the method asymptotically converges (in a probabilistic sense) to the global optimum. The convergence property is a reassuring one, but it is usually considered of little practical value, since its conditions would require impractical running times. However, the experience suggests that in many practical settings, very good solutions (often optimal) are actually found. The running time of the algorithm to obtain high-quality solutions, however, is problem dependent.

Tabu search Like simulated annealing, tabu search is a neighborhood search-based metaheuristic aimed at escaping local minima. Unlike simulated annealing, tabu search tries to keep the search biased toward good solutions.

The basic idea of tabu search is that the best solution in the neighborhood $\mathcal{N}$ of the current solution should be chosen as the new current solution, even if this implies increasing the cost. If we are in a local minimum, this means accepting a non-improving perturbation. The problem with this basic idea is that the possibility of cycling arises. If we try to escape from a local minimum by choosing the best solution in its neighborhood, it might well be the case that at the next iteration, we fall back into the local minimum, since this could be the best solution in the new neighborhood.

To prevent cycling, we must prevent revisiting solutions. One way would be to keep a record of the already visited solutions; however, this would be both memory- and time-consuming, since checking a candidate solution against the list of visited ones would require a substantial effort. A better idea could be to record only the most recent solutions. A practical alternative is to keep in memory only some *attributes* of the solutions or of the applied perturbations; such attributes are called *tabu*. For instance, the reverse of the selected perturbation at each step can be marked as tabu, restricting the neighborhood to be considered. Consider a pure integer program involving only binary variables; if we complement variable x_i, in the next few iteration we might forbid any perturbation complementing this variable again. As an alternative, a tabu attribute of a solution could be the value of the objective function. In practice, it is necessary to keep only a record of the most recent tabu attributes to avoid cycling; the data structure implementing this function is the *tabu list*.

The basic tabu navigation algorithm can be described as follows:

Step 1. Choose an initial current solution x_{cur}, a tabu list size; let $k = 1$, $\hat{f} = f(x_{\text{cur}})$, $\hat{x} = x_{\text{cur}}$.

Step 2. Evaluate the neighborhood $\mathcal{N}(x_{\text{cur}})$; update the current solution with the best non-tabu solution in the neighborhood; if necessary, update the current optimal solution $\hat{x}$ and the current optimal value $\hat{f}$.

Step 3. Add some attribute of the new solution or of the applied perturbation to the tabu list.

Step 4. If the maximum iteration number has been reached, stop; otherwise, set $k = k + 1$, and go to step 2.

Note that, unlike simulated annealing, this version of tabu search explores the *entire* neighborhood of the current solution; basic tabu search is a strategy of the *best-improving* rather than first-improving type. However, it is possible to restrict the neighborhood to reduce the computational burden.

There are several issues and refinements to consider in order to implement an effective and efficient algorithm. They are rather problem specific; this shows that, although local search metaheuristics are general-purpose, a certain degree of "customization" is necessary.

Genetic algorithms Unlike simulated annealing and tabu search, genetic algorithms work on a set of solutions rather than a single point. In this sense they are similar to the simplex search method of section 6.2.4. The idea is based on the survival-of-the-fittest mechanism of biological evolution. Each solution is represented by a string of numbers or symbols; strings are subject to random evolution mechanisms which change the current population. One evolution mechanism is mutation; an attribute of a string is randomly selected and modified using a neighborhood structure. Mutation is very similar to the usual local search mechanism, but there is another mechanism which is peculiar to genetic algorithms: crossover. In the crossover mechanism, two elements of the current set of solutions are selected and merged in some way.

Given two strings, we select a "breakpoint" position k and merge the strings as follows:

$$\left\{ \begin{array}{l} x_1, x_2, \ldots, x_k, x_{k+1}, \ldots, x_n \\ y_1, y_2, \ldots, y_k, y_{k+1}, \ldots, y_n \end{array} \right\} \Rightarrow \left\{ \begin{array}{l} x_1, x_2, \ldots, x_k, y_{k+1}, \ldots, y_n \\ y_1, y_2, \ldots, y_k, x_{k+1}, \ldots, x_n \end{array} \right\}.$$

Different variations are possible; for instance, a double crossover may be exploited, in which two breakpoints are selected for the crossover.

The set of solutions is updated at each iteration, selecting the "best" individuals for mutation and crossover and/or letting only the best individuals survive. Rather than selecting the best individuals deterministically, based on the value of the objective function, random selection mechanisms are employed to avoid freezing the population to a locally optimal solution. Genetic algorithms may be integrated with local search strategies; one idea is to use genetic mechanisms to find a set of initial points from which a local improvement search is carried out.

The idea of genetic algorithms certainly has a good potential for solving quite complex problems; the evident downside is that considerable experimentation may be needed to come up with the best strategy and the best setting of numerical parameters regulating the evolution mechanisms. The potential of this class of methods is also proved by the recent introduction of the Genetic Algorithm and Direct Search toolbox, which extends the functionalities of the MATLAB Optimization toolbox.

For further reading

In the literature

- A comprehensive reference on mixed-integer programming is [16]. A more recent treatment, including developments in automatic model strengthening, is [19].

- The use of mixed-integer programming models in portfolio management is the subject of an increasing number of papers including [2], [3], [4], [10], [11], and [17].

- For a textbook treatment, see also [18].

- The AMPL language is described in [5].

- Global optimization techniques for optimization of a fixed-mix portfolio are discussed in [12]; the model is extended and tackled by metaheuristics in [6].

- For a broader view of the principles behind global optimization algorithms see, e.g., [8].

- Tabu search is covered in depth by [7]. See, e.g., [1] for an application to global optimization.

- A textbook on genetic algorithms is [15]. An application to global optimization is described in [14].

On the Web

- The AMPL web site is http://www.ampl.com.

- See also http://www.ilog.com.

- Meta-heuristics are the algorithmic foundation of an optimization engine, OptQuest, which thanks to its flexibility has been integrated with many simulation packages. See http://www.optquest.com. The tool has also been applied to portfolio management problems, too.

- The Genetic Algorithm and Direct Search toolbox is described on The MathWorks' web site http://www.mathworks.com

REFERENCES

1. R. Battiti and G. Tecchiolli. The Continuous Reactive Tabu Search: Blending Combinatorial Optimization and Stochastic Search for Global Optimization. *Annals of Operations Research*, 63:153–188, 1996.

2. D. Bertsimas, C. Darnell, and R. Stoucy. Portfolio Construction through Mixed-Integer Programming at Grantham, Mayo, Van Otterloo and Company. *Interfaces*, 29:49–66, 1999.

3. D. Bienstock. Computational Study of a Family of Mixed-Integer Quadratic Programming Problems. *Mathematical Programming*, 74:121–140, 1996.

4. T.-J. Chang, N. Meade, J.E. Beasley, and Y.M. Sharaiha. Heuristics for Cardinality Constrained Portfolio Optimization. *Computers and Operations Research*, 27:1271–1302, 2000.

5. R. Fourer, D.M. Gay, and B.W. Kernighan. *AMPL: A Modeling Language for Mathematical Programming*. Boyd and Fraser, Danvers, MA, 1993.

6. F. Glover, J.M. Mulvey, and K. Hoyland. Solving Dynamic Stochastic Control Problems in Finance Using Tabu Search with Variable Scaling. In I.H. Osman and J.P. Kelly, editors, *Meta-Heuristics: Theory and Applications*, pages 429–448. Kluwer Academic, Dordrecht, The Netherlands, 1996.

7. F.W. Glover and M. Laguna. *Tabu Search*. Kluwer Academic, Dordrecht, The Netherlands, 1998.

8. R. Horst, P.M. Pardalos, and N.V. Thoai. *Introduction to Global Optimization*. Kluwer Academic, Dordrecht, The Netherlands, 1995.

9. H. Konno and H. Yamazaki. Mean-Absolute Deviation Portfolio Optimization Model and Its Application to Tokyo Stock Market. *Management Science*, 37:519–531, 1991.

10. M.S. Lobo, M. Fazel, and S. Boyd. Portfolio Optimization with Linear and Fixed Transaction Costs and Bounds on Risk. Unpublished manuscript (available at http://www.stanford.edu/~boyd), 1999.

11. R. Mansini and M.G. Speranza. Heuristic Algorithms for the Portfolio Selection Problem with Minimum Transaction Lots. *European Journal of Operational Research*, 114:219–233, 1999.

12. C.D. Maranas, I.P. Androulakis, C.A. Floudas, A.J. Berger, and J.M. Mulvey. Solving Long-Term Financial Planning Problems via Global Optimization. *Journal of Economic Dynamics and Control*, 21:1405–1425, 1997.

13. C.D. Maranas and C.A. Floudas. Global Minimum Potential Energy Conformations of Small Molecules. *Journal of Global Optimization*, 4:135–170, 1994.

14. Z. Michalewicz. Evolutionary Computation Techniques for Nonlinear Programming Problems. *International Transactions of Operations Research*, 1:223–140, 1994.

15. Z. Michalewicz. *Genetic Algorithms + Data Structures = Evolution Programs*. Springer-Verlag, Berlin, 1996.

16. G.L. Nemhauser and L.A. Wolsey. *Integer Programming and Combinatorial Optimization*. Wiley, Chichester, West Sussex, England, 1998.

17. J.K. Sankaran and A.A. Patil. On the Optimal Selection of Portfolios under Limited Diversification. *Journal of Banking and Finance*, 23:1655–1666, 1999.

18. B. Scherer and D. Martin. *Introduction to Modern Portfolio Optimization with NuOPT, S-Plus, and S^+Bayes*. Springer, New York, 2005.

19. L.A. Wolsey. *Integer Programming*. Wiley, New York, 1998.

Part V
Appendices

Appendix A
Introduction to MATLAB Programming

We give here a brief outline of the MATLAB basics, referring to the user manual for a full treatment. You may also type `demo` to see a demonstration of both MATLAB and the toolboxes you are interested in. Actual use of the features we describe is illustrated in the remainder of the book. A rich online documentation is available in the MATLAB environment; the reader should take advantage of this whenever a piece of code in the book is not clear.

A.1 MATLAB ENVIRONMENT

- MATLAB is an interactive computing environment. You may enter expressions and obtain an immediate evaluation:

```
>> rho = 1+sqrt(5)/2
rho =
    2.1180
```

By entering a command like this, you also define a variable `rho` which is added to the current environment and may be referred to in any other expression.

- There is a rich set of predefined functions. Try typing `help elfun`, `help elmat`, and `help ops` to get information on elementary mathematical functions, matrix manipulation, and operators, respectively. For each predefined function there is an online help:

```
>> help sqrt
 SQRT   Square root.
    SQRT(X) is the square root of the elements of X. Complex
    results are produced if X is not positive.

    See also sqrtm, realsqrt, hypot.

    Reference page in Help browser
       doc sqrt
```

The `help` command should be used when you know the name of the function you are interested in, but you need additional information. Otherwise, `lookfor` may be tried:

```
>> lookfor sqrt
REALSQRT Real square root.
SQRT   Square root.
SQRTM    Matrix square root.
```

We see that `lookfor` searches for functions whose online help documentation includes a given string. Recent MATLAB releases include an extensive online documentation which can be accessed by the command `doc`.

- MATLAB is case sensitive (`Pi` and `pi` are different).

```
>> pi
ans =
    3.1416
>> Pi
??? Undefined function or variable 'Pi'.
```

- MATLAB is a matrix-oriented environment and programming language. Vectors and matrices are the basic data structures, and more complex ones have been introduced in the more recent MATLAB versions. Functions and operators are available to deal with vectors and matrices directly. You may enter row and column vectors as follows:

```
>> V1=[22, 5, 3]
V1 =
    22    5    3
```

```
>> V2 = [33; 7; 1]
V2 =
    33
     7
     1
```

We may note the difference between comma and semicolon; the latter is used to terminate a row. In the example above, commas are optional, as we could enter the same vector by typing V1=[22 5 3].

- The who and whos commands may be used to check the user defined variables in the current environment, which can be cleared by the clear command.

```
>> who
Your variables are:
V1         V2
>> whos
  Name       Size                    Bytes  Class
  V1         1x3                        24  double array
  V2         3x1                        24  double array
Grand total is 6 elements using 48 bytes
>> clear V1
>> whos
  Name       Size                    Bytes  Class
  V2         3x1                        24  double array
Grand total is 3 elements using 24 bytes
>> clear
>> whos
>>
```

- You may also use the semicolon to suppress output from the evaluation of an expression:

```
>> V1=[22, 5, 3];
>> V2 = [33; 7; 1];
>>
```

Using semicolon to suppress output is important when we deal with large matrices (and in MATLAB programming as well).

- You may also enter matrices (note again the difference between ';' and ','):

```
>> A=[1 2 3; 4 5 6]
A =
```

```
           1       2       3
           4       5       6
>> B=[V2 , V2]
B =
          33      33
           7       7
           1       1
>> C=[V2 ; V2]
C =
          33
           7
           1
          33
           7
           1
```

Also note the effect of the following commands:

```
>> M1=zeros(2,2)
M1 =
           0       0
           0       0
>> M1=rho
M1 =
       2.1180
>> M1=zeros(2,2);
>> M1(:,:)=rho
M1 =
       2.1180    2.1180
       2.1180    2.1180
```

- The colon (:) is used to spot subranges of an index in a matrix.

```
>> M1=zeros(2,3)
M1 =
           0       0       0
           0       0       0
>> M1(2,:)=4
M1 =
           0       0       0
           4       4       4
>> M1(1,2:3)=6
M1 =
           0       6       6
           4       4       4
```

- The dots (...) may be used to write multiline commands.

```
>> M=ones(2,
??? M=ones(2,

Missing variable or function.
>> M=ones(2,...
2)
M =
     1     1
     1     1
```

- The zeros and ones commands are useful to initialize and preallocate matrices. This is recommended for efficiency. In fact, matrices are resized automatically by MATLAB whenever you assign a value to an element beyond the current row or column range, but this may be time consuming and should be avoided when possible.

```
>> M = [1 2; 3 4];
>> M(3,3) = 5
M =
     1     2     0
     3     4     0
     0     0     5
```

It should be noted that this flexible management of memory is a double-edged sword: It may increase flexibility, but it may make debugging difficult.

- [] is the empty vector. You may also use it to delete submatrices:

```
>> M1

M1 =
     0     6     6
     4     4     4
>> M1(:,2)=[]
M1 =
     0     6
     4     4
```

- Another use of the empty vector is to pass default values to MATLAB functions. Unlike other programming languages, MATLAB is rather flexible in its processing of input arguments to functions. Suppose we have a function f taking three input parameters. The standard call would be something like f(x1, x2, x3). If we call the function with one input arguments, f(x1), the missing ones are given default values. Of course this does not happen automatically; the function must be programmed that way, and the reader is urged to see how this is accomplished by opening predefined MATLAB functions with the editor.

Now suppose that we want to pass only the first and the third argument. We obviously cannot simply call the function like f(x1, x3), since x3 would be assigned to the second input argument of the function. To obtain what we want, we should use the empty vector: f(x1, [], x3).

- Matrices can be transposed and multiplied easily (if dimensions fit):

```
>> M1'
ans =
     0     4
     6     4
>> M2=rand(2,3)
M2 =
    0.9501    0.6068    0.8913
    0.2311    0.4860    0.7621
>> M1*M2
ans =
    1.3868    2.9159    4.5726
    4.7251    4.3713    6.6136
>> M1+1
ans =
     1     7
     5     5
```

The **rand** command yields a matrix with random entries, uniformly distributed in the (0,1) interval.

- Note the use of the dot . to operate element by element on a matrix:

```
>> A=0.5*ones(2,2)
A =
    0.5000    0.5000
    0.5000    0.5000
>> M1
M1 =
     0     6
     4     4
>> M1*A
ans =
     3     3
     4     4
>> M1.*A
ans =
     0     3
     2     2
```

```
>> I=[1 2; 3 4]
I =
     1     2
     3     4
>> I^2
ans =
     7    10
    15    22
>> I.^2
ans =
     1     4
     9    16
```

- Subranges may be used to build vectors. For instance, to compute the factorial:

```
>> 1:10
ans =
    1    2    3    4    5    6    7    8    9    10
>> prod(1:10)
ans =
     3628800
>> sum(1:10)
ans =
    55
```

You may also specify an optional increment step in these expressions:

```
>> 1:0.8:4
ans =
    1.0000    1.8000    2.6000    3.4000
```

The step can be negative too:

```
>> 5:-1:0
ans =
    5    4    3    2    1    0
```

- One more use of the colon operator is to make sure that a vector is a column vector:

```
>> V1 = 1:3
V1 =
    1    2    3
>> V2 = (1:3)'
V2 =
    1
    2
```

```
         3
>> V1(:)
ans =
     1
     2
     3
>> V2(:)
ans =
     1
     2
     3
```

The same effect cannot be obtained by transposition, unless one writes code using the function size to check matrix dimensions:

```
>> [m,n] = size(V2)
m =
     3
n =
     1
```

- Note the use of the special quantities Inf (infinity) and NaN (not a number):

```
>> l=1/0
Warning: Divide by zero.
l =
   Inf
>> l
l =
   Inf
>> prod(1:200)
ans =
   Inf
>> 1/0 - prod(1:200)
Warning: Divide by zero.
ans =
   NaN
```

- Useful functions to operate on matrices are: eye, inv, eig, det, rank, and diag:

```
>> eye(3)
ans =
     1     0     0
     0     1     0
     0     0     1
```

```
>> K=eye(3)*[1 2 3]'
K =
     1
     2
     3
>> K=inv(K)
K =
    1.0000         0         0
         0    0.5000         0
         0         0    0.3333
>> eig(K)
ans =
    1.0000
    0.5000
    0.3333
>> rank(K)
ans =
     3
>> det(K)
ans =
    0.1667
>> K=diag([1 2 3])
K =
     1     0     0
     0     2     0
     0     0     3
```

We should note a sort of dual nature in diag. If it receives a vector, it builds a matrix; if it receives a matrix, it returns a vector:

```
>> A = [1:3 ; 4:6 ; 7:9];
>> diag(A)
ans =
     1
     5
     9
```

- Some functions operate on matrices columnwise:

```
>> A = [1 3 5 ; 2 4 6 ];
>> sum(A)
ans =
     3     7    11
>> mean(A)
ans =
    1.5000    3.5000    5.5000
```

The last example may help to understand the rationale behind this choice. If the matrix contains samples from multiple random variables, and we want to compute the sample mean, we should arrange data in such a way that variables corresponds to columns, and joint realizations corresponds to rows. However, it is possible to specify the dimension along which these functions should work:

```
>> sum(A,2)
ans =
     9
    12
>> mean(A,2)
ans =
     3
     4
```

Another useful function in this vein computes cumulative sums:

```
>> cumsum(1:5)
ans =
     1    3    6   10   15
```

- Systems of linear equations are easily solved:

```
>> A = [3 5 -1; 9 2 4;  4 -2 -9];
>> b = (1:3)';
>> X = A\b
X =
    0.3119
   -0.0249
   -0.1892
>> A*X
ans =
    1.0000
    2.0000
    3.0000
```

- The efficiency of a function may be checked by using the commands tic and toc as follows:

```
>> tic, inv(rand(500,500));, toc
Elapsed time is 0.472760 seconds.
```

- We will see in section A.3 how MATLAB code can be developed in order to compute complicated functions. However, when the function is a relatively simple expression it may be preferable to define functions in a more direct way. One possibility is using the inline mechanism, which builds a function based on a string:

```
>> f = inline('exp(2*x).*sin(y)')
f =
     Inline function:
     f(x,y) = exp(2*x).*sin(y)
>> f(2,3)
ans =
    7.7049
```

Note the use of the dot operator to make sure the function works on vector inputs and how `inline` determines automatically the name and the order of the input arguments. If one wants to change that order, an explicit list of arguments can be given:

```
>> f = inline('exp(2*foo).*sin(fee)')
f =
     Inline function:
     f(fee,foo) = exp(2*foo).*sin(fee)
>> g = inline('exp(2*foo).*sin(fee)','foo','fee')
g =
     Inline function:
     g(foo,fee) = exp(2*foo).*sin(fee)
```

- An alternative approach to `inline` is based on the function handle operator @:

```
>> f = @(x,y) exp(2*x).*sin(y)
f =
     @(x,y) exp(2*x).*sin(y)
```

We see that the operator is used to "abstract" a function from an expression.[1] The @ operator is also useful to define anonymous functions which may be passed to higher-order functions, i.e., functions which receive functions as inputs (e.g., to compute integrals or to solve non-linear equations).

We may also fix some input parameters to obtain function of the remaining arguments:

```
>> g = @(y) f(2,y)
g =
     @(y) f(2,y)
>> g(3)
ans =
```

---

[1] Readers, like myself, with a little background in theoretical computer science or mathematical logic will notice some similarity with the notation used in λ-calculus.

7.7049

- In this book we will practically use only matrices, but MATLAB has included many more data structures over the years. We can deal with strings (delimited by quotes) and structures (called "structs") with arbitrary fields:

```
>> p.name = 'Donald Duck'
>> p.age = 55;
>> p
p =
    name: 'Donald Duck'
     age: 55
```

Structures are used by some functions to group output data in one structure, avoiding an excessive number of output arguments.

- Cell arrays may also be used to implements ragged arrays, i.e., arrays containing vectors with different lengths (which cannot be accomplished by traditional matrices):

```
>> M = cell(2,1);
>> M{1} = [ 1 2 3 ];
>> M{2} = [ 4 5 6 7 8 ];
>> M
M =
    [1x3 double]
    [1x5 double]
>> M{1}
ans =
     1     2     3
```

Note the use of braces rather than standard parentheses.

A.2 MATLAB GRAPHICS

- Plotting a function of a single variable is easy. Try the following commands:

```
>> x = 0:0.01:2*pi;
>> plot(x,sin(x))
>> axis([0 2*pi -1 1])
```

The axis command may be used to resize plot axes at will. There is also a rich set of ways to annotate a plot.

- Different types of plots may be obtained by using optional parameters of the plot command. Try with

  ```
  >> plot(0:20, rand(1,21), 'o')
  >> plot(0:20, rand(1,21), 'o-')
  ```

- To obtain a tridimensional surface, the surf command may be used.

  ```
  >> f = @(x,y) exp(-3*(x.^2 + y.^2)).*(sin(5*pi*x)+ cos(10*pi*y));
  >> [X Y] = meshgrid(-1:0.01:1 , -1:0.01:1);
  >> surf(X,Y,f(X,Y))
  ```

  Some explanation is in order here. The function surf must receive three matrices, corresponding to the x and y coordinates in the plane, and to the function value (the 'z' coordinate). A first requirement is that the function we want to draw should be encoded in such a way that it can receive matrix inputs; use of the dot operator is essential: Without the dots '.', input matrices would be multiplied row by column, as in linear algebra, rather than element by element. To build the two matrices of coordinates, meshgrid is used. To understand what this function accomplishes, let us consider a small scale example:

  ```
  >> [X,Y] = meshgrid(1:4,1:4)
  X =
         1     2     3     4
         1     2     3     4
         1     2     3     4
         1     2     3     4
  Y =
         1     1     1     1
         2     2     2     2
         3     3     3     3
         4     4     4     4
  ```

  We see that, for each point in the plane, we obtain matrices containing each coordinate.

- We may close this section with a more practical example: plotting the Black–Scholes price of a vanilla call option, for time to maturity T ranging from one year down to zero, initial price S_0 ranging from 30 to 70, strike price $K = 50$, risk-free rate $r = 0.1$, and volatility $\sigma = 0.4$. The following commands produce the surface in figure A.1:

  ```
  >> T = 1:-0.05:0;
  >> S0 = 30:70;
  >> K = 50;
  ```

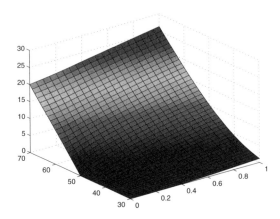

Fig. A.1 Price of a call option as a function of time to maturity and initial underlying asset price.

```
>> sigma = 0.4;
>> r = 0.1;
>> [X,Y] = meshgrid(T,S0);
>> f = @(time,price) blsprice(price, 50, 0.1, time, 0.4);
>> surf(X,Y,f(X,Y))
```

Of course we are relying on the fact that the blsprice function, available in the Financial toolbox, has been properly coded to handle matrix inputs.

A.3 MATLAB PROGRAMMING

- MATLAB toolboxes extend considerably the capabilities of the MATLAB core. They consist of a set of functions that are coded in the MATLAB programming language. They are contained in M-files, which are plain text files with default extension *.m. It is quite instructive to open some of these files in the MATLAB editor to see how a robust and flexible code is written.

- You may also write your own functions. You have simply to open the MATLAB editor and save the file in a directory which is on the MATLAB path.

- A simple function is displayed in figure A.2. The function consists of the function header, which specifies the input and output arguments. Note

```
function [xout, yout] = samplefile(x,y)
% a simple M-file to do some pointless computation
% this comment is printed by issuing the help samplefile
% command
[m,n] = size(x);
[p,q] = size(y);
z = rand(10,m)*x*rand(n,10) + rand(10,p)*y*rand(q,10);
xout = sum(z);
yout = sin(z);
```

Fig. A.2 Typical structure of a MATLAB function.

how multiple output arguments are expressed. The comments below the function heading are displayed if you ask for some help about the function:

```
>> help samplefile
   a simple M-file to do some pointless computation
   this comment is printed by issuing the help samplefile
   command
```

Then the function body is given, which may contain further comment lines and arbitrarily complex control structures.

- In general, you may write a function in which some input arguments are optional and are given default parameters. To see a simple example, try typing the following commands:

```
>> help mean
```

and

```
>> type mean.m
```

Alternatively, you may open mean.m within the MATLAB editor.

- The function body includes a sequence of instructions, which in turn are built by:
    - Using control structures common to any other programming language, such as **if**, **for**, **while**, etc.
    - Calling other predefined functions.

```
function p = myprimes(N)
found = 0;
trynumber = 2;
p = [];
while (found < N)
   if isprime(trynumber)
      p = [p , trynumber];
      found = found + 1;
   end
   trynumber = trynumber + 1;
end
```

Fig. A.3 MATLAB function to return the first N prime numbers.

- Building expressions based on the familiar arithmetic, relational, and logical operators.

• For instance, suppose you want to write a function that returns the first N prime numbers. MATLAB provides the user with two related functions, `primes` and `isprime`. The function `primes` returns the prime numbers that are less than or equal to an input number:

```
>> primes(11)
ans =
     2     3     5     7    11
```

whereas `isprime` returns 1 if the input number is prime, 0 otherwise:

```
>> isprime([3 4 5])
ans =
     1     0     1
```

Unfortunately, `primes` is not what we need, since we want the first N prime numbers. One way to accomplish our aim is illustrated in figure A.3. Note how the `if` statement treats 1 as "true" and 0 as "false."

```
>> myprimes(8)
ans =
     2     3     5     7    11    13    17    19
```

The function can and should be improved. To begin with, even numbers larger than 2 cannot be prime and should not be checked; furthermore, the vector p should be preallocated, rather than dynamically resized. These improvements are left as an exercise.

- A typical way to improve performance of MATLAB code is vectorization. This means that one should try to avoid for loops working on elements of vectors and matrices, which should be acted on as a whole block. As an example, we may write different functions to build a Hilbert matrix. This matrix is introduced in example 1.3 on page 18, and its elements are

$$\mathbf{H}_{ij} = \frac{1}{i+j-1}.$$

In figure A.4 we illustrate different functions to build a Hilbert matrix of order N: MyHilbDumb is based on two nested loops, without matrix preallocation; MyHilb is the same, but it preallocates the output matrix; MyHilbV is partially vectorized, as rows are built and assigned as vectors.

Let us compare the performance of the three functions:

```
>> tic, MyHilbDumb(1000);, toc
Elapsed time is 10.565729 seconds.
>> tic, MyHilb(1000);, toc
Elapsed time is 0.053242 seconds.
>> tic, MyHilbV(1000);, toc
Elapsed time is 0.063986 seconds.
>> tic, MyHilb(5000);, toc
Elapsed time is 1.245170 seconds.
>> tic, MyHilbV(5000);, toc
Elapsed time is 1.202888 seconds.
```

We see how fundamental preallocation is. Vectorization does not seem an impressive technique here (the reader is urged to check the performance of the built-in function hilb, which is fully vectorized). In older MATLAB versions vectorized code typically worked much better that non-vectorized code; improvements in the MATLAB interpreter have made this practice less important in some cases, but not always.

The following example shows that when the overhead of a function call is involved, vectorization may be useful:

```
>> prices = 30:0.1:70;
>> N = length(prices);
>> calls = zeros(N,1);
>> tic, calls = blsprice(prices,50,0.1,1,0.4);, toc
Elapsed time is 0.012505 seconds.
>> tic, ...
for i=1:N, calls(i)=blsprice(prices(i),50,0.1,1,0.4);, end, toc
Elapsed time is 0.397540 seconds.
```

```
function H = MyHilbDumb(N)
for i=1:N
    for j=1:N
        H(i,j) = 1/(i+j-1);
    end
end
```

```
function H = MyHilb(N)
H = zeros(N,N);
for i=1:N
    for j=1:N
        H(i,j) = 1/(i+j-1);
    end
end
```

```
function H = MyHilbV(N)
H = zeros(N,N);
for i=1:N
    H(i,:) = 1./(i:(i+N-1));
end
```

Fig. A.4 Three ways to build a Hilbert matrix.

- Useful operators to vectorize code are any and find:

```
>> V = [ 1 3 -4 9 -2 1]
V =
     1    3   -4    9   -2    1
>> any(V > 9)
ans =
     0
>> any(V >= 7)
ans =
     1
>> sum(V<0)
ans =
     2
>> find(V < 0)
ans =
     3    5
>> V(find(V<0))=[]
V =
     1    3    9    1
```

- When developing M-files, a most useful tool is the interactive debugger. We refer the reader to the manual for details.

Appendix B
Refresher on Probability Theory and Statistics

In this appendix we recall very briefly some basic facts about probability theory and parameter estimation. This is not meant as a substitute for a thorough treatment, for which we refer the reader to the references. We will not use measure theoretic concepts and will mostly rely on intuition. We also give information on some functions provided by the MATLAB Statistics toolbox.

B.1 SAMPLE SPACE, EVENTS, AND PROBABILITY

Probability is defined based on random events that take place within a sample space. A sample space S contains the possible outcomes of a random experiment or a sequence of random experiments. An event E is a subset of the sample space S. Which subsets are events may depend on the application, what we are interested in, and the available information on random outcomes. The empty set $\emptyset$ is a particular event. For any event E, we may consider its complement E^c; since the sample space S contains all the possible outcomes, we have $S^c = \emptyset$. Given any two events E_1 and E_2, we may consider their union $E_1 \cup E_2$ and their intersection $E_1 \cap E_2$; to ease notation, we will denote intersection by $E_1 E_2$. If the intersection of two events is empty, i.e., if $E_1 E_2 = \emptyset$, we say that the two events are mutually exclusive. More generally, we may consider the union and the intersection of an arbitrary number of events.

For each event E on a sample space S, we define a probability measure $P(E)$ which must satisfy the following three conditions:

1. $0 \leq P(E) \leq 1$.

2. $P(S) = 1$.

3. For any sequence of mutually exclusive events $E_1, E_2, E_3, \ldots$ (i.e., such that $E_i E_j = \emptyset$, for $i \neq j$), we have

$$P\left(\bigcup_{i=1}^{\infty} E_i\right) = \sum_{i=1}^{\infty} P(E_i).$$

Different properties may be proven as a consequence of these conditions. For instance, it can be shown that

$$P(E) + P(E^c) = 1$$

and that

$$P(E_1 \cup E_2) = P(E_1) + P(E_2) - P(E_1 E_2).$$

Often we are interested in the probability of an event E conditional on the occurrence of another event F, denoted by $P(E \mid F)$. It is natural to define the conditional probability as[1]

$$P(E \mid F) = \frac{P(EF)}{P(F)}.$$

This follows from the observation that if we know that the event F has occurred, the new sample space is F, so that probabilities must be adjusted accordingly. Finally, we say that two events are independent if

$$P(EF) = P(E)P(F),$$

which in turn implies that

$$P(E \mid F) = P(E).$$

So, for independent events, knowing that F has occurred tells us nothing about the probability of the occurrence of E. Note that mutually exclusive events are *not* independent; if we know that one has occurred, we know that the other cannot.

---

[1]This definition is not completely satisfactory: It does not work with events with zero probability. Conditioning is treated at a higher level in advanced probability texts, but we do not really need that machinery for this introductory textbook. Hence we will stick to this intuitive definition.

B.2 RANDOM VARIABLES, EXPECTATION, AND VARIANCE

When we associate numerical values of one or more variables to events, we obtain random variables. Random variables may be thought of as mappings from events to real or integer numbers. Usually, a random variable is denoted by a capital letter such as X; the value assumed by a random variable on a particular realization of the events is denoted by a lowercase letter such as x. A different notation is common in economics, and it may be preferable when dealing with the Greek alphabet: For instance, we may use $\tilde{\epsilon}$ for a random variable, and ϵ for its realizations. When X takes values on a finite or countable domain, such as non-negative integer numbers, we speak of a discrete random variable. For a discrete random variable, we define the **probability mass function** $p(\cdot)$ for each possible outcome value x_i:

$$p(x_i) = P\{X = x_i\}.$$

We have

$$\sum_{i=1}^{\infty} p(x_i) = 1.$$

We also define the (cumulative) **distribution function** $F(\cdot)$:

$$F(a) = P\{X \le a\} = \sum_{x_i \le a} p(x_i).$$

It is easy to see that the distribution function for a discrete random variable is a piecewise constant, nondecreasing function.

Example B.1 A typical example of discrete probability distribution is the Poisson random variable, with parameter λ. In this case the random variable X takes values in the set $\{0, 1, 2, 3, \ldots\}$, and its probability mass function is

$$p(i) = P\{X = i\} = e^{-\lambda} \frac{\lambda^i}{i!}, \qquad i = 0, 1, 2, \ldots.$$

We may check that this is indeed a probability mass function:

$$\sum_{i=0}^{\infty} p(i) = e^{-\lambda} \sum_{i=0}^{\infty} \frac{\lambda^i}{i!} = e^{-\lambda} e^{\lambda} = 1.$$

In practice, one usually works with a parameter λt, where λ is the rate at which certain events occurs over time and t is the length of the time interval we observe. For instance, this could model the number of shocks we observe over a time interval on the price of a stock or the credit rating of a bond issuer. □

If the random variable may take values over a continuous set, such as a bounded interval on the real line, say (a, b), or the entire line $(-\infty, +\infty)$, we have a continuous random variable. In this case, we cannot define a probability mass function; since the outcome values are infinite and uncountable, the probability that X takes a specific value will be zero.[2] We must define a non-negative **probability density function** $f(x)$ for $x \in (-\infty, +\infty)$ such that for a given subset B of real numbers,

$$P\{X \in B\} = \int_B f(x)\,dx.$$

Then we have

$$P\{a \leq X \leq b\} = \int_a^b f(x)\,dx,$$

and

$$\int_{-\infty}^{+\infty} f(x)\,dx = 1.$$

To understand what the probability density means, consider the following:

$$P\{X \in (x, x + \Delta x)\} = \int_x^{x+\Delta x} f(y)\,dy \approx f(x)\,\Delta x,$$

for a small Δx. So we see that density cannot be interpreted as a probability, but it does give a measure of how likely given values of the random variable are and is needed to define probability of sets. We may also define the distribution function:

$$F(a) = P\{X \leq a\} = \int_{-\infty}^a f(x)\,dx,$$

from which we obtain[3]

$$\frac{dF(x)}{dx} = f(x).$$

Given a random variable, we may compute its **expected value** using the probability mass function or the density function. In the discrete case we have

$$\mathrm{E}[X] \equiv \sum_i x_i p(x_i)$$

and, for the continuous case,

$$\mathrm{E}[X] \equiv \int_{-\infty}^{+\infty} x f(x)\,dx.$$

An important property of the expectation operator is

$$\mathrm{E}[aX + b] = a\,\mathrm{E}[X] + b.$$

---

[2] We are not considering mixed probability distributions, which are a hybrid between discrete and continuous distributions.
[3] The distribution function is not everywhere differentiable in the case of mixed distributions, which we do not consider.

Example B.2 Let us compute the expected value of a Poisson random variable. Applying the definition yields

$$\mathrm{E}[X] = \sum_{i=0}^{\infty} i e^{-\lambda} \frac{\lambda^i}{i!} = \lambda e^{-\lambda} \sum_{i=1}^{\infty} \frac{\lambda^{i-1}}{(i-1)!}$$

$$= \lambda e^{-\lambda} \sum_{k=0}^{\infty} \frac{\lambda^k}{k!} = \lambda.$$

This may be interpreted as follows. If events occur at a rate of λ events per time unit, the expected number of events over a unit interval is actually the rate λ. By the same token, the expected number of events occurring over an interval of length t is λt. □

The expected value of a random variable gives a measure of location of the entire distribution, but it does not tell anything about its dispersion. The typical measure of dispersion is **variance**:

$$\mathrm{Var}(X) \equiv \mathrm{E}[(X - \mathrm{E}[X])^2].$$

The variance of a random variable X is often denoted by σ_X^2. Unfortunately, variance has not the same unit of measure of the random variable itself; hence, the square root of variance, σ_X, called **standard deviation**, is often used. A couple of properties of the variance are the following:

$$\mathrm{Var}(X) = \mathrm{E}[X^2] - \mathrm{E}^2[X]$$
$$\mathrm{Var}(aX + b) = a^2 \mathrm{Var}(X).$$

We see immediately that, unlike the expectation, the variance operator is not linear. Indeed, it is *not* true in general that the variance of a sum of random variables is the sum of their variances (see later).

Example B.3 Consider a random variable X such that

$$\mathrm{E}[X] = \mu \quad \text{and} \quad \mathrm{Var}(X) = \sigma^2.$$

If we define another random variable

$$Z = \frac{X - \mu}{\sigma},$$

it is easy to see that the properties above imply

$$\mathrm{E}[Z] = 0 \quad \text{and} \quad \mathrm{Var}(Z) = 1.$$

□

It is also natural to define the expected value of a **function** $g(X)$ of a random variable:

$$E[g(X)] = \begin{cases} \sum_i g(x_i)p(x_i) & \text{for the discrete case} \\ \int_{-\infty}^{+\infty} g(x)f(x)\,dx & \text{for the continuous case.} \end{cases}$$

It is important to note that, in general,

$$E[f(X)] \neq f(E[X]).$$

If the function g is convex, then the following **Jensen's inequality** holds:

$$E[g(X)] \geq g(E[X]).$$

Another fundamental concept linked to probability distributions is the **quantile**. In the continuous case, the quantile q_β is linked to a probability level β as follows:

$$P\{X \leq q_\beta\} = \beta.$$

We see that the quantile is the solution of the equation

$$\int_{-\infty}^{q_\beta} f_X(y)\,dy = \beta.$$

If there are multiple solutions to this equation, we take the smallest one as the quantile. This does not happen in common probability distributions, as the distribution function is strictly monotonically increasing over the support of the distribution. In the discrete case, the cumulative distribution "jumps" and we may fail to find a solution to this equation. In this case we adapt the definition as follows: The quantile is the smallest number q_β such that

$$F_X(q_\beta) \geq \beta.$$

B.2.1 Common continuous random variables

Uniform random variable A random variable is distributed uniformly over the interval (a, b) if its density function is

$$f(x) = \begin{cases} 1/(b-a) & \text{if } x \in (a,b) \\ 0 & \text{otherwise.} \end{cases}$$

A typical case is the uniform distribution over the interval $(0, 1)$. It is easy to see that

$$E[X] = \int_a^b \frac{x}{b-a}\,dx = \frac{b^2 - a^2}{2(b-a)} = \frac{b+a}{2}$$

and

$$\begin{aligned}\text{Var}(X) &= \text{E}[X^2] - \text{E}^2[X] = \int_a^b \frac{x^2}{b-a}\,dx - \left(\frac{a+b}{2}\right)^2 \\ &= \frac{b^3 - a^3}{3(b-a)} - \frac{(b+a)^2}{4} = \frac{(b-a)^2}{12}.\end{aligned}$$

Exponential random variable The exponential random variable may only assume non-negative values, and its density is given by

$$f(x) = \begin{cases} \lambda e^{-\lambda x} & \text{if } x \geq 0 \\ 0 & \text{if } x < 0, \end{cases}$$

for some parameter $\lambda > 0$. The distribution function is

$$F(a) = \int_0^a \lambda e^{-\lambda x}\,dx = 1 - e^{-\lambda a}.$$

The expected value is

$$\text{E}[X] = \int_0^\infty x\lambda e^{-\lambda x}\,dx = \frac{1}{\lambda},$$

and the variance is $1/\lambda^2$. It is interesting to note that if the time elapsing between events is exponentially distributed with parameter λ, the events occur at a rate λ, and the distribution of the number of events over a time interval of length t is a Poisson random variable with parameter λt.

Normal random variable The normal random variable has an infinite support, i.e., it may take values over the whole real line, and its density function is the bell-shaped function:

$$f(x) = \frac{1}{\sqrt{2\pi}\,\sigma} e^{-\frac{1}{2}\left(\frac{x-\mu}{\sigma}\right)^2}, \quad -\infty < x < +\infty,$$

for given parameters μ and σ^2. The distribution function for the normal distribution is not known in closed form, but it can be computed by numerical approximations (see section 3.3.1). With some calculations it can be shown that the parameters μ and σ have indeed a precise meaning:

$$\text{E}[X] = \mu, \quad \text{Var}[X] = \sigma^2.$$

We use the notation $X \sim \mathcal{N}(\mu, \sigma^2)$ to say that X has normal distribution with given expected value and variance. A variable $Z \sim \mathcal{N}(0,1)$ is called a **unit** or **standard normal** variable.

Example B.4 The parameter μ influences where the maximum of the density is located, whereas the variance σ^2, or the standard deviation σ, tells how

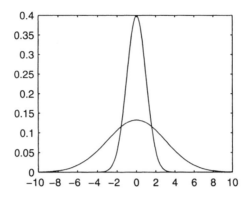

Fig. B.1 Normal density functions for $\mu = 0$ and $\sigma = 1$ or $\sigma = 3$.

stretched the function is. We may plot the density functions for two normal distributions with $\mu = 0$ and $\sigma = 1$ or $\sigma = 3$.

```
>> x=-10:0.1:10;
>> plot(x, normpdf(x,0,1))
>> hold on
>> plot(x, normpdf(x,0,3))
```

The result is plotted in figure B.1. The Statistics toolbox includes functions to compute the probability functions for all of the main probability distributions.
□

An important property of the normal distribution is that if X is normally distributed with parameters μ and σ^2, then $\alpha X + \beta$ is normally distributed with parameters $\alpha\mu + \beta$ and $\alpha^2 \sigma^2$. In particular, $Z = (X - \mu)/\sigma$ is a standard normal.

The importance of the standard normal distribution is apparent if we think of computing the distribution function or the quantiles for a generic normal variable. By working with standard normal variables, we are actually able to deal with the more general case. For instance, to compute the distribution function for an arbitrary normal variable, it is sufficient to come up with an approximation for the standard case:

$$N(x) = \frac{1}{\sqrt{2\pi}} \int_{-\infty}^{x} e^{-z^2/2}\, dz.$$

Let z_β be the β-quantile for the standard normal:

$$P\{Z \leq z_\beta\} = N(z_\beta) = \beta.$$

Knowing z_β, it is easy to find the β-quantile for a normal variable $X \sim \mathcal{N}(\mu, \sigma^2)$:

$$\beta = P\{X \leq q_\beta\}$$

$$= P\left\{\frac{X-\mu}{\sigma} \le \frac{q_\beta - \mu}{\sigma}\right\}$$
$$= P\left\{Z \le \frac{q_\beta - \mu}{\sigma}\right\},$$

from which we get
$$q_\beta = \mu + z_\beta \sigma.$$

In statistics, we are typically interested in quantiles of the form $z_{1-\alpha}$, where α is a relatively small number, such as 0.01 or 0.05. Quantiles and values of $N(x)$ are tabulated or computed using suitable approximations.

Example B.5 The function normcdf(x,sigma,mu) yields the distribution function. To compute the probability that a standard normal variable lies in the interval $(-2, 2)$:

```
>> p = normcdf([-2 2]);
>> p(2) - p(1)
ans =
    0.9545
```

Similarly:

```
>> p = normcdf([-3 3]);
>> p(2)-p(1)
ans =
    0.9973
```

from which we see that for a normal distribution, the probability of falling outside the interval $(\mu - 3\sigma, \mu + 3\sigma)$ is quite small. In fact, the normal distribution is a debatable model for asset returns, as in practice these exhibit fat tails, i.e., the occurrence of extreme values is more likely than it should be with the normal distribution.

You may also invert the distribution function. Compare x and xnew in the following:

```
>> x=[-3:0.2:0.3];
>> xnew=norminv(normcdf(x,0,1),0,1);
```
☐

The importance of normal variables, apart from their many properties, stems from the **central limit theorem**. Roughly speaking, it states that if we sum many identically distributed and independent random variables, their sum tends to have a normal distribution as the number of summed random variables goes to infinity.

Lognormal random variable Due to the central limit theorem, a normal random variable may be thought of as the limit of a sum of random variables. The lognormal variable may be thought of as the limit of a *product* of random variables. Formally, we say that a random variable Z is lognormally distributed if $\log Z$ is normally distributed; put another way, if X is normal, then e^X is lognormal.

The following formulas illustrate the relationships between the parameters of a normal and a lognormal distribution. If $X \sim \mathcal{N}(\mu, \sigma^2)$ and $Z = e^X$, then

$$\begin{aligned} \mathrm{E}[Z] &= e^{\mu + \sigma^2/2} \\ \mathrm{Var}(Z) &= e^{2\mu + \sigma}(e^{\sigma^2} - 1). \end{aligned}$$

In particular, we see that

$$\mathrm{E}[e^X] = e^{\mu + \sigma^2/2} \geq e^\mu = e^{\mathrm{E}[X]}.$$

Since the exponential is a convex function, this is a consequence of Jensen's inequality.

B.3 JOINTLY DISTRIBUTED RANDOM VARIABLES

When considering jointly distributed random variables, we may follow the same route as in the scalar case. We illustrate in the bidimensional case, as the generalization is straightforward. Given two random variables X and Y, we may define the joint distribution function:

$$F(x, y) = P\{X \leq x, Y \leq y\}.$$

In the discrete case we also consider the probability mass function:

$$p(x, y) = P\{X = x, Y = y\},$$

whereas continuous variables are characterized by a density $f(x, y)$ such that, for a region D in the plane,

$$P\{(X, Y) \in D\} = \iint_D f(x, y)\, dy\, dx.$$

From the joint distribution we may derive the marginal distributions for the single variables. For instance

$$\begin{aligned} P\{X \in A\} &= P\{X \in A, Y \in (-\infty, +\infty)\} = \int_A \int_{-\infty}^{+\infty} f(x, y)\, dy\, dx \\ &= \int_A f_X(x)\, dx, \end{aligned}$$

where

$$f_X(x) = \int_{-\infty}^{+\infty} f(x, y)\, dy$$

is the marginal density for the random variable X; the other density $f_Y(y)$ may be defined similarly.

The computation of expected values is quite similar to the scalar case. Given a function $g(X, Y)$ of the two random variables, we have

$$E[g(X,Y)] = \begin{cases} \sum_i \sum_j g(x_i, y_j) p(x_i, y_j) & \text{in the discrete case} \\ \int_{-\infty}^{+\infty} \int_{-\infty}^{+\infty} g(x,y) f(x,y) \, dy \, dx & \text{in the continuous case.} \end{cases}$$

From the linearity of these operations, it is easy to see that the expected value of a linear combination of random variables,

$$Z = \sum_{i=1}^{n} \lambda_i X_i,$$

is the same linear combination of the expected values:

$$E[Z] = \sum_{i=1}^{n} \lambda_i E[X_i].$$

However, a similar result does not hold, in general, for variance. Similarly, for jointly distributed variables it is *not* true in general that

$$E[g(X)h(Y)] = E[g(X)]E[h(Y)].$$

To investigate this matter we must deal with the dependence or independence between the random variables.

B.4 INDEPENDENCE, COVARIANCE, AND CONDITIONAL EXPECTATION

Two random variables X and Y are independent if the two events $\{X \le a\}$ and $\{Y \le b\}$ are independent, i.e.,

$$F(a,b) = P\{X \le a, Y \le b\} = P\{X \le a\} P\{Y \le b\} = F_X(a) F_Y(b).$$

This in turn implies that

$$p(x,y) = p_X(x) p_Y(y), \qquad f(x,y) = f_X(x) f_Y(y),$$

for discrete and continuous variables, respectively. If the variables are independent, it is easy to show that

$$E[g(X)h(Y)] = E[g(X)]E[h(Y)]$$

holds.

If there is some degree of dependence between random variables, we should try to measure it somehow. One measure of mutual dependence is the **covariance**:

$$\text{Cov}(X, Y) = \text{E}[(X - \text{E}[X])(Y - \text{E}[Y])] = \text{E}[XY] - \text{E}[X]\text{E}[Y].$$

If X and Y are independent, their covariance is zero (but the converse is not necessarily true, as the covariance is only one measure of dependence). If $\text{Cov}(X, Y) > 0$, Y tends to be large when X is, and small when X is. More precisely, when X is above its expected value, then Y is too, and when X is below its expected value, then Y is too. As a result, the expected value of $(X - \text{E}[X])(Y - \text{E}[Y])$ is positive because the two factors tend to have the same sign. A similar observation holds when covariance is negative. The following properties of the covariance are useful:

- $\text{Cov}(X, X) = \text{Var}(X)$,
- $\text{Cov}(X, Y) = \text{Cov}(Y, X)$,
- $\text{Cov}(aX, Y) = a\,\text{Cov}(Y, X)$,
- $\text{Cov}(X, Y + Z) = \text{Cov}(X, Y) + \text{Cov}(X, Z)$.

Using these properties (or the definitions), it can be shown that

$$\begin{aligned}\text{Var}(X + Y) &= \text{Var}(X) + \text{Var}(Y) + 2\,\text{Cov}(X, Y),\\ \text{Var}(X - Y) &= \text{Var}(X) + \text{Var}(Y) - 2\,\text{Cov}(X, Y).\end{aligned}$$

More generally,

$$\text{Var}\left(\sum_{i=1}^{n} X_i\right) = \sum_{i=1}^{n} \text{Var}(X_i) + 2\sum_{i=1}^{n}\sum_{j<i} \text{Cov}(X_i, X_j).$$

Thus, for mutually independent variables, the variance of a sum is the sum of the variances.

Example B.6 We often have to work with multivariate normals. Let

$$\mathbf{X} = \begin{bmatrix} X_1 \\ X_2 \\ \vdots \\ X_n \end{bmatrix}$$

be a vector of normal random variables with expected value $\boldsymbol{\mu}$ and covariance matrix

$$\Sigma = \text{E}[(\mathbf{X} - \boldsymbol{\mu})(\mathbf{X} - \boldsymbol{\mu})'].$$

Then the joint density function is given by

$$f(\mathbf{x}) = \frac{1}{(2\pi)^{n/2} |\Sigma|^{1/2}} e^{-\frac{1}{2}(\mathbf{x}-\mu)'\Sigma^{-1}(\mathbf{x}-\mu)},$$

where $|\Sigma|$ is the determinant of the covariance matrix. If the normal variables are mutually uncorrelated, then both the matrix Σ and its inverse are diagonal. This implies that the density function may be factorized into separate components, one for each X_i; hence, uncorrelated normal variables are also independent.

Another property of jointly normal variables is that they may combined linearly to yield other jointly normal variables. Given a matrix $\mathbf{T} \in \mathbb{R}^{m,n}$, $\mathbf{TX}$ is a vector of m jointly normal variables. □

The value of the covariance depends on the magnitude of the random variables involved. Often, a normalized measure of dependence is preferred, the **coefficient of correlation**:

$$\rho_{XY} = \frac{\text{Cov}(X,Y)}{\sqrt{\text{Var}(X)}\sqrt{\text{Var}(Y)}}.$$

It can be shown that $\rho_{XY} \in [-1, 1]$.

Example B.7 Correlation is often used in finance. However, it is important to realize its limitations. Consider the following example.

```
>> x = -1:0.001:1;
>> y = sqrt(1-x.^2);
>> cov(x,y)
ans =
    0.3338    0.0000
    0.0000    0.0501
```

Here we have a random variable X which is distributed uniformly on $(-1, 1)$, and a random variable Y which is deterministically linked to X, as

$$Y = \sqrt{1 - X^2}.$$

However, the covariance and the correlation are zero, since

$$\text{Cov}(X, Y) = \text{E}[XY] - \text{E}[X]\text{E}[Y],$$

but $\text{E}[X] = 0$, and (because of symmetry)

$$\text{E}[XY] = \int_{-1}^{1} x \frac{1}{2}\sqrt{1 - x^2}\, dx = 0.$$

The key issue is that the correlation is a measure of *linear* dependence. Here the dependence is non-linear, as the points (X, Y) lie on the upper half of the unit circle $X^2 + Y^2 = 1$. □

If two variables are not independent, then knowing something about the value taken by one of them can give us valuable information about the other one. This leads us to investigate conditioning. Just as we have defined conditional probabilities for events, we may define **conditional expectation**. This means that we want to know how an event like $(Y = y)$ influences the distribution of X. For discrete random variables we have

$$E[X \mid Y = y_j] = \sum_i x_i P\{X = x_i \mid Y = y_j\} = \frac{\sum_i x_i P\{X = x_i, Y = y_j\}}{P\{Y = y_j\}}.$$

Similarly, for continuous variables

$$E[X \mid Y = y] = \frac{\int x f(x, y) \, dx}{\int f(x, y) \, dx}.$$

Conditioning is a useful way to solve many problems. A fundamental property is the following:

$$E[X] = E[E[X \mid Y]]. \tag{B.1}$$

In practice, this may be used when fixing the value of a random variable makes working with another one easier. Equation (B.1) may be rewritten, in concrete, as

$$E[X] = \begin{cases} \sum_j E[X \mid Y = y_j] P\{Y = y_j\} & \text{in the discrete case} \\ \int E[X \mid Y = y] f_Y(y) \, dy & \text{in the continuous case.} \end{cases}$$

We may also define a conditional variance:

$$\text{Var}(X \mid Y) = E\left[(X - E[X \mid Y])^2 \mid Y\right].$$

The following formula may be proved for the conditional variance:

$$\text{Var}(X) = E[\text{Var}(X \mid Y)] + \text{Var}(E[X \mid Y]). \tag{B.2}$$

This formula may be used to compute variance by conditioning, but it also implies that

$$\text{Var}(X) \geq E[\text{Var}(X \mid Y)]$$
$$\text{Var}(X) \geq \text{Var}(E[X \mid Y]),$$

since variance is a non-negative quantity by definition. These properties may be used for variance reduction in Monte Carlo simulation (see section 4.5).

We would like to close this section by pointing out that our treatment of conditioning, apart from being very brief by necessity, has followed the classical lines of basic textbooks on probability theory. A solid understanding of conditioning and of the role of information in probability requires advanced tools which are beyond the scope of this book (see the references).

B.5 PARAMETER ESTIMATION

In the theory of probability, we assume a lot of knowledge about a set of random variables, and we ask questions about the probability of some events, about expected values of functions of those variables, etc. However, the knowledge required to get those answers, i.e., the whole probability distribution, is a scarce commodity. Even the expected value and variance are typically unknown, and must be estimated on the basis of samples. The sample data might come from the real world (e.g., stock prices) or from a Monte Carlo simulation. Typical parameters we want to estimate are the expected value, the variance, or the covariance matrix; furthermore, we would also like to quantify the reliability of the estimate.

A random sample should be thought as a set $X_1, X_2, \ldots, X_n$ of independent and identically distributed random variables, drawn from the same underlying distribution. Say that the expected value of the underlying population is μ and the variance is σ^2; these parameters are unknown, and we would like to come up with a reasonable estimate of them. An intuitive way to estimate μ is to use the **sample mean**:

$$\bar{X} = \frac{1}{n} \sum_{i=1}^{n} X_i.$$

Note that the expected value is an unknown number, whereas the sample mean is a random variable. It is a reasonable estimator, in the sense that it is unbiased:

$$E[\bar{X}] = \mu.$$

The more samples we get, the better, in the sense that the variance of the estimator decreases:

$$\text{Var}(\bar{X}(n)) = \frac{1}{n^2} \text{Var}\left(\sum_{i=1}^{n} X_i\right) = \frac{1}{n^2} \sum_{i=1}^{n} \text{Var}(X_i) = \sigma^2/n.$$

It is fundamental to understand that in this derivation we have assumed the independence of the samples; if the samples are not independent, reasoning this way may lead to underestimate the uncertainty in the estimate.[4] We see from the last formula that, if n goes to infinity, the variance of the estimator goes to zero. So, in some sense, the sample mean should "tend" to the unknown expected value. To state this in a mathematically precise way, we should introduce concepts of stochastic convergence. In fact, the *law of large numbers* comes in two forms, weak and strong, depending on the kind of stochastic convergence we use. We will not consider this issue and settle for an intuitive understanding. Another interesting property of the sample mean

---

[4]See, e.g., [2] for a clear discussion of this point.

stems from the central limit theorem. Roughly speaking, when the number of samples grow, $\bar{X}$ tends to be distributed normally. More precisely, we have that the distribution of

$$\frac{\bar{X} - \mu}{\sigma/\sqrt{n}} \qquad (B.3)$$

tends to the standard normal distribution. How many samples it takes to have an approximately normal distribution depends on the distribution of the X_i. If they are normal, then the sample mean is always normal. If it is symmetric, a few samples may be enough; if it is strongly asymmetric (skewed), then many samples may be needed. This is not an issue in this book, as we apply these ideas to Monte Carlo simulation, where many thousands of samples are taken. We should recall again that all of the ideas above rely on the assumption of *independence* between samples.

Another difficulty derives form the fact that the variance is typically unknown too. If we knew μ, we could estimate σ^2 by averaging squared deviations:

$$\frac{1}{n} \sum_{i=1}^{n} [X_i - \mu]^2.$$

Since we must use an estimate of μ, the estimator of σ^2 is the **sample variance**:

$$S^2 = \frac{1}{n-1} \sum_{i=1}^{n} [X_i - \bar{X}]^2.$$

Note the $1/(n-1)$ factor, which is essentially due to the use of an estimate of μ. It can be shown that this factor is needed to make the estimator unbiased ($E[S^2] = \sigma^2$). By a similar expression we may estimate the covariance between two random variables X and Y:

$$S_{XY} = \frac{1}{n-1} \sum_{i=1}^{n} \left(X_i - \bar{X}\right)\left(Y_i - \bar{Y}\right).$$

It can be shown that $E[S_{XY}] = \text{Cov}(X, Y)$. We can also estimate the correlation coefficient:

$$r_{XY} = \frac{\sum_{i=1}^{n} \left(X_i - \bar{X}\right)\left(Y_i - \bar{Y}\right)}{\sqrt{\sum_{i=1}^{n} \left(X_i - \bar{X}\right)^2} \sqrt{\sum_{i=1}^{n} \left(Y_i - \bar{Y}\right)^2}}.$$

These tasks are accomplished by MATLAB functions. The basic versions are available in the MATLAB core; some advanced functionalities are included only in the Statistics toolbox.

Example B.8 The function `mean` yields the sample mean. For instance, let us use the `normrnd` function to generate a set of independent normally distributed data values[5]:

```
>> randn('state',0)
>> x=normrnd(2,3,1000,2);
>> mean(x)
ans =
    1.8708    2.1366
```

The first two parameters of `normrnd` are the expected value and the standard deviation of the normal variable; the remaining two are optional and define the size of the matrix to generate. The matrix, which here has 1000 rows and two columns, is interpreted columnwise, as 1000 realizations of two random variables. This is why two means are estimated, one per column of the data matrix. The function `cov(x)` estimates the covariance matrix (assuming a column-oriented data matrix).

```
>> randn('state',0)
>> x=normrnd(10,2,10000,4);
>> cov(x)
ans =
    4.0091    0.0480    0.0204   -0.0457
    0.0480    4.0291    0.0374   -0.0050
    0.0204    0.0374    4.0390    0.0193
   -0.0457   -0.0050    0.0193    4.0464
```

Note that the values on the diagonal are close to the "correct" variance $\sigma^2 = 4$ for each of the four independent variables; off-diagonal elements should be zero, as the samples should be independent. Given the limited number of samples, it is not surprising that the results do not match exactly what we would expect in theory.

In practice, estimating parameters may be a tough problem. Consider drawing 100 samples from a normal distribution with known parameters and checking if the sample mean corresponds to the known expected value. Let us repeat ten of these experiments:

```
>> randn('state',0)
>> x=normrnd(0.3,2,100,10);
>> mean(x)'
ans =
    0.3959
```

---

[5]We use the instruction `randn('state',0)` to make sure you will get the same numbers shown here. Otherwise, the numbers you get may differ from the following ones, depending on the current state of the random number generator; the issue is explained in section 4.3. Furthermore, if you repeat the experiment, you will get different outcomes.

0.0460
0.1437
0.2803
0.0048
0.1646
0.4143
0.1915
0.4961
0.0013

You see that the estimated mean value may be quite different from the correct value $\mu = 0.3$. Actually, if you repeat the experiment a few times, you will even get negative sample means. This is due to the fact that the expected value is small with respect to the variance of the data; if you think of estimating stock returns over short periods, using historical data when volatility is high, you will realize that this is not a hypothetical circumstance. This phenomenon, called *mean blur*, is described, e.g., in [3, chapter 8]. Another point worth mentioning is that if you use historical data, you might question the validity of the old data; however, using only the recent ones may lead to unreliable estimates. The Financial toolbox includes a more sophisticated function (ewstats) to compute a covariance matrix by applying a "forgetting factor," reducing the weight of the old data. □

Given the remarkable amount of variability in the estimator, which is evident in the last example, it is clear that we need some way to measure the reliability of our estimate. Consider (B.3) and assume we know the $(1 - \alpha/2)$-quantile from the standard normal distribution, i.e., the number $z_{1-\alpha/2}$ such that

$$P\{Z \leq z_{1-\alpha/2}\} = \frac{1}{\sqrt{2\pi}} \int_{-\infty}^{z_{1-\alpha/2}} e^{-y^2/2} \, dy = 1 - \alpha/2,$$

where $Z \sim \mathcal{N}(0,1)$. Then, given the symmetry of the standard normal distribution, we see that

$$P\left\{-z_{1-\alpha/2} \leq \frac{\bar{X} - \mu}{\sigma/\sqrt{n}} \leq z_{1-\alpha/2} \leq\right\} \approx 1 - \alpha.$$

This is only approximately true unless the X_i are normal, but given the central limit theorem, it will be a good approximation when a large number of samples are taken. Rearranging the above inequality, we see that, with probability close to $1 - \alpha$, we have

$$\bar{X} - z_{1-\alpha/2} \frac{\sigma}{\sqrt{n}} \leq \mu \leq \bar{X} + z_{1-\alpha/2} \frac{\sigma}{\sqrt{n}}.$$

In other words, we may build a **confidence interval** that, with a suitable degree of confidence, will contain the unknown parameter μ. Unfortunately, this is not really true, since we have to estimate σ^2 by the sample variance. Hence, we should consider the distribution of the random quantity:

$$\frac{\bar{X} - \mu}{S/\sqrt{n}}.$$

It turns out that the distribution is not really standard normal. If the X_i are normal, then this ratio is distributed according to a Students's t distribution with $n-1$ degrees of freedom. This distribution is qualitatively similar to a standard normal distribution, as it is bell shaped and symmetric around the origin, but it has fatter tails. In practice, in building the confidence interval, we should use the quantiles $t_{n-1,1-\alpha/2}$ from this distribution, where $t_{n-1,1-\alpha/2} > z_{1-\alpha/2}$. This basically means that the confidence interval should be wider, which makes sense given the need to estimate more parameters. It turns out that when n is large, the t distribution tends to the standard normal distribution. Again, all of this is only approximately true in general, since the samples are not necessarily normal themselves. However, when the number of samples is large, thanks to the central limit theorem, we may use the following approximate confidence interval:

$$\bar{X} \pm z_{1-\alpha/2} \frac{S}{\sqrt{n}}.$$

The idea is that if we repeat the sampling and estimation procedure over and over, the percentage of cases in which the "true" value falls within this interval should be $100 \times (1-\alpha)$. Typical values of α are 0.05 and 0.01.

Example B.9 Calling the function [muhat, sigmahat, muci, sigmaci] = normfit(x) yields an estimate of the expected value and the standard deviation and the respective 95% confidence intervals.

```
>> randn('state',0)
>> x=normrnd(1,2,100,1);
>> [mu,s,mci,sci] = normfit(x)
mu =
    1.0959
s =
    1.7370
mci =
    0.7512
    1.4405
sci =
    1.5251
    2.0178
```

This function assumes normal samples and uses the quantiles from the t distribution. Keeping the above warnings in mind, we may use this function to build confidence intervals for parameters we estimate by Monte Carlo simulation. It is possible to specify a different confidence level by calling the function with an optional parameter: normfit(x,alpha). □

B.6 LINEAR REGRESSION

Linear regression by the method of least squares is a two-fold technique. On the one hand, we may consider it as a function approximation technique. Say that we have a set of n data points (x_i, y_i), $i = 1, \ldots, n$. We may assume a functional form $y = f(x)$ linking the data, and we look for the function $f(\cdot)$ that yields the best fitting. Linear regression is the case in which we assume a linear form:

$$y = f(x) = a + bx.$$

If we define the residual e_i as

$$e_i = y_i - f(x_i) = y_i - (a + bx_i), \tag{B.4}$$

we may look for the optimal parameters a and b minimizing the sum of squared residuals:

$$e = \sum_{i=1}^{n} e_i^2 = \sum_{i=1}^{n} (y_i - a - bx_i)^2. \tag{B.5}$$

Straightforward calculus yields

$$a = \frac{1}{n} \sum_{i=1}^{n} y_i - b \frac{1}{n} \sum_{i=1}^{n} x_i = \bar{y} - b\bar{x}, \tag{B.6}$$

where $\bar{x}$ and $\bar{y}$ are formally equivalent to sample means, and

$$b = \frac{n \sum_{i=1}^{n} x_i y_i - \sum_{i=1}^{n} x_i \cdot \sum_{i=1}^{n} y_i}{n \sum_{i=1}^{n} x_i^2 - \left(\sum_{i=1}^{n} x_i \right)^2}. \tag{B.7}$$

All of this has nothing to do with statistics, and it is just a simple case of the more general problem of function approximation (see section 3.3). However, the expression for b looks suspiciously like the ratio of a sample covariance over a sample variance. The following manipulations show that this interpretation is not unreasonable:

$$b = \frac{\sum_{i=1}^{n} x_i (y_i - \bar{y}) - \sum_{i=1}^{n} \bar{x} (y_i - \bar{y})}{\sum_{i=1}^{n} x_i (x_i - \bar{x}) - \sum_{i=1}^{n} \bar{x} (x_i - \bar{x})}$$

$$= \frac{\sum_{i=1}^{n} (x_i - \bar{x})(y_i - \bar{y})}{\sum_{i=1}^{n} (x_i - \bar{x})(x_i - \bar{x})} = \frac{\frac{1}{n-1} \sum_{i=1}^{n} (x_i - \bar{x})(y_i - \bar{y})}{\frac{1}{n-1} \sum_{i=1}^{n} (x_i - \bar{x})^2}$$

$$= \frac{S_{xy}}{S_x^2}. \tag{B.8}$$

Here we have somewhat misused the notations S_{xy} and S_x^2, since we have no statistical interpretation of these quantities. A statistical interpretation can be given if we assume that our data come from a statistical model. One possible model is

$$Y_i = \alpha + \beta x_i + \epsilon_i, \qquad i = 1, \ldots, n, \tag{B.9}$$

where

- the parameters α and β are (in practice) unknown *numbers*;
- ϵ_i is a random variable such that

$$\mathrm{E}[\epsilon_i] = 0, \qquad \mathrm{Var}(\epsilon_i) = \sigma^2, \qquad i = 1, \ldots, n;$$

this implies that the *errors* ϵ_i are identically distributed;

- the random variables ϵ_i are mutually independent and do not depend on the associated value of x_i;
- the values x_i are given numbers.

The last observation makes sense when the x_i is under our control; hence, Y_i is random due to the impact of the random error, but x_i is not. In other statistical models, we consider random variables X_i, but the general approach does not change that much.

Under these hypotheses, it can be shown that the regression coefficients a and b are unbiased estimators of the parameters α and β. Note that the regression coefficients are random because they are influenced by the errors. Under additional assumptions on the distribution of the errors, which are typically assumed normal, we may build confidence intervals for the estimates.

Example B.10 The Statistics toolbox offers a function to perform *multiple* linear regression, i.e., linear regression where there are multiple "x" variables. It is interesting to carry out a little experiment to understand the nature of the problem. Let us assume a linear model:

$$Y = 10 + 5x + \epsilon$$

where $\epsilon \sim \mathcal{N}(0, 4)$. We consider ten values of x:

$$x_i = 1 + 0.2 \times i, \qquad i = 0, 1, \ldots, 9,$$

and generate ten random samples as errors. Then we check if the estimates we get are close to the known values:

```
>> randn('state',0)
```

```
>> errors=normrnd(0,2,10,1);
>> x = 1 + 0.2*(0:9)'
x =
    1.0000
    1.2000
    1.4000
    1.6000
    1.8000
    2.0000
    2.2000
    2.4000
    2.6000
    2.8000
>> y = 10 + 5*x + errors
y =
   14.1349
   12.6688
   17.2507
   18.5754
   16.7071
   22.3818
   23.3783
   21.9247
   23.6546
   24.3493
>> v = regress(y, [ones(10,1), x])
v =
    7.2801
    6.4328
```

What we get, $a = 7.2801$ and $b = 6.4328$, is fairly distant from what we know. This is due to the amount of noise, but there is another factor. Let us repeat the experiment with different x values:

```
>> x = (1:10)';
>> y = 10 + 5*x + errors;
>> v = regress(y, [ones(10,1), x])
v =
    8.4264
    5.2866
```

Here the estimates look a bit better. The reason is that the values of x are more widespread, and the errors have a smaller impact. If we could reduce noise, we would get really close to the correct value:

```
>> y = 10 + 5*x + normrnd(0,1,10,1);
>> v = regress(y, [ones(10,1), x])
v =
   10.6117
```

4.9308

Of course, we have cheated and this is not what happens in a real setting, and confidence intervals for the estimates should be derived. □

We will only use regression in pricing American-style options, and this is why we just give this very sketchy overview of an important topic. However, we should at least mention the following *caveats* about linear regression:

- Regression describes association, not causation: we tend to interpret x as a cause and Y as an effect, but this need not be true.

- Due to sampling variability we may "see" relationships which are not really supported by data.

- On the other hand, since the b parameter is linked to covariance, and covariance is only a measure of linear association (see example B.7), linear regression may not properly account for more complex, non-linear, associations.

For further reading

There are many excellent books on probability theory, ranging from the elementary to the very sophisticated.

- An introductory book characterized by a remarkable clarity, plenty of insightful examples, and a wide range of topics is [5], which does not rely on measure-theoretic concepts.

- If you are interested in a more advanced treatment, based on rigorous axiomatic foundations, see, e.g., [6].

- A less encyclopedic, but perhaps more readable, treatment can be found in [1].

- Apart from good statistics books, such as [4], a quick and readable introduction to parameter estimation may be found in simulation books such as [2].

REFERENCES

1. M. Capiński and T. Zastawniak, editors. *Probability through Problems*. Springer-Verlag, Berlin, 2000.

2. A.M. Law and W.D. Kelton. *Simulation Modeling and Analysis (3rd ed.)*. McGraw-Hill, New York, 1999.

3. D.G. Luenberger. *Investment Science*. Oxford University Press, New York, 1998.

4. S. Ross. *Introduction to Probability and Statistics for Engineers and Scientists (2nd ed.)*. Academic Press, San Diego, CA, 2000.

5. S. Ross. *Introduction to Probability Models (8th ed.)*. Academic Press, San Diego, CA, 2002.

6. A.N. Shiryaev. *Probability (2nd ed.)*. Springer-Verlag, New York, 1996.

Appendix C
Introduction to AMPL

In this brief appendix, we want to introduce the basic syntax of AMPL. We use AMPL only in the last chapters on optimization models, and the syntax is almost self explanatory. Hence, we will just describe a few basic examples, so that the reader can get a grasp of the basic language elements. The reader is referred to the original reference [1], written by the developers of AMPL. Unlike MATLAB, AMPL is not a procedural language. There is a part of the language which is aimed at writing scripts, which behave like any program based on a sequence of control statements and instructions. But the core of AMPL is a *declarative* syntax to describe a mathematical programming model and the data to instantiate it. The optimization solver is separate: You can write a model in AMPL, and solve it with different solvers, possibly implementing different algorithms. Actually, AMPL interfaces have been built for many different solvers; in fact, AMPL is more of a language standard which has been implemented and is sold by a variety of providers.

A demo version is currently available on the web site http://www.ampl.com. The reader with no access to a commercial implementation can get the student demo and install it following the instructions. This student demo comes with two solvers: MINOS and CPLEX. MINOS is a solver for linear and nonlinear programming models with continuous variables, developed at Stanford University. CPLEX is a solver for linear and mixed-integer programming models. Originally, CPLEX was a university product, but it is now developed and distributed by ILOG. Recent CPLEX versions are able to cope with quadratic programming models, both continuous and mixed-integer. All the examples in this book have been solved using CPLEX.

Clearly, software choice is a very subjective matter. I personally work a lot integrating MATLAB and ILOG AMPL/CPLEX. But for the sake of fairness, alternative modeling languages are listed in the references.

C.1 RUNNING OPTIMIZATION MODELS IN AMPL

Typically, optimization models in AMPL are written using two separate files.

- A **model** file, with standard extension *.mod, contains the description of parameters (data), decision variables, constraints, and the objective function.

- A separate **data** file, with standard extension *.dat, contains data values for a specific model instance. These data must match the description provided in the model file.

Both files are normal ASCII files which can be created using any text editor, including MATLAB editor (if you are using word processors, be sure you are creating plain text files, with no hidden control characters for formatting). It is also possible to describe a model in one file, but separating structure and data is a good practice, enabling to solve multiple instances of the same model easily.

When you start AMPL, you get a DOS-like window[1] with a prompt like:

ampl:

To load a model file, you must enter a command like:

ampl: model mymodel.mod;

where the semicolon must not be forgotten, as it marks the end of a command (otherwise AMPL waits for more input by issuing a prompt like ampl?).[2] To load a data file, the command is

ampl: data mymodel.dat;

Then we may solve the model by issuing the command:

ampl: solve;

---

[1]The exact look of the window and the way you start AMPL depend on the AMPL version you use.
[2]Here we are assuming that the model and data files are in the same directory as the AMPL executable, which is not good practice. It is much better to place AMPL on the DOS path and to launch it from the directory where the files are stored. See the manuals for details.

To change data without loading a new model, you should do something like:

```
ampl: reset data;
ampl: data mymodel.dat;
```

Using reset; unloads the model too, and it must be used if you want to load and solve a different model. This is also important if you get error messages because of syntax errors in the model description. If you just correct the model file and load the new version, you will get a lot of error messages about duplicate definitions.

The solver can be select using the option command. For instance, you may choose

```
ampl: option solver minos;
```

or

```
ampl: option solver cplex;
```

Many more options are actually available, as well as ways to display the solution and to save output to files. We will cover only the essential in the following. We should also mention that the commercial AMPL versions include a powerful script language, which can be used to write complex applications in which several optimization models are dealt with, whereby one model provides input to another one.

C.2 MEAN VARIANCE EFFICIENT PORTFOLIOS IN AMPL

The best way to get acquainted with AMPL syntax is by considering a simple but relevant example. We describe the theory of mean-variance efficient portfolios in section 2.4.2. This framework leads to the solution of the following quadratic program:

$$\begin{aligned}
\min \quad & \mathbf{w}'\Sigma\mathbf{w} \\
\text{s.t.} \quad & \mathbf{w}'\bar{\mathbf{r}} = \bar{r}_T \\
& \sum_{i=1}^{n} w_i = 1 \\
& w_i \geq 0.
\end{aligned}$$

AMPL syntax for this model is given in figure C.1. First we define model parameters: the number of assets NAssets, the vector of expected return (one per asset), the covariance matrix, and the target return. Note that each declaration must be terminated by a semicolon, as AMPL does not consider end of line characters. The restriction NAssets > 0 is *not* a constraint of the model: It is an optional consistency check that is carried out when data are loaded, *before* issuing the solve command. Catching data inconsistencies as

```
param NAssets > 0;
param ExpRet{1..NAssets};
param CovMat{1..NAssets, 1..NAssets};
param TargetRet;

var W{1..NAssets} >= 0;

minimize Risk:
      sum {i in 1..NAssets, j in 1..NAssets} W[i]*CovMat[i,j]*W[j];

subject to SumToOne:
      sum {i in 1..NAssets} W[i] = 1;

subject to MinReturn:
      sum {i in 1..NAssets} ExpRet[i]*W[i] = TargetRet;
```

```
param NAssets := 3;
param ExpRet :=
    1 0.15
    2 0.2
    3 0.08;
param CovMat:
       1         2         3       :=
1    0.2000    0.0500   -0.0100
2    0.0500    0.3000    0.0150
3   -0.0100    0.0150    0.1000;

param TargetRet := 0.1;
```

Fig. C.1 AMPL model (MeanVar.mod) and data (MeanVar.dat) files for mean-variance efficient portfolios.

early as possible may be very helpful. Also note that in AMPL it is typical (but not required) to assign long names to parameters and variables, which are more meaningful than the terse names we use in mathematical models.

Then the decision variable W is declared; this variable must be non-negative to prevent short-selling, and this bound is associated to the variable, rather than being declared as a constraint. Finally, the objective function and the two constraints are declared. In both cases we use the sum operator, with a fairly natural syntax. We should note that braces ({}) are used when declaring vectors and matrices, whereas squares brackets ([]) are used to access elements. Objectives and constraints are always given a name, so that later we can access information such as the objective value and dual variables. Expressions for constraints and objective can be entered freely. There is no natural order in the declarations: We may interleave any type of model elements, provided what is used has already been declared.

In the second part of figure C.1 we show the data file. The syntax is fairly natural, but you should notice its basic features:

- Blank and newline characters do not play any role: We must assign vector data by giving both the index and the value; this may look a bit involved, but it allows quite general indexing.

- Each declaration must be closed by a semicolon.

- To assign a matrix, a syntax has been devised that allows to write data as a table, with rows and columns arranged in a visually clear way.

Now we are ready to load and solve the model, and to display the solution:

```
ampl: model MeanVar.mod;
ampl: data MeanVar.dat;
ampl: solve;
    CPLEX 9.1.0: optimal solution; objective 0.06309598494
    18 QP barrier iterations; no basis.
ampl: display W;
    W [*] :=
        1   0.260978
        2   0.0144292
        3   0.724592
    ;
```

We see that a barrier solver is used, hence, no basis is available; see section 6.4 to understand this point. We can also evaluate expressions based on the output from the optimization models, as well as checking dual variables of constraints:

```
ampl: display Risk;
    Risk = 0.063096
ampl: display sqrt(Risk);
```

```
  sqrt(Risk) = 0.251189
ampl: display MinReturn.dual;
  MinReturn.dual = -0.69699
ampl: display sum {k in 1..NAssets} W[k]*ExpRet[k];
  sum{k in 1 .. NAssets} W[k]*ExpRet[k] = 0.1
```

C.3 THE KNAPSACK MODEL IN AMPL

We have considered the knapsack model as a trivial model for capital budgeting (example 1.2 on page 15). This is a pure binary programming model:

$$\max \quad \sum_{i=1}^{n} R_i x_i$$

$$\text{s.t.} \quad \sum_{i=1}^{N} C_i x_i \leq W$$

$$x_i \in \{0, 1\}.$$

The corresponding AMPL model is displayed in figure C.2. Again, the syntax is fairly natural, and we should just note a couple of points:

- The decision variables are declared as `binary`.

- In the data file, the two vectors of parameters are assigned at the same time to save on writing; you should compare carefully the syntax used here against the syntax used to assign a matrix (see the covariance matrix in the previous example).

Now we may solve the model and check the solution (we must use `reset` to unload the previous model):

```
ampl: reset;
ampl: model Knapsack.mod;
ampl: data Knapsack.dat;
ampl: solve;
  CPLEX 9.1.0: optimal integer solution; objective 34
  3 MIP simplex iterations
  0 branch-and-bound nodes
ampl: display x;
x [*] :=
1  1
2  0
3  0
4  1
;
```

In this case, branch and bound is invoked (see chapter 12). In fact, if you are using the student demo, you cannot solve this model with MINOS; CPLEX must be selected using

```
param NItems > 0;
param Value{1..NItems} >= 0;
param Cost{1..NItems} >= 0;
param Budget >= 0;

var x{1..NItems} binary;

maximize TotalValue:
      sum {i in 1..NItems} Value[i]*x[i];

subject to AvailableBudget:
      sum {i in 1..NItems} Cost[i]*x[i] <= Budget;
```

```
param NItems = 4;

param: Value Cost :=
    1      10    2
    2       7    1
    3      25    6
    4      24    5;

param Budget := 7;
```

Fig. C.2 AMPL model (Knapsack.mod) and data (Knapsack.dat) files for the knapsack model.

```
ampl: option solver cplex;
```

If you use MINOS, you will get the solution for the continuous relaxation of the model above, i.e., a model in which the binary decision variables are relaxed: $x \in [0,1]$, instead of $x \in \{0,1\}$. The same can be achieved in ILOG AMPL/CPLEX by issuing appropriate commands:

```
ampl: option cplex_options 'relax';
ampl: solve;
  CPLEX 9.1.0: relax
  Ignoring integrality of 4 variables.
  CPLEX 9.1.0: optimal solution; objective 36.2
  1 dual simplex iterations (0 in phase I)
ampl: display x;
  x [*] :=
  1  1
  2  1
  3  0
  4  0.8
;
```

Here we have used the `relax` option to solve the relaxed model. We may also use other options to gain some insights on the solution process:

```
ampl: option cplex_options 'mipdisplay 2';
ampl: solve;
CPLEX 9.1.0: mipdisplay 2
MIP start values provide initial solution with objective 34.0000.
Clique table members: 2
MIP emphasis: balance optimality and feasibility
Root relaxation solution time =    0.00 sec.

        Nodes                                Cuts/
Node  Left   Objective  IInf  Best Integer  Best Node   ItCnt    Gap

   0     0     36.2000     1       34.0000    36.2000       1   6.47%
              cutoff               34.0000    Cuts:  2       2   0.00%

Cover cuts applied:   1
CPLEX 9.1.0: optimal integer solution; objective 34
2 MIP simplex iterations
0 branch-and-bound nodes
```

To interpret this output, the reader should have a look at chapter 12, where the branch and bound method is explained.

```
param NBonds >0, integer;
param TimeHorizon >0, integer;
param BondPrice{1..NBonds};
param CashFlow{1..NBonds, 1..TimeHorizon};
param Liability{1..TimeHorizon};

var x{1..NBonds} >= 0;

minimize PortfolioCost:
    sum {i in 1..NBonds} BondPrice[i]*x[i];

subject to MeetLiability {t in 1..TimeHorizon}:
    sum {i in 1..NBonds} CashFlow[i,t]*x[i] >= Liability[t];
```

Fig. C.3 AMPL model file for simple cash flow matching.

C.4 CASH FLOW MATCHING

As a final example, we consider a cash flow matching model (see section 2.3.2)

$$\min \quad \sum_{i=1}^{N} P_i x_i$$
$$\text{s.t.} \quad \sum_{i=1}^{N} F_{it} x_i \geq L_t \quad \forall t$$
$$x_i \geq 0.$$

The only new point here, with respect to previous models, is the constraint which must be replicated for each time period within the planning horizon. How this can be accomplished is illustrated in the AMPL model of figure C.3. Also note that a few parameters have been restricted to integer variables; the integer keyword can also be used to specify general integer decision variables.

For further reading

In the literature

- AMPL was introduced in [1] by its developers.
- There are many other modeling languages. A notable one is GAMS, which are similar in spirit to AMPL, in the sense that it is not linked to a specific solver. See http://www.gams.com. GAMS is probably more

familiar to people in Economics, and it is also used in [2, 3] to develop financial optimization models.

On the Web

- The AMPL student version and additional material can be found on http://www.ampl.com. There you may also see the list of solvers compatible with AMPL.

- For the commercial ILOG AMPL version and the CPLEX solver, see http://www.ilog.com.

- MINOS and other optimization solvers from Stanford University are described in http://www.sbsi-sol-optimize.com.

- We should mention that there are other languages such as LINGO. This is a more of a "proprietary" system, as it is linked to a specific optimization library. See http://www.lindo.com.

REFERENCES

1. R. Fourer, D.M. Gay, and B.W. Kernighan. *AMPL: A Modeling Language for Mathematical Programming.* Boyd and Fraser, Danvers, MA, 1993.

2. S. Zenios, editor. *A Library of Financial Optimization Models.* Blackwell Publishers, Oxford, 2006.

3. S. Zenios. *Practical Financial Optimization.* Blackwell Publishers, Oxford, 2006.

Index

acceptance–rejection method, 233, 235, 237, 247, 276, 458
active set method, 365, 378
ADI, *see* Alternating Direction Implicit method
algorithm
    polynomial, 145
Alternating Direction Implicit method, 319
antithetic
    sampling, 244, 447, 547
arbitrage, 71, 103, 126, 415, 486, 550
    opportunity, 39, 104, 129, 551
arc (in a network), 497
arithmetic
    finite precision, 15
asset allocation, 73, 77, 506
asset–liability management, 530, 534, 556
augmented Lagrangian method, 351

backsubstitution, 154
barrier
    function, 349, 375
    logarithmic, 375
    monitoring, 122
    option, *see* option, barrier
base
    binary, 138
    decimal, 138
    in Halton sequence, 270
basic
    feasible solution, 369
    solution, 369
    variable, 369
basis, 370
    function, 174, 204, 503, 512, 517
    monomial, 175
Bayesian statistics, 26
Bellman equation, 502, 510
bias, 259, 512
biased low estimator, 259
bid–ask spread, 24
binary
    decision variable, 565
binomial
    lattice, *see* lattice, binomial

657

model, 26
bisection method, 192, 410
Black–Derman–Toy (BDT) model, 127
Black–Scholes
    equation, 290, 292, 307, 511
    formula, 110, 173, 224
bond, 30
    above par, 31
    at par, 31
    below par, 31
    callable, 31, 125
    convexity, 59
    coupon, 30
    coupon rate, 30
    embedded options, 31
    face value, 30
    option, 125
    par value, 30
    portfolio, 380
    pricing, 52
    yield, 53
    zero-coupon, 30, 49, 124, 128
boundary
    condition, 110, 292, 477
    free, 486
    Neumann condition, 478
Box–Muller method, 236, 247, 276, 458
branch and bound, 572, 578, 584
    LP-based, 584
branching, 582, 586
branching factor, 27
Brownian bridge, 440, 462
Brownian motion, *see* geometric Brownian motion
butterfly spread, 248
buy and hold, 88

C++, 11
calibration
    Cox, Ross, and Rubinstein (CRR), 405, 411
    Jarrow–Rudd, 405
    lattice, 417

    model, 9
    of a binomial lattice, 403
canonical form (of LP problem), 526
caplet, 125
cash flow
    matching, 55, 655
central limit theorem, 236, 241, 631
central path, 377
certainty equivalent, 68
chance constraint, 82, 510
chance-constrained model, 526
Chebyshev
    node, 180
    polynomial, 183
Cholesky
    factor, 238, 444
    factorization, *see* factorization
clean price, 381
code vectorization, 434, 436
collocation method, 511
combination
    convex, 390
    linear, 369
combinatorial optimization, 495
common random numbers, 251, 470, 578
compact model formulation, 540
complementary slackness, 354, 355, 360, 374, 377
complexity, 144, 155
    exponential, 145, 377
    polynomial, 368, 377
concave
    function, 334, 391, 530, 567
    optimization problem, 334
condition number, 142, 150
conditional
    density, 266
    distribution, 504
    expectation, 504, 509, 512, 636
    Monte Carlo, 447, 448
    probability, 530
    Value at Risk (CVaR), 87
    variance, 255, 636
conditioning, 20, 142, 255

INDEX 659

confidence
    interval, 240, 640
    level, 83
consistency, 320
consistent numerical scheme, 323
constraint
    active, 333, 354, 366
    bounding, 333
    dualization, 373
    dualized, 358
    equality, 333, 347, 357, 381
    inactive, 333, 354
    inequality, 333, 347, 358, 381
    integrality, 337
    qualification, 351, 353
consumption–saving problem, 500
continuation
    region, 118
    value, 117, 414, 419
continuous-time
    dynamic system, 332
contraction mapping, 161
control variate, 253, 447, 455
convection term, 304
convection–diffusion equation, 304
convergence, 168
    global, 204
    linear, 143, 193
    quadratic, 143, 195
    rate of, 143
convex
    combination, 169, 300, 309, 390, 394, 569
    function, 334, 390, 528, 567
    hull, 342, 368, 390, 394, 570, 578
    optimization problem, 334
    problem, 527, 528
    set, 334, 335, 390
convex hull, 578, 591
convexity, 63, 113, 334, 359, 389
    bond, 380
correlation, 73, 82, 86, 253, 417
    coefficient, 638
    coefficient of, 635

instantaneous, 101, 444
negative, 244
positive, 252
cost-to-go, 498
covariance, 73, 337, 451, 548, 634
    matrix, 238, 337, 444
cover inequality, 590
covered position, 435
Cox–Ingersoll–Ross (CIR) model, 126
Crank–Nicolson method, 313, 485, 488
cumulative distribution function, *see* distribution, function
cut generation, 591
cutting plane, 557
cycling, 371

decision variable, 329
    binary, 565
    semicontinuous, 567, 571
decomposition, *see* factorization
    LU, 483
default, 31, 87
delta-hedging, 435
derivative, 4, 33
    Over the Counter (OTC), 30
descent direction, 338
diagonal dominance, 163, 168
differentiable function, 334
diffusion
    partial differential equation, 292
    term, 304
direction number, 281
discounted gain, 552
discrepancy, 269
discrete-time
    model, *see* model
    system, 500
distribution
    t, 241
    beta, 234
    conditional, 440
    discrete empirical, 232
    exponential, 231

660 INDEX

function, 110, 173, 230, 233, 625, 626
    joint, 632
    lognormal, 89, 100, 219, 403, 431
    marginal, 548, 632
    multivariate normal, 238, 548
    normal, 80, 82, 235
    standard normal, 110, 173, 638
    Student's t, 641
    symmetric, 82
    uniform, 228
dividend, 31, 112, 417
    yield, 112, 417
domain of influence, 302
drift, 84, 112, 435
dual
    feasibility, 374
    function, 358, 360, 373
    problem, 358, 373, 552, 557
    variable, 352, 554
duality, 372
    strong, 359, 360
    theory, 358
    weak, 359
duration, 57, 63, 86, 113, 125, 380
    Macauley, 58
    modified, 58
dynamic programming, 210, 332, 401, 415, 539, 558
    average cost, 501
    discounted, 501
    discrete state, 510
    finite horizon, 502
    infinite horizon, 510
    stochastic, 504

early exercise, 117, 414, 486
    boundary, 117
efficient frontier, 74, 75, 383, 571
eigenvalue, 148, 162, 163, 239, 311, 312, 393, 581
eigenvector, 149
epigraph, 391, 392, 578
equation
    linear system, 18, 483
    non-linear, 142, 191, 410
    polynomial, 47, 142, 191
equilibrium, *see* market, equilibrium
    Cournot, 201
    pricing, 37
error, 150
    absolute, 162, 241
    approximation, 174
    discretization, 430
    function, 178
    relative, 140, 141, 241
    roundoff, 140, 173
    sampling, 430
    truncation, 140, 212
estimator
    biased, 469
    high-biased, 519
    low-biased, 519
    unbiased, 637, 643
Euler scheme, 431
event, 623
    independent, 624
excess return, 506
expected return, *see* return
expected value, 626
    of a function, 209, 628, 633
explicit method, 305
extreme
    point, 368, 370, 378, 394, 558
    ray, 368, 394, 558

factorization
    Cholesky, 159, 238
    LU, 157
    QR, 366
Faure sequence, 472
feasible
    region, 365, 564
    set, 329
feedback control, 505
Feynman–Kač formula, 111, 129
finite difference, 251, 468
    Alternating Direction Implicit, 319

INDEX   661

backward, 294, 476
central, 294, 470, 476
Crank–Nicolson method, 313
explicit method, 305, 478
forward, 294, 476, 482
fully implicit method, 482
implicit method, 309
method, 402, 496
stability, 423
symmetric, see finite difference, central, see finite difference, central
first-order optimality condition, 334
fixed point, 510
fixed-charge problem, 566, 571
fixed-income
    portfolio, 102
    security, 30
fixed-mix portfolio, 576
fixed-point iteration, 161
floorlet, 125
Fortran, 11
forward contract, 33, 35, 103
Fourier analysis, 302
free boundary, 118, 293, 511
function
    affine, 335, 363
    approximation, 173, 505
    concave, 67, 361
    distribution, 84, 178
    indicator, 210, 266
    interpolation, 175
    inverse demand, 201
    non-convex, 581
    non-differentiable, 334, 340
    piecewise linear, 545, 567, 574
    Runge, 180, 186
    strictly concave, 530
    strictly convex, 390, 393
    utility, 176, 530
function approximation, 642
functional equation, 498
future contract, 34

Gauss–Seidel method, 488

Gaussian
    elimination, 154
    quadrature, see quadrature
gearing, 36
genetic algorithm, 389, 596
geometric Brownian motion, 98, 116, 126, 430, 441, 477, 482, 504
    bidimensional, 443
global optimization, 564
gradient
    conjugate, 173
    method, 387
graph optimization, 496
Gray code, 283
Greek, see option, sensitivity
grid, 476
    notation, 295, 476

Halton sequence, 269, 276, 458
heat equation, 292, 303
    bidimensional, 314
    physical interpretation, 304
hedging, 33, 108, 435
heuristic method, 591
homotopy, 205
    continuation, 206, 377
Hull–White model, 127

ill-conditioning, 151
immunization, 63, 125
importance sampling, 261, 450, 547
inadmissibility form, 372
independent increment, 92, 108
indicator function, 447
infinite-time horizon, 501
initial condition, 292
inner product, 188, 215
integer programming, see programming, integer
integration
    numerical, 448
interest
    accrued, 61
    compound, 43

662   INDEX

continuously compounded, 44
    simple, 43
interest rate, 43
    spot, 52, 56
    term structure, 52, 64
interest-rate
    cap, 125
    derivative, 9, 124
    dynamics, 126
    floor, 125
    risk management, 125
    swap, 124
interior point method, 375, 378
internal rate of return, 47, 53
interpolation, 212, 503
    linear, 479
intrinsic value, 117, 414, 419
inverse transform method, 230, 234, 237, 247, 277, 458
iterative method, 488, 511
Ito
    lemma, 96, 128, 417, 431
    multidimensional lemma, 101
    stochastic differential equation, 430
    stochastic integral, 95

Jacobian
    determinant, 236
    matrix, 197
Jensen's inequality, 628, 632
jump
    in asset price, 6

Kelley's cutting planes algorithm, 364, 556
knapsack problem, 16, 144, 337, 495, 564, 587, 652
Kuhn–Tucker conditions, 351
kurtosis, 548

L-shaped decomposition, 365, 540, 555, 557
Lagrange
    multiplier, 352, 356, 366, 375, 535
    polynomial, 212
Lagrangian
    function, 351, 354, 358, 375
    multiplier, *see* Lagrange, multiplier
large numbers
    strong law of, 222
lattice, 214
    binomial, 28, 489, 510
    implied, 426
    method, 496
    recombining, 28
    structure in LCG, 228, 237
    trinomial, 422, 481
law of large numbers, 637
law of one price, 51
Lax's equivalence theorem, 323
LCG, *see* linear congruential generator
least squares, 147, 175, 388, 503, 513, 548, 642
leverage, 36
liability, 54, 57, 530
    uncertain, 544
LIBOR, 130
likelihood ratio, 262, 450
limited liability (assets), 31, 32
line search, 339
linear congruential generator, 226, 267
linear programming, *see* programming, linear
    canonical form, 367
    duality, 553
    standard form, 367
linear regression, 147, 388, 512, 642
local improvement, 592
local search, 591, 592
    best-improving, 592
    first-improving, 592
low-discrepancy sequence, 269, 458, 547
    Halton, 269
    Sobol, 281
lower bound, 572, 581, 583

INDEX

LU decomposition, 310

marginal density, 633
market
    complete, 107
    efficiency, 89
    equilibrium, 126
    incomplete, 117, 128
    model, 130
Markovian
    dynamic system, 498
    state property, 501
martingale, 25
matrix
    block-diagonal, 556
    diagonal, 451
    diagonally dominant, 163
    Hessian, 334, 341, 356, 392, 393, 581
    Hilbert, 18, 150, 191, 619
    orthogonal, 366
    permutation, 156
    positive definite, 239, 341, 393
    positive semidefinite, 337, 392, 393
    singular, 151
    sparse, 80, 160, 556
    triangular, 366
    tridiagonal, 159, 308, 310, 483
maturity
    bond, 30
    option, 4, 35
mean absolute deviation, 573
mean blur, 640
mean reversion, 102, 127
mean-variance
    efficient portfolio, 383, 649
    framework, 577
    portfolio optimization, 571
metaheuristic, 389, 578, 591
metamodel, 388
method
    direct, 144, 154
    Gauss–Seidel, 168, 169
    iterative, 143, 161

    Jacobi, 163
Microsoft Excel, 11
MILP, 584
minimizer, 329
minimum
    variance portfolio, 383
minimum lot, 573
model
    binomial, 39, 105
    calibration, 39, 117
    continuous-state, 29
    continuous-time, 29
    discrete-state, 26
    discrete-time, 27
modulus
    in LCG, 226
moment matching, 214, 401, 510, 548
monomial, 190
monomial basis, 512
Monte Carlo
    integration, 222
    sampling, 528, 547
Monte Carlo sampling, 505
multiplier
    in LCG, 226

naked position, 435
neighborhood structure, 592
Nelder–Mead method, 342
network optimization, 496
Newton's method, 195, 197, 204, 377, 511
    for optimization, 341
no-arbitrage
    argument, 103
    principle, 50, 51, 106
node (in a network), 496
non-anticipativity, 256
    condition, 532, 576
    constraint, 533
non-convex
    function, 387
    problem, 577
    set, 567

non-differentiability, 361
norm
    L_p, 146
    compatible, 148, 163
    Euclidean, 146, 172
    Frobenius, 148
    matrix, 147
    spectral, 148, 150
    subordinate, 149
    vector, 146
normal
    multivariate, 159
    standard distribution, 84
    variate, 247
normed linear space, 188
not-a-knot condition, 184, 188
null space, 365
numeraire
    good, 38
numerical
    instability, 19, 169, 300, 480
    stability, 307, 320
numerical integration, *see* quadrature, 504

objective function, 329
    separable, 501
optimal control, 332, 500
optimal stopping, 414
optimality principle, 501
optimization, 198
    discrete, 390
    global, 576
    problem, 172
optimization method
    active set, 366, 368
    gradient, 339
    interior point, 535
    steepest descent, 339
    subgradient, 340
    trust region, 341
optimization problem
    concave, *see* convex
    constrained, 333, 346
    convex, *see* convex

dual, 358
finite-dimensional, 329
infeasible, 329, 331
infinite-dimensional, 332
non-smooth, 347
relaxed, 358, 581
unbounded, 329, 331
unconstrained, 333, 338
optimizer, 329
optimum
    global, 329, 334
    local, 329, 334
option, 4, 35
    American, 4, 35, 117, 256, 478, 496, 510
    American call, 8
    American put, 414, 486
    American put , 488
    American spread, 417
    as-you-like-it, 255
    Asian, 35, 109
        arithmetic, 454
        arithmetic average, 123
        average rate, 454
        geometric, 457
        geometric average, 123
    at-the-money, 35
    barrier, 119, 446, 478, 486
    Bermudan, 35, 511
    call, 35
    chooser, 255, 519
    continuation value, 117
    delta, 108, 109, 115, 478
    down-and-in put, 119, 447
    down-and-out put, 119, 446, 485
    European call, 4, 110, 219, 242, 247, 276, 406, 435, 477, 615
    European put, 110, 477
    exchange, 443
    exotic, 35, 119
    expiration, 117
    gamma, 115
    Greek, 111, 210

in-the-money, 35, 117, 266, 435, 477, 513
intrinsic value, 117, 486
lookback, 123, 425
multidimensional, 417
on a bond, 125
out-of-the-money, 35, 266, 435, 477
path dependent, 123
pay-later, 410
put, 35
sensitivity, 251, 468, 479
spread, 444
weakly path-dependent, 446
orthogonal
elements, 189
matrix, 366
polynomial, 191, 215, 512
projection, 190
system, 189
orthonormal
polynomial, 217
system, 189
overrelaxation
method, 168, 488
parameter, 491

parity
for barrier options, 119
put-call, 104
partial differential equation, 109
elliptic, 291
first-order, 291
hyperbolic, 291
linear, 291
order of, 291
parabolic, 291, 307
quasilinear, 291
second-order, 290
path generation, 430
pathwise estimator, 471
PDE, *see* partial differential equation
Peaceman-Rachford method, 319
penalty function, 346, 375

barrier, 349
exact, 347
exterior, 347
interior, 349, 375
perturbation analysis, 389
pivoting, 156
point
extreme, 558
Poisson
distribution, 625
process, 6, 100, 232
random variable, 627
polar
coordinate, 236
rejection, 237, 276
polyhedron, 393
bounded, 394
polynomial
function, 329
interpolating, 212
interpolation, 179
Lagrange, 179, 183
primitive, 281
polytope, 394
portfolio
cardinality-constrained, 571
efficient, 74, 385
management, 380
mean-variance optimization, 337
optimization, 15, 40, 71
rebalancing, 534
position
covered, 435
long, 33
naked, 435
short, 33
power utility, 509
predecessor node, 540
present value, 44, 45, 52, 59
price
clean, 61
dirty, 62
spot, 103
pricing
linearity of, 51

risk-neutral, 104
primal
    feasibility, 374
    optimization problem, 358
    variable, 352
prime number, 270, 460, 618
principal (amount), 43
principal (notional), 124
probability
    conditional, 624
    density, 233, 632
    density function, 626
    mass function, 625, 632
    measure, 261, 555, 624
process
    continuous-time, 88
    discrete-time, 88
    generalized Wiener, 93
    standard Wiener, 91
    Wiener, 29, 91, 108, 126
        non-differentiability of, 93
programming
    integer, 16, 83, 564, 580
    linear, 147, 335, 380, 647
    mixed-integer, 337, 565, 647
    mixed-integer quadratic, 571
    non-linear, 261, 336, 385
    nonlinear, 647
    pure binary, 565, 652
    pure integer, 337
    quadratic, 337, 355, 648, 649
    stochastic, 28, 65, 82, 365, 387, 496
        with recourse, 527
    stochastic multistage, 530
protective put, 36
pseudorandom numbers, 225

QR factorization, 366
quadrature
    adaptive, 220
    composite formula, 213
    formula, 211
        closed, 211
        open, 212
    Gauss–Hermite, 215, 219, 506
    Gaussian, 215, 505, 509, 547
    Newton– formula, 220
    Newton–Cotes formula, 212
    product rule, 220
    recursive, 220
    trapezoidal rule, 213
quantile, 83, 84, 236, 241, 628, 630, 640
quasi-Monte Carlo simulation, 269, 458
quasi-Newton method, 201
    for optimization, 342
    in optimization, 388
quasirandom sequence, 269

Radon–Nikodym derivative, 262
random variable, 232, 625
    continuous, 626
    discrete, 625
    exponential, 629
    function of, 628, 633
    independent, 633
    jointly distributed, 632
    lognormal, 631
    multivariate normal, 634
    normal, 215, 629
    standard normal, 236, 432, 629, 630
    uniform, 628
random variate generator, 225
rare event, 450
rate of return
    internal, 191
ray, 394
    extreme, 394, 558
recourse
    fixed, 527
    function, 363, 527, 556, 558
    relatively complete, 556
    variable, 527
recursive equation, 498, 502
reduced
    cost, 371
    gradient, 365

relaxation
    continuous, 585
    of an optimization problem, 580
replicating portfolio, 127
replication, 107
replication (in simulation), 240
residual, 175, 513, 642
response surface, 388
return
    expected, 40, 73, 383
    rate of, 32, 73
    total, 32
Richardson extrapolation, 519
risk, 73, 383, 573
    aversion, 66, 74, 80, 415, 509, 530, 577
    coefficient of absolute aversion, 69
    coefficient of relative aversion, 69
    coherent measure, 88
    credit, 31, 54
    interest rate, 54
    market price of, 117, 129
    measure, 41, 83
    minimization, 574
    neutrality, 526
    premium, 68
    reinvestment, 31
risk-free
    asset, 72, 552
    interest rate, 39
    rate, 224, 500
risk-neutral
    dynamics, 112
    measure, 111, 117, 128, 129, 224, 266, 481, 552
    probability, 107, 482
    probability measure, 555
    valuation, 262
Romberg integration, 220
roots
    of a polynomial, 48, 142, 191, 217

sample
    covariance, 642
    mean, 221, 240, 637
    path, 430
    space, 623
    variance, 240, 638, 641, 642
scenario, 532
    aggregation, 535
    bushy tree, 27
    fan, 26
    generation, 546
    tree, 528, 546
search tree, 582
second-order optimality condition, 334
seed (in random number generation), 227
semicontinuous variable, 567
shift
    in LCG, 226
short rate, 126
short-selling, 32, 73, 103
short-squeezing, 33
shortest path problem, 496
shortfall, 530, 556
significance loss, 140, 156
simplex, 342
    method, 370, 378
    search, 342, 344, 372, 389
Simpson's rule, 214, 220
simulated annealing, 593
simulation, 578
    stochastic, 338
simulation-based optimization, 387
skewness, 548
Sobol sequence, 281, 462
solution
    basic, 369
    unbounded, 368, 551
spectral radius, 148, 163, 166, 169, 311
spline, 479
    cubic, 184
    linear, 183
    natural, 184

668   INDEX

shape-preserving, 188
split-variable model formulation, 532
St. Petersburg paradox, 41
stability, 141
    condition, 302, 309, 315
stability analysis
    matrix theoretic, 307
    Von Neumann, 302, 323
standard deviation, 74, 83, 383, 627
state equation, 500
stationarity, 357
stationarity condition, 334, 338, 357, 376
statistical model, 643
step
    length, 339
stochastic
    differential, 91
    differential equation, 90
    integral, 91, 93
    optimization, 210, 256
stochastic programming, *see* programming, stochastic, 363
stop-loss hedging, 435
stopping
    time, 117
stratification, 440
stratified sampling, 260, 547
    optimal allocation of the samples, 261
strike price, 4, 35
subadditivity, 87
subdifferentiable
    function, 335
subdifferential, 334, 392
subgradient, 335, 340, 362, 392
support hyperplane, 364, 392

tabu
    list, 596
    search, 389, 578, 595
Taylor expansion, 97, 115, 173, 195, 293, 365, 392
term structure
    equation, 129

terminal cost, 332
trajectory cost, 332
transaction cost, 35, 39, 51, 65, 82, 111, 127, 544, 567, 571, 574
transition probability, 504
transport equation, 296
truncation error, 294, 323
trust region method, 341

unbounded
    problem, 368
unconstrained
    optimization problem, 330
unit hypercube, 220, 223, 269
utility
    Cobb–Douglas function, 37
    CRRA, 70, 72, 508
    DARA, 70
    expected, 41, 338, 545
    from bequest, 501
    function, 37, 66, 500, 506, 530, 545
    IARA, 70
    logarithmic, 70, 71, 506
    power, 509
    quadratic, 70, 80, 577
    theory, 80
    Von Neumann–Morgenstern, 66

value
    intrinsic, 486
Value at Risk, 83, 115, 262, 266, 472
value function, 498, 558
value of the stochastic solution, 529
Van der Corput sequence, 270
VAR, *see* vector autoregressive model
VaR, *see* Value at Risk
variable
    artificial, 372
    binary, 565, 571
    dual, 554
    slack, 367
variance, 83, 627

of a sum of variables, 634
  reduction, 244
    by conditioning, 447
Vasiček model, 126
vector autoregressive model, 546
vectorization (of MATLAB code), 619
Visual Basic, 11
volatility, 84, 112
  historical, 116
  implied, 116, 426
  role in barrier options, 121
  stochastic, 102, 446
Von Neumann
  stability analysis, 302, 323
VSS, *see* value of the stochastic solution

well-posed problem, 293, 323
Wiener process, 432, 439, 462
  bidimensional, 445

yield, 59, 86, 381
  curve, 57
  required, 54

STATISTICS IN PRACTICE

Human and Biological Sciences

Brown and Prescott · Applied Mixed Models in Medicine
Ellenberg, Fleming and DeMets · Data Monitoring Committees in Clinical Trials: A Practical Perspective
Lawson, Browne and Vidal Rodeiro · Disease Mapping With WinBUGS and MLwiN
Lui · Statistical Estimation of Epidemiological Risk
*Marubini and Valsecchi · Analysing Survival Data from Clinical Trials and Observation Studies
Parmigiani · Modeling in Medical Decision Making: A Bayesian Approach
Senn · Cross-over Trials in Clinical Research, *Second Edition*
Senn · Statistical Issues in Drug Development
Spiegelhalter, Abrams and Myles · Bayesian Approaches to Clinical Trials and Health-Care Evaluation
Whitehead · Design and Analysis of Sequential Clinical Trials, *Revised Second Edition*
Whitehead · Meta-Analysis of Controlled Clinical Trials

Earth and Environmental Sciences

Buck, Cavanagh and Litton · Bayesian Approach to Interpreting Archaeological Data
Glasbey and Horgan · Image Analysis in the Biological Sciences
Helsel · Nondetects and Data Analysis: Statistics for Censored Environmental Data
McBride · Using Statistical Methods for Water Quality Management: Issues, Problems and Solutions
Webster and Oliver · Geostatistics for Environmental Scientists

Industry, Commerce and Finance

Aitken and Taroni · Statistics and the Evaluation of Evidence for Forensic Scientists, *Second Edition*
Brandimarte · Numerical Methods in Finance and Economics: A MATLAB-Based Introduction, *Second Edition*
Chan and Wong · Simulation Techniques in Financial Risk Management
Lehtonen and Pahkinen · Practical Methods for Design and Analysis of Complex Surveys, *Second Edition*
Ohser and Mücklich · Statistical Analysis of Microstructures in Materials Science

*Now available in paperback.